Springer Series in Reliability Engineering

For further volumes:
http://www.springer.com/series/6917

Wallace R. Blischke · M. Rezaul Karim
D. N. Prabhakar Murthy

# Warranty Data Collection and Analysis

Springer

Prof. Wallace R. Blischke
5401 Katherine Avenue
Sherman Oaks, CA 91401-4922
USA
e-mail: wrblischke@sbcglobal.net

Prof. M. Rezaul Karim
Department of Statistics
University of Rajshahi
Rajshahi
Bangladesh
e-mail: mrkarim@ru.ac.bd

Prof. D. N. Prabhakar Murthy
School of Mechanical and Mining Engineering
The University of Queensland
Brisbane, QLD 4072
Australia
e-mail: p.murthy@uq.edu.au

ISSN 1614-7839
ISBN 978-0-85729-646-7 e-ISBN 978-0-85729-647-4
DOI 10.1007/978-0-85729-647-4
Springer London Dordrecht Heidelberg New York

British Library Cataloguing in Publication Data
A catalogue record for this book is available from the British Library



*Cover design:* eStudio Calamar, Berlin/Figueres

Printed on acid-free paper

Springer is part of Springer Science+Business Media (www.springer.com)

*Dedicated to*

*Douglas Paul Satterblom, in Memorium*

WRB

*My wife Tahmina and daughters Nafisa and Raisa*

MRK

*The Memory of my parents, Seethamma and Narsimha Murthy*

DNPM

# Preface

Warranty is a critical element in the marketing of products. It provides assurance to customers that the manufacturer will provide compensation, through repair, replacement or refund, for purchased items that do not perform satisfactorily. This has become more critical with the increase in consumer expectations that has occurred over recent years and the passage of legislation demanding better customer protection.

Offering warranty has serious implications for manufacturers. Although offering better warranty terms may give a manufacturer a marketing advantage over competitors, this entails an additional cost, namely that associated with the servicing of warranty claims. Depending on the product and the manufacturer, these costs typically vary from 1 to 10% of the sale price of the item, and may have serious implications with regard to the manufacturer's reputation and the profitability of the business.[1]

Warranty costs depend on the number of warranty claims and the servicing strategy used by the manufacturer. The number of claims depends on the field reliability of the product, which is influenced by both consumer actions (such as operating environment, usage intensity, maintenance, due care, etc.) and decisions of the manufacturer (design, development, production, testing, etc.).

Manufacturers must make decisions with regard to product launch based on the limited information gathered during the design and development stages of the new product development process. Warranty data provide useful information for assessing product reliability and detecting reliability problems (e.g., those associated with design, production, component suppliers, etc.), as well as problems associated with the servicing of warranty claims. Warranty data consist of (1) claims data and (2) supplementary data. Claims data are data that are collected during the servicing of warranty claims. Supplementary data are additional data and information needed for proper analysis of the claims data. These data are

[1] See *Warranty Week*—a weekly electronic newsletter—for reports on warranty costs in different industry sectors.

obtained from various sections of the manufacturing business as well from many other sources external to the manufacturer, and can be used to reduce not only the costs of servicing warranty for current products on the market, but, equally importantly, for products to be developed and launched in the future.

This book deals with warranty data collection and analysis, and the use of these results in improving business performance for both current and future products. The intent is to develop an appropriate framework for this purpose. The book consists of an introductory chapter (Chap. 1) and six parts (Parts I–VI), with each part consisting of two or more chapters. The six parts are as follows:

- Part I (2 chapters) gives an introduction and overview of reliability theory and product warranty.
- Part II (2 chapters) deals with warranty data collection. It looks at the issues and challenges associated with the collection of claims and supplementary data.
- Part III (5 chapters) looks at tools and techniques. The first two chapters deal with modeling of warranty claims and predicting warranty costs for several one-dimensional (1D) and two-dimensional (2D) warranties. The next three chapters deal with a review of some of the basic tools of statistical analysis and statistical inference.
- Part IV (4 chapters) deals with the analysis of warranty data using the tools and techniques of Part III. Several different data scenarios are considered and the methods are illustrated by application to claims and supplementary data for several types of products.
- Part V (2 chapters) looks at the framework for the improvement process. The first chapter deals with improvements for current (or existing) products and the second with new products.
- Part VI (2 chapters) examines two real case studies. The first case deals with air-conditioner data sold with a 1D warranty and the second with a component of an automobile sold with 2D warranty.

Effective management of reliability and warranty in a manufacturing business requires an interdisciplinary team consisting of engineers, scientists, operations researchers, statisticians, IT and management experts. The book is aimed at all of these groups of practitioners. Some practitioners want only to be trained to use a method. A book written for that purpose would be a training manual. This book is intended to educate the practitioner and provide an understanding of the underlying concepts and the issues involved. As such, the book should appeal to practitioners who want more than a manual. The concepts, tools and techniques are highlighted through examples and case studies in order to emphasize proper techniques for data collection and analysis and to show how the results may be used for effective decision making.

The book is also intended for researchers in industry and academia. We raise new and challenging issues that would be of interest to researchers and, in the process, bridge the gap between theory and practice. This book would be of interest to analytically oriented practitioners and to researchers wanting to tackle complex real world warranty problems of importance to manufacturers. As indicated earlier, the book is structured in a manner to facilitate both readers, in that we aim for a middle ground that will have practical content for the practitioner and discussion of unsolved or partially solved problems that will interest the researcher. In the process, we give both a realistic notion of the strengths as well as the limitations of the current state of knowledge.

It is assumed that the reader has some knowledge of statistics, mathematics, operations analysis, and warranty, though reviews of many basic concepts in these areas are included in the book.

The authors bring an interdisciplinary perspective to the book through their backgrounds, research into various aspects of reliability and warranty, and their interaction with many businesses in Asia, Australia, Europe and the USA.

The authors are grateful to their ex-students and other researchers with whom them they have collaborated over many years (ranging from 15 to nearly 40). These include

- Professor John Eccleston, Dr. Richard Wilson and Dr. Michael Bulmer (The University of Queensland, Australia)
- Professor Kazuyuki Suzuki (The University of Electro-communications, Japan)
- Professor Jaiwook Baik (Korea National Open University, South Korea)
- Professor Bermawi Iskandar (ITB, Indonesia)
- Professor Renyan Jiang (Changsha University, China)

We are grateful to the staff at Springer Verlag for their support. We especially want to thank Anthony Doyle for his early interest and encouragement, and Claire Protherough, who provided much valuable guidance in the preparation of the manuscript and much patience and understanding during several unavoidable delays in completion of the project.

Sherman Oaks, CA, USA — Wallace R. Blischke
Rajshahi, Bangladesh — M. Rezaul Karim
Brisbane, Australia — D. N. Prabhakar Murthy

# Contents

## Part III Models, Tools and Techniques

# Chapter 1
# An Overview

## 1.1 Introduction

A salient feature of modern industrial societies is that new products are appearing on the market at an ever increasing pace. This is due to (i) rapid advances in technology and (ii) constantly increasing demands of customers, with each a driver of the other. As a result, products are becoming more complex and their performance capabilities are increasing with each new generation.

Customers need assurance that a product will perform satisfactorily over its designed useful life. Product performance depends on the reliability of the product, which, in turn, depends on decisions made during its design, development and production. One way in which manufacturers can assure customers of satisfactory product performance is through warranty. A warranty is a contractual obligation that requires the manufacturer to remedy any problems or failures that occur during a specified period of warranty coverage. The base warranty on an item is part of the initial transaction. In contrast, extended warranties are optional additional warranty coverages that are purchased separately by the customer.

Offering warranty results in additional costs to the manufacturer. This cost depends on the reliability of the product (which is at least partially under the control of the manufacturer) and on operating environment, usage mode, and usage intensity (all of which are determined by the users). The costs associated with the base warranty are factored into the sale price and, as mentioned earlier, the customer pays extra to cover the costs associated with extended warranties. For large manufacturers, annual warranty costs often run into billion of dollars and as a fraction of the sale price they typically vary from 2 to 10%. As a consequence, warranty costs have a significant impact on overall profits.

Manufacturers arrive at decisions about reliability based on their understanding of customer behavior and on engineering judgment during product design, development, and production. In the early stages of this process, there will be limited data and other information available on product performance. As a result,

W. R. Blischke et al., *Warranty Data Collection and Analysis*,
Springer Series in Reliability Engineering, DOI: 10.1007/978-0-85729-647-4_1,

there will be considerable uncertainty in predictions of item reliability and hence in assessment of warranty costs based on these estimates.

Warranty claims data are data that are collected during the processing of claims under warranty. Together with supplementary data, they provide valuable information for assessing product reliability and the impact of operating environment, usage mode, and usage intensity on the performance of the product. Through proper analysis, one can obtain better estimates of product reliability and of warranty costs. These, in turn, help manufacturers take appropriate actions relating to product reliability over the remainder of the product life cycle and to address warranty issues from an overall business perspective.

There are several possible problems associated with the use of warranty data. These make data analysis and extraction of useful information more difficult and challenging. This book deals with issues relating to warranty data collection and analysis. In this chapter we give an overview of the book.

The outline of the chapter is as follows. In Sect. 1.2, we discuss various issues regarding products themselves. Section 1.3 deals with product performance and introduces the notion of product failure. In Sects. 1.4 and 1.5, we look at various issues relating to product warranty and product reliability, respectively. Section 1.6 discusses some aspects of warranty data. The objectives of the book are discussed in Sect. 1.7 and an outline of the book is given in Sect. 1.8.

## 1.2 Products

According to [4]:

> A product can be tangible (e.g., assemblies or processed materials) or intangible (e.g., knowledge or concepts), or a combination thereof. A product can be either intended (e.g., offering to customers) or unintended (e.g., pollutant or unwanted effects)

In this book, we are concerned with tangible products, specifically manufactured goods sold with warranty. This includes most consumer goods, with the exception of food and other agricultural products. It also includes many products that are acquired in commercial transactions and government procurement. The following is an example of a typical consumer product:

*Example 1.1* [Cell Phone] A cell phone has the capacity to send and receive radio signals. This provides a seamless transmission of voice as long as the telephone sending the message and the one receiving the message are within geographical range of the communication satellite involved. Millions of people around the world use cellular phones. A modern cell phone may perform an array of functions in addition to sending and receiving voice messages. These include[1]:

---

[1] This list is not intended to be exhaustive and will quite likely be out of date by the time this book is in print.

- Accessing information (such as news, entertainment, stock quotes) from the internet
- Sending and receiving text mail
- Sending and receiving e-mails
- Storing contact information details
- Serving as an alarm clock
- Receiving FM radio
- Taking digital pictures
- Sending and receiving digital pictures
- Making task and do lists
- Keeping a calendar of appointments and reminders
- Carrying out simple arithmetic calculations
- Providing a games console
- Integrating other devices, such as MP3 players

Worldwide sales of these devices have been growing exponentially.

### *1.2.1 Product Classification*

Products can be classified in many different ways.[2] Two of these are the following:

#### 1.2.1.1 Classification 1

1. *Consumer non-durables and durables*: These are products that are bought by households. Non-durables differ from durables in the sense that the life of a nondurable item (for example, food) is relatively short, and the item is less complex than a durable item (for example, television, automobile).
2. *Industrial and commercial products*: These are products used by businesses for their operations. The technical complexity of such products can vary considerably. The products may be either complete units (for example, trucks, pumps) or components (for example, batteries, bearings, disk drives).
3. *Specialized products*: Specialized products (for example, military and commercial aircraft, ships, rockets) are usually complex and expensive, often involve state-of-the-art technology, and are usually designed and built to the specific needs of the customer. Still more complex are very large systems that are collections of many inter-linked products (for example, power stations, communication networks, chemical plants).

[2] See [6] for more details.

#### 1.2.1.2 Classification 2

1. *Standard products*: These are manufactured in anticipation of a subsequent demand. Standard products are manufactured based on previous products of the same type, competing goods, and the results of market surveys. They include all consumer non-durables and durables and most commercial and industrial products.
2. *Custom-built products*: These are manufactured in response to a specific request from a customer. They include airliners, ships, railroad equipment, apartment and office buildings, refineries, specialized defense products, and many other items.

### *1.2.2 Product Complexity*

The complexity of products has been increasing with technological advances. As a result, a product must be viewed as a system consisting of many elements and capable of decomposition into a hierarchy of levels, with the system at the top level and parts at the lowest level. There are many ways of describing this hierarchy. One such is the nine-level description shown in Table 1.1 and based on a hierarchy given in [1]:

*Example 1.2* [Automobile] The modern automobile is a complex system consisting of over 15,000 components. This system can be decomposed into the following sub-systems[3]:

- Body (passenger compartment)
- Engine (power source)
- Chassis (for supporting the engine and body)
- Transmission (for transmitting power from the engine to the wheels through shafts and gears)
- Controls (for accelerating, braking, steering, etc.)
- Cooling (for cooling the engine, providing comfort to passengers)
- Electrical (battery, starting motor, lights, logic controllers)
- Safety (seat belts, air bags, locks)
- Lubrication (oil pump, grease)
- Fuel (tank, carburetor, filters, fuel lines)
- Exhaust system (muffler, catalytic converter)
- Others (seats, doors, windows, radio, etc.)

[3] See http://auto.howstuffworks.com for a discussion of the principles of how these sub-systems work.

**Table 1.1** Multilevel decomposition of a product

| Level | Characterization |
|---|---|
| 0 | System |
| 1 | Sub-system |
| 3 | Assembly |
| 4 | Sub-assembly |
| 5 | Module |
| 6 | Sub-module |
| 7 | Component |
| 8 | Part |

*Comment* The number of levels needed to describe a product from the system level down to the part level depends on the complexity of the product

Each of these, in turn, can be decomposed into assemblies, sub-assemblies, and so forth, down to the part level.

To illustrate the complexity of the system, the following is a partial list of the components and parts for the engine sub-system: cylinder block, cylinder heads, pistons rings, connecting rods, bearings, crankshaft main bearings, camshaft bearings, cam followers, timing chain or belt; timing gears, guides, rocker arms, rocker shaft, rocker bushings, cylinder head valves, valve guides, valve lifters, valve springs, valve seals, valve retainers, valve seats, push rods, water pump, oil pump and oil pump housing, oil pan, intake and exhaust manifolds, valve covers, engine mounts, turbocharger/supercharger housing seals and gaskets.

## 1.3 Product Performance

### *1.3.1 Concept and Notions*

We begin with a basic definition of performance [8]:

> Performance, n. The accomplishment, execution, carrying out, working out of anything ordered or undertaken; the doing of any action or work; working, action (personal or mechanical); spec. the capabilities of a machine or device, now esp. those of a motor vehicle or aircraft measured under test and expressed in a specification

Many different definitions of performance can be found in the technical literature. The following two are a small illustrative sample:

> Performance is the measure of function and behavior—how well the device does what it is designed to do. [9]
>
> How well a product implements its intended functions. Typical product performance characteristics are speed, efficiency, life, accuracy, and noise. [10]

In general, product performance is a measure of the functional aspects of the product. It is *a vector of variables*, where each variable is *a measurable property of the product or its elements*. The performance variables can be:

- Functional properties (e.g., power, throughput, and fuel consumption)
- Reliability related (defined in terms of failure frequency, mean time to failure [MTTF], etc.)

Products are designed for a specified set of conditions such as the usage mode, usage intensity, and operating environment. When the conditions differ significantly from those specified, the performance of the product is affected. Product performance is also influenced by the skills of the operator and other factors.

### *1.3.2 Product Failure*

Product failure is closely linked to product function, as can be seen from the two following two definitions of failure:

> The termination of the ability of an item to perform a required function [3]
> Equipment fails if it is no longer able to carry out its intended function under the specified operational conditions for which it was designed [7]

### *1.3.3 Consumer Perspective*

Consumers view a product in terms of its attributes. According to [5],

> To a potential buyer a product is a complex cluster of value satisfactions.

Reference [2] states:

> Consumers seek benefits rather than products per se.

As a result, we have the following relationship:

$$\text{Attributes (Features)} \rightarrow \text{Bundle of benefits} \rightarrow \text{Value to the customer}$$

A successful new product

1. Satisfies new (or earlier unsatisfied) needs, wants or desires.
2. Possesses superior performance in such need satisfactions, compared to other products on the market.

Products are becoming more complex in order to meet ever-growing consumer needs and expectations.

When a failure occurs, no matter how benign, its impact is felt. For customers, the consequences of failures may range from a mere nuisance value (for example, failure of air-conditioner) to actual economic loss (for example, loss of food due to failure of a freezer) to something resulting in serious damage to the environment and/or loss of life (for example, brake failure in a car). All of these lead to customer dissatisfaction with the product, at the very least.

When the customer is a business enterprise, failures may lead to downtimes. This affects the production of services and goods, which, in turn, affects both the goodwill of clients as well the bottom line of the balance sheet of the seller.

## 1.4 Product Warranty

Consumers are often uncertain about product performance. As a result, they look for assurance that the product will perform satisfactorily over it's the useful life. One way of providing this assurance is through product warranty. This is a service element bundled with the product. Different types of warranties are offered, depending on the product and the anticipated buyer.

### *1.4.1 Warranty Concept*

A warranty is a manufacturer's assurance to a buyer that a product or service is or shall be as represented. It may be considered to be a contractual agreement between buyer and manufacturer (or seller) that is entered into upon sale of the product or service.

In broad terms, the purpose of a warranty is to establish liability of the manufacturer in the event that an item fails or is unable to perform its intended function when properly used. The contract specifies both the performance that is to be expected and the redress available to the buyer if a failure occurs or the performance is unsatisfactory. The warranty is intended to assure the buyer that the product will perform its intended function under normal conditions of use for a specified period of time.

The terms warranty and guarantee are often used synonymously. The distinction is that a guarantee is defined to be a pledge or assurance of something; a warranty is a particular type of guarantee, namely a guarantee concerning goods or services provided by a seller to a buyer. Another related concept is that of a service contract or "extended warranty." The difference between a warranty and a service contract is that the latter is entered into voluntarily and is purchased separately—the buyer may even have a choice of terms—whereas the basic warranty is a part of the product purchase and is an integral part of the sale.

### *1.4.2 Manufacturer's Perspective*

From the manufacturer's point of view, warranties also serve another important role, as buyers typically compare characteristics of comparable models of competing brands in making their purchase decisions. When competing brands are nearly identical, it is very difficult, in many instances, to choose a particular product solely on the basis of product-related characteristics such as price, special features, perceived product quality and reliability, financing offered by the manufacturer, and so on. In such situations, post-sale factors—warranty, parts availability and cost, service, maintenance, and so forth—take on added importance in product choice. Of these, warranty is a factor that is known (or at least potentially known) to the buyer at the time of purchase, and buyers may reasonably interpret longer warranties as a signal of a more reliable and durable product. As such, warranty can become an effective promotional tool.

### *1.4.3 Warranty Costs*

Offering warranty results in additional costs to the manufacturer. This cost depends on several factors, the most important of which are shown in Fig. 1.1.

Warranty terms offered by a manufacturer are influenced by warranty legislation and by the warranties offered by competitors. Customer usage can vary across the customer population and is outside of the control of the manufacturer. As a result, product reliability is the only factor affecting claims that is solely under the control of the manufacturer, and it has a significant impact on warranty claims and costs. Warranty costs are high, typically varying from 2 to 10% of the sale price. Table 1.2 shows 2006 warranty costs for a sample of US companies.[4] Viewed as a fraction of profits, this number can vary by an order of magnitude.

Warranty costs are inversely related to product reliability. It follows that one way of reducing warranty costs is to improve reliability. This, however, can be a costly exercise, as it involves considerable investment in research and development. The optimum reliability program is one that achieves a sensible trade-off between the cost of building in reliability and the cost of warranty.

## 1.5 Product Reliability

Product reliability is determined primarily by decisions made during the early stages (design and development) of the product life cycle and it has implications for later stages (marketing and post-sale support) because of the impact of unreliability on sales and on warranty costs.

Assessment of product reliability prior to launch of the product on the market is based on limited information—data supplied by vendors, subjective judgment of design engineers during the design stage, and data collected during the

[4] See Warranty Week (http://www.warrantyweek.com) for detailed lists of warranty costs for companies in different industry sectors in the USA.

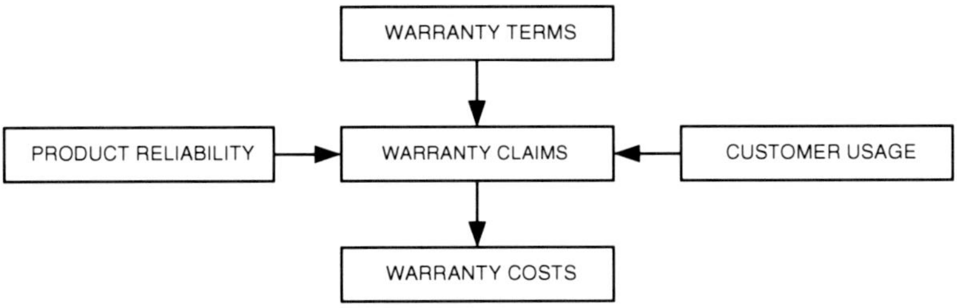


**Fig. 1.1** Factors influencing warranty costs

**Table 1.2** Warranty Costs for 2006 for a sample of US Companies (extracted from Warranty Week, April 13, 2007)

| Company | Claims (Millions, $) | Claims rate (%) |
|---|---|---|
| Apple Inc. | 244 | 0.9 |
| General Electric Co. | 665 | 1.0 |
| General Motors Corp. | 4,463 | 2.6 |
| Caterpillar Inc. | 745 | 1.9 |
| Hewlett-Packard Co. | 2,346 | 3.2 |
| IBM Corp. | 762 | 3.4 |
| Motorola Inc. | 891 | 2.1 |
| Black & Decker Corp. | 118 | 1.8 |
| Cummins Inc. | 292 | 2.6 |
| Deere & Co. | 509 | 2.6 |
| Ford Motor Co. | 4,106 | 2.9 |
| Boeing Co. | 206 | 0.7 |
| General Dynamics | 59 | 1.4 |
| Dell Inc. | 1,775 | 4.0 |

development stage. Data from field failures are needed to assess the actual reliability and compare it with the predicted reliability. If the actual reliability is significantly lower than the predicted value, it is essential that the manufacturer identify the cause or causes—design, production, materials, storage, or other factors. Once this is done, actions can be initiated to improve reliability. On the other hand, if the actual reliability is significantly above the predicted value, then this information can be used to make changes to the marketing strategy—increasing the warranty period and/or lowering the price—which will likely result in an increase in total sales.

The collection of field data is typically costly and time consuming. Warranty data provide at least a partial alternative to obtaining field data. With warranty periods becoming longer, tracking products through this longer time frame provides much additional information that may be of significant value in the new product development process.

## 1.6 Warranty Data Collection and Analysis

### *1.6.1 Types and Sources of Data*

Warranty data are the data that are needed for effective management of warranty for both existing and new products. Warranty data are categorized into the following two groups:

1. Warranty claims data: These are data collected during the processing of claims and servicing of repairs under warranty. The data are obtained from the post-sale support system for data collection. We shall call this "*claims data*" in the remainder of the book.
2. Supplementary data for warranty analysis: These are data collected from all other sources—either internal or external to the manufacturing organization. The data are obtained from internal sources such as design and development, production and marketing, and external sources such as vendors and others. We shall call this "*supplementary data*" in the remainder of the book.

The various kinds of data that may be collected are grouped into the following categories:

- Product related—technical: component failed, mode of failure, age and usage at failure, etc.
- Customer related: Operating mode, usage intensity, operating environment, maintenance, etc.
- Service agent related: Rectification actions (repair, replacement, refund, etc.), quality of repair/service, cost of repair, etc.
- Market related: Competitor's product performance, price, warranty terms, etc.

### *1.6.2 Warranty Data Analysis*

The objectives or goals of warranty data analysis may include one or more of the following:

- To extract information for assessing product reliability
- To detect problems in design, production, transport, storage or servicing
- To evaluate and control costs
- To aid in new product development
- To compare field performance with the desired design performance
- To evaluate and, if necessary, change warranty policies
- To make appropriate decisions for implementation of any needed changes
- To provide valuable information that is necessary for continuous improvement at all levels of the business

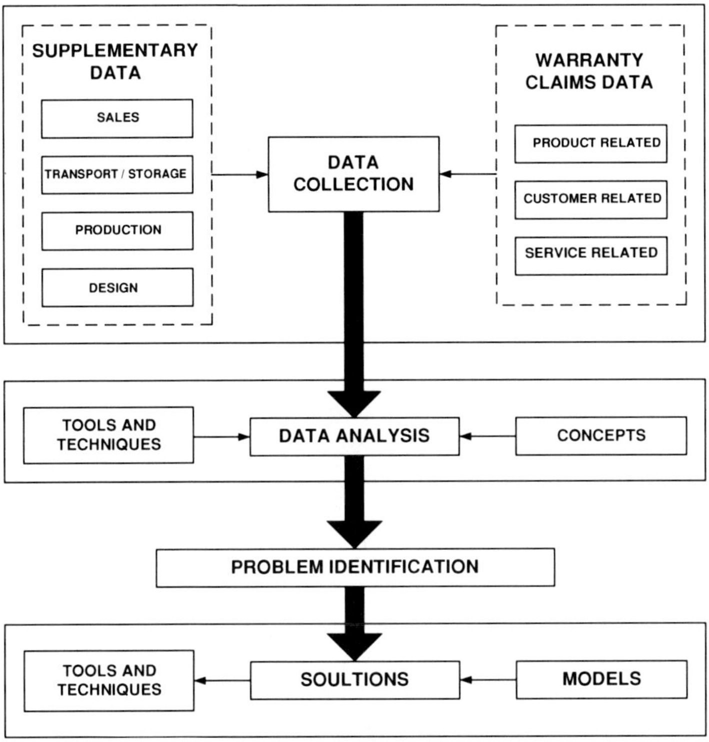


**Fig. 1.2** Data collection, analysis and use

Depending on the context, the analysis can be at the product, component or some intermediate level (assembly, module, etc.), and can be qualitative and/or quantitative.

Proper analysis requires additional (or supplementary) data obtained from other sources, such as production, marketing, design and development. Quantitative analysis involves many different types of models and a wide array of tools and techniques, as indicated in Fig. 1.2. These topics are discussed in detail in the remainder of the book.

### *1.6.3 Challenging Issues*

The four key issues are:

- What kinds of warranty data are currently collected?
- What kinds of warranty data should be collected?
- What needs to be done to collect the data properly?

- What needs to be done to properly and adequately analyze the data?

The following are a few of the many data-related problems that may occur:

- Data and information are recorded only for failed items, with little or no information obtained on un-failed items
- Data are not obtained on failures outside the warranty period
- Data are often incomplete or missing
- Delays in reporting are common
- Some data may be pooled or aggregated (For example, total sales in different time periods rather than individual sale dates for each item)
- Data may be censored in one or more ways
- Data may be reported incorrectly

Improper collection of warranty data may result in a significant loss of information, which may, in turn, seriously affect the decision making process. Furthermore, the analysis of warranty data not collected properly is more difficult and challenging. Effective approaches are needed to compensate for the loss of information and to arrive at credible inferences based on the data.

## 1.7 Objectives of the Book

As indicated in the previous section, warranty claims data as typically collected are very messy. Extracting the maximum amount of information requires sophisticated statistical techniques, and the use of this information to make proper and effective decisions requires building suitable models. In this book, we deal with both of these topics and provide many specific examples of applications, including data, results of the analyses, interpretation of the results, and illustrations of their usefulness in engineering and business decision making. Along the way, we highlight areas needing further research.

## 1.8 Outline of the Book

The book consists of an introductory chapter (Chapter 1) and six parts (Part I–VI), with each part consisting of several chapters. The six parts are as follows:

- Part I: Warranty and Reliability
- Part II: Warranty Data Collection
- Part III: Models, Tools and Techniques
- Part IV: Warranty Data Analysis
- Part V: Warranty Management
- Part VI: Case Studies

The chapter titles and brief descriptions of their contents are as follows:

Part I: Warranty and Reliability

*Chapter 2: Product Warranty.* Warranty is a contractual agreement between manufacturer and buyer. The chapter discusses the classification of warranties and describes several one- and two-dimensional warranty policies. Servicing of warranty results in additional costs to the manufacturer. The role of data in warranty cost management is discussed.

*Chapter 3: Product Reliability.* Warranty servicing costs depend on product reliability. Various aspects of reliability are discussed, including (i) notions of reliability, ranging from design to field reliability; (ii) reliability modeling, from simple models to complex models that capture the many factors that affect reliability; and (iii) the link between product reliability and warranty.

Part II: Warranty Data Collection

*Chapter 4: Warranty Claims Data.* Warranty claims data, collected during the servicing of claims are of four types—product related, customer related, service agent related and cost related. Details of each of these and factors that affect their collection of warranty are discussed. The use of warranty claims data for effective management of warranty is emphasized.

*Chapter 5: Supplementary Data for Warranty Analysis.* Warranty claims data alone are not adequate for estimation of product reliability, prediction of future claims, costs, and so forth. For these, supplementary warranty data are required. This chapter deals with censored data and data from various internal and external sources.

Part III: Models, Tools and Techniques

*Chapter 6: Cost Models for One-Dimensional Warranties.* When a manufacturer offers a warranty, claims under warranty must be serviced, resulting in additional costs. The expected number of claims depends on the reliability of the product. Here we focus on models for prediction of warranty cost as a function of product reliability for various one-dimensional warranties.

*Chapter 7: Cost Models for Two-Dimensional Warranties.* We consider two-dimensional warranties under which the warranty expires when the item reaches an age $W$ or the usage reaches a level $U$, whichever comes first. Failures are random points scattered over the two-dimensional warranty region. We discuss three approaches for modeling failures and warranty claims. Cost models for some simplified cases are derived.

*Chapter 8: Preliminary Data Analysis.* The objectives of preliminary data analysis are to edit the data to prepare it for further analysis, describe the key features of the data, and summarize the results. This chapter deals with quantitative and qualitative approaches, including scales of measurement, types of data, graphical methods, and basic descriptive statistics.

*Chapter 9: Basic Statistical Inference.* We look at a number of key statistical techniques that are used in inference problems regarding reliability and warranty. These include (i) estimation (point estimation, and confidence intervals), (ii) hypothesis testing, (iii) nonparametric methods, (iv) tolerance intervals, and (v) rank correlation.

*Chapter 10: Additional Statistical Techniques*. This chapter covers tests for outliers; goodness-of-fit tests, tests for comparing means of two or more normal populations; basic linear regression and correlation analysis; estimation of functions of parameters, including the coefficient of variation, and cost and reliability functions; and tests of assumptions.

Part IV: Warranty Data Analysis

*Chapter 11: Nonparametric Approach to Analysis of 1-D Warranty Data*. In this chapter, the nonparametric approach to inference based on distribution functions, renewal functions, mean cumulative functions, etc., is discussed in the context of 1-D warranty data. The estimators depend on the data structures and scenarios discussed in Chap. 5.

*Chapter 12: Parametric Approach to Analysis of 1-D Warranty Data*. Different types of one dimensional warranty data are analyzed by a number of commonly used parametric models. The method of maximum likelihood estimation is used to estimate the parameters of the models. The selected model for a given data scenario is used for predicting and drawing inference about reliability and warranty related quantities.

*Chapter 13: Complex Models for Analysis of 1-D Warranty Data*. When quality variations in production occur and/or customers vary with regard to usage intensity, operating environment, etc., more complex reliability models are required for analysis the data. Here we consider competing risk, mixture, AFT, PH and parametric regression models for modeling 1-D warranty claims data.

*Chapter 14: Parametric Approach to Analysis of 2-D Warranty Data*. For 2-D warranty policies, the unavailability of information on censored items leads to difficulties in estimation of the life distribution of the items. We look at three approaches to modeling failures—conditioning on usage rate, composite scale, and bivariate distribution function. The models are applied to forecasting of warranty claims and warranty costs using 2-D warranty data.

Part V: Warranty Management

*Chapter 15: Use of Warranty Data for Improving Current Products and Operations*. Proper analysis of warranty data allows a manufacturer to evaluate various performance measures at product and business levels, assess these relative to the design targets, and determine what corrective actions, if any, are needed. The use these results in the context of the TQM approach to continuous improvement is discussed.

*Chapter 16: Use of Warranty Data in New Product Development*. Because of the importance of warranty and its potential cost implications, warranty issues must be addressed during the early stages of the new product development process. This chapter deals with a framework for implementing this approach and looks at the role of warranty data in this context.

Part VI: Case Studies

*Chapter 17: Case Study 1 [Analysis of Air-conditioner Claims Data]*. The first 18 months of warranty claims data on 15 main components of a room AC unit covered under a free-replacement warranty are analyzed. Multiple failure modes are modeled and the results used to predict future costs and to compare alternative warranty policies.

*Chapter 18: Case Study 2* [*Analysis of Automobile Components Warranty Claims Data*]. This chapter presents an analysis of 2-D warranty claims data on a subsystem of an automobile. The objectives of the study are to (i) present a preliminary analysis of the claims data, (ii) investigate the relationship between age and usage, (iii) select suitable probability models for age and usage, (iv) model two variables, age and usage, using a single (composite scale) variable, and (v) discuss the managerial implications for cost-benefit analysis.

## References


1. Blischke WR, Murthy DNP (2000) Reliability. Wiley, New York
2. Day GS, Shocker AD, Shrivastava RK (1978) Consumer oriented approaches to identifying product markets. J Mark 43(Fall):9–19
3. IEC 50 (191) (1990) International electrotechnical vocabulary (IEV) Chapter 191—Dependability and quality of service. International electrotechnical commission, Geneva
4. ISO 8402 (1994) Quality vocabulary. International standards organization, Geneva
5. Levitt T (1980) Marketing success through differentiation of anything. Harv Bus Rev 1980(Jan–Feb):83–91
6. Murthy DNP, Rausand M, Osteras T (2008) Product reliability–performance and specifications. Springer, London
7. Nieuwhof GWE (1984) The concept of failure in reliability engineering. Reliab Eng 7:53–59
8. Oxford Dictionary (1989) Oxford english dictionary. Oxford University Press, Oxford
9. Ullman DG (1992) The mechanical design process. McGraw Hill, London
10. Ulrich KT, Eppinger SD (1995) Product design and development. McGraw-Hill, New York

# Part I
# Warranty and Reliability

# Chapter 2
# Product Warranty

## 2.1 Introduction

In Chap. 1, we defined warranty as a contractual agreement between manufacturer (or seller) and buyer that is entered into upon sale of the product. This contract defines the compensation available to the buyer if the performance of the product is found to be unsatisfactory. There are many different types of warranties that have been reported in the literature and warranties have been studied by researchers from many different and diverse disciplines. In this chapter, we provide a brief overview of the literature on warranty, with particular emphasis on those aspects that are important in the context of warranty management.

The outline of the chapter is as follows: Sect. 2.2 lists the various disciplines that have studied different aspects of warranty. In Sect. 2.3, we briefly discuss three perspectives on warranty, those of the seller, the buyer, and society as a whole. Section 2.4 provides a classification of the many types of warranties that may be used in the sale of products. Some specific classes of warranties along with illustrative warranty policies are discussed in Sects. 2.5–2.9. The servicing of warranty is discussed in Sect. 2.10. Offering warranties result in additional costs to the manufacturer; this issue is discussed in Sect. 2.11. Section 2.12 deals with methods of reducing the cost of warranty and other issues in warranty management.

## 2.2 The Study of Warranty

Because of the diversity of purpose and application, product warranty has received the attention of researchers from many diverse disciplines.[1] The following is a list of these along with some of the important issues that have been dealt with by each:

---

[1] Reference [5] lists over 1500 papers on warranty that have appeared prior to 1995. Reviews of subsequent recent literature on warranty can be found in [11] and [9].

W. R. Blischke et al., *Warranty Data Collection and Analysis*, Springer Series in Reliability Engineering, DOI: 10.1007/978-0-85729-647-4_2, 

- Historical: origin and use of the notion of a warranty
- Legal: court actions, dispute resolution, product liability
- Legislative: Magnusson–Moss Act; Federal Trade Commission, Warranty requirements in government acquisition (particularly military) and TREAD Act in the USA and similar legislation in the EU and governments throughout the world
- Economic: market equilibrium, social welfare
- Behavioral: buyer reaction, influence on purchase decisions, perceived role of warranty, claims behavior
- Consumerist: product information, consumer protection
- Engineering: design, manufacturing, quality control, testing
- Statistics: data acquisition and analysis, data-based reliability analysis
- Operations Research: cost modeling, optimization
- Accounting: tracking of costs, time of accrual
- Marketing: assessment of consumer attitudes, assessment of the marketplace, use of warranty as a marketing tool, warranty and sales
- Management: integration of many of the previous items, determination of warranty policy, warranty servicing decisions
- Societal: public policy issues

As a consequence, the literature on warranty is very large. Reference [2] deals with these issues in detail. Administration of warranties in the context of government acquisition is discussed in [3]. Reference [8] deals with warranty management in the context of new product development.

## 2.3 Three Perspectives on Warranty

Warranties are an integral part of nearly all commercial transactions and of many government transactions that involve product purchases as well. As discussed in Chap. 1, the buyer (individual, corporation, or government agency) point of view of a warranty is different from that of the manufacturer (or distributor, retailer, and so forth). Yet another perspective on warranty is that of society. From a societal point of view, warranty is dealt with by many groups, including legislators, consumer affairs groups, the courts, and public policy decision-makers.

Civilized society has always taken a dim view of the damage suffered by its members that is caused by someone or some activity, and has demanded a remedy or retribution for offences against it. Consequently, manufacturers are required to provide compensation for any damages resulting from failures of an object. This has serious implications for manufacturers of engineered objects. Product liability laws and warranty legislation are signs of society's desire to ensure fitness of products for their intended use and compensation for failures.

During the twentieth century, consumer movements had an impact on warranty. In fact, there were three important consumer movements during this period [4], the third of which began after the end of World War II and gained momentum in the 1960s. The result of this social pressure was a growing concern for buyers' protection, and the notion of express warranty was augmented by another concept, "implied warranty," which basically states that a product must be capable of performing its intended function when used properly and under normal operating conditions. By 1952, every state in the United States except Louisiana adopted what is termed the Uniform Commercial Code (UCC).

Several forms of legislation have been enacted during the past few decades to regulate warranties on various products, the most notable such legislation being the Magnuson–Moss Warranty-Federal Trade Commission Improvement Act of 1975 and the TREAD Act of 1999.[2]

## 2.4 Classification of Warranties

### 2.4.1 *Implied Warranties*

Implied warranties are unspoken, unwritten promises. In the US, these are created by state law. Implied warranties are based on the common law principle of "fair value for money spent." There are two types of implied warranties that occur in consumer product transactions, the implied warranty of merchantability, and that of fitness. These are briefly defined as follows:

1. The implied warranty of merchantability is a merchant's basic promise that the goods sold will do what they are supposed to do and that there is nothing significantly wrong with them. In other words, it is an implied promise that the goods are fit to be sold.
2. The implied warranty of fitness is a promise that a product can be used for a particular purpose and is applicable when the customer relies on the manufacturer or seller's advice that this is the case.

### 2.4.2 *Express Warranties*

Section 2-313 of the UCC covers the various actions that a seller may take to create an express warranty. Specifically, Section 2-313 states [12]:

[2] See "A Businesspersons Guide to Federal Warranty Law," available at http://www.ftc.gov/bcp/conline/buspubs/warranty.htm, for an excellent discussion of express and implied warranties, the Magnuson–Moss Act and related issues."

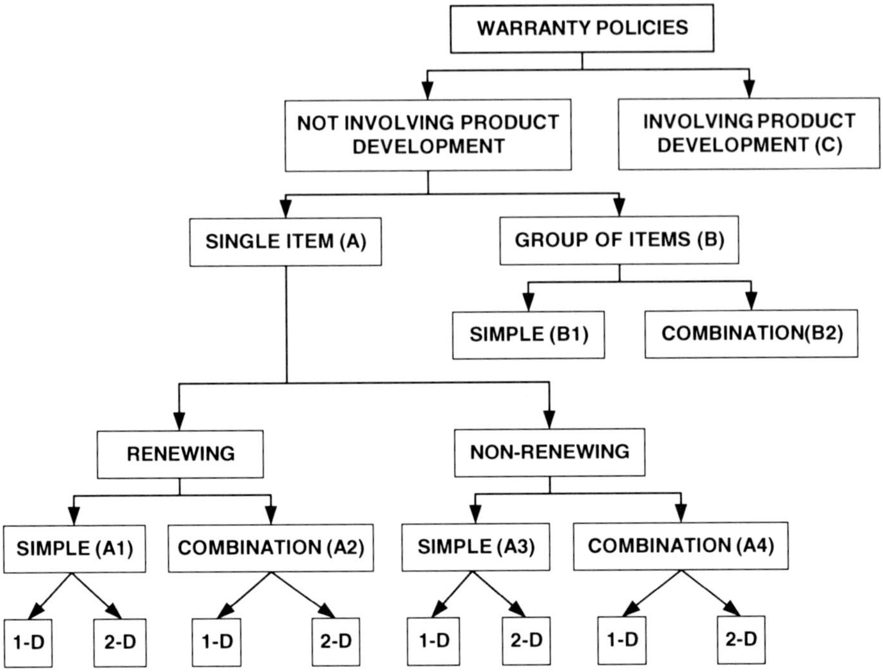


**Fig. 2.1** Warranty taxonomy

Express warranties by the seller are created as follows:

(a) Any affirmation of fact or promise made by the seller to the buyer which relates to the goods and becomes part of the basis of the bargain creates an express warranty that the goods shall conform to the affirmation or promise.

(b) Any description of the goods which is made part of the basis of the bargain creates an express warranty that the goods shall conform to the description.

(c) Any sample or model which is made part of the basis of the bargain creates an express warranty that the whole of the goods shall conform to the sample or model.

### *2.4.3 Classification of Express Warranties*

A taxonomy of warranty policies is given in Fig. 2.1. The first criterion for classification of express warranties is whether or not the warranty contract includes a provision that the manufacturer may be required to carry out further product development (for example, to improve product reliability) subsequent to the sale of the product. Policies which do not involve such further product development can be further divided into two groups—Group A, consisting of policies applicable for single item sales, and Group B, policies used in the sale of groups of items (called lot or batch sales).

Policies in Group A are divided into two sub-groups, based on whether the policy is renewing or non-renewing (i.e., whether or not the warranty term begins anew on replacement or repair of a failed item). A further subdivision comes about in that warranties may be simple or combination. The basic free replacement and pro-rata warranties discussed in the next section are simple policies. A combination policy is a simple policy combined with some additional features or a policy that combines the terms of two or more simple policies. The resulting four different types of policies under category A are labeled A1–A4 in Fig. 2.1. Each of these four groupings can be further subdivided into two sub-groups based on whether the policy is one-dimensional [1-D] or two-dimensional [2-D]. Policies belonging to Group B can also be sub-divided into simple or combination, labeled B1 and B2 in Fig. 2.1. As in group A, B1 and B2 can be further subdivided based on whether the policy is one-dimensional or two-dimensional.

Finally, policies that involve product development subsequent to the sale are labeled Group C. Warranties of this type are typically part of a service maintenance contract and are used principally in commercial applications and government acquisition of large, complex items—for example, aircraft or military equipment. Nearly all such warranties involve time and/or some function of time as well as a number of characteristics that may not involve time, for example, fuel efficiency.

The number of warranty policies offered by manufacturers and/or reported in the literature is large. In the remainder of this chapter we discuss a few of these. Interested readers should consult [1] for details regarding the policies not discussed in this book.

In describing the policies, we use the following notation:

$W$, $U$: Warranty parameters
$WP$: Warranty period (duration of coverage till the warranty expires)
$C_s$: Sale price
$T$: Age at failure
$X$: Usage at failure
$T_i$: Lifetime of item $i$
$S_n$: Lifetime of $n$ items $\left(=\sum_{i=1}^{n} T_i\right)$

## 2.5 One-Dimensional Warranties

In the one-dimensional case, a policy is characterized by an interval defined in terms of a single variable—time or age.[3] Some commonly used renewing and non-renewing one-dimensional warranties are defined below.

[3] The variable can also be usage—for example, number of copies made in the case of a photocopier, numbers of hours flown in the case of jet engines.

## 2.5.1 Non-Renewing Warranties

Policy 1: Non-renewing Free-Replacement Warranty (FRW) Policy

The seller agrees to repair or provide replacements for failed items free of charge up to a time $W$ from the time of the initial purchase. The warranty expires at time $W$ after purchase.

*Comment*: The warranty period is $WP = W$.

*Example 2.1* [Titan Sales Corporation—Electric heater for home use] This electric heater is warranted against defects in material and workmanship for one year from the date of original purchase. Titan will repair or replace, at its option, any defective unit delivered prepaid to an authorized service station during the warranty period.

The warranty does not apply to commercial use, unreasonable use, or damage to the product (not resulting from a defect or malfunction) while in the possession of the consumer.

Policy 2: Non-renewing Pro-rata Warranty (PRW) Policy

The seller agrees to refund an amount $\alpha(T)C_s$ if the item fails at age $T$ prior to time $W$ from the time of purchase, where $\alpha(T)$ is a non-increasing function of $T$, with $0 < \alpha(T) < 1$.

*Comments*:

1. Since failures occur in an uncertain manner, the refund in Policy 2 is uncertain.
2. The warranty period is $WP = W$.

*Example 2.2* [Sears warranty for small electric appliances] For one year from the date of purchase, Sears will replace this electric appliance, free of charge, if defective in material or workmanship. Warranty service is available by returning the appliance to the nearest Sears store throughout the United States. The warranty gives you specific legal rights, and you may also have other rights which vary from state to state.

## 2.5.2 Renewing Warranties

Policy 3: Renewing Free-Replacement Warranty (FRW) Policy

Under this policy, the manufacturer agrees to either repair or provide a replacement free of charge up to time $W$ from the initial purchase. Whenever there is a replacement, the failed item is replaced by a new one with a new warranty whose terms are identical to those of the original warranty.

Policy 4: Renewing Pro-rata Warranty (PRW) Policy

Under this policy, the manufacturer agrees to provide a replacement item, with warranty, at prorated cost $[1 - (T/W)]C_s$, for any item (including the item

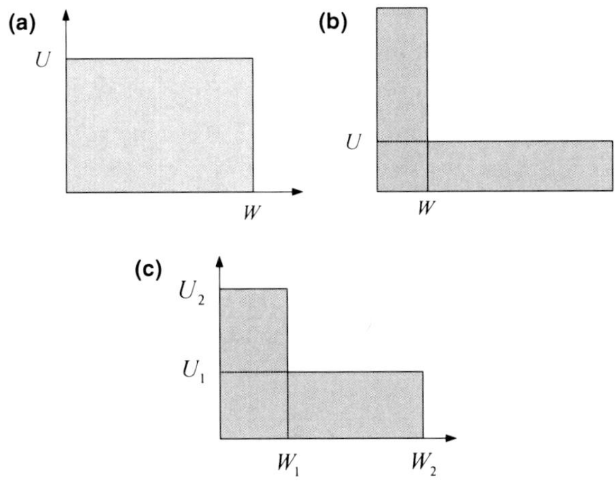


**Fig. 2.2** Three two-dimensional warranty regions

originally purchased and any replacements made under warranty) which fails to achieve a lifetime of at least $W$.

*Comment*: In contrast to the non-renewing case, here the warranty period $WP$ (the period from the instant of sale to the expiry of warranty) is uncertain, with $WP \geq W$.

## 2.6 Two-Dimensional Warranties

In the one-dimensional case, a policy is characterized by an interval that is defined in terms of a single variable—for example, time, age, or usage. In the case of two-dimensional warranties, the warranty is characterized by a region in a two-dimensional plane, usually with one axis representing time or age and the other representing item usage. Three possibilities with regard to the region of coverage are shown in Fig. 2.2.

For the region shown in Fig. 2.2a, the warranty policy is characterized by two parameters, $W$ and $U$, which are the respective limits on the maximum time and the maximum usage for the buyer. The region shown in Fig. 2.2b is also characterized by two parameters, which are the assured minimum time of coverage and minimum usage. The region in (a) tends to favor the manufacturer because it limits both the maximum time and the maximum usage for the buyer. For a buyer who is a heavy user, the warranty expires before time $W$ because usage has reached $U$. Similarly, for a buyer who is a light user, the warranty expires at time $W$ with total usage below $U$. In contrast, the region shown in (b) favors the buyer. Here, a heavy user is covered for a time period $W$, by which time the usage may have well exceeded the limit $U$, and a light user is covered well beyond $W$, for the policy expires only when the total usage reaches $U$.

The region shown in Fig. 2.2c is a compromise between (a) and (b) and involves four parameters. In this case, the buyer is provided warranty coverage for a minimum time period $W_1$ and for a minimum usage $U_1$. At the same time, the manufacturer is obliged to cover the item for a maximum time period $W_2$ and for a maximum usage $U_2$.

The following policy utilizes the warranty region given in Fig. 2.2a:

Policy 5: Two-dimensional Non-renewing Free-Replacement Warranty (FRW) Policy

The seller agrees to repair or provide a replacement for failed items free of charge up to a time $W$ or up to a usage $U$, whichever occurs first, from the time of the initial purchase. $W$ is called the warranty period and $U$ the usage limit.

*Comment*: If the usage is heavy, the warranty can expire well before $W$, and if the usage is very light, then the warranty can expire well before the limit $U$ is reached. Should a failure occur at age $T$ with usage $X$, it is covered by warranty only if $T$ is less than $W$ and $X$ is less than $U$. If the failed item is replaced by a new item, the replacement item is warranted for a time period $W - T$ and for usage $U - X$. Nearly all auto manufacturers offer this type of policy, with usage corresponding to distance driven.

*Example 2.3* [Automobile] The warranty for the whole automobile (called bumper-to-bumper) covers all parts of the automobile except for certain items (e.g., tires, radio, battery) covered by suppliers. Warranty coverage varies from manufacturer to manufacturer and by brand of the car. The power train (which includes the engine, transmission and other parts of the drive train only) is usually covered by a longer warranty. Warranty terms have changed significantly over time, with some manufacturers now offering 5 years and 50,000 miles for bumper-to-bumper coverage and 10 years and 100,000 miles for the power train.

Policy 6: Two-dimensional Non-renewing Pro-rata Warranty (PRW) Policy

The seller agrees to refund the buyer a fraction of the original sale price should the item fail at age $T < W$ with usage $X$ at failure less than $U$. The fraction refunded is a function of $W - T$ and/or $U - X$.

## 2.7 Group Warranties

Many commercial and governmental transactions involve purchase of a group of items. In such cases, it may be desirable to provide warranty coverage for the group as a whole. The rationale for such a policy is as follows: The advantage to the buyer is that multiple-item purchases can be dealt with as a unit rather than having to deal with each item individually under a separate warranty contract. The advantage to the manufacturer is that fewer warranty claims may be expected because longer-lived items can offset early failures.

Policies of this type, often called *cumulative* warranties, are conceptually straightforward extensions of the non-renewing free-replacement and pro-rata warranties discussed earlier. Under a cumulative warranty, the lot of $n$ items is warranted for a total time of $nW$, with no specific service time guarantee for any individual item. The following is an illustrative example.

Policy 7: Group Free-Replacement Warranty (FRW) Policy

A lot of $n$ items is warranted for a total aggregate period $nW$. The $n$ items in the lot are used one at a time. If $S_n < nW$ free-replacement items are supplied, also one at a time, until the first instant when the total lifetimes of all failed items plus the service time of the item then in use is at least $nW$.

Details of other policies of this type can be found in [1].[4]

## 2.8 Reliability Improvement Warranties

The basic idea of a reliability improvement warranty (RIW) is to extend the notion of a basic consumer warranty (usually the FRW) to include guarantees on the reliability of the item and not just on its immediate or short-term performance. This is particularly appropriate in the purchase of complex, repairable equipment that is intended for relatively long use. The intent of reliability improvement warranties is to negotiate warranty terms that will motivate a manufacturer to continue improvements in reliability after a product is delivered.

Under RIW, the contractor's fee is based on his ability to meet the warranty reliability requirements. These often include a guaranteed MTBF (mean time between failures) as a part of the warranty contract. The following is an illustrative example:

Policy 8: RIW Policy [6]

Under this policy, the manufacturer agrees to repair or provide replacements free of charge for any failed parts or units until time $W$ after purchase. In addition, the manufacturer guarantees the mean time between failures (MTBF) of the purchased item to be at least $M$. If the computed MTBF is less than $M$, the manufacturer will provide, at no cost to the buyer, (1) engineering analysis to determine the cause of failure to meet the guaranteed MTBF requirement, (2) engineering change proposals, (3) modification of all existing units in accordance with approved engineering changes, and (4) consignment spares for buyer use until such time as it is shown that the MTBF is at least $M$.

## 2.9 Extended Warranties

An extended warranty (which is also sometimes referred to as a "service contract") is a related concept. The difference between a warranty and a service contract is that the latter is entered into voluntarily and is purchased separately—the buyer may

---

[4] For more on cumulative warranties, see [7].

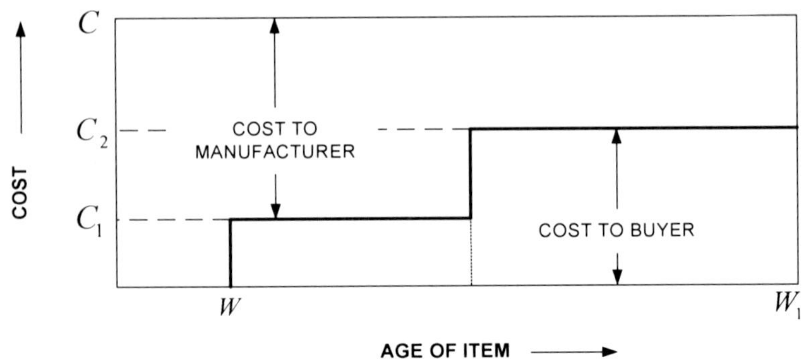


**Fig. 2.3** An illustrative example of cost-sharing extended warranty

even have a choice of terms, whereas a warranty is part of product purchase and integral to the sale.

With an extended warranty, the warranty coverage (for the non-renewing case) is $W_1$ $(> W)$, with $W_1 - W$ being the duration of the extended warranty and $W$ the duration of the base warranty. The terms of the extended warranty can be the same as those for the base warranty provided by the manufacturer for a new product at no additional cost to the buyer, or they may differ in the sense that the extended warranty may include features such as (i) cost sharing (see Fig. 2.3), (ii) exclusions (labor cost to be borne by the buyer), (iii) limits on individual claims and the total claim under warranty, and (iv) deductibles. Extended warranties are currently offered on a wide range of products, including automobiles, electronics, appliances, and many other items.

## 2.10 The Warranty Servicing Process

The warranty servicing process is shown in Fig. 2.4. The manufacturer's ability to service warranty is affected by the geographical distribution of customers and by the level of demand for prompt response. The manufacturer ordinarily requires a dispersed network of service facilities that store spare parts and provide a base for field service. The service delivery network requires a diverse collection of human and capital resources and careful attention must be paid to both the design and the control of the service delivery system. This involves several strategic and operational issues.

The strategic issues are (i) the number of service centers and their location, (ii) the capacity and manning of each service center to ensure adequate response time for customer satisfaction, and (iii) whether to own these centers or out-source them, in which case the servicing is carried out by an independent agent. The operational issues are (i) transportation of the material needed for warranty servicing, (ii) spare parts inventory management, (iii) scheduling of jobs, and (iv) optimal repair/replace decisions. For more on this, see [10].

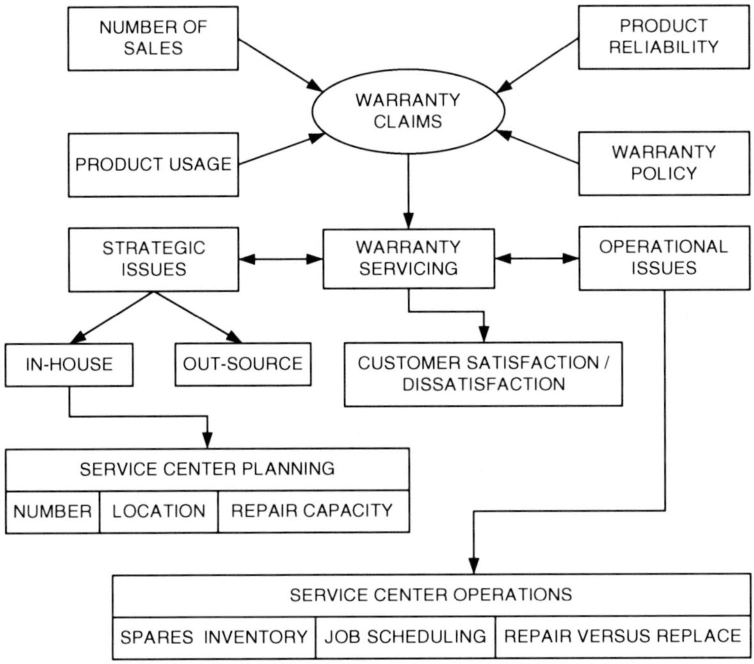


**Fig. 2.4** Warranty servicing process

## 2.11 Warranty Costs

As mentioned in Chap. 1, the cost of warranty as a percentage of the sale price can typically vary from 1 to 10%, depending on the product and the manufacturer. The total annual warranty cost for the automobile industry around the world in 2009 was around 30 billion US dollars.[5]

Whenever an item is returned under warranty, the manufacturer incurs various costs (handling, material, labor, facilities, etc.). These costs are random, i.e., unpredictable quantities. The following three costs are of importance to both consumers and manufacturers:

1. Warranty cost per unit sale
2. Life cycle cost per unit sale
3. Life cycle cost over repeat purchases

In the following, we look at each of these.

[5] See Warranty Week (http://www.warrantyweek.com) for warranty costs for the automotive and other industry sectors.

### 2.11.1 Warranty Cost per Unit Sale

The basic cost of warranty to the manufacturer is the sum of the costs associated with the servicing of an item that fails under warranty. These costs include the cost of replacement items or parts, the cost of testing to determine that the item has failed and how, the costs of repair personnel, the repair facility, spares warehousing, administrative costs, and many other possible cost elements.

The type of warranty determines the warranty period over which the item is covered, as well as the compensation to the buyer in the event of failure of the item. For a renewing warranty, the warranty cost is the cost of servicing all warranty claims for an item over the total warranty period, including the original item and all of its replacements under warranty, continuing until an item reaches age $W$ without failure. For a non-renewing warranty, claims occur on failure of the item and continue, if additional failures occur, until the total lifetime of the item and its replacements reaches $W$. The warranty cost is the total cost of servicing these claims. Thus in both cases warranty cost is the sum of a random number of individual costs.

The cost to the manufacturer per unit sold is important in the context of pricing the product. The sale price must exceed the sum of this cost and the cost of production in order to ensure that, in the long run, the manufacturer will not incur a loss as a result of the warranty.

### 2.11.2 Life Cycle Cost per Unit Sale

The life cycle cost of an item is of particular interest to the buyer. Many products are used for long periods of time. For example, aircraft and locomotives may be often used for 30–50 years, automobiles for 10–15 years, and so forth. Over the lifetime $L$ of the product, one or more components will ordinarily need to be replaced more than once. This is true, for example, of aircraft engines, automobile tires and batteries, and so forth. Many component replacements occur after the original warranty has expired, and the replacements are covered by a separate warranty. As a result, we have repeat purchases of a component over the period $L$. The time between repeat purchases of a component is uncertain. It depends on the time of first failure of the component outside the original warranty period in the case of non-repairable components, and on consumer replacement decisions in the case of repairable components.

Warranty costs over the life cycle for the consumer are different from those for the manufacturer. For both, however, the costs are uncertain and depend on $L$, $W$, and other factors, such as product reliability, consumers' replacement decisions, etc. The cost to the manufacturer includes the production costs associated with units sold at full price and the cost of servicing claims under warranty over the life cycle.

### 2.11.3 Life Cycle Cost over Multiple Purchases

From a marketing perspective, the product life cycle is the period (usually also denoted $L$) from the instant a new product is launched to the instant it is withdrawn from the market because of obsolescence and/or replacement by another product. Over the product life cycle, product sales (first and repeat purchases) occur dynamically over time. The manufacturer must service the warranty claims associated with each such sale.

Warranty claims occur over a period that is greater than $L$ and depends on the type of warranty. In the case of products sold with one-dimensional non-renewing warranty, this period is simply $L + W$. If the warranty is renewing, the period is somewhat longer, extending until the lifetime of the last item covered by warranty exceeds $W$.

The expected number of warranty claims per unit time also changes dynamically since it is a function of sales over time, product reliability, and other factors such as usage intensity, the usage environment, and so forth. This number is needed for planning of spares, repair facilities, and other service elements. The expected warranty cost per unit time, needed for determining warranty reserve requirements, also changes dynamically over time.

## 2.12 Warranty Management

### 2.12.1 Stages of Management

Warranty management is an important issue for manufacturers. Three stages in the evolution of warranty management have been recognized [8]. These are:

*Stage 1 [Administration]*. For companies in the first stage of warranty management, the focus is the on the administration of warranty claims. The aim is to control warranty servicing costs through detection of fraud (by customers and/or service agents) and efficient servicing of valid claims.

*Stage 2 [Operational Improvement]*. At the second stage of warranty management, the focus has moved to understanding the causes that lead to warranty claims and the resulting costs and customer dissatisfaction. Warranty data collected from service centers are used for improvements that lead to reduced warranty servicing costs and increased customer satisfaction.

In both of these stages, warranty is viewed as an afterthought and warranty management is neither integrated into the overall new product management process nor is it strategic in focus. Warranty impacts the commercial side of a business, both marketing and financial, since sales and revenue generated are influenced by warranty terms and the cost of servicing warranty claims affects overall profits. Warranty claims depend on product reliability and are thus influenced by the technical side of the business as well, since reliability is a function of

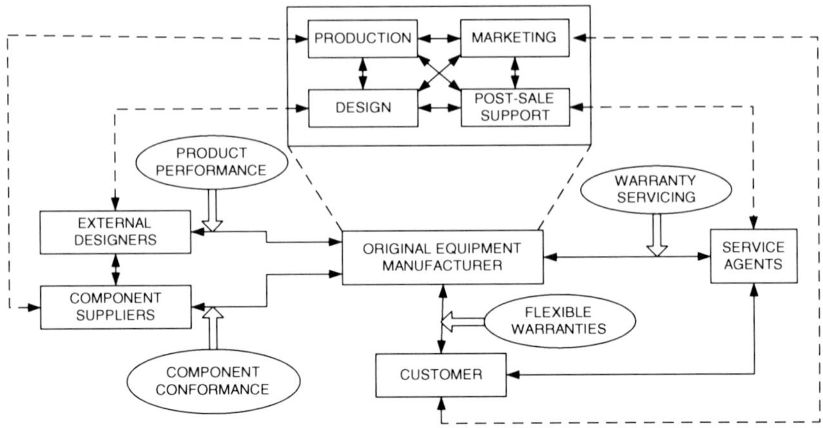


**Fig. 2.5** Warranty management

design and production. Warranty can no longer be viewed as an afterthought. Instead, it should be viewed as an important element of all new products, and in that context, it is important that it be managed strategically.

*Stage 3 [Strategic Warranty Management].* At the highest level of warranty management, the manufacturer looks at warranty from a strategic management perspective. Product warranty management must be done in the overall product life cycle context. This implies defining a warranty strategy, in conjunction with all other technical and commercial strategies, so as to achieve the overall business goals.

Figure 2.5 shows the interactions between the different key elements of the process. The overall system can be viewed as defining a *warranty chain* involving several external parties. A manufacturing process is comprised of several units or sections. Four of these that are important in the context of warranty management are shown in the box at the top of the figure. (There are many other units, such as legal, human resources, and so forth that are not shown in the box.) The dashed lines in Fig. 2.5 show the links between four of these units and the different external parties involved.

Some of the important issues in the context of external parties involved are:

1. *Product Performance*: This depends on the decisions made by the designer, and may have a significant impact on warranty costs. When external designers are involved, how does one apportion the warranty costs resulting due to design problems? If the problem is due to poor design from an external designer, the costs should be borne by the external designer, but it is often difficult to establish whether the internal or the external designers are responsible for a particular design problem.
2. *Component Conformance*: Non-conforming items result in higher warranty cost. When a warranty claim occurs and the reason for it is identified as non-

conformance of a component supplied by an external party, how are the warranty costs to be shared?
3. *Warranty Servicing*: The service agent might not deliver the appropriate level of service quality. This impacts customer satisfaction and subsequent sales. How are the costs to be shared between the two parties (manufacturer and service agent)? Increased monitoring will minimize the risk of the service agent shirking on the quality of service. However, this adds to the total cost to the manufacturer.
4. *Flexible Warranties*: Warranties should be designed to meet the needs of the customer rather than those of the manufacturer. In addition, this is a potential revenue-generating source, since customers (especially industrial and commercial) are willing to pay extra for better warranty service.

Through proper contracts, the manufacturer can ensure that external parties carry their share of the warranty costs as well as the indirect costs resulting from product failures attributable to the external parties involved.

Strategic warranty management deals with all of these and other issues in warranty, in an integrated manner, taking into account all of the implications of the decisions and actions of all of the parties involved.

### 2.12.2 Role of Warranty Data in Management

Warranty data play a very critical role in the effective management of warranty for both existing and new products. As indicated in Chap. 1, warranty data can be categorized into two groups (i) warranty claims data and (ii) supplementary warranty data. These are discussed in Chaps. 4 and 5, respectively, and their use in improving operations for existing products and in new product development are discussed in Chaps. 15 and 16, respectively.

## References

1. Blischke WR, Murthy DNP (1994) Warranty cost analysis. Marcel Dekker, New York
2. Blischke WR, Murthy DNP (eds) (1996) Product warranty handbook. Marcel Dekker, New York
3. Brennan JR (1994) Warranties: planning analysis and implementation. McGraw-Hill, New York
4. Burton JR (1996) Warranty protection: a consumerist perspective. In: Blischke WR, Murthy DNP (eds) Product warranty handbook. Springer, New York
5. Djamaludin I, Murthy DNP, Blischke WR (1996) Bibliography on warranties. In: Blischke WR, Murthy DNP (eds) Product warranty handbook. Springer, New York
6. Gandara A, Rich MD (1977) Reliability improvement warranties for military procurement. RAND Corp, Santa Monica Report No. R-2264-AF

7. Guin L (1984) Cumulative warranties: conceptualization and analysis. Doctoral dissertation, University of Southern California, Los Angeles, CA
8. Murthy DNP, Blischke WR (2005) Warranty management and product manufacture. Springer, London
9. Murthy DNP, Djamaludin I (2002) Product warranty—a review. Int J Prod Econ 79:231–260
10. Murthy DNP, Solem M, Roren T (2003) Product warranty logistics: issues and challenges. Eur J Oper Res 156:110–126
11. Thomas MU, Rao SS (1999) Warranty economic decision models: a summary and some suggested directions for future research. Oper Res 47:807–820
12. White JJ, Summers RS (1972) Uniform commercial code. West Publishing Company, St. Paul

# Chapter 3
# Reliability

## 3.1 Introduction

Offering warranty results in additional costs to the manufacturer due to the servicing of claims resulting from product failures under warranty. Product failures are depend on product reliability and this, in turn, is influenced by several factors, some under the control of the manufacturer (decisions made during the design and production phases) and others under the control of the customer (operating environment, usage mode and intensity, and so forth).

During the design phase, an assessment of product reliability is made based on product design and available knowledge of component reliability (often supplied by vendors). This, in combination with limited test data collected during the development phase, forms the basis for deciding whether or not to launch the product. This decision must be made at an early stage because building in reliability is costly but the consequence of not having adequate reliability can be costlier (due to higher warranty costs, product recall, etc.) Warranty data provide a valuable source of information for assessing the reliability of an item in operation (called the "field reliability") and to make decisions regarding the reliability improvements needed to control the consequences of unreliability.

A good understanding of reliability theory is essential for designing proper systems for the collection and analysis of warranty data. These provide essential information for making effective management decisions. In this chapter, we briefly discuss some topics from reliability theory that will be used in later chapters.

The outline of the chapter is as follows. We begin with a brief discussion of some basic concepts in Sect. 3.2. It is important that product reliability be viewed from a product life perspective. This is discussed in Sect. 3.3, where we consider the life cycle of both standard and custom-built products. This provides a framework for characterization of the different notions of product reliability that are discussed in Sect. 3.4. Reliability modeling is important for a variety of reasons, including estimation of reliability based on parametric models, and prediction of warranty

W. R. Blischke et al., *Warranty Data Collection and Analysis*,
Springer Series in Reliability Engineering, DOI: 10.1007/978-0-85729-647-4_3,

costs. These issues are discussed further in later chapters of the book. Section 3.5 looks at the modeling process in general. The modeling of first failure needs to done differently from that of subsequent failures, since the latter depend on actions taken to rectify failures. Sections 3.6 and 3.7 deal with the modeling of first and subsequent product failures, respectively. Even a simple product is comprised of several components. In Sect. 3.8, we discuss the linking of product reliability to component reliabilities. Finally, in Sect. 3.9, we look at the relationship between warranty and reliability. The use of reliability models to predict warranty costs is discussed in Chaps. 6 and 7.

## 3.2 Basic Concepts

### *3.2.1 Product Deterioration*

All products degrade with age and/or usage. When product performance falls below a desired level, the product is deemed to have *failed*. Failures occur in an uncertain manner and are influenced by factors such as design, manufacture or construction, maintenance, and operation. In all of these, the human factor is an important element.

Failure is often a result of the effect of deterioration. The deterioration process leading to a failure is a complicated process that varies with the type of product and the material used. The rate at which deterioration occurs is a function of time and/or usage intensity.

### *3.2.2 Fault*

A *fault* is the state of the system characterized by its inability to perform its required function. (Note: This excludes situations arising from preventive maintenance or any other intentional shutdown period during which the system is unable to perform its required function.) A fault is therefore a state resulting from a failure.

It is important to differentiate between failure or fault and error. The International Electrotechnical Commission defines an error to be a "discrepancy between a computed, observed or measured value or condition and the true, specified or theoretically correct value or condition." [7] As a result, an error is not a failure, because it is within the acceptable limits of deviation from the desired performance (target value). An error is sometimes referred to as an incipient failure [19].

### 3.2.3 Failure Modes

A *failure mode* is a description of a fault. It is sometimes referred to as fault mode (for example, in [7]). Failure modes are identified by studying the performance of the item. A classification scheme for failure modes is shown in Fig. 1.7 of [2] and a brief description of the different failure modes is as follows:

1. *Intermittent failures*: Failures that last for only a short time. A good example of this is a switch that sometimes does not make proper contact.
2. *Extended failures*: Failures that continue until some corrective action rectifies the failure. These can be divided into the following two categories:

   – *Complete Failures* which result in total loss of function.
   – *Partial Failures* which result in partial loss of function.

   Each of these can be further subdivided into the following:

1. *Sudden failures*: Failures that occur without any warning.
2. *Gradual failures*: Failures that occur with signals to warn of the occurrence of a failure.

A complete and sudden failure is called a *catastrophic failure* and a gradual and partial failure is designated a *degraded failure*.

### 3.2.4 Failure Causes and Classification

According to IEC 50 (191), *failure cause* is "the circumstances during design, manufacture or use which have led to a failure". Failure cause is useful information in the prevention of failures or their reoccurrence. Failure causes may be classified based on the causes of failure as follows:

1. Design Failure: Due to inadequate design.
2. Weakness failure: Due to weakness (inherent or induced) in the system so that the system cannot stand the stress it encounters in its normal environment.
3. Manufacturing failure: Due to non-conformity during manufacturing.
4. Aging failure: Due to the effects of age and/or usage.
5. Misuse failure: Due to misuse of the system (operating in environments for which it was not designed).
6. Mishandling failures: Due to incorrect handling and/or lack of care and maintenance.

### 3.2.5 Failure Mechanism

According to IEC 50 (191), a *failure mechanism* is "the physical, chemical or other processes that may lead to a failure". There are other causes as well, such as human errors.

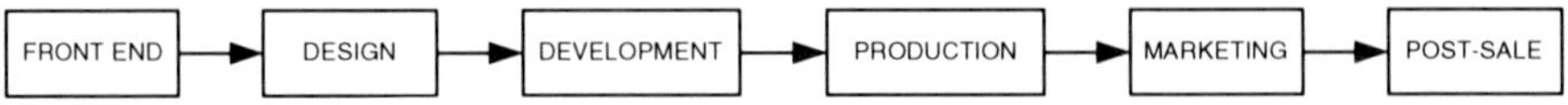


**Fig. 3.1** Product life cycle (standard product)

Mechanisms of failure can be divided into two broad categories, (1) overstress mechanisms, and (2) wear-out mechanisms [6]. In the former case, an item fails only if the stress to which the item is subjected exceeds the strength of the item. If the stress is below the strength, the stress has no permanent effect on the item. In the latter case, however, the stress causes damage that usually accumulates irreversibly. The accumulated damage does not disappear when the stress is removed, although sometimes annealing is possible. The cumulative damage does not cause any performance degradation as long as is it below the endurance limit. Once this limit is reached, the item fails. The effects of stresses are influenced by several factors—geometry of the part, constitutive and damage properties of the materials, manufacturing, and operational environment.

## 3.3 Product Life Cycle

The life cycle of a product is basically the period of time during which it is in existence, either conceptually or physically, and may be defined in various ways. Below we look at the product life cycles for standard and custom-built products. These differ somewhat, and both depend on the point of view taken—buyer, manufacturer, seller, and so forth.

### *3.3.1 Standard Products*

A product life cycle for a standard consumer durable or an industrial product, from the point of view of the manufacturer, is the time from initial concept of the product to withdrawal of the product from the marketplace. The life cycle involves several stages, as indicated in Fig. 3.1.

The process begins with the idea of building a product to meet some customer requirements, such as performance targets, including reliability. This is usually based on a study of the market and the potential demand for the product being planned. The next step is to carry out a feasibility study. This involves determining if it is possible to achieve the targets within specified cost limits. If this analysis indicates that the project is feasible, an initial product design is undertaken. A prototype is then developed and tested.

It is not unusual at this stage to find that achieved performance levels of the prototype product are below the target values. In this case, further product

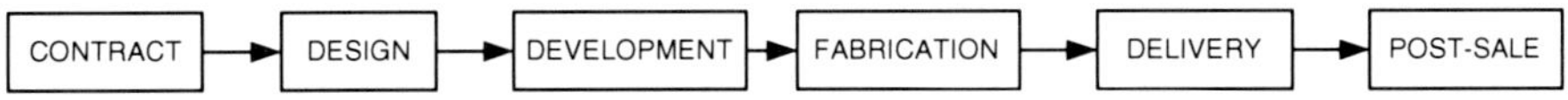


**Fig. 3.2** Product life cycle (custom built product)

development is undertaken to overcome the problem. Once this is achieved, the next step is to carry out trials to determine performance of the product in the field and to start a pre-production run. This is required because the manufacturing process must be fine-tuned and quality control procedures established to ensure that the items produced have the same performance characteristics as those of the final prototype.

After this, the production and marketing efforts begin. The items are produced and sold. Production continues until the product is removed from the market because of obsolescence and/or the launch of a new product. Post-sale support of the product continues at least until expiration of the warranty on the last item sold, but can continue beyond this point in terms of spare parts, service contracts, etc.

### *3.3.2 Custom Built Products*

The life cycle for a custom built product is slightly different and is as shown in Fig. 3.2. Here the product requirement is supplied by the customer and then jointly agreed upon by the customer and manufacturer. The manufacturer builds the product to these specifications under a negotiated contract. The process then follows basically the same steps as those for standard products.

## 3.4 Product Reliability

### *3.4.1 Concept and Definition*

Reliability of a product conveys the concept of dependability, successful operation or performance, and the absence of failures. It is an external property of great interest to both manufacturer and consumer. Unreliability (or lack of reliability) conveys the opposite. More technical definitions of reliability are the following:

1. The ability of an item to perform a required function, under given environmental and operational conditions and for a stated period of time. [8]
2. The reliability of a product (system) is the probability that the product (system) will perform its intended function for a specified time period when operating under normal (or stated) environmental conditions. [4]

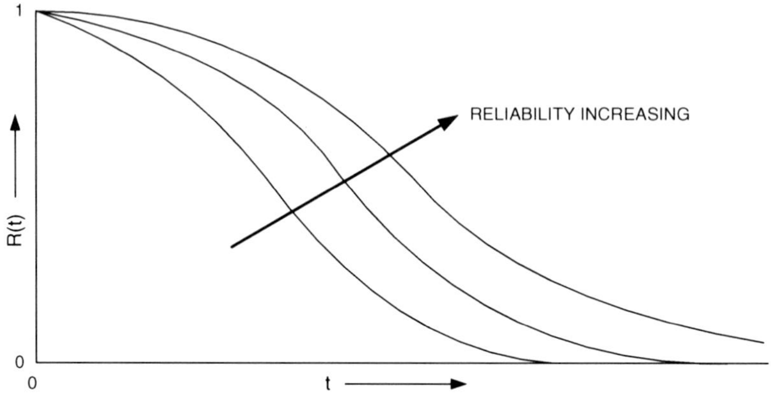


**Fig. 3.3** Plots of reliability functions

The reliability of product is given by a function $R(t)$ with the following properties:

1. $R(t)$ is a non-increasing function of $t$, $0 \leq t < \infty$
2. $R(0) = 1$ and $R(\infty) = 0$

Typical plots of $R(t)$ are shown in Fig. 3.3.

## 3.4.2 Product Life Cycle Perspective

From a product life cycle perspective, there are several different notions of reliability. Figure 3.4 [13] shows how these are sequentially linked and the factors that affect them. We briefly discuss four reliability concepts.

### 3.4.2.1 Design Reliability

At the design stage, the desired product reliability is determined through a tradeoff between the cost of building in reliability and the consequences of failures. This trade-off is discussed in detail in [13]. From this, one derives the reliability specification at the component level. One then evaluates the design reliability.[1]

### 3.4.2.2 Inherent Reliability

For standard products produced in volume, the reliability of the produced item can differ from the design reliability because of assembly errors and component

[1] The linking of component reliabilities to product reliability is discussed in Sect. 3.8.

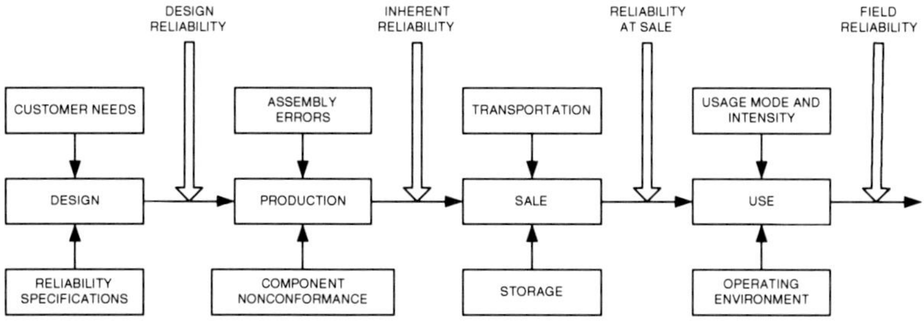


**Fig. 3.4** Different notions of reliability (standard product)

non-conformance. The reliability of produced items is the "inherent reliability" of the product.

#### 3.4.2.3 Reliability at Sale

After production, the product must be transported to the market and is often stored for some time before it is sold. The reliability of a unit at sale depends on the mechanical load (resulting from vibrations during transport) and impact load (resulting from mishandling) to which it has been subjected, the duration of storage, and the storage environment (temperature, humidity, etc.). As a result, the reliability at sale can differ from the inherent reliability. Once an item is sold, it may either be stored for an additional time (if the unit has been purchased for later use or is used as a spare), or it may be put into operation immediately. The additional storage time may again affect the reliability of the unit.

#### 3.4.2.4 Field Reliability

The reliability performance of a unit in operation depends on the length and environment of prior storage and on operational factors such as the usage intensity (which determines the load—electrical, mechanical, thermal, chemical—on the unit), usage mode (whether used continuously or intermittently), and operating environment (temperature, humidity, vibration, pollution, etc.) and, in some instances, on the human operator. The reliability performance of an item in operation is often referred to as "field reliability."

*Example 3.1* Washing machines are designed to some nominal functional and reliability requirements. The functional requirements might be, for example, a nominal load of 12 pounds per wash and a usage intensity of 6 washes per week. The reliability requirement might be, for example, that not more than one washer per thousand fails in the first year when the machine is operated under normal load and usage intensities. This defines the design reliability.

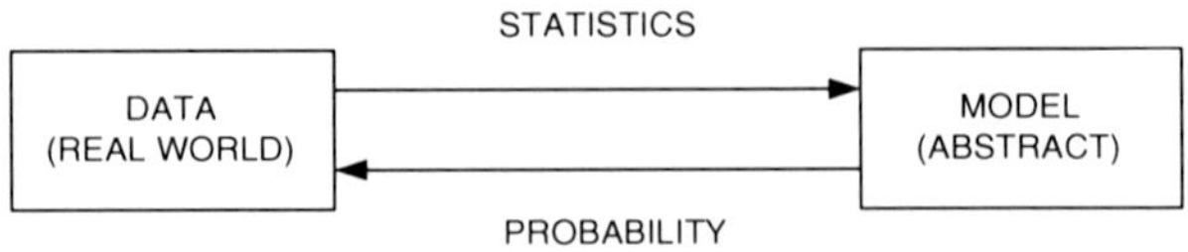


**Fig. 3.5** Link between real world and model

Due to variations in manufacturing, the inherent reliability can differ from the design reliability. If some of the bearings are defective, for example, they can wear faster, causing washing machines with these defective bearings to fail earlier.

In practice, the load will vary from wash to wash. If the load per wash is significantly greater than 12 pounds, then it can affect the performance and the reliability of components such as bearings, motor, etc. This may also occur if the usage intensity is significantly higher than the nominal value.[2]

## 3.5 Models and Modeling Process

### *3.5.1 The Role of Models*

Models play an important role in solving a variety of problems. A model is a representation of the real world that is relevant to the problem. There are many different types of models. Some of these are physical models and others abstract. We confine our attention to mathematical models.[3]

A mathematical model is an abstract representation involving a mathematical formulation. When uncertainty is a significant feature of the real world (as is the case, for example, in the time to failure of an item), then concepts from probability theory and statistics, as well as data from the real world, play an important role in linking the model to reality, as indicated in Fig. 3.5.[4]

### *3.5.2 Modeling Process*

Building a model is an iterative process involving several steps, as indicated in Fig. 3.6.[5]

---

[2] Note that for some products of this type designed for domestic use, the warranty becomes null and void if used in a commercial context (e.g., in a laundromat).

[3] There are many books that discuss models. See for example, [11] and the references cited therein.

[4] In this chapter, we confine our attention to models for product failures. In later chapters we deal with models for other purposes, such as estimating warranty costs, etc.

[5] There are many books that discuss the modeling process in detail; see for example [11] and the references cited therein.

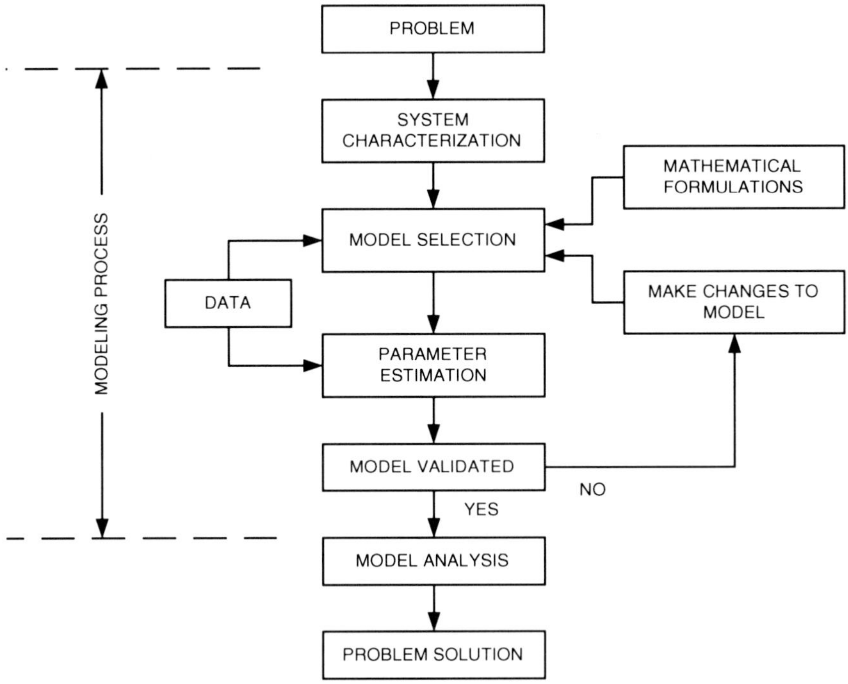


**Fig. 3.6** Modeling process

In the following, we discuss the key steps in the modeling process. These principles will be applied to reliability modeling in the following section.

Step 1: Defining the Problem

Problem definition depends on the context. In this chapter, the problem is to predict product failures over time.

Step 2: System Characterization

Characterization of a system details the salient features of the system that are relevant to the problem under consideration. This generally involves a process of simplification. The variables used in the system characterization and the relationships between them are problem dependent. If the problem were to understand product failures, then the system characterization would involve reliability theory; if the problem were to study the impact of warranty on sales, then one would use theories from marketing; and so forth. The characterization of the cause–effect relationship between the variables can be done in several ways. A common approach is to use diagrams with nodes representing variables and directed arcs indicating the cause–effect relationships.

Step 3: Model Selection

There are two approaches to model selection. These are:

*Empirical (black-box) Approach*: Model selection is based solely on the data available.

*Physics-based (white-box) Approach*: Model selection is based on relevant theories (for example, the different theories for component failures).

The kind of mathematical formulation to be used depends on the system characterization and the approach used. For modeling product failures based on the black-box approach, distribution functions are used to model the time to first failure and counting processes are used to model subsequent failures.

Step 4: Parameter Estimation

The model will involve one or more unknown parameters, and numerical values for these are needed. These are obtained by means of a statistical methodology called parameter estimation. The approach used depends on the type and amount of data available. This is discussed in Chap. 9.

Step 5: Model Validation

Validation involves testing whether or not the model selected (along with the assigned parameter values) models the real world sufficiently adequately to yield a meaningful solution to the problem of interest. The approach used can vary from a visual comparison between model predictions and observed data to statistical methods such as hypothesis testing and goodness-of-fit. These procedures are discussed in Chap. 10.

Step 6: Model Analysis

One can use several different approaches to analysis of the model. These include analytical methods (which yield closed form results as functions of the model parameters), computational methods, and simulation.

## 3.6 Modeling First Failure and Reliability

### 3.6.1 Basic Results

Let $T$ be a continuous random variable denoting the time to failure of an item. This is modeled by a *distribution function* $F(t;\theta)$ (also called a *cumulative distribution function* or *CDF*), which characterizes the probability that the item fails before $t$. The CDF is given by

$$F(t;\theta) = P\{T \leq t\}. \tag{3.1}$$

*Comment*: For notational ease, the dependence on the parameter $\theta$ is often suppressed and $F(t)$ is used instead of $F(t;\theta)$. We follow this convention in the remainder of the chapter.

$F(t)$ is called the *failure distribution* function. When $F(t)$ is differentiable, the result is called the *failure density* function, and denoted $f(t)$. This is given by

$$f(t) = \frac{dF(t)}{dt}. \tag{3.2}$$

The *reliability* function $R(t)$ (sometimes denoted $\bar{F}(t)$),[6] is defined to be the probability that the item survives for at least a period $t$, so that

$$R(t) = P\{T > t\} = 1 - F(t). \tag{3.3}$$

The conditional probability that the item will fail in the interval $[t, t + \delta t)$, given that it has not failed prior to $t$, is given by

$$F(\delta t | T > t) = \frac{F(t + \delta t) - F(t)}{R(t)} \tag{3.4}$$

The *hazard function* (or *failure rate function*) $h(t)$ associated with $F(t)$ is defined as

$$h(t) = \lim_{\delta t \to 0} \frac{F(\delta t | T > t)}{\delta t} = \frac{f(t)}{R(t)} \tag{3.5}$$

The hazard function $h(t)$ can be interpreted as the probability that the item will fail in $[t, t + \delta t)$, given that it has not failed prior to $t$. In other words, it characterizes the effect of age on item failure more explicitly than $F(t)$ or $f(t)$.

The *cumulative hazard function*, $H(t)$, is defined as

$$H(t) = \int_0^t h(t')dt' \tag{3.6}$$

$H(t)$ is also called the *cumulative failure rate function.*

Appendix A provides a list of distributions that have been used extensively in reliability modeling.

*Example 3.2* [Two-parameter Weibull Distribution] The two-parameter Weibull distribution is used extensively in reliability modeling. The CDF for this distribution is

$$F(t;\theta) = 1 - e^{-(t/\alpha)^{\beta}} \tag{3.7}$$

---

[6] We will use both notations throughout the book.

for $t \geq 0$. The parameter set is $\theta = \{\alpha, \beta\}$, with $\alpha > 0$ and $\beta > 0$. $\alpha$ is a scale parameter and $\beta$ is a shape parameter. The failure density and hazard functions are given by

$$f(t;\theta) = \frac{\beta t^{(\beta-1)} e^{-(t/\alpha)^\beta}}{\alpha^\beta} \tag{3.8}$$

and

$$h(t;\theta) = \frac{\beta t^{(\beta-1)}}{\alpha^\beta} \tag{3.9}$$

The shape of the hazard functions depend on the shape parameter and can have one of the following three shapes:

1. Increasing failure rate (IFR) when $\beta > 1$
2. Decreasing failure rate (DFR) when $\beta < 1$
3. Constant failure rate when (CFR) $\beta = 1$.

Figure 3.7 shows plots of the density and hazard functions for $\beta = 0.5$, 1, and 2. These values of the shape parameter illustrate the three regions indicated above.

### 3.6.2 Design Reliability

Let $F_0(t)$ denote the design failure distribution. Let $R_0(t)$, $f_0(t)$ and $h_0(t)$, denote, respectively, the reliability function, the density function and the hazard function associated with $F_0(t)$. The hazard function $h_0(t)$ is IFR (curve A in Fig. 3.8),[7] which reflects the effect of ageing. Good design requires that the hazard function be below some specified value over the useful life of the product.

### 3.6.3 Effect of Quality Variations in Manufacturing

Two causes of variations are (1) assembly error and (2) component non-conformance.

#### 3.6.3.1 Assembly Errors

Even a simple product consists of several components that are assembled in production. The type of assembly operation depends on the product. For an

---

[7] Figure 3.8 shows four plots (A–D). Plot A is the designed hazard function. Plots B–D indicate how this is affected as a result of quality variations, as discussed in the next subsection.

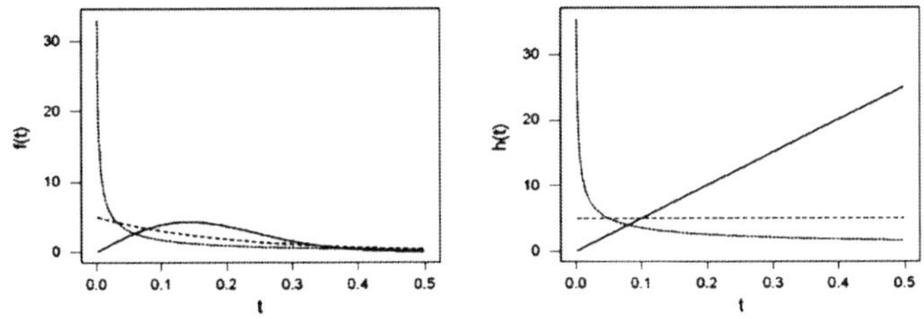


**Fig. 3.7** Plots of Weibull density and hazard functions for $\beta = 0.5$, 1, and 2 (*top*, *middle*, and *bottom curves*, respectively, along left axis in both plots)

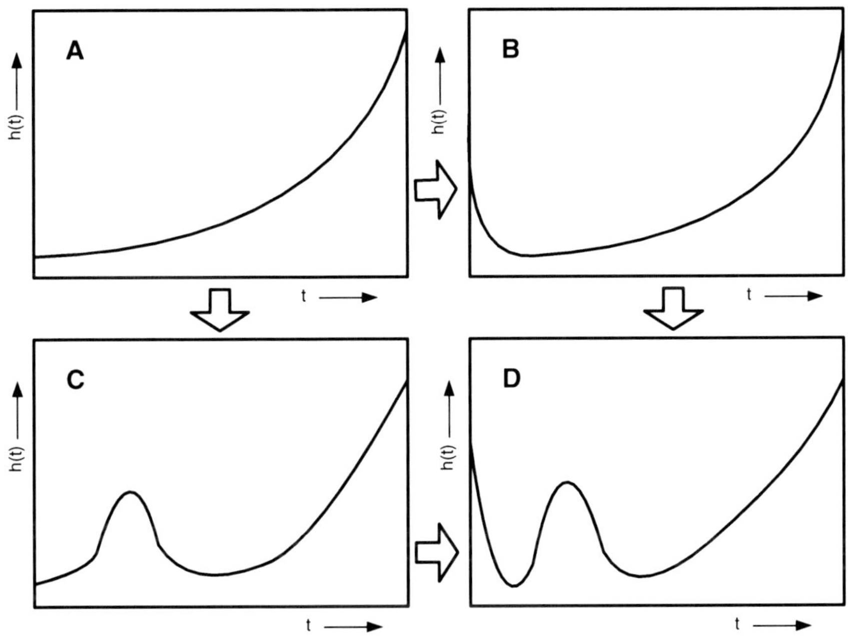


**Fig. 3.8** Shapes of hazard function with quality variations

electronic product, one of the assembly operations is soldering. If the soldering is not done properly (called dry solder), then the connection between the components can break within a short period, leading to a premature failure. For a mechanical component, a premature failure can occur if the alignment is not correct or the tolerances are violated.

Failures resulting from assembly errors can be viewed as a new mode of failure that is different from other failure modes that one examines during the design process. Let $F_1(t)$ denote the distribution function associated with this new failure mode, and $R_1(t)$, $f_1(t)$ and $h_1(t)$ the survivor function, density function and failure rate function associated with $F_1(t)$. The failure rate $h_1(t)$ is a decreasing function of $t$, implying that failure will occur sooner rather than later, and that the mean time to failure (*MTTF*) under this new failure mode is much smaller than the design *MTTF*.

Not all items are affected by assembly errors. Let $q$, $0 \le q \le 1$, denote the probability that an item has an assembly error. The reliability of produced items can be modeled by a modified competing risk model [12] given by[8]

$$R_a(t) = R_0(t)[1 - qF_1(t)] \tag{3.10}$$

*Comments*: (1) If $q = 0$, then $R_a(t) = R_0(t)$. If $q = 1$, then $R_a(t) = R_0(t)R_1(t)$, which is the standard competing risk model (see Appendix A). (2) The hazard function $h_a(t)$ associated with $F_a(t)$ is the sum of the design hazard function (which is increasing) and the hazard function for the new failure mode (which is decreasing). As a result, $h_a(t)$ has a bathtub shape (curve B in Fig. 3.8).

#### 3.6.3.2 Component Non-Conformance

Because of variations in quality, some components do not meet design specifications. Suppose, in particular, that their *MTTF* is much smaller than intended. Items that are produced with such nonconforming components will also tend to have an *MTTF* that is much smaller than the intended design value. To model this situation, we proceed as follows. Let $F_2(t)$ denote the failure distribution of items that have nonconforming components, and $R_2(t)$, $f_2(t)$ and $h_2(t)$denote, respectively, the survivor, density, and failure rate functions associated with $F_2(t)$. $h_2(t)$ is an increasing function of $t$, with $h_2(t) > h_0(t)$ for all $t$.

Let $p$, $0 \le p \le 1$, denote the probability that an item produced has nonconforming components, so that its failure distribution is given by $F_2(t)$. Then $(1 - p)$ is the probability that the item is conforming and has failure distribution $F_0(t)$. As a result, the reliability of the items produced is given by

$$R_n(t) = (1 - p)R_0(t) + pR_2(t) \tag{3.11}$$

*Comments*: (1) This is a standard mixture model involving two distributions (see Appendix A). If $p = 0$, then $R_n(t) = R_0(t)$, as to be expected, and if $p = 1$, then $R_n(t) = R_2(t)$, as all items have nonconforming components. (2) The hazard function $h_n(t)$ associated with $F_n(t)$ has an N-shape [increasing followed by decreasing and ultimately increasing (curve C in Fig. 3.8)].

---

[8] [5] deals with this model, which they call the "general limited failure population model." They give an interpretation of the model in the context of reliability theory where an item failure is due to one of two competing causes—common cause and another, called special cause. The time to failure due to common cause failure (for example, wear-out) has a distribution function $F_0(t)$ and a proportion $q$ can fail due to the other cause (for example, infant mortality) with a distribution function $F_I(t)$.

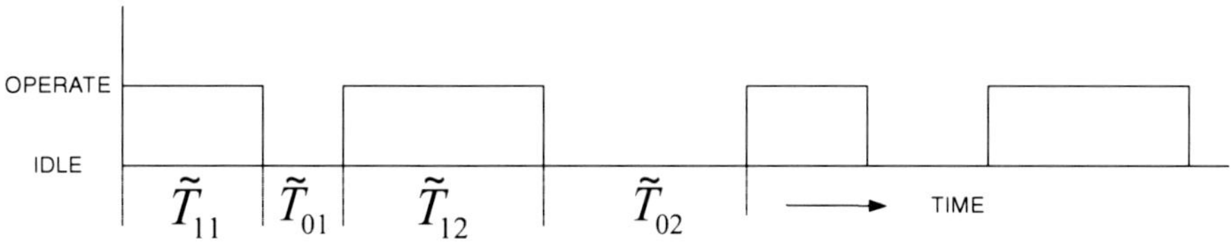


**Fig. 3.9** Intermittent usage time history

#### 3.6.3.3 Modeling the Combined Effect

With both assembly errors and component nonconformance may occur, the reliability of the items produced is given by

$$R_q(t) = [(1-p)R_0(t) + pR_2(t)](1 - qF_1(t)) \tag{3.12}$$

In this case, the hazard function $h_q(t)$ associated with $F_q(t)$ $[= 1 - R_q(t)]$ has a $W$-shape (curve D in Fig. 3.8).

*Comment*: The plots shown in Fig. 3.8 provide a basis for identifying quality variation problems from empirical plots of the hazard function based on warranty data.

### *3.6.4 Usage Mode*

Products are often used intermittently, resulting in usage pattern such as that shown in Fig. 3.9. Intermittent usage involves a cyclic change from the "Operate" state to the "Idle" state in an uncertain manner. Here $\tilde{T}_{1j}$ denotes the time in operating state and $\tilde{T}_{0j}$ the time in the idle state during the $j$th cycle.

Let $R_0(t)$ denote the reliability of the product when it is used continuously and $R_i(t)$ the reliability when used intermittently. In order to link the two, we need to model operate and idle times.

*Special Case* We assume the following:

1. $\tilde{T}_{1j}$ is a sequence of independent and identically distributed (iid) random variables from a distribution $G_1(t)$
2. $\tilde{T}_{0j}$ is a sequence of iid random variables from a distribution $G_0(t)$
3. There is no degradation when an item is in its idle state

Then it can be shown [14] that

$$R_i(t) = R_0(t) + \int_0^t R_0(z)h(z,t)dz \tag{3.13}$$

where

$$h(z,t) = g_1(z)\bar{G}_0(t-z) + \int_0^z \left[ \int_0^{t-x} h(z-x, t-x-y) g_0(y) dy \right] g_1(x) dx. \quad (3.14)$$

Since $h(z,t) > 0$, we have from (3.13) that $R_i(t) \geq R_0(t)$, as would be expected.

### 3.6.5 Usage Intensity (Operating Load)

A product is designed for some nominal usage intensity (for example, the number of washes per week and/or size of loads washed in a washing machine; the number of miles travelled per year in an automobile). Usage intensity can vary considerably across the customer population. When the usage intensity is higher (lower) than the nominal usage intensity, the degradation (due to higher wear and/or increased stresses on the components) is faster (slower). As a result, the actual field reliability can be lower or higher than the design reliability.[9]

We use the term "operating environment" to cover all of these. Let $s$ denote the stress on the components in operation. Let $s_0$ denote the stress (electrical, mechanical and/or thermal, depending on the product) on the components under nominal usage intensity. Define $\tilde{s} = s/s_0$. Let $R_e(t)$ denote the field reliability (which takes into account the influence of the operating environment) and $R_0(t)$ the design reliability. The two well known models linking field reliability to design reliability are the following:

- Model 1: Accelerated Failure Time (AFT) Model [16]
- Model 2: Proportional Hazard (PH) Model [10]

#### 3.6.5.1 AFT Model

Let $T_s$ denote the time to failure under stress $s$ and $T_0$ the failure time under nominal stress. The AFT model assumes the following

$$T_s = T_0 \phi(\tilde{s}) \quad (3.15)$$

where $\phi(\tilde{s})$ is a non-negative and monotonically increasing function with

$$\phi(\tilde{s}) \begin{cases} >1 & \text{when } \tilde{s} > 1 \\ =1 & \text{when } \tilde{s} = 1 \\ <1 & \text{when } \tilde{s} < 1 \end{cases} \quad (3.16)$$

[9] The same is true regarding the operating environment—for example, road conditions in the case of an automobile, operating temperature in the case of an electronic product.

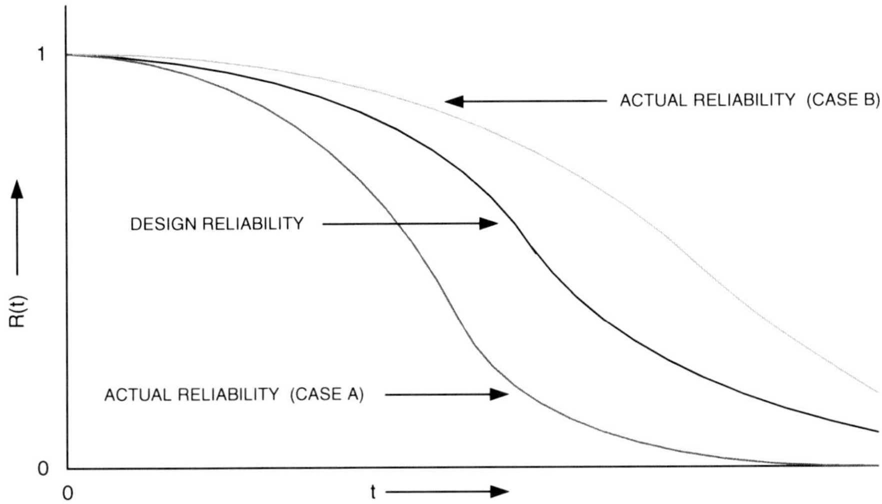


**Fig. 3.10** Design and actual (field) reliabilities

As a result, $R_e(t)$ has the same form as $R_0(t)$ and the two scale parameters are linked by a relationship similar to that in (3.15). The scale parameter for $R_e(t)$ decreases [increases] as $\tilde{s}$ increases [decreases].

Figure 3.10 shows the effect of $\phi(\tilde{s})$ on the field reliability, with case A corresponding to $s > s_0$ and case B corresponding to $s < s_0$.

#### 3.6.5.2 PH Model

Let $h_e(t)$ $[h_0(t)]$ denote the hazard function associated with $R_e(t)$ $[R_0(t)]$. The PH model assumes that

$$h_e(t) = h_0(t)\phi(\tilde{s}) \tag{3.17}$$

where $\phi(\tilde{s})$ is as in the AFT Model. As a result, $R_e(t) = [R_0(t)]^{\phi(\tilde{s})}$.

### 3.6.6 Other Notions of Usage

In addition to intermittent usage discussed in Sect. 3.6.4, one can define two other notions of usage, namely:

1. *Number of times an item is used*: Let $N(t)$ denote the number of times an item is used is over the interval $[0, t)$. Typical examples are (a) the landing gear used in the landing of an aircraft, and (b) number of loads done in a washing machine.

2. *Output of an item*: Let $U(t)$ denote the usage up to time $t$. The output is some measurable quantity. Typical examples of this are (a) miles an automobiles is driven, and (b) copies made on a photocopier.

In these cases, the item degradation and failure depend on the age and usage of the product. This can be modeled in several different ways. Approaches to modeling are discussed in Chap. 6.

## 3.7 Modeling Failures over Time

When a repairable item fails, it can either be repaired or replaced by a new item. In the case of a non-repairable item, the only option is to replace the failed item by a new one. Since failures occur in an uncertain manner, the number of failures over a time interval is a non-negative random variable. The distribution of this variable depends on the failure distribution of the item, the actions (repair or replace) taken after each failure, and the type of repair.

In this section, we model the number of failures over the interval $[0, t)$, starting with a new item at $t = 0$, for several different scenarios. Let $N(t)$ denote the number of failures over $[0, t)$. This is a counting process (see Appendix A). Let $p_j(t)$ denote the probability that $N(t) = j$, $j = 1, 2, \ldots$. Models for repairable and non-repairable items are as follows:

### 3.7.1 *Non-Repairable Product*

In the case of non-repairable product, every failure results in the replacement of the failed item by a new item. We assume that all new items are statistically similar, with distribution function $F(t)$. If the failures are detected and replaced immediately with replacement time negligible, then $N(t)$ is an ordinary renewal process, and we have the following results (see Appendix B):

$$p_j(t) = P\{N(t) = j\} = F^{(j)}(t) - F^{(j+1)}(t), \tag{3.18}$$

where $F^{(j)}(t)$ is the $j$-fold convolution of $F(t)$ with itself, and the expected number of failures over $[0, t)$ is given by

$$M(t) = F(t) + \int_0^t M(t - t')f(t')dt' \tag{3.19}$$

In general, it is difficult to obtain an analytical expression for $M(t)$ and computational approaches must be used to evaluate it [3].

## 3.7.2 Repairable Product

In this case, the characterization of the number of failures over time depends on the type of repair. The two types of repair are as follows:

### 3.7.2.1 Minimal Repair

Here the failure rate after repair is essentially the same as that if the item had not failed [1]. This is appropriate for complex products for which the product failure is due to failure of one or few of its components. The equipment becomes operational by replacing (or repairing) the failed components. This action ordinarily has very little impact on the reliability characteristics of the product.

If the failures are statistically independent, then $N(t)$ is a non-stationary Poisson process with intensity function $\lambda(t) = h(t)$, the failure rate associated with $F(t)$ [15]. As a result, we have the following (see Appendix B):

$$p_j(t) = P\{N(t) = j\} = \frac{e^{-\Lambda(t)}\{\Lambda(t)\}^j}{j!} \tag{3.20}$$

where

$$\Lambda(t) = \int_0^t \lambda(t')dt', \tag{3.21}$$

and the expected number of failures over $[0, t)$ is given by

$$E[N(t)] = \Lambda(t) \tag{3.22}$$

### 3.7.2.2 Imperfect Repair

Here the failure rate changes (in either direction) after repair. Many different types of imperfect repair models have been proposed [17]. The two that have been used extensively are the following:

*Reduction in failure rate*: If the repair time is negligible, then $h(t^+) = h(t^-) - \delta$, where $t$ is the time at which the failure occurs and $\delta$ is the reduction, subject to the constraint $0 \leq \delta < h(t^+) - h(0)$.

*Reduction in age*: This involves the notion of virtual age [9]. Let $A(t)$ denote the virtual age at time $t$. If the repair time is negligible, then $A(t^+) = A(t^-) - x$ if the failure occurs at time $t$ and the reduction in age is $x$, subject to the constraint $0 \leq x < A(t^-)$.

*Comment*: $\delta = 0$ and $x = 0$ imply minimal repair.

#### 3.7.2.3 Repaired Items Different from New

Here, the failed item is subjected to a major overhaul which results in the failure distribution of the repaired items being $F_r(t)$, say, which is different from the failure distribution, $F(t)$, for new items. Since repaired items are assumed to be inferior to new ones, the mean time to failure for a repaired item is taken to be smaller than that for a new item.

## 3.8 Linking Product Reliability and Component Reliabilities

Even simple products are built using many components, and the number used increases with the complexity of the product. As such, a product can be viewed as a system of interconnected components. In Chap. 1, we discussed a decomposition of a product or system involving several levels. The number of levels that is appropriate depends on the product. The performance of the product depends on the state of the system (working, failed, or in one of several partially failed states) and this in turn depends on the state (working/failed) of the various components.

The two approaches for linking product reliability to component are (1) reliability block diagrams and (2) fault tree analysis. We discuss these briefly below. For additional details, see [4].

### *3.8.1 Reliability Block Diagrams*

In a reliability block diagram, each component is represented by a block with two end points. When the component is in its working state, there is a connection between the two end points. This connection is broken when the component is in a failed state. A multi-component system can be represented as a network of such blocks, each with two end points. The system is in working state if there is a connected path between the two end points. If no such path exists, then the system is in a failed state. Systems may be of the following types:

*Series Structure*: This represents the case where the system is in its working state only when all the components are in working states.

*Parallel Structure*: This represents the case where the system is in a failed state only when all of the components are in failed states.

*General Structure*: This is a combination of series and parallel sub-structures and is needed for modeling more complex products.

### *3.8.2 Fault Tree Analysis (FTA)*

A fault tree is a logic diagram that displays the relationship between a potential event affecting system performance and the reasons or underlying causes for this event. The reason may be failures (primary or secondary) of one or more components of the system, environmental conditions, human errors, and other factors.

A fault tree illustrates the state of the system (denoted the TOP event) in terms of the states (working/failed) of the system's components (denoted basic events). The connections are done using *gates*, where the output from a gate is determined by the inputs to it. A special set of symbols (for gates and basic events) is used for this purpose.[10]

### 3.8.3 Structure Function and Product Reliability

Let $X_i(t),\ 1 \le i \le n$, denote the state of component $i$, at time $t$, with

$$X_i(t) = \begin{cases} 1 & \text{if component } i \text{ is in working state at time } t \\ 0 & \text{if component } i \text{ is in failed state at time } t \end{cases} \tag{3.23}$$

Let $\underset{\sim}{X}(t) = (X_1(t), X_2(t), \ldots, X_n(t))$ denote the state of the $n$ components at time $t$, and $X_S(t)$ (a binary random variable) denote the state of the system at time $t$. Then from FTA one can derive an expression of the form

$$X_S(t) = \phi(\underset{\sim}{X}(t)), \tag{3.24}$$

which links the component states to the system state. $\phi(\cdot)$ is called the structure function.[11]

Let $R_S(t)$ and $\underset{\sim}{R}(t) = (R_1(t), R_2(t), \ldots, R_n(t))$ denote the reliability of the system and of the set of reliabilities of the $n$ components, respectively. If the component failures are independent, then

$$R_S(t) = \phi(\underset{\sim}{R}(t)) \tag{3.25}$$

so that we have the system reliability in terms of the component reliabilities. Results for the two simplest systems are:

*Series Structure*

$$R_S(t) = \prod_{i=1}^{n} R_i(t) \tag{3.26}$$

[10] For more on the construction and analysis of fault trees, see [4] and [18].

[11] The details can be found in many books on reliability; see, for example, [4, 18].

*Parallel structure*

$$R_S(t) = 1 - \prod_{i=1}^{n} (1 - R_i(t)) \tag{3.27}$$

*Example 3.3* Suppose a system is constructed based on three components as shown by the following diagram.

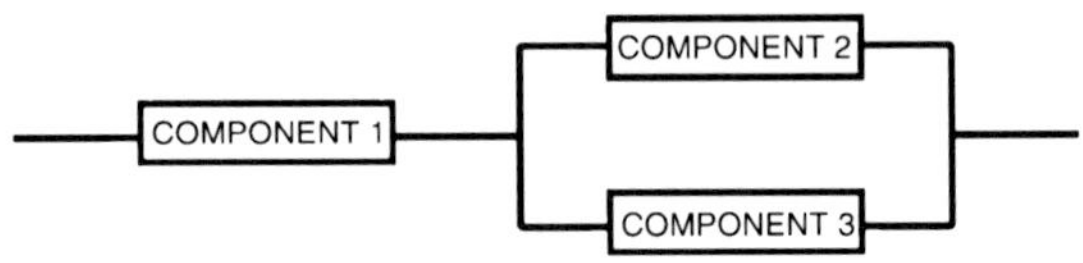


If the lifetimes of components 1, 2 and 3 all follow exponential distributions (A.22) with $\lambda = 0.001$, 0.002 and 0.003 failures per hour, respectively, then the reliability of the system for ten hours ($t = 10$) can be computed as follows:

Components 2 and 3 are a subsystem in parallel structure. The reliability of this subsystem at $t = 10$ (based on (3.27)) is

$$\begin{aligned} R_{2,3}(t = 10) &= 1 - \{(1 - R_2(10))(1 - R_3(10))\} = R_2(10) + R_3(10) - R_2(10)R_3(10) \\ &= e^{-0.002\times 10} + e^{-0.003\times 10} - e^{-0.002\times 10}e^{-0.003\times 10} = 0.99941 \end{aligned}$$

Component 1 and the sub-system with components 2 and 3 are in series structure. From (3.26), the reliability of the system at $t = 10$ is

$$R_s(t = 10) = R_1(10)R_{2,3}(10) = e^{-0.001\times 10} \times 0.99941 = 0.98947.$$

## 3.9 Warranty and Reliability

As mentioned in Chap. 1, offering warranty results in additional costs to the manufacturer. The various factors that affect these costs are shown in Fig. 3.11.

The key factors are:

1. Design reliability
2. Inherent reliability
3. Operating environment
4. Servicing strategy

In this chapter we have focused on (1)–(3). The effect of these on warranty costs are discussed in Chaps. 6 and 7.

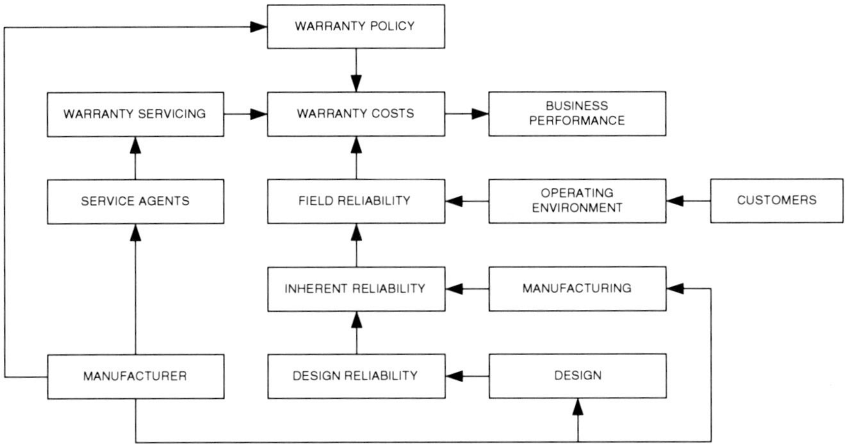


**Fig. 3.11** Reliability and warranty

## References


1. Barlow RE, Hunter L (1961) Optimum preventive maintenance policies. Oper Res 8:90–100
2. Blache K, Shrivastava AB (1994) Defining failure of manufacturing machinery and equipment. In: Proceedings of annual reliability and maintainability symposium, 69–75
3. Blischke WR, Murthy DNP (1994) Warranty cost analysis. Marcel Dekker, New York
4. Blischke WR, Murthy DNP (2000) Reliability. Wiley, New York
5. Chan V, Meeker WQ (1999) Failure-time model for infant-mortality and wearout failure modes. IEEE Trans Reliab 48:377–387
6. Dasgupta A, Pecht M (1991) Material failure mechanisms and damage models. IEEE Trans Reliab 40:531–536
7. IEC 50 (191) (1990) International Electrotechnical Vocabulary (IEV)—Chapter 19: Dependability and Quality of Service. IEC, Geneva
8. ISO 8402 (1986) Quality vocabulary. ISO, Geneva
9. Kijima M (1989) Some results for repairable systems with general repair. J Appl Probab 26:89–102
10. Kumar D, Klefsjo B (1994) Proportional hazards model: A review. Reliab Eng Sys Saf 29:177–188
11. Murthy DNP, Page NW, Rodin Y (1990) Mathematical modelling. Pergamon Press, Oxford, England
12. Murthy DNP, Xie M, Jiang R (2003) Weibull Models. Wiley, New York
13. Murthy DNP, Rausand M, Osteras T (2008) Product reliability—performance and specifications. Springer, New York
14. Murthy DNP, Wilson RJ (2009) Field reliability (under preparation)
15. Nakagawa T, Kowada M (1983) Analysis of a system with minimal repair and its application to a replacement policy. Eur J Oper Res 12:253–257
16. Nelson W (1990) Accelerated testing. Wiley, New York
17. Pham H, Wang H (1996) Imperfect maintenance. Euro J of Op Res 94:425–438
18. Rausand M, Høyland A (2004) System reliability theory: models, statistical methods, and applications (2nd ed.). Wiley, Hoboken
19. Rausand M, Oien K (1996) The basic concept of failure analysis. Reliab Eng Sys Saf 53:73–83

# Part II
# Warranty Data Collection

# Chapter 4
# Warranty Claims Data

## 4.1 Introduction

As discussed in Chap. 1, warranty data are the data that are needed for effective management of warranty for both existing and new products. Such data can be categorized into two classes—(1) warranty claims data and (2) warranty supplementary data. Warranty claims data (which we shall often refer to as "claims data") are the data collected during the processing of claims and servicing of repairs under warranty and are the focus of this chapter. We look at various issues relating to the collection and analysis of claims data. The outline of the chapter is as follows.

Section 4.2 provides a general discussion on data, information and knowledge. An understanding of this is important for the design of an effective data collection system. Section 4.3 deals with structured and unstructured data. In Sect. 4.4, we look at the overall warranty process. This involves two sub-processes—the claim process and the servicing process. Section 4.5 deals with warranty claims data collection. The data collected can be classified into several different categories; this is the focus of Sect. 4.6. There are a number of potential problems with claims data. These are examined in Sect. 4.7. The goals for warranty data collection are discussed in Sect. 4.8. In Sect. 4.9 we comment on the current practice in claims data collection.

## 4.2 Data, Information and Knowledge

Data and information are two terms used either interchangeably as synonyms or with only slight differences. Generally, "data" represents a collection of realizations of a measurable quantity such as component failure times, component material property, load on the component, etc. "Information" is extracted from data through analysis. There is considerable discussion about this topic in the literature, as illustrated by the following:

W. R. Blischke et al., *Warranty Data Collection and Analysis*,
Springer Series in Reliability Engineering, DOI: 10.1007/978-0-85729-647-4_4,

1. Data represents a fact or statement of an event without relation to other things. Information embodies the understanding of the relationship of some sort, possibly cause and effect. [2]
2. Data are raw facts that have not been organized or cannot possibly be interpreted. Information is data that are understood. Information comes from the relationship between pieces of data. [3]
3. ... they do not mean the same thing. Information is the wider of the two concepts, not only statements of facts but also explanatory discourse or discussion, whereas data is the plural of datum, defined as a thing known or granted. [6]

Knowledge is the ability of individuals to understand the information and the manner in which the information is used in a specific context, as illustrated by the following:

1. Knowledge represents a pattern that connects and generally provides a high level of predictability as to what is described and what will happen next. [2]
2. Data gets transformed into information through an understanding of the relationships, and information yields knowledge through an understanding of the patterns. [2]

The link between data, information and knowledge can be characterised through the DIKW (Data, Information, Knowledge, and Wisdom) hierarchy—a term attributed to [1]. According to this, the content of the human mind can be classified into five categories:

- *Data*: symbols
- *Information*: data that are processed to be useful; provides answers to "who", "what", "where", and "when" questions
- *Knowledge*: application of data and information; answers "how" questions
- *Understanding*: appreciation of "why"
- *Wisdom*: evaluated understanding.

It is worth noting that knowledge includes theories, models, tools and techniques, standards, and so forth.

## 4.3 Structured and Unstructured Data

### *4.3.1 Information Technology (IT) Perspective*[1]

Structured data have a well defined format that requires close-ended answers—a choice from a finite set of choices. In contrast, unstructured data are usually in the form of a text with no specified set of choices. Two common sources of

[1] A different perspective is the statistical perspective. To a statistician, structured data are data that are collected under controlled conditions (e.g., a designed experiment or sample survey), while unstructured data are those collected haphazardly under conditions not under the control of the experimenter. The latter case is often called an "observational study". In this sense, claims data would almost always be unstructured.

unstructured data in the context of warranty claims data are customer descriptions of problems and technicians comments. Examples are given below.

1. Customer statements describing problems:
   - Brakes are sluggish
   - The air conditioner is not cooling adequately
   - TV picture flickers after being in use for some time
   - Computer is slow to respond
   - Cell phone loses its signal
2. Technicians' comments on fault identification:
   - Brake pedal stiff
   - Compressor is not functioning properly
   - Loose contact wire
   - Bolt not tightened sufficiently
   - Motor exceptionally noisy

### *4.3.2 Converting Unstructured Data into Structured Data*

For analysis (either qualitative or quantitative), it is necessary to convert unstructured data into structured data. In the context of unstructured warranty claims data, this involves a natural language processing technique called *named-entity extraction* (also referred to as *text tagging and annotation*).

Named-entity extraction consists of identifying the names of entities in free-form or unstructured text. Some of the common types of entities in the context of warranty claims data are the technician action, car part, location of a defect, reason of failure, effect of failure, defect type, condition under which defect occurred and customer action that caused the defect [13]. Tagging and annotation are based on hand-crafted rules and look-up in tables containing domain terms and clue words or phrases. Common approaches to named-entity recognition are based on using statistical modeling or machine learning.

For more on data mining and converting unstructured data to structured data, see [5, 7, 9, 13, 14].

## 4.4 Warranty Process

The warranty process consists of two sub-processes—(1) the claim process and (2) the servicing process. The main factors that influence the warranty process are indicated in Fig. 4.1 and discussed in the following sections.

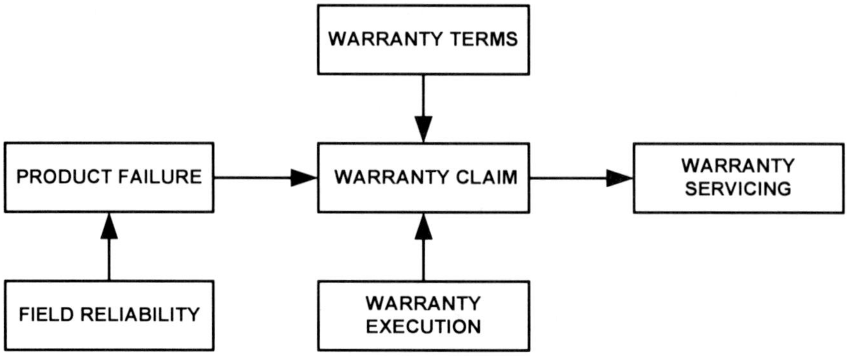


**Fig. 4.1** Warranty process

## 4.4.1 Claim Process

The starting point of a claim process is item failure (real or perceived). In Chap. 3, we defined an item failure as the inability of an item to function as required when operated properly. Most failures can be defined in an objective manner (e.g., pollution levels of an engine exhaust exceeding regulatory limits), but others are subjective (e.g., the tonal quality of a musical instrument).

As discussed in Chap. 3, product failure depends on field reliability and this, in turn, depends on several other factors, some under the control of the manufacturer (such as decisions made during the design and production of the product) and others under the control of customers (such as operating environment, usage mode and intensity, maintenance, etc.).

### 4.4.1.1 Warranty Execution

Not every failure under warranty leads to a warranty claim. Some reasons for this are that a customer:

1. may develop a strong dissatisfaction with the product and switch brands rather than exercising the claim;
2. may not be happy with the warranty service provided;
3. might feel that exercising the claim is not worth the hassle; and
4. may simply not be aware of or have forgotten the warranty coverage.

As a result, whether or not a warranty claim is executed is uncertain. It depends on, among other things, the age of the item at failure. One way of modeling this is through a warranty execution function $\phi(t)$, where $t$ is the age of the item at failure. $\phi(t)$ is the probability that a failure at age $t$ will result in a warranty claim. Typically, $\phi(t) = 1$ for an interval $0 < t \leq W_1$ $(< W)$ and is a decreasing function of $t$ for $W_1 < t \leq W$. Many different forms for $\phi(t)$ have been proposed [10].

### 4.4.2 Servicing Process

The servicing process is a complicated, multi-step process, as indicated in Fig. 4.2.[2] It begins once a customer (or representative of the customer) decides to lodge a warranty claim. For items that need to be serviced on site (for example, consumer durables such as washing machines, refrigerators, etc., and most commercial industrial products such as lifts, pumps, etc.), the first step is to contact the warranty handling personnel (either a call center or at the warranty service center). Other items are usually brought to a warranty service center.

In describing the data that are obtained in the servicing process, we use the following notation:

- D-I: Data and information
- D-I-1: Data and information collected by the service agent (Data from other sources are denoted D-I-k, $k = 2, 3, 4$ and 5. These are discussed in Chap. 5.)
- D-I-1.j, $j = 1, 2$ and 3, denote the data generated during various stages of the warranty servicing process, as shown in Fig. 4.2.

The first step in the servicing process is the collection of relevant data and information regarding the failed item brought to the center. The information and data collected, denoted D-I-1.1 in Fig. 4.2, can involve varying levels of interaction between the customer and the warranty handling personnel. Specific details of D-I-1.1 are discussed in a later section.

The second step in the process is to decide whether or not the claim is valid. A claim may be invalid for one or more of the following reasons:

- The claim is fraudulent (e.g., because the warranty has expired or because the item has not failed).[3]

---

[2] Notes for Fig. 4.2

Can be either a call center or receptionist at a warranty service centre (either retailer or independent agent).

- The skills and competencies may vary.
- The repair technician can either be trained by the manufacturer (for repairing specific products) or have general competency to carry out repairs. If the failed item is serviced on site, then a technician must be dispatched to the site.
- The report for the service agent can differ from that for the vendor of the failed components and/or the manufacturer.
- The ability to transfer data depends on the compatibilities of the different warranty systems. Some information can be lost in the transfer.

[3] Fraudulent claims from customers (and service agents) that go undetected account for about 10–15% of manufacturers' warranty costs [4].

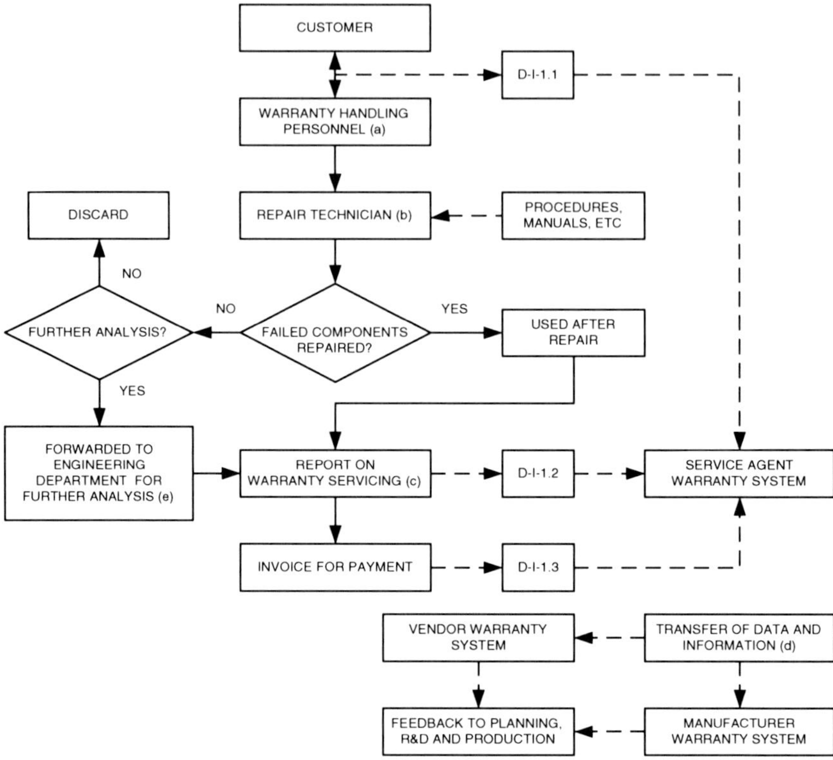


**Fig. 4.2** Warranty servicing process

- The warranty is invalidated due to resale of the item.[4]
- The item was used in a manner that voids the warranty.[5]

If a claim is not rejected, the next step is the servicing of the failed unit by a repair technician. As mentioned earlier, for some products the repair has to be carried out on site, and for others the failed item is brought to a warranty servicing center.

In completing the servicing, the repair technician must: (1) determine whether or not the problem reported can be observed or reproduced during testing[6];

[4] In the automobile industry, Chrysler was the first to introduce the transfer of warranty with resale before the original warranty expired (*Warranty Week* February 3, 2003). As a result, the resale values of such cars increased significantly.

[5] The conditions under which a warranty contract becomes null and void must be stated explicitly in the warranty document.

[6] There can be several reasons for not being able to observe or reproduce the failure (for example, intermittent failures not observed at the service depot). The "trouble not identified" phenomenon in automotive electronics is another example of this [15].

(2) identify the failure cause if there is a problem, and (3) carry out the repair to fix the failed item. The repair can involve different types of actions, for example re-soldering to fix a broken contact, repair/replacement of a failed components, etc.

If a failed component is replaced by a new item, then the failed unit is either discarded or sent back to the manufacturer. The latter action can be taken for a number of reasons: (1) to prevent fraud by the service agent; (2) for more detailed analysis of the failed component; (3) for possible refurbishment and resale, etc.

The actions taken by the repair technician generate a large amount of data and information (denoted D-I-1.2 in Fig. 4.2). Finally, the service agent, in invoicing the manufacturer for payment will generate additional data (D-I-1.3 in Fig. 4.2). D-I-1.2 and D-I-1.3 are discussed further in later sections of the chapter.

The data and information collected (D-I-1.1 through D-I-1.3) constitute the warranty claims data. This is ordinarily stored in the service agent's warranty system. Some of the data may need to be transferred to the warranty systems of the manufacturer and/or the vendors for subsequent use. This topic is discussed further in Chap. 15.

## 4.5 Warranty Claim Data Collection

Table 4.1 shows the different events and activities associated with the servicing of warranty claims and some typical warranty claims data that are generated in the process for the case where the warranty servicing is done on site.

The process for the case where the failed item is brought to a warranty depot (or an authorized agent) is very similar, except that there is no need for dispatching service technicians and hence no travel time.

## 4.6 Classification of Warranty Claim Data

The warranty claims data identified as D-I-1.1–D-I-1.3 in Fig. 4.2 can be grouped into the following four categories:

- Product related
- Customer related
- Service agent related
- Cost related

Some warranty claims data are structured and others unstructured (indicated by NS). Similarly some are objective and others subjective (indicated by SU). In the following, we indicate the typical elements of D-I-1.1–D-I-1.3, classified into the four categories listed above.

**Table 4.1** Warranty claims data (Failed item serviced on site)

| Events | Activities | Typical warranty claims data |
|---|---|---|
| Customer reports failure | Logging warranty claim | Unstructured description of symptoms |
| Dispatch of service technician | Manuals, spares, tools to be taken | Time to travel to site |
| Start of diagnosis | Testing<br>Trouble shooting | No fault detected<br>Fault detected (a)<br>Cause of failure (b)<br>Age (usage) of failed component(s)<br>Action initiated (c) |
| Ordering spares (d) | Order spares | Cost of spares ordered<br>Time to delivery |
| Start of repair | Repair | Time to travel to site (e)<br>Components replaced<br>Time for repair<br>Further action regarding disposal of failed components |
| Follow up action with customer | | Customer satisfaction<br>Need for further action |

Notes: (a) Failure may not be completely understood, in which case the technician may make an educated guess.
(b) The cause of failure being reported will depend on the knowledge and experience of the technician.
(c) This can depend on the ulterior motives of the technician. Some codes for failure may offer longer times for repair or may be less questioned by the manufacturer.
(d) This occurs only when one or more of the needed spares have to be ordered.
(e) The repair activity can only recommence after the spare has been delivered.

### *4.6.1 Product Data*

D-I-1.1

- Make
- Model
- Purchase date
- Identification number[7]
- Retailer name and details (if relevant)[8]
- Usage at failure (if appropriate)
- Failure date
- Type of warranty (base, extended, terms, etc.)
- Customer statements (NS)

[7] In the case of automobiles, each vehicle has a unique identification number (referred to as VIN).

[8] This information is needed for traceability of components to batch numbers in production and is discussed in Sect. 5.9.

- Symptoms prior to failure (NS)
- Other relevant information (some of which may be NS)

D-I-1.2

- Diagnostic testing and outcomes
- Fault not found—testing procedures followed
- Fault found (defect codes)
  - Minor fix (alignment, adjustment, etc.)
  - Components that failed
- Description of the condition of failed components (structured and NS)
- Listing possible causes leading to failure (NS)
- Actions to rectify
  - Failed components repaired
  - Failed components replaced by new
- Actions with regards components not repaired
  - Discard
  - Send to manufacturer for further analysis

### *4.6.2 Customer Data*

D-I-1.1

- Name
- Address
- Contact details
- Usage mode
- Usage intensity
- Operating environment

Comment: Data and information relating to the last three items are often categorical (for example, usage intensity high, average or low).

### *4.6.3 Service Data*

D-I-1.2

- Identification number of servicing agent[9]

[9] The servicing agent can be a retailer servicing products of a single manufacturer, as is typically the case for products such as automobiles, or a retailer selling brands of various manufacturers (for example, departmental stores), or an authorized independent agent.

- Name(s) of warranty handling personnel
- ID number
- Assessment of customer usage mode and intensity (NS and SU)
- Level of customer dissatisfaction with product performance (SU)
- Claim ID number (for tracking)
- Identification number of repair technician
- Time expended for servicing—can be divided into times for different activities, such as transport, diagnostic, disassembly, repair actions, assembly, etc.
- Reasons for servicing time exceeding its target value (e.g., unavailability of spares, tools, etc.)

### 4.6.4 Cost Related Data

D-I-1.3

Servicing costs may be direct or indirect costs. Each of these categories, in turn, can be divided into various subgroups, depending on the product.[10]

*Direct Expense (DE)* Direct expenses are those incurred while dealing directly with the shipping, repairing/replacing or refurbishing of a product, and typically include the following:

- DE-1: RMA (return of material authorization) and transactional administration process expenses
- DE-2: Warranty depot repair/replace/refurbish expenses
- DE-3: Field services repair/replace/refurbish expenses

*Indirect Expenses (IE)* Indirect expenses include the following:

- IE-1: Vendor related recovery (credits and debits) expenses
- IE-2: Warranty related call center activity expenses
- IE-3: Warranty related inventory expenses
- IE-4: Warranty registration expenses
- IE-5: Warranty analysis process expenses
- IE-6: End of life disposal expenses

## 4.7 Problems in Dealing with Warranty Claims Data

A number of problems may be encountered in dealing with warranty claims data. The most important of these are the following: (1) delays in reporting, and (2) unreported failures. These and some additional problems are discussed below.

---

[10] The material for this section is based on [12].

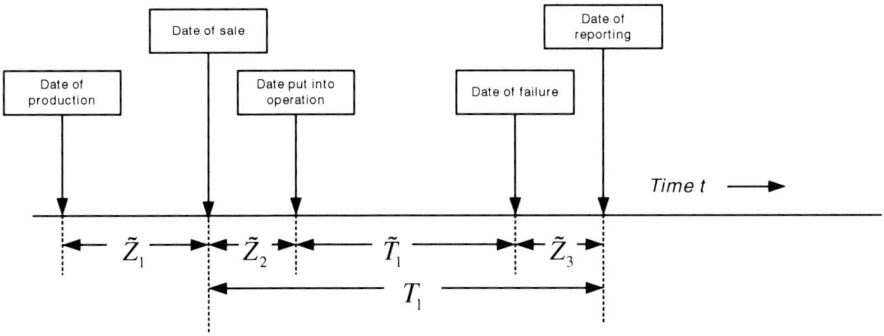


**Fig. 4.3** Delays affecting failure data

### 4.7.1 Delays in Reporting

Figure 4.3 shows various delays (denoted by $\tilde{Z}_i$, $i = 1, 3$) that can result in the observed failure time ($T_1$) being greater than the actual failure time ($\tilde{T}_1$).

Some examples of reasons for these delays are as follows:

1. Delay $\tilde{Z}_1$: This delay may result from (a) the time needed for transport (which can be several weeks if the items are transported across continents or overseas), and (b) the wait in a warehouse or retailer's shop before the unit is sold (which can vary from days to several months or possibly years).
2. Delay $\tilde{Z}_2$: This typically arises when a unit is bought as a spare and not put into use immediately.[11]
3. Delay $\tilde{Z}_3$: This depends on the product and on several other factors, such as criticality of failure, environmental conditions, usage mode, etc. For example, in the case of an automobile, the failure of windshield wiper controls would be more critical during the rainy season and less so during the dry season. As a result, the delay in reporting can be longer during the dry season.[12]

Of particular importance are $\tilde{Z}_2$ and $\tilde{Z}_3$, neither of which is observable. As a result of these delays, the observed time to failure $T_1$ can be greater than the actual time to failure $\tilde{T}_1$. A consequence of this is that the analysis and inferences based on the observed failure time will yield reliability estimates that are, on average, greater than the true values.

[11] In general, customers may not collect this information. Even when it is collected, it may or may not be communicated to the service agent when a warranty claim is exercised.

[12] Customers may or may not know the exact time of failure. Through a process of interaction, however, the service agent may, on occasion, obtain an estimate of this delay.

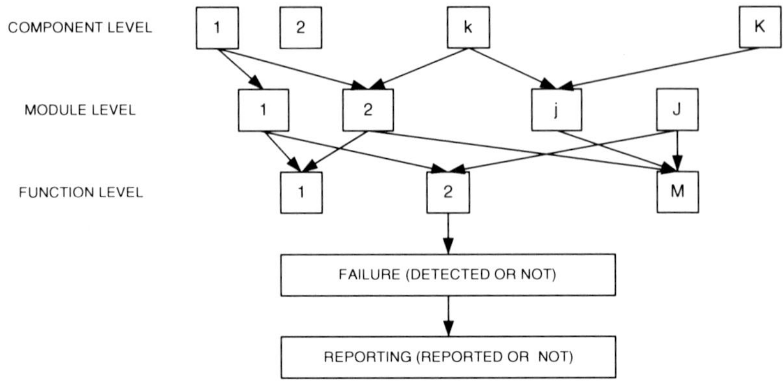


**Fig. 4.4** Failures not detected or not reported

### *4.7.2 Failure Not Reported*

Many products perform more than a single function. A typical example is a cell phone that provides several functions, such as sending voice signals, texts, pictures, etc. Such products involve several components that are interconnected to form several different modules, and these are then interlinked to provide the different functions, as indicated in Fig. 4.4.

If a customer does not use a functional mode, then failure of that mode is not recognized. Even if the failure is detected (for example, inability to send pictures), the customer may not report it. This may occur, for example, when the main usage of the cell phone is for some other function and the failed function mode is viewed as not being relevant.[13]

### *4.7.3 Other Problems*

Some other problems that may be encountered in dealing with warranty claims data are the following:

- Aggregated data[14]
- Data reported incorrectly (due to fraud or accidental error)
- Incomplete data

---

[13] Another example is the modern microwave oven. A customer who uses it only for warming food or boiling water might not detect a failure in some other mode (such as thawing, sensor cooking, etc.).

[14] An example of this is a communications company that bought a batch of microwave antennas (for ground-to-ground and ground-to-satellite transmission) for operation in a remote region. Some were put in use and others kept as spares. Failed units were replaced by the spares and claims under warranty would involve returning all failed units on a periodic basis [8].

### 4.7.4 *Loss of Information*

The maximum information that a service agent can collect while servicing a warranty claim is typically the following:

- Date of sale (to ensure if item still covered by warranty)
- Age of item at failure
- Usage of item at failure (if relevant)
- Reason for failure—failure mode (component causing the failure, assembly errors, etc.)
- Usage mode, intensity, operating environment (unstructured data from customers)
- Symptoms prior to failure (unstructured data from customers)
- Actions taken to rectify the failure (replace or repair of failed components, condition of other components, etc.)

There is loss of information when some of the data and information on the above list are not collected. This loss increases as the amount of data collected increases.

## 4.8 Use of Warranty Claims Data

Warranty claims data alone are useful mainly for identifying and solving certain type of problems at Stage 1 of warranty management and for few problems at Stage 2 of warranty management.[15]

### 4.8.1 *Stage 1 of Warranty Management*

Two of the problems that warranty claims data can be used for are (1) detecting fraud and (2) detecting over-servicing by service agents. This requires doing an analysis on pooled data from all the service agents to evaluate different performance metrics (e.g., fraction of items failing due to the failure of a particular component, time to fix a particular fault, and so on) and then doing the same analysis for each agent separately. Any significant deviation between the metrics for a particular service agent and the pooled results should trigger further investigations to determine if there are any valid reasons (such as the mean travel time by repair technician to service items on site varying with region, etc.). To carry out a proper analysis requires additional or supplementary warranty data.

[15] Stages in the evolution of warranty management are discussed in Sect. 2.12.1.

### *4.8.2 Stage 2 of Warranty Management*

Items returned to the engineering department for further analysis provide insights into failure mechanisms and causes. This needs to be fed back to various other departments, such as design, R&D, and production, as indicated at the bottom of Fig. 4.2. The data are mainly useful for qualitative analysis, such as failure mode analysis. At this point, it is not possible to do any meaningful quantitative analysis, such as estimating reliabilities at product and component levels, based solely on warranty claims data. For this purpose, supplementary warranty data are required for proper analysis. This will be discussed further in Chap. 5.

## 4.9 Current Industry Practice

The collection and the use of warranty claims data varies significantly from industry to industry and across each industry. The majority of manufacturers have warranty systems and collect warranty claims data that are appropriate for Stage 1 of warranty management. A few (mainly large manufacturers in the automotive, computer and some other industries) have warranty systems and collect warranty claims data that are appropriate for Stage 2 of warranty management. As examples, we briefly discuss the automotive industry and then outline the warranty claims data collection process at Intron, a global manufacturer of wireless data acquisition and communication products for electric, gas, and water utilities.

### *4.9.1 Automotive Industry*[16]

We focus on some issues that are relevant for the warranty claims process.

*Part Identification/Defect Analysis and Codification* Warranty data coding of failures varies considerably from dealer to dealer and from technician to technician within a manufacturer's dealer network. Some of the reasons for this are:

- The fact that a component failed might be obvious, but the cause of failure may not be readily apparent.
- The repair technician may need to make rapid decisions (for entry into the service agent's warranty system) because of time and cost constraints.
- Lack of experience on the part of the technician.

[16] This section is adapted from a report [11] involving interviews with representatives from three major automotive manufacturers, four automotive suppliers and one automobile dealer in the USA.

*Manufacturer's Warranty System* The internal process for handling warranty data varies from manufacturer to manufacturer. The input to the system is the transfer of data from the different service agents. The starting point is a review of dealership claims in order to monitor dealer repair work (Stage 1 of warranty management). This also serves as a first defense for identifying potential warranty problems (Stage 2 of warranty management).

The type of data that is transferred from a manufacturer to its component suppliers (vendors) differs significantly among manufacturers. Three different types of data transferred are:

- Incident-based data (limited to claims and counts)
- Rate-based data (based on production/sales)[17]
- Warranty data with month of production/months in service (MOP/MIS) data included.

The date of production is critical for traceability of the batch in production.

*Challenging Issues* According to CAR,[18] the automotive industry handles 100 million warranty claims per year and each claim includes numerous fields and several lines of text. Proper analysis requires efficient text data mining approaches.

### 4.9.2 Itron Inc.

Itron Inc. is a global supplier of wireless data acquisition and communication products for electric, gas, and water utilities. Its systems are installed at more than 2,000 utilities worldwide. One such product is a system called ERT, which is a sophisticated encased module that has many components, including a small computer and radio. ERTs that are purchased directly from Itron are covered by an FRW policy, with all failed ERTs returned during the warranty period replaced free of charge.

The warranty process begins with the customer contacting "Itron Customer Care" to obtain a "Returned Material Authorization number" (RMA#). The customer will be asked to provide the following information:

1. Customer (Utility) name and shipping address.
2. Contact name, phone number and e-mail address.
3. Shipping date and carrier name.
4. Quantity and type of ERT modules.

The failed units are sent to Itron along with a completed form (FormCF73 REV A 05/09/03), which provides the following data and information for each unit:

[17] This requires supplementary warranty data, as discussed in Chap. 5.

[18] Center for Automotive Research (CAR): For more details, visit www.cargroup.org

- ID Number
- Model
- Install Date
- Install Reading
- Symptom Code
  - No response. Describe verifying system.
  - ERT read doesn't match meter index.
  - Can't program. Describe programming tool.
  - ERT type, such as Type 15.
  - Physical damage, please describe. (Text data)
  - Mechanical failure, please describe. (Text data)
- Failure Code
  - Unit failed at/before time of installation
  - Unit was working after installation
- Meter Location
  - Inside
  - Outside
- Removal Date
- Removal Reading
- Further Description

These data are then used as an important tool for effective warranty management.

## References

1. Ackoff RL (1989) From data to wisdom. J Appl Sys Anal 16:3–9
2. Bellinger G, Castro D, Mills A (1977) Data, information, knowledge, and wisdom. From http://www.outsights.com/systems/dikw/dikw.htm
3. Benyon D (1990) Information and data modeling. Alfred Waller, Heneley-on-Thames
4. Byrne PM (2004) Making warranty management manageable. Logist Manag, August 1, 2004
5. Cios KJ, Pedrycz W, Swiniarski RW, Kurgan LA (2007) Data mining. a knowledge discovery approach. Springer Science, NY
6. Holstrom JE (1971) Personal filing and indexing of design data. Proc information systems for designers, University of Southampton, Paper No. 1
7. Jeske DR, Liu RY (2007) Mining and tracking massive text data: classification, construction of tracking statistics, and inference under misclassification. Technometrics 49:116–128
8. Lyons KF, Murthy DNP (1996) Warranty data analysis: a case study. Proceedings of the 2nd Australia Japan workshop, Gold Coast, July 17–19
9. McCallum A (2005) Information extraction: distilling structured data from unstructured text. Soc Comput 3(9):48–57
10. Patankar JG, Mitra A (1996) Warranty and consumer behavior: warranty execution. In: Blischke WR, Murthy DNP (eds) Product warranty handbook. Marcel Dekker, New York

11. Smith BC, Miller RT (2005) The warranty process flow within the automotive industry: an investigation of automotive warranty processes and issues. Report prepared for the centre for automotive research. From www.cargroup.org
12. Sparker G (2006) Warranty financial management: Part 1—Defining warranty management expenses. Warranty week, Sept 12
13. Sureka A, De S, Varma K (2008) Mining automotive warranty claims data for effective root cause analysis. Database systems for advanced applications (Lecture notes in computer science), vol 4947. pp 621–626
14. Tan AH (1999) Text mining: the state of the art and the challenges. In: Zhong N, Zhou L (eds) PAKDD 1999. LNCS (LNAI), vol 1574. pp 65–70
15. Thomas DA, Ayers K, Pecht M (2002) The "trouble not identified" phenomena in automotive electronics. Microelectron Reliab 42:641–651

# Chapter 5
# Supplementary Warranty Data

## 5.1 Introduction

In Chap. 4, we classified warranty data into two categories: (i) warranty claims data and (ii) supplementary warranty data. The former deals with data generated during the servicing of claims under warranty and are not sufficient for estimating field reliability. Supplementary warranty data are the additional data that are needed for estimation and for many other warranty related analyses. In this chapter, we focus on supplementary warranty data. There are two different notions of supplementary warranty data—narrow and broad. In this chapter we look at both notions and discuss some related issues. The outline of the chapter is as follows:

In Sect. 5.2, we discuss the two notions of supplementary warranty data. The narrow definition deals with data that are censored as a result of expiration of the warranty. This topic is discussed in detail in Sect. 5.3, where we highlight the uncertainties that result in the case of two-dimensional warranties. Section 5.4 deals with the broad definition and looks at all relevant data from a product life cycle perspective. The data are categorized into three groups (i) pre-production, (ii) production and, (iii) post-production. These are discussed further in Sects. 5.5–5.7, respectively. In Sect. 5.8, we look at the different uses of both warranty claims and supplementary data from the warranty management perspective. A critical factor for effective management is traceability to the component level. This is discussed in Sect. 5.9. In Sect. 5.10, we look at some of the problems in the collection of warranty supplementary data. We comment on current practice in industry in Sect. 5.11 and conclude with discussions of various data structures and possible scenarios in which these may be encountered in Sects. 5.12 and 5.13, respectively. These structures and scenarios will be used is Chaps. 11–14, where we deal with warranty data analysis.

W. R. Blischke et al., *Warranty Data Collection and Analysis*,
Springer Series in Reliability Engineering, DOI: 10.1007/978-0-85729-647-4_5,

## 5.2 Supplementary Data Needed

Estimating field reliability based solely on warranty claims data will not yield correct estimates because this approach ignores information relating to items that are under warranty and still operational as well as information on the age/usage of items after warranty expiration. This additional information can be obtained from the sale date or the last replacement date under warranty for items sold with one-dimensional warranties. In the case of two-dimensional warranties, one can obtain partial information, namely the censored data that have been referred to as supplementary data. We define this as the narrow notion of that term.

The broad notion of supplementary data includes data from throughout the product life cycle that are of relevance to effective warranty management. As indicated in Fig. 3.1, the product life cycle involves several phases. At each phase, a great deal of data and information that are of relevance to warranty and reliability analysis are generated. We divide the data into the following three groups:

- Pre-production data
- Production (Quality Assurance) data
- Post-production data

These are discussed in Sects. 5.5–5.7, respectively.

## 5.3 Censored Data

In this section, we look at the structure of the data collected in the warranty process and the censoring that comes about as a result of the warranty. Let $t = 0$ denote the instant the product is launched on the market. Sales occur over time. Let $t_{i0}$ denote the instant when the $i$th item is sold, $i = 1, 2, \ldots, I$. We assume that items are immediately put into operation. Let $t_{ik}$ denote the time of the $k$th failure, $k = 1, 2, \ldots$, for the $i$th item. We also assume that there is no lag between the failure and the time that the claim is made. Let $[0, t)$ denote the time interval over which warranty data are collected. We consider both one- and two-dimensional warranties.

### *5.3.1 One-Dimensional Warranties*

Let $W$ denote the parameter of the one-dimensional warranty. For $0 < t < W$, all the items sold are still under warranty and for $t \geq W$ some of the items sold are no longer under warranty. Note that the warranty period $WP$ is $W$ in the case of non-renewing warranties and can be greater than $W$ in the case of renewing warranties.

Figure 5.1 shows the time histories of two items sold with non-renewing warranty. For item $i$, we have $(t - t_{i0}) > W$, implying that the item is no longer under warranty by time $t$. Also, since one failure has occurred, we have one failure

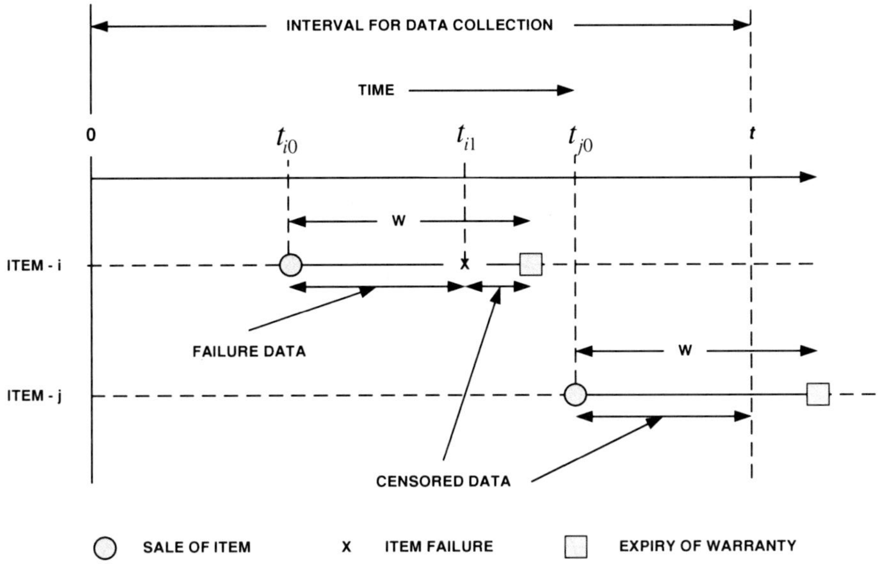


**Fig. 5.1** Example of censoring in a one-dimensional non-renewing warranty

observation (given by $t_{i1} - t_{i0}$, the age at failure) and *a right-censored* observation (given by $W - t_{i1} - t_{i0}$).[1] For item $j$, we have $(t - t_{j0}) < W$, implying that the item is still under warranty. Since this item has had no failure, the censored value observed is $t - t_{j0}$.

For the case of a renewing warranty, the censored observation for an item no longer under warranty is simply $W$, and the failure data are all the ages at earlier failures that resulted in the warranty being renewed. If an item is still under warranty, the censored observation is the age of the item at time $t$.

### *5.3.2 Two-Dimensional Warranties*

In the case of two-dimensional warranties, failures are random points in a two-dimensional plane, with one axis representing age and the other usage (mileage, copies made, etc.). We confine our attention to two-dimensional warranties characterized by a rectangle, as indicated in Fig. 5.2.

The warranty can cease due to either the age or the usage limit being exceeded. Which of these occurs depends on the usage rate, as indicated in Fig. 5.2. As a result, knowing the sale date does not provide any information about the censored

---

[1] If the interval over which data are collected is given by $[\tau, t)$, $\tau > 0$, then one can have left-censored data for items sold before $\tau$ that fail in the interval $[\tau, t)$ and are still under warranty.

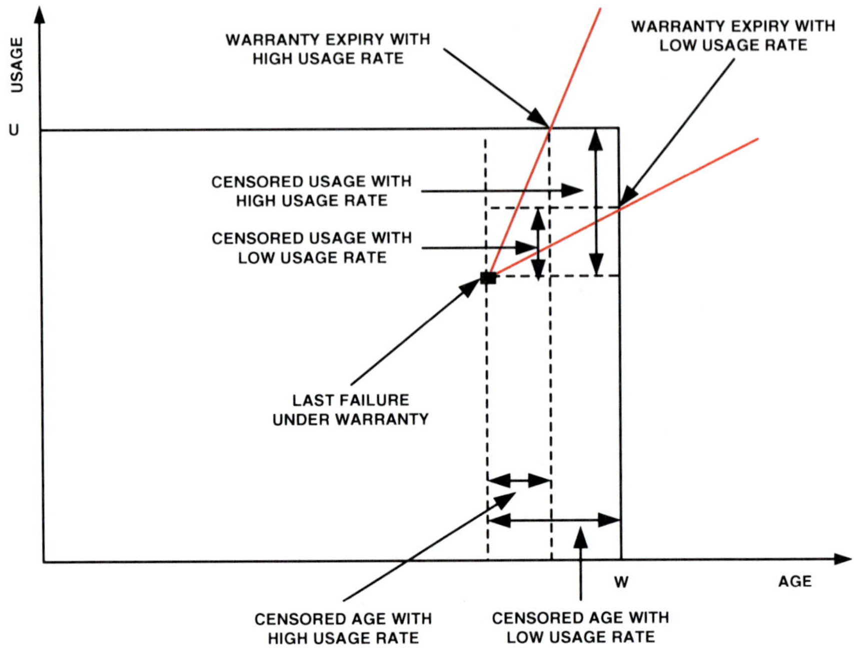


**Fig. 5.2** Censoring in two-dimensional warranty

age and usage when the warranty expires. This uncertainty poses extra challenges in estimating product reliability. However, if one assumes that the usage rate is constant (but varying from customer to customer), then one can predict the censored age and usage based on the last failure and knowledge of the usage rate.

### *5.3.3 Types of Censoring*

In analysis of incomplete data, the structure of the data (i.e., the nature of the incompleteness) must be taken into account for a proper analysis. Here we are concerned with data that are incomplete in the sense that the period of observation is terminated prior to the failure of all items under observation. Such data are called *censored* data.

Censored data are encountered in reliability and claims data for a number of reasons. Claims data are typically censored on the right at time $W$ after sale, since little or no additional data are obtained on items whose lifetimes exceed the length of the warranty period. Thus times to failure are observed only for items and replacements that fail while under warranty. For the remaining items, two types of right-censoring occur: (1) items whose lifetime exceeds $W$ (For these, we know only that $Y \geq W$); and (2) items sold at some time $t$ that is less than $W$ from the time of observation (For these we know only that $Y > t$). Occasionally left-censored data are obtained as well. This occurs when the period of observation begins at a time after some of the items have been sold and put into service.

In reliability testing, two forms of censoring commonly occur:

1. Type I Censoring: $n$ items are put on test and testing stops at some predetermined time $T$.
2. Type II Censoring: $n$ items are put on test and testing stops when a predetermined number $r$ of items has failed.

Under Type I censoring, we know the lifetimes of the items that fail prior to $T$. For the remaining items, we know only that $Y > T$. Under Type II, we know the lifetimes of the first $r$ items to fail, and know only that the lifetimes of remaining items exceed that of the last item to fail.

Data may be *singly* or *multiply* censored. The forms of reliability data just discussed are singly right-censored. Claims data may be singly or multiply censored, depending on the time perspective taken. (This will be discussed in detail in Chap. 11.) There are, in fact, many types of censored data. Data may be right- or left-censored, or both. They may be randomly censored (e.g., with Type II censoring), there may be combinations of different types of censoring, and so forth. Here we look only at right-censored data. The results are easily extended to left-censored data and to data censored on both the right and the left. Additional results will be given in later chapters for other types of censoring and other types of incompleteness found in claims and related data.

Censoring significantly affects the form of the likelihood function (See Appendix D), which is essentially the joint distribution of the data. For complete data, the likelihood function is typically simply a product of densities or discrete probabilities. In the case of censored data, it is a product of density functions for failed items and CDF's for censored items.

## 5.4 Life Cycle Data

Warranty costs result from failures under warranty. The rate at which these occur depends on the field reliability of the item, as discussed in Chap. 3. Field reliability, in turn, depends on decisions made during the design, development and production phases of the product life cycle. Failures have an impact on customer satisfaction and sales. As such, data from all the different phases of the product life cycle play an important role in effective warranty management. These are the supplementary data in the broad sense, as discussed earlier.

### *5.4.1 Data Sources*

Typical sources for supplementary data are the various units within the organization that monitor or implement functions that generate data (such as engineering,

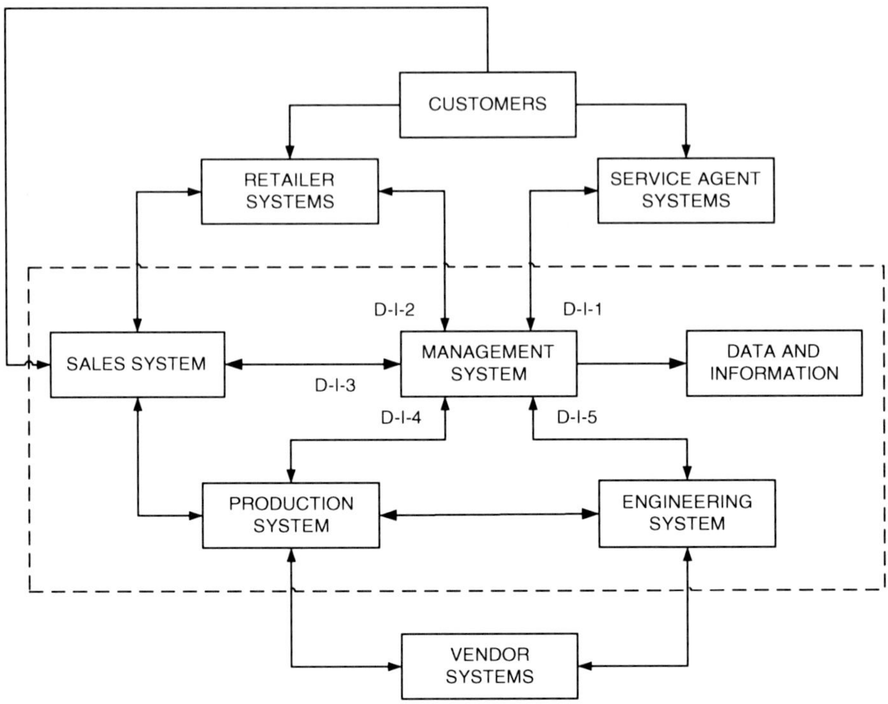


**Fig. 5.3** Data collection systems needed for warranty management

production, marketing, etc.) as well as units outside the organization (such as component vendors, service agents, retailers and customers). Each unit, both within the organization and external to it, can have a different data collection system. Effective warranty management requires that the manufacturer have a management system that collects all of these data into a central data bank, as indicated in Fig. 5.3.

The management system essential to the process is the system used by the organization to manage the overall business. It must contain several sub-systems for collection and dissemination of data and information and for making decisions. The system must have the capability of dealing with inputs from units within the organization (indicated by the units inside the dotted rectangle) as well as inputs from units external to the organization.[2]

[2] One such sub-system is that responsible for collecting the data needed for decision making by top management in the front-end phase of the product life cycle.

### 5.4.2 Data Classification

Life cycle data can be classified in many different ways.[3] A scheme to classify the life cycle data and information (denoted D-I in the remainder of the book) that is appropriate for effective warranty management involves the following six categories:

D-I-1: Data and information from service agent systems. These include warranty claims and servicing data and are discussed in detail in Chap. 4.[4]

D-I-2: Data and information from retailer systems. These include sales at the retail level, customer related information, etc.

D-I-3: Data and information from the marketing unit of the manufacturer. The data relate to the promotion of the product, sales and revenue at the wholesale level, etc.[5]

D-I-4: Data and information from the production unit of the manufacturer. The data from the production system also include data and information from vendor's production systems. The data relate to materials and processes used in the production phase.

D-I-5: Data and information from the engineering unit of the manufacturer. The engineering system contains data and information relating to the design (conceptual and detailed) and development phases of the product development process.

D-I-6: Data and information collected for decision making during the feasibility (or front-end) phase of the product life cycle. (This is usually a subsystem of the management system.)

Note that data and information from external units come through one or more of the above systems and include the vendor system and information from customers (usually relating to product performance and satisfaction), as indicated in Fig. 5.3.

Some of the data and information are generic (common to all manufacturers and to all products), while others are manufacturer and/or product specific. It is impossible to list all the data; it is extensive and, to some extent, product specific. In the next three sections we list some data and information that are highly relevant in the context of effective warranty management.

[3] In Sect. 4.6, we proposed a scheme for the classification of data. A scheme used in the automotive industry involves four domains—(i) Computer Aided Design (CAD), (ii) Computer Aided Manufacturing (CAM), (iii) Computer Aided Engineering (CAE), and (iv) Service. For details, see [1], where the different data types in each domain are listed.

[4] In some organizations D-I-1 and/or D-I-2 are fed to the management system through the marketing system.

[5] This can also include sales at the retail level if the manufacturer sells directly to final end-customers.

## 5.5 Pre-production Data

Pre-production data are data from the first three phases (feasibility, design, and development) of the product life cycle.

### 5.5.1 *Feasibility Phase Data [D-I-6]*

Important data and information include the following:

- Warranty (claims + supplementary) data for earlier generations and for similar products (from the management system)
- Customer needs and usage profiles (through surveys)
- Potential sales (through surveys, expert panels, etc.)
- Information regarding competitors' products currently on the market and any available information regarding planned future products (from industry magazines, competitors' advertising, government statistics bureaus, former employees of competitors and other sources.)

### 5.5.2 *Design Phase Data [D-I-5]*

Important data and information include the following:

- Standards used
- Details of conceptual design
- Design reliability[6]
- Engineering drawings of detail design
- Design reliability specifications at the component level
- Material specifications for components

The main sources for data and information are:

- Standards (Industry, National and International)
- Vendors' brochures and manuals
- Design details of earlier generations of the product
- Regulatory standards and requirements
- Technical books, journals and magazines

---

[6] The overall reliability of the product must take into account the implications for warranty costs, customer satisfaction level, etc., all of which are determined during the feasibility phase, as well as the implications for development and production costs. This issue is discussed in detail in [2].

### *5.5.3 Development Phase Data [D-I-5]*

Important data and information include the following:

- Standards used
- Test plans at component and intermediate levels
  - Types of tests (e.g., different operating environments)
  - Stress levels (electrical, mechanical, thermal, etc.) in the case of accelerated testing
  - Duration of test
  - Number of items to be tested
- Details of test facilities
- Test data—stress levels, test times, failure and censored data, etc.

The main sources for data and information are:

- Standards (Industry, National and International)
- Test reports

## 5.6 Production Data [D-I-4]

Important data and information include the following:

- Details of the production process
- Flow of materials
- Details of quality assurance schemes
  - Plans for testing at component, product and intermediate levels
  - Types of testing
  - Quality control schemes
  - Acceptance sampling schemes for components from external vendors
- Testing to detect
  - Assembly errors
  - Component non-conformance
- Burn-in testing (to improve reliability)
- Details of vendor's process operation
- Vendor's quality assurance schemes

The main sources for data and information are:

- Process control systems
- Results of testing
- Vendor systems

## 5.7 Post-production Data

Post-production data are the data from marketing to post-sale support, the last two phases of the product life cycle.

### *5.7.1 Marketing Phase Data [D-I-3]*

Important data and information include the following:

- Product support services (warranties, extended warranties, service contracts, customization, etc.)
- Bundling of product and product support
- Price and promotion
- Information about competitors' products

The main sources for data and information are:

- Retailer feedback
- Customer feedback, surveys, etc.
- Market surveys
- Industry magazines

### *5.7.2 Data from Retailer [D-I-2]*

Most products are sold through a chain involving wholesalers and retailers. The retailer is the organization that sells the manufactured products to consumers. In general, retailers are independent businesses. In some case, the retailer sells the products of a single manufacturer under a franchise agreement (e.g., most car dealerships). In other cases, the retailer sells products from more than one manufacturer (e.g., department or specialized stores).

Since retailers are the people dealing directly with customers, they are in a unique position for obtaining data and information relating to customer needs, choice preferences, etc.

### *5.7.3 Warranty [Extended Warranty] Data [D-I-1]*

Important data and information include the following:

- Warranty claims
- Product failure data

- Customer usage profiles
- Preventive maintenance actions under warranty

The main sources for data and information are:

- Service agent systems
- Retailer systems (feedback from customers)
- Direct feedback from customers

### 5.7.4 Usage Data

In Sect. 3.6.4, we defined three types of usage. For many products, there is a counter to display cumulative usage (total miles in the case of an automobile, total number of copies made in the case of photocopier, total number of landings in the case of an aircraft). These data are usually collected from items that fail during the warranty.

For some products (e.g., large photocopiers), the counter is read on a regular basis through remote sensing and the data are transmitted electronically to the manufacturer. In other cases, the data are collected manually by asking customers to mail the usage at set times. This provides additional information about usage over time, in contrast to usage data recorded only at failure times.

### 5.7.5 Post-warranty Data

Once a warranty or extended warranty expires, there is no need for customers to have failures fixed by an authorized service agent. Repairs may be performed by independent repair facilities or by the customer himself. In both cases, failure data will not be collected.

Customers who use an authorized service agent in the post-warranty period are a potential source for failure data if the manufacturer has an agreement with the service agent that requires the collection of such data. Other sources for getting these data are

- Maintenance service contracts (provided that the service agent is willing to collect and share the data)
- Sales of spares (only for proprietary components where the manufacture or the component vendor is the sole provider of spares)
- Consumer choice magazines and reports
- Follow up surveys

## 5.8 Use of Warranty Data

Both warranty claims data and supplementary warranty data are needed for making decisions in all three stages of warranty management.[7] We discuss the role of data in each stage.

### 5.8.1 Stage 1 of Warranty Management

At the time at which a manufacturer releases a product into the marketplace, reliability estimates are based on design calculations and limited testing at component and product levels during development. At this stage, knowledge of product reliability (design, inherent or field) is limited. Warranty claims data and censored data (using sales data D-I-2) allow the manufacturer to obtain a better estimate of field reliability. At the product level, this information is important for obtaining (i) better estimates of future warranty costs (and hence managing the warranty reserves needed) and (ii) better planning of warranty logistics. At the component level, this information becomes critical for the sharing of warranty costs between the manufacturer and the component suppliers. The contract between the manufacturer and component supplier defines how these costs are to be shared.

### 5.8.2 Stage 2 of Warranty Management

In Stage 2 warranty management, the focus is on continuous improvement. This requires root cause analysis of failed components (carried out by the engineering department). The warranty data needed for this purpose are shown in Fig. 5.4[8]

---

[7] In the automotive industry, the Automotive Industry Action Group (AIAG) determined that the four basic uses of (retained) data are (i) product design re-use (including machine and tooling re-use), (ii) service parts management, (iii) legal, and (iv) historical.

[8] Notes for Fig. 5.4

(a) Data and information from OEM management system

- Strategic decisions relating to the product

(b) Data and information from OEM engineering system

- Component design data—reliability specifications, material selection, etc.
- Test data relating to the component during development

(c) Data and information from OEM production system

- Assembly of component into the product
- Linking failed components to batch numbers
- Other data relating to the batches under consideration (such as vendors if more than one vendor is used)

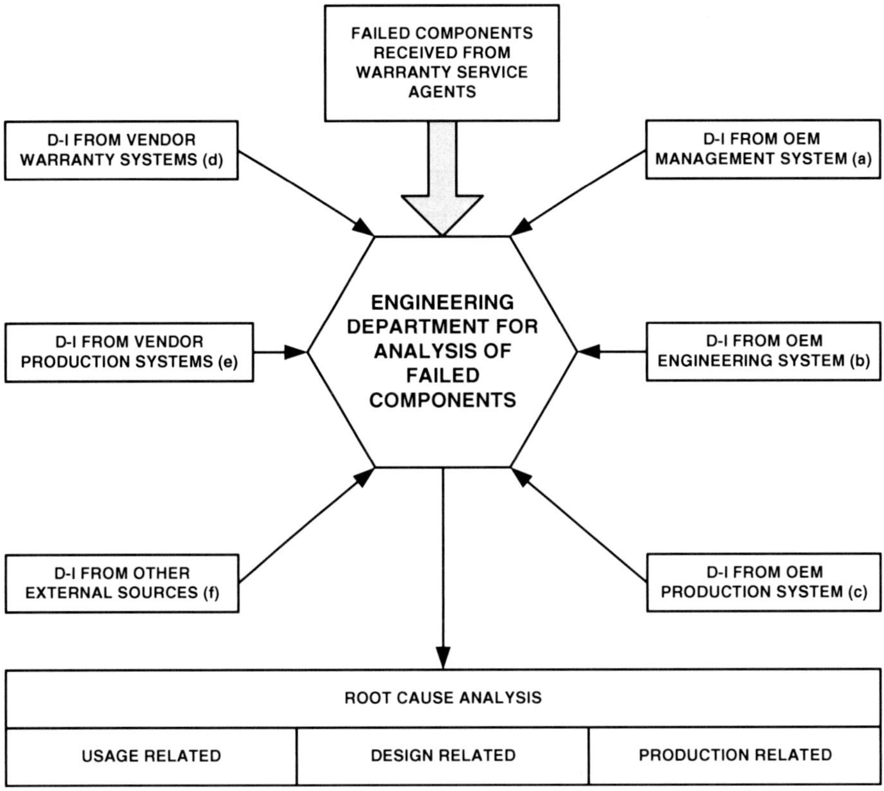


**Fig. 5.4** Root cause analysis for continuous improvement processes

(Footnote 8 continued)
(d) Data and information from different service agent warranty systems

- D-I-1 of Fig. 5.3

(e) Data and information from different vendor production systems

- Linking failed components to batch numbers
- Other data relating to the batches under consideration (such as material used)
- Process control data
- QC data—inspection, testing, etc., for the batch under consideration

(f) Data and information from other external sources

- Technical journals
- Reports from research laboratories
- Research reports from academic institutions
- Trade journals (industry specific)

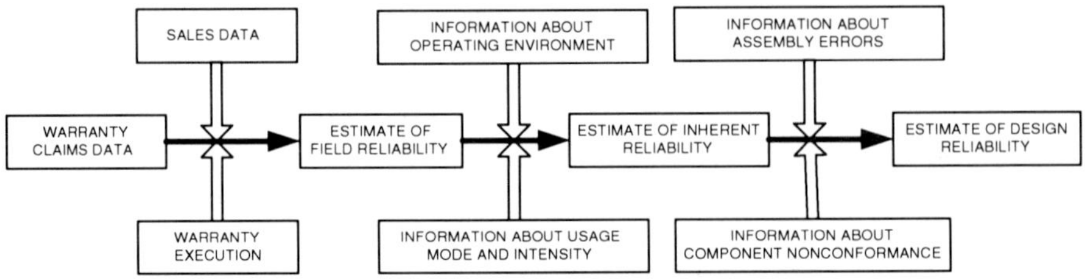


**Fig. 5.5** Assessing field, inherent and design reliabilities

Some of the data needed are indicated in (a)–(f) in the notes for the figure. The root cause can be one or more of the following:

- Usage Related: This arises when customers use the product in a mode for which it is not designed.
- Design Related: An example of this is a component not having sufficient strength due to improper size specification or wrong selection of material.
- Production Related: The two most common problems are assembly errors and component nonconformance.

#### 5.8.2.1 Assessing Reliability

In Sect. 3.4, we defined four different notions of reliability: (i) design reliability, (ii) inherent reliability, (iii) reliability at sale, and (iv) field reliability. Assessing these reliabilities involves warranty claims data as well as supplementary data, as shown in Fig. 5.5.

The estimates of these reliabilities provide information regarding problems (design, production, etc.) and are important in the context of continuous improvement processes.

The use of warranty data for continuous improvement of the product and the manufacturing process is discussed further in Chap. 15.

### *5.8.3 Stage 3 of Warranty Management*

Stage 3 warranty management advocates warranty management from a strategic and overall business perspective. Here warranty issues for new products are addressed during the front-end phase, taking into account both commercial aspects (sales, price, etc.) and technical considerations (product reliability, development effort needed, etc.) in the various phases of the product life cycle.

Warranty data from earlier generations provide useful information for decision making in the front-end phase. This topic is discussed further in Chap. 16.

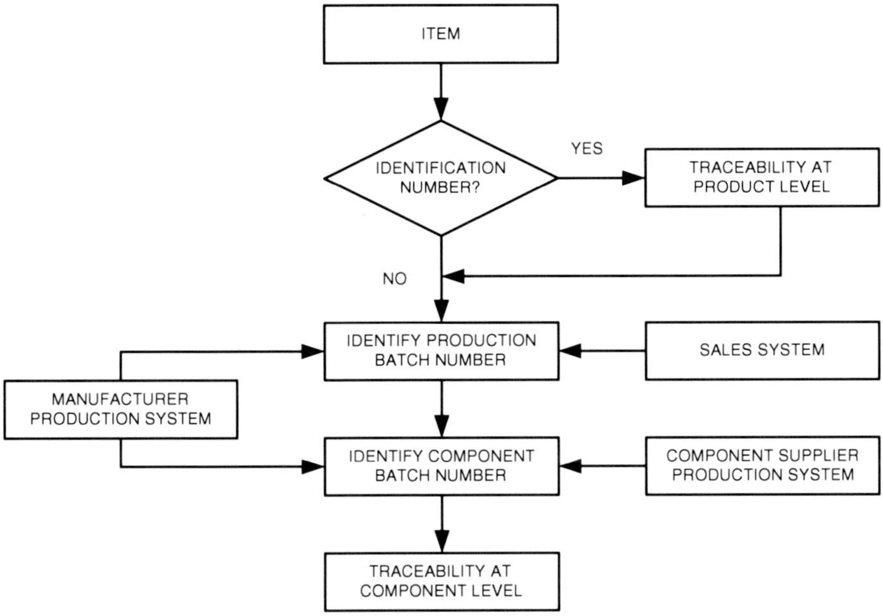


**Fig. 5.6** Traceability at product and component levels

## 5.9 Traceability

Most consumer durables, industrial products and commercial products are produced using either continuous or batch production. In the former case, by dividing the time into suitable intervals (e.g., 8 h shifts, days, etc.), one can view the continuous production as a batch production.

Similarly, components, whether produced internally or bought from external vendors, are produced in batches. Due to variations in materials and/or production, the quality of components produced can vary from batch to batch. This, combined with variations in assembly of the components, can result in quality variations, at the product level, from batch to batch.

For proper root cause analysis, it is necessary to identify the batch number for each failed item at both the product and component levels. The ability to perform this identification is called *traceability*.

For certain products (e.g., automobiles, personal computers), each unit at the product level has a unique identification number [called Vehicle Identification Number (VIN) in the case of automobiles]. For others (e.g., household appliances), there are no such identification numbers. In either case, there is ordinarily no numbering at the component level. Traceability at component and product levels requires supplementary data, as indicated in Fig. 5.6.

Once the batch numbers of failed products or components are identified, the analysis for assessing quality variation from batch to batch can be performed and the results related to other variables of the production process in order to control or reduce quality variations. This topic is pursued further in Chap. 15.

## 5.10 Problems with Supplementary Data

As in the case of claims data, discussed Chap. 4, there can be several possible problems associated with supplementary data. These include the following:

- Aggregated (pooled) data as opposed to item level data
- Missing data
- Errors in reporting
- Data not collected for a variety of reasons, such as time, cost, etc.

## 5.11 Current Practice in Industry

The collection of supplementary data varies from industry to industry and from manufacturer to manufacturer within an industry sector. There are several reasons for this, including the following:

1. Businesses that are in Stage 1 of warranty management have warranty systems that collect data mainly for controlling fraud and warranty costs. In contrast, businesses in Stage 2 of warranty management collect some or all of the supplementary data that are required for effective warranty management.
2. Data collection is done at the functional unit level (e.g., departmental), and the data collected are mainly for managing the operations of the unit involved. As a result, data of relevance to overall business management are often not collected.
3. The different systems used for data collection are often incompatible, resulting in difficulties in transferring data between different units of the organization.
4. There is no one person at the very senior management level with the overall responsibility for data collection across the organization and for ensuring that proper data are collected and used effectively for making management decisions.

### *5.11.1 Automotive Industry*[9]

The Automotive Industry Action Group (AIAG) has considered various issues and problems with automotive product data and has served as an automotive voice in national and international standards communities.

[9] This section is adapted from [1]

**Table 5.1** Data types in the automotive industry

| Automotive product—data types | PLC perspective |
|---|---|
| Advanced engineering information (1) | D-I-1 |
| Equipment maintenance and repair (2) | D-I-2 |
| Feasibility information (3) | D-I-6 |
| Field service information (4) | D-I-1 |
| Fixture and tooling information (5) | D-I-4 |
| Inspection and test information (6) | D-I-4 |
| Manufacturing source list (7) | D-I-4 |
| Manufacturing process design information (8) | D-I-4 |
| Part termination information (9) | D-I-4 |
| Plant layout information (10) | D-I-4 |
| Process control information (11) | D-I-5 |
| Product design information (geometry) (12) | D-I-5 |
| Product design information (non-geometry) (13) | D-I-5 |
| Product feature information (14) | D-I-6 |
| Product regulatory information (15) | D-I-4 |
| Product test information (16) | D-I-4 |
| Product warranty information (17) | D-I-1 |
| Program cycle information (18) | D-I-6 |
| Purchasing bill of material (19) | D-I-4 |
| Repair/problem reports (20) | D-I-1 |
| Service tools and equipment information (21) | D-I-1 |
| Technical vehicle owner information (22) | D-I-1 |

The AIAG has categorized automotive data into the following domains:

1. *CAD Domain*: Part design and its associated data
2. *CAM Domain*: Some paper artifacts, but mostly application software (tool design or drawings developed using a standard package)
3. *CAE Domain*: CAE data not tied to a specific production part or assembly in the same way that a CAD design is associated with a specific production part.
4. *Service Domain*: Data needed to meet the operational service needs of the vehicle

The different product data types are listed in the first column of Table 5.1 (adapted from Table 5 of [1]). The second column shows the connection to data classifications D-I-1 to D-I-6 of Sect. 5.4.2.

The definitions of the various types of data and information indicated by numbers 1–22 in Table 5.1 are as follows:

1. Information on new and advanced methods, processes, and concept studies and project files for product engineering.
2. Equipment operation, repair and maintenance documentation; includes check sheets, routine preventive maintenance, calibration/verifications, return information on equipment.
3. Other types of reports, documents and data used in the design and development of vehicle products.

4. Documents related to customer product problems or defects, which may include the plan, notification, announcements, customer lists, etc.
5. Information which defines the fixtures and tooling required for manufacturing parts and vehicles; includes CAD/CAM files and drawings, specifications, production, feasibility, status, etc.
6. General manufacturing quality and reliability records used to manage product quality; does not include agency/regulatory testing records.
7. Documentation identifying the acceptance of quality material.
8. Records that demonstrate the development and verification of the design of the manufacturing processes.
9. Examples of procurement or purchasing types of records using product data and requiring retention.
10. Plant layout drawings, machine specs, reliability, production capacity, maintenance schedules, equipment history, etc., to specify the manufacture of components, subsystems, and vehicles.
11. Process control direction and documentation that directly supports the manufacture of a product, assembly or part, such as control plans, SPC charts, etc.
12. Part drawing and/or geometry data for released production parts.
13. Records of product specifications and design from initiation to verification and validation, through ongoing improvement.
14. Documents that provide product or program direction.
15. Records showing compliance to regulatory standards; i.e., environment/ regulatory test records, military standards, special test requirement records, certificate of compliance, third party testing records.
16. Documents test data required to design, authorize and conduct requested/ required tests.
17. Documentation of product warranties and claims.
18. Documents product and program management data.
19. Procedures and manuals used to manufacture test products, assemblies, and parts.
20. Documents related to problems encountered by customers and subsequent statistics.
21. Design, analysis, and related data used to support the development and specification of service related tools, equipment and processes.
22. Material that promotes or describes services and products and provides customers with guidance and direction.

## 5.12 Characterization of Data Structures

The following three data structures will be considered:

- Structure 1 Data: The data consist of the sale dates and failure times for each item.
- Structure 2 Data: The data consist of counts of numbers of failures for each item over non-overlapping time intervals.

- Structure 3 Data: The data consist of aggregated numbers of failures (combining failures of several items or failure modes) over discrete time intervals.

From the reliability analysis point of view Structure 1 Data have the maximum information and Structure 3 Data the least. Ideally, service agents should use Data Structure 1 to collect data and most do so. However, a few use either Structure 2 or Structure 3 for data collection. We introduce notation and format for the three data structures before proceeding to analysis of the data.

For warranty claims data, time can be measured relative to two clocks: (i) calendar (or time) clock and (ii) age clock. Under the calendar clock, the data (e.g., the dates of sale, failures, censoring, etc.) are given in terms of the calendar clock, with time zero being the same for all items. On the other hand, under the age clock, the data (failures, censoring, etc.) are based on a clock where time zero differs from item to item in that it corresponds to the instant when an item is put into use or sold. We use $t$ to denote time as measured by the calendar clock and $a$ to denote time as measured by age clock.

### 5.12.1 Structure 1 Data (Detailed Data)

The following notation and conventions are used to describe the data:

- Customers: $I$ (numbered 1, 2, …., $I$)
- Subscripts $i$ and $j$ denote the customer number and failure number, respectively

Calendar Clock

- $t_{i0}$ denotes the sale time for the item bought by customer $i$ (based on the calendar clock)
- $t_{ij}$ denotes the calendar time for the $j$th failure experienced by customer $i$, $i = 1, 2, \ldots, I$, $j = 1, 2, \ldots, J$
- $\tilde{t}_i$ denotes the censored data for customer $i = 1, 2, \ldots, I$.

Age Clock

- $a_{i0} = 0$
- $a_{ij}$, $i = 1, 2, \ldots, I$, $j = 1, \ldots$, denotes the age of the item at the time of the $j$th failure experienced by customer $i$ (based on age clock).
- $\left[a_{ij} = t_{ij} - t_{i,j-1}, i = 1, 2, \ldots, I, j = 1, 2, \ldots\right]$
- $\tilde{a}_i$ denotes the censored observation for customer $i$, $i = 1, 2, \ldots, I$.

Number of Failures

- $n_i$ ($\geq 0$) denotes the number of failures under warranty experienced by customer $i$, $i = 1, 2, \ldots, I$.

Figure 5.7 is an illustrative example of warranty data collected over the interval $[0, t)$ with $I = 3$ and the items sold with a non-renewing warranty with warranty period $W$.

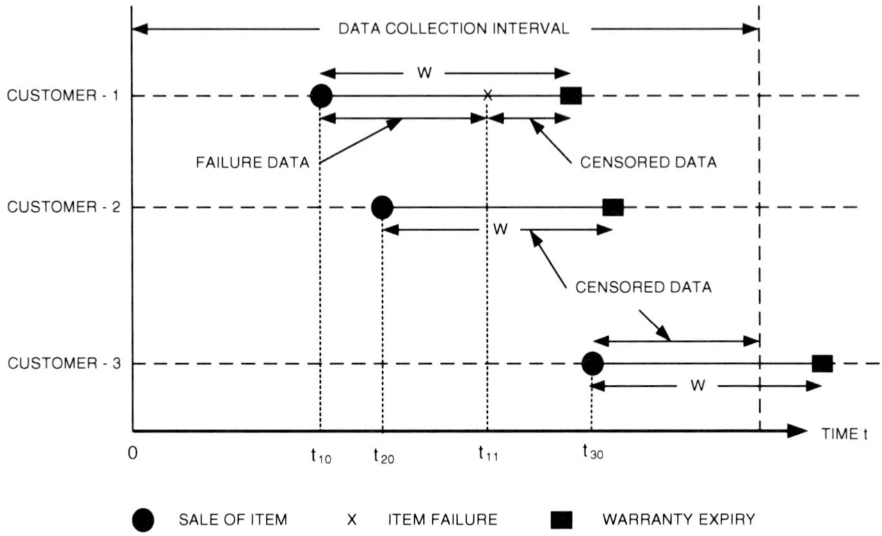


**Fig. 5.7** Claims history for Data Structure 1

Note that censoring can occur in the following two ways:

1. Due to expiration of the warranty: The censored observation for customer $i$ is given by $W$ if the item experienced no failure ($n_i = 0$) and by $\tilde{t}_i = W - t_{n_i}$ when $n_i > 0$. The censored data for customers 2 and 1 in Fig. 5.7 illustrate the two cases.
2. Due to reaching the end of the data collection interval: In this case the censored observation for customer $i$ is given by $\tilde{t}_i = t - t_{n_i}$, when $n_i \geq 0$. Customer 3 in Fig. 5.7 illustrates this.

The data are best stored in a tabular form for analysis. Table 5.2 illustrates this for the data of Fig. 5.7.

In Table 5.2, each customer has one or more rows. The first row contains the following information in Columns 1–3 and 6.

C1: Customer number
C2: Date of sale
C3: Warranty status ($Y$ indicates still under warranty and $N$ indicates no longer under warranty)
C6: Censored data (based on the calendar clock)

The subsequent rows (one for each failure) contain the following information in columns 4 and 5 and columns 7 onwards.

C4: Failure number
C5: Failure time (based on the calendar clock)
C7 onwards: Other data (such as failed component, failure mode, time to repair, cost of repair, etc.)

**Table 5.2** Tabular representation of Data Structure 1

| C1 | C2 | C3 | C4 | C5 | C6 | C7 | C8 |
|---|---|---|---|---|---|---|---|
| 1 | $t_{10}$ | $N$ | | | $\tilde{t}_1$ | | |
| | | | 1 | $t_{11}$ | | | |
| | | | 2 | $t_{12}$ | | | |
| 2 | $t_{20}$ | $N$ | | | $\tilde{t}_2$ | | |
| 3 | $t_{30}$ | $Y$ | | | $\tilde{t}_3$ | | |

*Comment*: One can store failure and censored times based on the age clock instead of the calendar clock. Which is used depends on the purpose of the analysis.

### 5.12.2 Structure 2 Data (Count Data)

The following notation and conventions are used to describe the data structure:

- $\Delta$ is the width of the interval for discretization of time
- Period $j$ ($j = 1, 2, \ldots,$) corresponds to the time interval $[(j-1)\Delta, j\Delta)$ (measured using the age clock)
- $n_{ij}$ denotes the number of failures experienced by customer $i$ in period $j$
- $N_{ij}$ denotes the number of failures experienced by customer $i$ in the first $j$ periods $\left(N_{ij} = \sum_{m=1}^{j} n_{im}\right)$.

Figure 5.8 is an illustrative example of warranty data over 8 time intervals for three customers.

A tabular form of the data for analysis is one in which each row corresponds to a customer and the columns contain the following information:

C1: Customer number $(i)$
C2: $n_{i1}$
C3: $n_{i2}$
⋮
$C(j+1)$: $n_{ij}$ (number of failures in period $j$)

Table 5.3 is the tabular form for the data from Fig. 5.8.

### 5.12.3 Structure 3 Data (Aggregated over Discrete Time Intervals)

This data structure is useful for certain types of analysis.[10] Time is divided into discrete intervals as in Data Structure 2. Usually the time period is a month, so that

[10] Chapter 16 uses this structure in the context of improvements to product and operations.

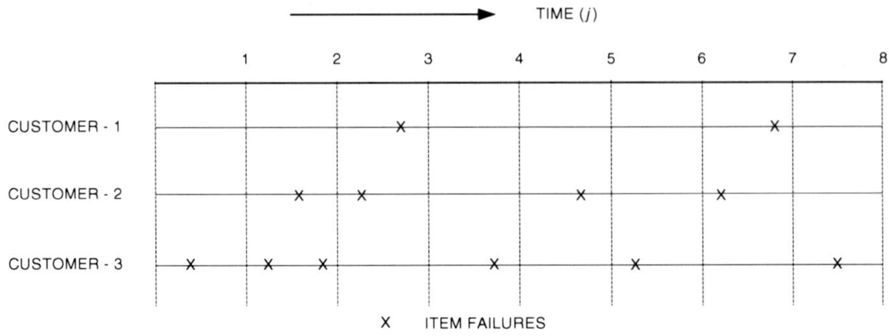


**Fig. 5.8** Claims history for Data Structure 2

**Table 5.3** Tabular representation of Data Structure 2

| Customer number ($i$) | Time interval ($j$) | | | | | | | |
|---|---|---|---|---|---|---|---|---|
| | 1 | 2 | 3 | 4 | 5 | 6 | 7 | 8 |
| 1 | 0 | 0 | 1 | 0 | 0 | 0 | 1 | 0 |
| 2 | 0 | 1 | 0 | 0 | 1 | 0 | 1 | 0 |
| 3 | 1 | 2 | 0 | 1 | 0 | 1 | 0 | 1 |

one is looking at aggregated monthly data.[11] The following notation and conventions are used to describe the data structure:

- $\Delta$: length of time interval for discretization
- $j$, $k$, $t$: indices denoting time intervals
- $S_j$ : number of units sold in interval $j$ ($j = 1, 2, \ldots, J$)
- $n_{jk}$: number of warranty claims in interval $k$ ($k = 1, 2, \ldots, K$) units sold in interval $j$ ($j \leq k$)
- MOS: Month of Sale (indexed by subscript $j$)
- MIS: Months in Service before failure (duration for which the item is in use—indexed by subscript $a$)
- MOP: Month of Production
- $\tilde{n}_{ja}$ : Number of items from MOS $j$ that fail with age at failure in the interval $[(a-1)\Delta, a\Delta)$

Figure 5.9 is an illustrative example of warranty data collected over 8 months for items sold over 5 months. The time interval is $\Delta = 1$ month and $J = 5$. The items are sold with a non-renewing warranty with warranty period $W$ equal to 3 months.

A tabular form to store the data for analysis would involve one row for each month of sale. The entries in the columns of the table are as follows:

[11] The period can be shorter—for example, a week or a day. If there are two or more in a workday, it can be a shift within a day.

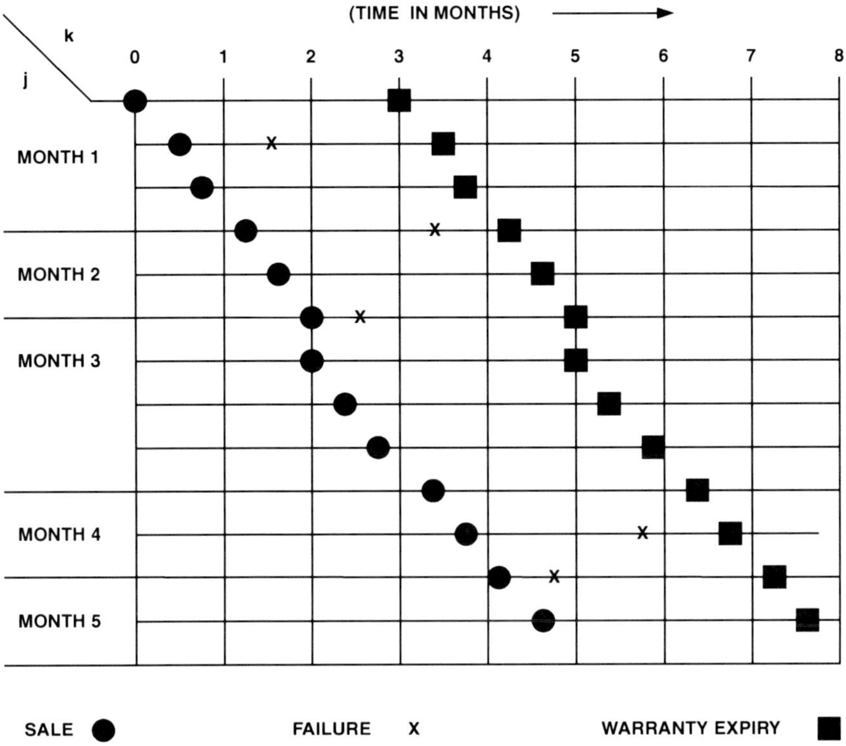


**Fig. 5.9** Claims history for Data Scenario 3

C1: Month of sale (MOS), indexed by $j$

C2: Number $s_j$ of items sold in interval $j$

C3: Number of warranty claims in the first time interval, $(n_{j1})$, $j = 1, 2, \ldots, J$

⋮

C$(k+2)$: Warranty claims in interval $k$ $(n_{jk})$, $j = 1, 2, \ldots, J$

⋮

Table 5.4 is one such tabular form for the data from Fig. 5.9. Note that items sold in month $j$ can result in warranty claims in intervals $j$ to $j + W = j + 3$.

From Table 5.4 one can create a table in which the age clock instead of the calendar clock. Failures in interval $k$ for items with MOS equal to $j$ would correspond to months in service before failure given by $a = k - j + 1$, $(j \leq k)$. The table will have one row for each month of sale and entries in the columns as follows:

**Table 5.4** Tabular representation of Data Structure 3 (MOS–MIS Table based on calendar clock)

| ($j$) | Monthly sales ($S_j$) | $\{n_{jk}\}$ in interval ($k$) | | | | | | | |
|---|---|---|---|---|---|---|---|---|---|
| | | 1 | 2 | 3 | 4 | 5 | 6 | 7 | 8 |
| 1 | 3 | 0 | 1 | 0 | 0 | – | – | – | – |
| 2 | 2 | – | 0 | 0 | 1 | 0 | – | – | – |
| 3 | 4 | – | – | 1 | 0 | 0 | 0 | – | – |
| 4 | 2 | – | – | – | 0 | 0 | 1 | 0 | – |
| 5 | 2 | – | – | – | – | 1 | 0 | 0 | 0 |

**Table 5.5** Tabular representation for data structure 3 (MOS–MIS table based on age clock)

| MOS ($j$) | Monthly sales ($S_j$) | MIS ($a$) | | |
|---|---|---|---|---|
| | | 1 | 2 | 3 |
| 1 | 3 | 1 | 0 | 0 |
| 2 | 2 | 0 | 1 | 0 |
| 3 | 4 | 0 | 0 | 0 |
| 4 | 2 | 0 | 1 | 0 |
| 5 | 2 | 1 | 0 | 0 |

C1: Month of sale (MOS) indexed by $j$

C2: Number of items sold in interval $j$ ($S_j$)

C3: Warranty claims in the first MIS interval $(n_{j1})$, $j = 1, 2, \ldots, J$

⋮

C($k + 2$): Warranty claims in the $k$th MIS interval $(n_{jk})$, $j = 1, 2, \ldots, J$

⋮

The maximum value for $k$ is $[W/\Delta]$.[12] This table is called the *MOS–MIS* table. Table 5.5 is the MOS–MIS table based on the data of Table 5.4.

If all the items sold in a month are from a batch produced from the same month, then the month of production (MOP) is the same as the month of sale (MOS). In this case the above table can be called a *MOP–MIS* table.

Often items produced in a given month are sold over several months subsequent to the month of production. In this case the construction of the MOS–MIS table is a bit more involved and is possible only if items sold can be linked to the month of production. This issue is discussed in detail in Chaps. 11 and 14.

## 5.13 Scenarios for Data Analysis

A proper analysis depends on the objectives of the analysis and the structure of the data. The two broad categories of data that we have identified are warranty claims data, discussed in Chap. 4, and supplementary data, discussed in this

[12] [x] is the smallest integer equal to or greater than x.

chapter. Other issues that are relevant are the data collection interval, the data structure as discussed in Sect. 5.12, and the warranty servicing policies and procedures.

As a result, several different scenarios for analysis of data are possible. We begin with a brief discussion of the above issues before proceeding to a description of the data scenarios.

*Data Collection Interval*. Service agents collect data on a continuous basis and the data are transmitted to the manufacturer either in real time or on a regular basis. The manufacturer normally analyzes the data on a periodic basis, with the period being month, quarter, half yearly or yearly. As a result, the interval of data collection varies with time.

*Warranty Servicing*. In the case of a non-renewing PRW rebate policy, the customer is refunded a fraction of the sale price on failure of an item under warranty. Under a renewing PRW policy, the customer is provided with a new item at a discounted price. In the case of the FRW (non-renewing and renewing) policy, the manufacturer has a choice of either repairing or replacing a failed item. Repair may be done in a number of ways, each modeled differently.[13]

For analysis of warranty data, the replace/repair policy, and, in the latter case, the type of repair must be known. When this information is missing, assumptions regarding the service actions must be made. The two assumptions commonly used (which are appropriate for most warranty data) are the following:

1. A failed item is always replaced by a new item.
2. A failed item is always repaired minimally.

*Additional Terminology and Notation*. Each new sale is identified with a unique customer number. When a customer buys more than one unit, each unit has a different number and we use the term "customer" to correspond to the identification number. As a result, each customer buys only one unit and for the renewing PRW policy and the FRW policy with replacement by new, a customer can receive one or more new units as replacements during the warranty period. We use the following notation:

$I$ : Total number of customers who have bought the product over the data collection interval

$I_1$: Number of customers who have one or more warranty claims over the data collection interval

$I_2$: Number of customers who have had no warranty claim over the data collection interval

[13] The different kinds of repair are discussed in Sect. 3.7.2. Optimal servicing strategy can involve both repair and replace actions [3, 4]. The analysis of warranty data for these cases follows along similar lines, but can be considerably more involved.

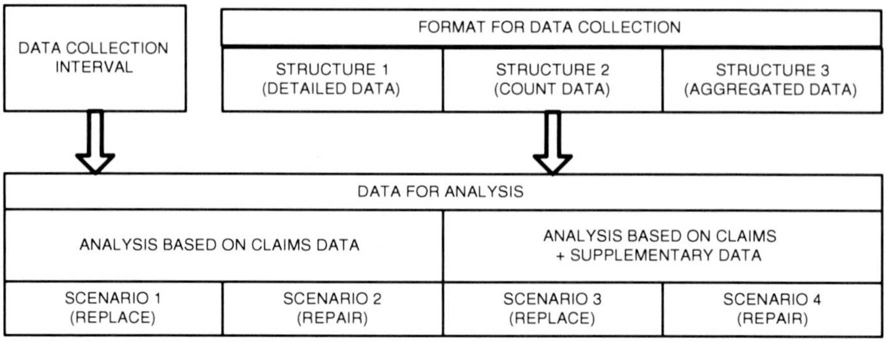


**Fig. 5.10** Different scenarios for warranty data analysis

*Comment*: $I_1 + I_2 = I$.

*Data Scenarios*. For each data structure, we have four different scenarios, as indicated in Fig. 5.10. Thus we have a total of twelve different scenarios.

In the remainder of this section, we describe the data format for the twelve scenarios. The dependence of the analysis on the data format will be discussed in later chapters.

## *5.13.1 Scenarios for Structure 1 Data*

### 5.13.1.1 Scenario 1.1

The data are the warranty claims data and consist of the ages of items that failed under warranty and were replaced by new items. These data are given by the set[14] $I_{11} : \{a_{ij},\ i = 1, 2, \ldots I_1,\ j = 1, 2, \ldots\}$, with $a_{ij}$ denoting the age at $j$th failure for customer $i$.

Let $n_1$ denote the total number of failed items (includes the items sold as well as those used as replacements) and $t_i$, $1 \le i \le n_1$, denote the ages at failure. As a result, the data, in a form more appropriate for analysis, are given by the set

$$I_{11} : \{t_i,\ 1 \le i \le n_1\}.$$

*Comments*:

1. In the case a non-renewing PRW policy, $j$ cannot exceed one.
2. In the case of a renewing PRW and both non-renewing and renewing FRW policies, $j$ can be greater than 1.

---

[14] We use the first sub-script to denote the structure and the second to denote the scenario so that the set $I_{uv}$ denotes data set for Scenario $v$ (varying from 1 to 4) under Structure $u$ (varying from 1 to 3).

3. Since failed items are replaced by new ones and it is assumed that the time to replace is negligible and all new items are statistically similar, the failure data are a set of *iid* variables. For each customer, failures over time are modeled by a renewal process.[15]

#### 5.13.1.2 Scenario 1.2

The data consist of the age at failure $a_{ij}, i = 1, 2, \ldots I_1,\ j = 1, 2, \ldots$, as discussed in Sect. 5.12, with $a_{ij}$ denoting the age of the item in service at the time of the $j$th failure for customer $I_1$, with failed items repaired minimally. Let $n_i$ denote the number of failures experienced by customer $i$ (or repairs carried out by the service agent for customer $i$) over the period during which the data are collected. The data for analysis are given by the set

$$I_{12} : \{(a_{ij},\ j = 1, 2, \ldots n_i),\ i = 1, 2, \ldots I_1\}.$$

*Comments*:

1. This scenario is applicable only for the non-renewing FRW policy.
2. $j$ can be 1 or greater than 1.
3. Since failed items are minimally repaired and the repair time negligible, the failures can be viewed as a counting process characterized through the mean cumulative function (MCF).[16]
4. In Scenario 1.1 all $a_{ij}$'s are statistically similar. In Scenario 1.2, only $a_{i1}$'s are statistically similar, whereas the $a_{ij}$'s are not for $j > 1$ and all $i$.

#### 5.13.1.3 Scenario 1.3

The data consist of the ages of items that failed under warranty (as in Scenario 1.1) and the censoring times for all customers, given by $\tilde{a}_i,\ i = 1, 2, \ldots I$, with $\tilde{a}_i$ denoting the censored item age for customer $i$. As in Scenario 1.1, the alternative representation, more appropriate for analysis, is given by the data set

$$I_{13} : \{t_i,\ 1 \le i \le n_1;\ \tilde{t}_i,\ 1 \le i \le n_2\}$$

where $n_1$ and $t_i,\ 1 \le i \le n_1$ are as in Scenario 1.1; $n_2 \le I$ and $\tilde{t}_i = \tilde{a}_i,\ i = 1, 2, \ldots, I$.

---

[15] See Chap. 11 and Appendix B.

[16] See Chap. 11 and Appendix B.

*Comments*:

1. In the case of a non-renewing PRW policy, $j$ cannot exceed one and $n_2 = I - n_1$.
2. For a renewing PRW and both non-renewing and renewing FRW policies, $j$ can be greater than 1 and $n_2 = I$.
3. Since failed items are replaced by new ones and it is assumed that all new items are statistically similar, the failed and censored data are a set of *iid* variables. For each customer, failures over time are modeled by a renewal process.

#### 5.13.1.4 Scenario 1.4

The data consist of the ages at failures as in Scenario 1.2 and the additional set $\{\tilde{a}_i, i = 1, 2, \ldots I_1\}$, where the $\tilde{a}_i$, $i = 1, 2, \ldots, I$, are the censored ages (see Comment 2 below), with $\tilde{a}_i = W$ if the warranty of customer $i$ expired within the data collection period. If the warranty had not expired, $\tilde{a}_i$ is the age of the item at the end of interval. The data for analysis are given by the set

$$I_{14} : \{(a_{ij},\ j = 1, \ldots n_i),\ i = 1, 2, \ldots I_1;\ \tilde{a}_i, i = 1, 2, \ldots, I\}.$$

*Comments*:

1. The first three comments of Scenario 1.2 are applicable in this case.
2. $\tilde{a}_i$ may be viewed as a censored observation in the sense that the time between failures $n_i$ and $(n_i + 1)$ is greater than $\tilde{a}_i$ for customer $i$.

### *5.13.2 Scenarios for Structure 2 Data*

#### 5.13.2.1 Scenario 2.1

The data consist of the number of replacements under warranty for each customer over different time intervals during the data collection period. We use the calendar clock and let $n_{ij}$, $i = 1, 2, \ldots I_1$, $j = 1, 2, \ldots$, denote the replacements for customer $i$, where period $j$ is the time interval $[(j-1)\Delta, j\Delta)$, as discussed in Sect. 5.12. The data for analysis are given by the set

$$I_{21} : \{(t_{i0}, n_{ij},\ j = 1, \ldots),\ i = 1, 2, \ldots I_1\},$$

where $t_{i0}$, $i = 1, 2, \ldots I_1$, denotes the time interval during which customer $i$ bought the product.

*Comments*:

1. In the case of a non-renewing PRW policy, for each $i$, $n_{ij} = 1$ for only one time interval.
2. For the non-renewing PRW and for both renewing and non-renewing FRW policies, $n_{ij} \geq 0$ for all $i$ and $j$, with at least one interval having a value greater than zero.
3. Failures for each customer occur according to a renewal process.

#### 5.13.2.2 Scenario 2.2

This is the same as Scenario 2.1, except that failed items are repaired instead of being replaced. As such, the data for analysis are given by the set

$$I_{22} : \{(t_{i0}, n_{ij}, j = 1, 2 \ldots), i = 1, 2, \ldots I_1\},$$

which is the same as the set for Scenario 2.1.

*Comments*:

1. Data of this type are generated by FRW policies.
2. Failures for each customer modeled by a counting process characterized by the mean cumulative function (MCF).
3. From the sales data it is possible to calculate the number of sales that led to warranty claims in each time interval. The sum of these equal to $I_1$.

#### 5.13.2.3 Scenario 2.3

The warranty claims data consist of the number of replacements under warranty for each customer over different time intervals during the data collection period as in Scenario 2.1. Supplementary data consist of the number of customers who had no warranty claims and the interval for the sale of these units. As such, the data for analysis are given by the set

$$I_{23} : \{t_{i0}, i = 1, 2, \ldots I; (n_{ij}, j = 1, \ldots), i = 1, 2, \ldots I_1\}.$$

*Comment*: The comments of Scenarios 1.3 and 2.1 are applicable here.

#### 5.13.2.4 Scenario 2.4

The warranty claims data consist of the number of repairs under warranty for each customer over different time intervals during the data collection period as in Scenario 2.2. The supplementary data are the number of customers who had no warranty claims and the interval for the sale of these units. As such, the data for analysis are the set

$$I_{24}:\ \{t_{i0}, i=1,2,\ldots I;\ (n_{ij},\ j=1,2,\ldots),\ i=1,2,\ldots I_1\}$$

*Comment:* The comments of Scenario 2.2 also apply here.

## *5.13.3 Scenarios for Structure 3 Data*

### 5.13.3.1 Scenario 3.1

The data consist of the total number of replacements in each time interval aggregated over all items sold in the interval. We use the calendar clock and the warranty claims data for analysis are given by the set

$$I_{31}:\{n_{jk},\ j=1,2,\ldots;k=j,j+1,\ldots\},$$

where $n_{jk}$ is the number of items that were replaced in the interval $k$ for items that were sold in interval $j$.

*Comments*:

1. The number of replacements in calendar clock interval $k$ is given by

$$n_{\bullet k}=\sum_j n_{jk}.$$

2. The number of replacements in age clock interval $a$ (or MIS $a$) can be expressed as $n_{\bullet a}=\sum_j n_{j,j+a-1}$.

### 5.13.3.2 Scenario 3.2

The data consist of the total number of repairs in each time interval aggregated over all items sold in the interval. We use the calendar clock and the warranty claims data for analysis are given by the set

$$I_{32}:\{n_{jk},\ j=1,2,\ldots;k=j,j+1,\ldots\},$$

where $n_{jk}$ is the number of items that were repaired in the interval $k$ for items that were sold in interval $j$.

*Comments*:

1. The number of repairs in interval $k$ is given by $n_{\bullet k}=\sum_j n_{jk}$.
2. The number of repairs in age clock interval $a$ (or MIS $a$) can be expressed as $n_{\bullet a}=\sum_j n_{j,j+a-1}$.

#### 5.13.3.3 Scenario 3.3

The data consist of the total number of replacements aggregated over all items sold in the same interval as well as sales in each period. Let $S_j$ denote the number of items sold in interval $j$. The data available are given by the set

$$I_{33} : \{S_j, n_{jk},\ j = 1, 2, \ldots; k = j, j + 1, \ldots\}$$

Additional information may include the number of customers having items still under warranty at the start of each interval.

#### 5.13.3.4 Scenario 3.4

This is similar to Scenario 3.3, except that $n_{jk}$ is the number of failed items repaired in the interval $k$ for items that were sold in interval $j$, so that

$$I_{34} : \{S_j, n_{jk},\ j = 1, 2, \ldots; k = j, j + 1, \ldots\}$$

## References

1. Bsharah F, Less M (2000) Requirements and strategies for the retention of automotive product data. Comput Aided Des 32:145–158
2. Murthy DNP, Rausand M, Osteras T (2008) Product reliability—performance and specifications. Springer, London
3. Murthy DNP, Jack N (2007) Warranty servicing. In: Kenett F, Faltin FW R (eds) Encyclopedia of statistics in quality and reliability. Ruggeri. Wiley, New York
4. Yun WY, Murthy DNP, Jack N (2008) Warranty servicing strategies with imperfect repair. Int J Prod Econ 111:159–169

# Part III
# Models, Tools and Techniques

# Chapter 6
# Cost Models for One-Dimensional Warranties

## 6.1 Introduction

When a manufacturer offers a warranty on a product, then all legitimate claims under warranty must be serviced by the manufacturer or his agent. The frequency of failures leading to claims depends on the reliability of the product. Servicing of the claims results in additional costs to the manufacturer. As discussed in Sect. 2.11, there are several different notions of warranty costs. In this chapter, we focus on models for predicting these costs as a function of product reliability. Models will be given for several one-dimensional warranties. These models play a critical role in warranty management.

The outline of the chapter is as follows: In Sect. 6.2, we look at system characterization. This is Step 1 of the model-building process discussed in Sect. 3.5.1. Here we apply to process to the problem of building cost models for one-dimensional warranties. Modeling of the various elements of the system is discussed in Sect. 6.3. Section 6.4 deals with models for warranty cost per unit sale for several one-dimensional warranty policies. In Sects. 6.5 and 6.6, we look cost models for two different notions of life cycle.

## 6.2 System Characterization for Warranty Cost Analysis

In Sect. 2.11, we defined three different notions of warranty costs.[1] These are:

- Warranty cost per unit sale
- Life cycle cost per unit sale (which we shall denote LCC-I)
- Cost over product life cycle (which we shall denote LCC-II)

---

[1] Other cost bases that might be considered are cost per unit of time, other life cycle concepts, and so forth. For further discussions of these, see [2, 3].

W. R. Blischke et al., *Warranty Data Collection and Analysis*,
Springer Series in Reliability Engineering, DOI: 10.1007/978-0-85729-647-4_6,

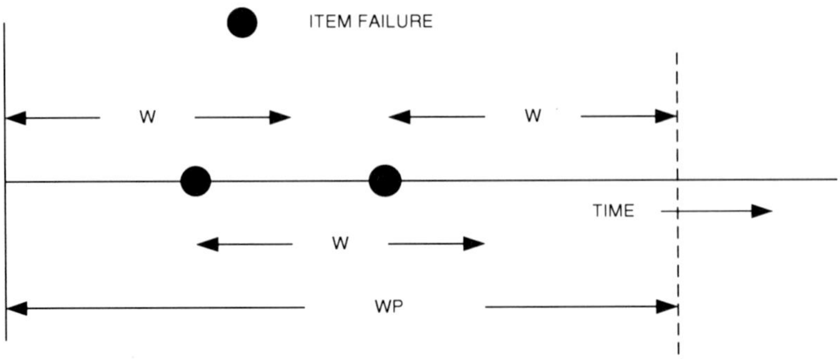


**Fig. 6.1** Claims history for a renewing warranty

The costs associated with these are uncertain since claims and the cost to rectify failed items are uncertain. In fact, all of these depend on many other factors, including product reliability, warranty execution, servicing strategy, etc.

The characterization of any system depends on the level of detail desired. For example, the cost to service a warranty claim is comprised of several cost elements, as discussed in Sect. 4.6.4. In the simplest characterization, one looks at the aggregate of these different cost elements and treat this as a deterministic variable. In a more detailed characterization, this would be treated as an uncertain variable, with the uncertainty modeled by a distribution function. A still more detailed characterization of the system would be to characterize each of the cost elements involved separately.

The system characterization for warranty cost per unit sale is different from that for life cycle costs, as will be indicated later in the chapter.

### 6.2.1 Warranty Period

The one-dimensional warranty is characterized by a single parameter $W$. The period for which an item is covered under warranty (which we denote $WP$) depends on whether the warranty is non-renewing or renewing. In the former case, $WP = W$. In the latter case the warranty is renewed each time an item fails under warranty. As a result, the warranty period is uncertain and can assume any value in the interval $[W, \infty)$, depending on the claims generated. Figure 6.1 is an illustration of this in which the initial item sold and the item used as a replacement fail within warranty, whereas the last item survives for a period $W$. The warranty period $WP$ equals $W$ plus the lifetimes of the first two items.

### 6.2.2 Characterization of Cost per Unit Sale

The system characterization for modeling warranty cost per unit sold is indicated in Fig. 6.2.

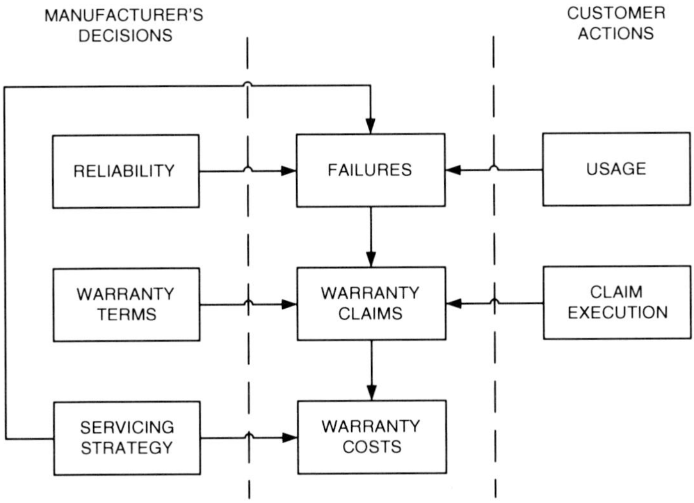


**Fig. 6.2** System characterization for cost per unit sale

As can be seen, some of the elements involved are under the control of the manufacturer and others are influenced by the actions of customers. The characterization of these elements is discussed in the next section.

### *6.2.3 Characterization of Life Cycle Costs*

We discuss system characterization for LCC-I and LCC-II.

LCC-I: Let $L_l$ denote the useful life of the product. As discussed in Sect. 2.11.2, one or more components may need to be replaced more than once during this period. Many component replacements occur after the original warranty has expired, and the replacements are covered by separate warranties. A typical history of repeat purchases for a component sold with a non-renewing warranty policy with warranty period $W$ is shown in Fig. 6.3. The times between repeat purchases of a component (indicated by $Y_i$, $i = 1, 2, \ldots$) are uncertain. These depend on the times of first failure outside the original warranty period (indicated by $Z_i$, $i = 1, 2, \ldots$) in the case of non-repairable products, and on consumer replacement decisions in the case of repairable products.

Life cycle costs are different for the manufacturer and the customer. For the manufacturer, these costs include the cost of servicing warranty claims associated with the product initially sold as well as claims associated with subsequent component sales. For the customer, the cost is the initial purchase price plus the costs associated with subsequent purchases of replacements for components that fail outside warranty.

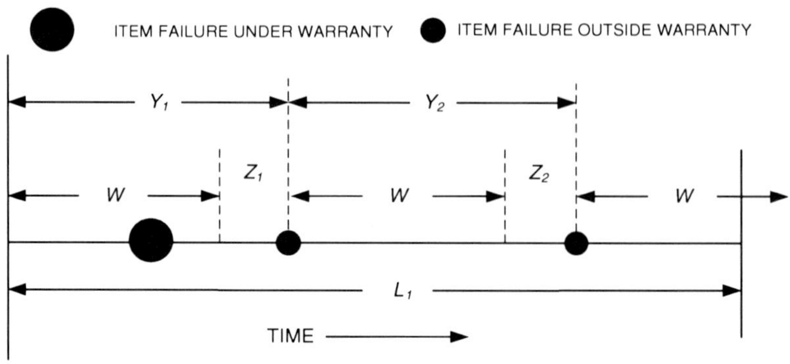


**Fig. 6.3** Typical history of repeat purchases over period $L_1$

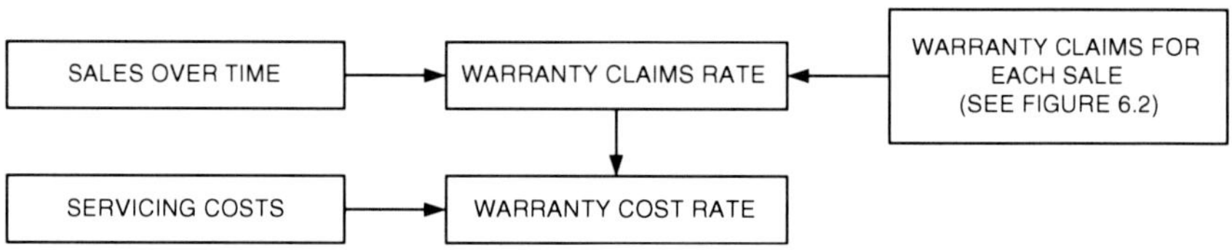


**Fig. 6.4** System Characterization for Warranty Cost

LCC-II: For a manufacturer, the product life cycle (from a marketing perspective) extends from the time that the product is introduced into the marketplace until the time it is withdrawn. Over this period, there are ordinarily several first purchase sales. If the useful life (as defined in LCC-I) is large compared to the product life cycle, then there are no repeat purchases. On the other hand, if the useful life is small relative to the product life cycle, then there may be repeat purchases from satisfied customers. The manufacturer must service all warranty claims for each item sold. The cost of this is referred to as the life cycle cost and is relevant in the context of decision making at the front-end phase of the new product development process.

Let $L$ denote the length of the product life cycle. The system characterization for warranty costs over $L$ is indicated in Fig. 6.4. The characterization of warranty claims with each sale is as indicated in Fig. 6.2.

## 6.3 Modeling for Warranty Cost Analysis

In practice, characterization of each element in Figs. 6.2 and 6.4 is needed in order to formulate a working cost model. The modeling efforts may be synthesized as follows. In essence, the direct cost of warranty is primarily a result of two key determinants: (1) the structure of the warranty policy and (2) the failure pattern of the product. The latter depends on the life distribution of the product. Although it is recognized that each of these determinants is a function of many factors, the cost models that will be discussed in this chapter require precisely these two inputs.

The first of these is directly controlled by the manufacturer through choice of the policy to be offered, although this choice may be influenced by market and other factors. We will look at cost models for a few of the common consumer warranties discussed in Chap. 2. The objectives here are (1) to provide some insight into the cost modeling process, (2) to evaluate warranty costs as a function of warranty length in realistic applications, (3) to compare costs of different warranty policies, and (4) to assess, to a limited extent, the sensitivity of the results to some of the assumptions made.

The second important input to the analysis, the failure pattern of the item, is only partly controlled by the manufacturer and, in any case, is far more difficult to determine precisely. The failure pattern is influenced not only by the design of the product and the manufacturing process, but also by the raw materials used, the failure patterns of components received from suppliers, the type and intensity of usage by the purchaser, and many additional uncontrolled (and usually uncontrollable) factors. As a result, careful attention must be paid to two important interrelated ingredients of this aspect of the modeling effort, (1) selection of an appropriate, realistic probabilistic model, and (2) acquisition and analysis of as much relevant data as possible.

Some details of the characterization and modeling of the different elements of Figs. 6.2 and 6.4 are discussed in this section.

### *6.3.1 Servicing Strategy*

For a non-repairable product, the only option available is to replace a failed unit with a new one. For a repairable product, there are several strategies for dealing with a failed item, including the following:

- Always replace
- Always repair minimally
- Imperfect repair
- Cost limit repair
- Repair or replace

Minimal and imperfect repair are discussed in Sect. 3.7.2. Cost limit repair and repair versus replace decisions are discussed below. Two other elements of importance in defining a servicing strategy are service time and service costs. These are also discussed below.

*Cost limit repair* In a cost repair limit strategy, an estimate of the cost to repair a failed product is made and by comparing it with some specified limit, the failed item is either repaired or replaced by a new one.

*Repair versus replace* There are many strategies that utilize average repair cost in making repair versus replace decisions. In all of these, the warranty period is divided into distinct intervals for repair and replacement. In [18], the warranty period is split into a replacement interval followed by a repair interval. In [8], the authors propose a strategy under which the warranty period is divided into three distinct intervals – [0, x), [x, y] and (y, WP]. The first failure in the middle interval is remedied by replacement and all other failures are minimally repaired.

*Service Time* As discussed in Sect. 4.5 service time consists of several components, including investigation time (time needed to locate the fault), time needed to carry out the actual repair, and testing time after repair. It may also include the waiting times that can result because of lack of spares or because of other failed items awaiting remedial actions. This time is dependent on the inventory of spares and on staffing of the repair facility. If the failed item must be serviced on site, travel time is an additional element.

Some of these times can be predicted precisely, while others (e.g., time to carry out the actual repair) may be highly variable, depending on the item and the type of failure. The easiest approach is to aggregate all the above times into a single repair time $T_r$ modeled as a random variable with CDF $F_r(t) = P\{T_r \le t\}$. We assume that $F_r(t)$ is differentiable and let $f_r(t) = dF_r(t)/dt$ denote the corresponding density function. Analogous to the concept of the failure rate function, we can define a repair rate function $\rho(t)$ given by

$$\rho(t) = \frac{f_r(t)}{1 - F_r(t)} \tag{6.1}$$

$\rho(t)\delta t$ is interpreted as the probability that the repair will be completed in $[t, t + \delta t)$, given that it has not been completed in $[0, t)$. In general, $\rho(t)$ would be a decreasing function of $t$ [14], indicating that the probability of a repair being completed in a short time increases with the duration of the service. In other words, $\rho(t)$ has a 'decreasing repair rate', a concept analogous to that of a decreasing failure rate.[2]

*Service Cost* In Sect. 4.6.4, we discussed the different cost elements that constitute the cost of servicing a warranty claim. This cost is uncertain since it depends as the service time needed and on material costs, which may vary. Let $C_r$ denote the random cost of servicing a warranty claim, with distribution function $G(c) = P\{C_r \le c\}$. For analysis of expected cost, it is often sufficient to deal with the average cost of repair, which we denote $c_r = E[C_r]$.

### 6.3.2 Effect of Usage

As discussed in Sect. 3.6.5, a product is designed for some nominal usage intensity (for example, the number of washes per week and/or size of loads washed in a washer; the number of Km travelled per year in an automobile). The actual usage

---

[2] Reference [12] suggests that the lognormal distribution is appropriate for modeling the repair times for many different products.

intensity may vary across customers. When the usage intensity is higher [lower] than the nominal usage intensity, the degradation (due to higher wear and/or increased stress on the components) is faster [slower]. As a result, the actual (field) reliability can be lower or higher than the design reliability.

The two well known models linking field reliability to design reliability are the following:

- Accelerated Failure Time (AFT) Models
- Proportional Hazard (PH) Models

These models are discussed in Sect. 3.6.5, and references to additional sources of information are given.

### 6.3.3 Warranty Execution

Not all customers may exercise the warranty when the item fails within the warranty period. There are several reasons for this, including the following:

- The customer may develop dissatisfaction for the product and switch to a competitor's product.
- The effort involved in exercising the warranty claim may not be worthwhile in relation to the benefits derived. This is especially true when failures occur near the end of the warranty period.
- The item may be sold before the warranty expires and the warranty terms invalidate the warranty on change of ownership.[3]

Warranty execution is uncertain. One can model this through a warranty execution function $\phi(\cdot)$ which is a non-negative and non-increasing function. If the age at failure is $x$ $(< W)$, then the probability that the warranty is executed is given by $\phi(x)$, with $\phi(0) = 1$ and $0 \leq \phi(W) < 1$. Figure 6.5 is illustrates some of the different forms $\phi(x)$ may assume.[4]

### 6.3.4 Sales

In analyzing sales, it is necessary to differentiate between first and repeat purchases. If the product life cycle ($L$) is less than the useful product life ($L_{\mathrm{l}}$), then there are no repeat purchases. If not, there may be one or more repeat purchases.

---

[3] This is true for most manufacturers. Some manufactures allow for transfer of warranty. In the case of automobiles, this has resulted in higher resale values.

[4] See [20] for more on this.

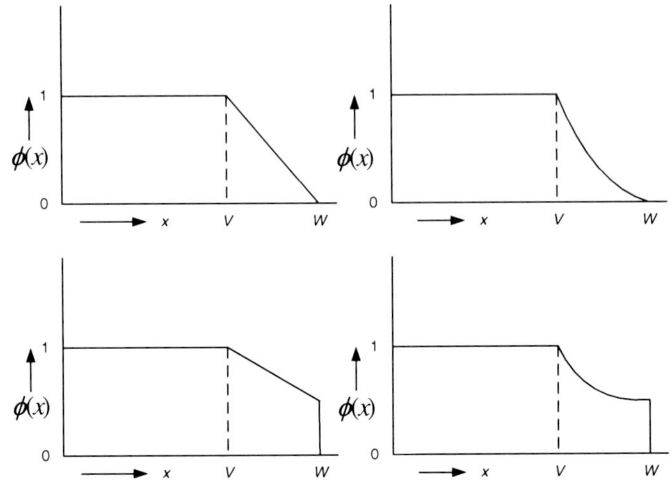


**Fig. 6.5** Warranty execution functions

Below, we consider first and repeat purchases separately, in the former case looking at both static and dynamic models.

### 6.3.4.1 First Purchase Sales

Total first purchase sales may be modeled either over the life of the product (static model) or over time. In the latter case, the key component of the model is a sales rate function (dynamic model).

*Static Model*. Let $S_m$ denote the maximum first purchase sales over the product life cycle. $S_m$ may be modeled by means of a formulation called a Cobb–Douglas model, given by[5]

$$S_m = k_1 C_b^{\alpha}(W + k_2)^{\beta}, \tag{6.2}$$

where $C_b$ is the sale price, $W$ is the parameter of the warranty policy, the parameters $(k_1, k_2, \beta)$ are all positive and $\alpha$ is negative. $\alpha$ is defined to be the cost elasticity and is given by $[\partial S_m/S_m]/[\partial C_b/C_b]$. $\beta$ is the warranty elasticity and is given by $[\partial S_m/S_m]/[\partial W/W]$). Note that total sales (1) decrease as $\alpha$ increases in magnitude and (2) increase as $W$ increases.[6]

*Dynamic Model*. The sales rate $s_1(t)$, $0 \le t \le L$, can be modeled in many different ways. One well known model is the Bass Diffusion model given by

---

[5] For more on the Cobb-Douglas model, see [6].

[6] Warranty elasticity for Chrysler is claimed to be 0.14, which is regarded as typical of those reported in the literature [19].

$$s_1(t) = \frac{dS_1(t)}{dt} = (a + bS_1)(S_m - S_1(t)), \quad S_1(0) = 0 \tag{6.3}$$

with $a > 0$ denoting the effect of advertising and $b > 0$ denoting the word-of-mouth effect.[7]

The total number of first purchase sales over the life cycle, $S_m$, is given by

$$S_m = \int_0^L s_1(t)dt \tag{6.4}$$

#### 6.3.4.2 Repeat Purchase Sales

This aspect of the analysis is relevant only when the useful-life of product is smaller than the length of the product life cycle. Let $v_j$ denote the probability that a customer who has bought the product for the $j$th time buys the product again when the purchased item reaches the end of its useful life. An upper limit on the number of repeat purchases is given by the largest integer less than $[L/L_1]$.

Let $s_j(t)$ denote the sales rate for the $j$th purchase. Then

$$s_{j+1}(t) = \frac{dS_{j+1}(t)}{dt} = v_j s_j(t - L_1), \quad jL_1 \le t < L, \ j \ge 1 \tag{6.5}$$

with $s_{j+1}(t) = 0$ for $t \le jL_1$. The rate of first purchase sales, $s_1(t)$, is given by (6.3). The total sales rate at $t$ is given by

$$s(t) = \sum_{j=1}^{[L/L_1]} s_j(t). \tag{6.6}$$

### 6.3.5 Warranty Claims

Let $N(t)$ denote the number of warranty claims over $[0, t)$.[8] The total number of claims over the warranty period is given by $N(WP)$. This quantity depends on the design reliability, usage intensity, warranty execution and the servicing strategy

---

[7] This is the simple diffusion model first proposed in [1]. Since then the basic model has been extended to take into account other factors, e.g., advertising effort, negative and positive word-of-mouth effects, etc.). Details of these can be found in [13].

[8] We assume that every failure results in an instantaneous claim and that all claims are valid. This results in a simplified model. One can relax these assumptions and the resulting model is more complex.

used. For the most general case, modeling of $N(t)$ involves very complex stochastic point processes (see Appendix B). By making some simplifying assumptions, however, the model becomes manageable. We discuss this further in the next section.

In principle, it should be possible to obtain the complete probabilistic characterization of $N(t)$. However, this is difficult even for the simplified cases. As a result, it is often necessary to characterize $N(t)$ using only its first and second moments.

### 6.3.6 Warranty Costs

The cost of warranty, i.e., the cost of servicing all warranty claims for an item over the total warranty period, is a sum of a random number of random individual costs, since both the individual costs and the number of claims over the warranty period are random.

Let $\tilde{C}_i$ denote the cost of servicing the $i$th warranty claim. The costs are random we assume that they are iid with distribution function $G(\cdot)$. The total cost of warranty over an interval $[0, t)$ is given by

$$C(t) = \sum_{i=1}^{N(t)} \tilde{C}_i \qquad (6.7)$$

A complete probabilistic characterization of $C(t)$ is extremely difficult, even for simplified cases.[9] Again, the characterization is often done by means of the first and second moments. To determine the first moment, let $c_r = E[C_r]$ denote the average cost of servicing a warranty claim. Then the expected warranty cost is given by

$$E[C(t)] = E[N(t)]E[\tilde{C}_i] = E[N(t)]c_r \qquad (6.8)$$

The variance is obtained similarly.

### 6.3.7 Some Comments on Analysis

Once the warranty process and the failure pattern of the item are modeled properly, the cost analysis is carried out (at least conceptually) using appropriate mathematical and statistical techniques. A number of approaches may be used, depending on the complexity of the model and analysis and the information that is available to the analyst.

In the simplest cases, the models lend themselves to straightforward mathematical analysis and numerical results are easily obtained. More complex models

9 For a complete probabilistic analysis of the warranty costs for some very simple cases, see [21].

(which occur even for the simplest warranties) may lead to considerable computational complexities, but are very doable on modern computer facilities. Complex mathematical problems may also be encountered. Intractable mathematical models are often dealt with through simulation studies. These require modeling of the warranty process, simulating part and system failures, repair activities, and elements that affect costs, and repeating the simulation under various choices of input parameters to obtain a detailed profile of predicted warranty costs under different scenarios. A well-designed warranty simulation will enable the user to assess the effect of changing warranty policies, lengths, and other terms.[10]

Another aspect of the analysis concerns the type and amount of information available. In the analysis as described, it is assumed that the models are correct. Implementation of the models requires, in addition, knowledge of the model parameters, including those of the failure distribution. In some cases, all of this information may be obtained from engineering analysis or from a long history of dealing with like products. In other cases, particularly for new products, many of the required model inputs may not be known. Lack of the required information introduces uncertainty into the results. In order to deal with this problem, data (experimental, historical, and others) are required. This introduces additional uncertainty, and statistical methods are employed in the decision making process. These are discussed in Chaps. 8–14, and include confidence intervals to express uncertainty in statistical estimates, test of assumptions and models, and so forth.

In data-based warranty cost analysis, it is important to investigate the uncertainty of the results. One way of doing this is to use confidence intervals in the cost calculations. This may be done by calculating costs using the nominal estimated parameter values and then repeating the calculations using both the upper and lower confidence limits as inputs. (With multiple parameters, the process may be somewhat more complicated.) One can then reasonably conclude that the true cost will lie somewhere in the range of values obtained. This is a type of *sensitivity study*, and much more detailed studies of this type, varying, for example, the probability model, cost elements, warranty periods and other important drivers of cost, may be done as well. This is particularly important for new products and any other situations involving substantial uncertainty, and proper management requires that resources for such studies be provided.

### 6.3.8 Notation

We use the following notation in the remainder of the chapter.

$F(t)$: Failure distribution of product
$T_0$: Age at failure for the unit sold
$T_i$ $(i \geq 1)$: Age of item $i$ supplied under warranty servicing

[10] For an example of such a program, see [7].

$C_m$: Unit manufacturing cost
$C_s$: Cost to the manufacturer of supplying a new item for use in warranty servicing ($C_s \geq C_m$).[11]
$C_b$: Cost to the customer of purchasing a unit (sale price per unit)
$C_r$: Cost of minimal repair (a random variable)
$c_r$: Average cost of repair ($\equiv E[C_r]$)
$C_m(W)$: Warranty cost to the manufacturer (a random variable)[12]
$\tilde{C}_b(I_{i+1}, T_i)$: Cost to buyer for the purchase of item $(i + 1)$ under renewing PRW Policy

## 6.4 Warranty Cost Analysis: Cost per Unit Sale

We look at cost models for the following common consumer warranties[13]:

- Non-renewing and renewing free replacement warranty (FRW) policies
- Non-renewing and renewing pro-rata warranty

We consider both repairable and non-repairable products. In the former case, we consider two simple servicing strategies—always replace or always minimal repair. In the latter case, this involves either a refund or replacement a by new item, depending on the warranty policy offered.[14]

Cost models for all but the simplest of these warranties are mathematically complex. We omit details and provide some numerical results.

In order to simplify the building and analysis of the model, we make the following assumptions:

1. All consumers are alike in their usage. One can relax this assumption by dividing consumers into two or more groups based on usage intensity.[15]
2. All items are statistically similar. One can relax this assumption by including two types of items (conforming and non-conforming) to take into account quality variations in manufacturing.[16]

---

[11] This includes materials, labor, testing, inventory, administration, and any other costs.

[12] The sale price must exceed the sum of the unit manufacturing cost and the expected warranty costs in order to ensure that, in the long run, the manufacturer will not incur a loss.

[13] The cost analysis of many other one-dimensional warranty policies can be found in [2]. See also [15] for subsequent warranty cost models.

[14] Cost analysis for other servicing strategies involves more complex model formulation and analysis. Details may be found in the following sources:

- Cost limit policy: [2, 16, 24]
- Repair versus replace: [9, 10, 17, 18, 22]
- Imperfect repair: [5, 4, 23]

[15] Cost analysis with heterogeneous usage intensity is discussed in [11].

[16] For more on this, see [2, 3, 15].

3. Whenever a failure occurs, it results in an immediate claim. Relaxing this assumption involves modeling the delay time between failure and claim.
4. All claims are valid. This can be relaxed by assuming that a fraction of the claims are invalid, either because the item was used in a mode not covered by the warranty or because it was a bogus claim.
5. The time to rectify a failed item (either through repair or replacement) is sufficiently small in relation to the mean time between failures that it can be approximated as being zero.
6. The manufacturer has the logistic support (spares and facilities) needed to carry out the rectification actions without any delays.

We omit the details of the derivation and present only the final results.

### 6.4.1 Non-renewing FRW Policy

Under this policy, any item that fails during the warranty period is repaired or replaced free of charge and the warranty period is $WP = W$. We first consider the case where the item is non-repairable, so that any failures during warranty require replacement of the failed item by a new item.

#### 6.4.1.1 Non-repairable Product

Failures are modeled by the renewal process associated with the distribution function $F(t)$, as discussed in Sect. 3.7.1.

*Warranty Claims*. The probability function and the first and second moments for the number of failures $N(W)$ during the warranty period are as follows:

$$P[N(W) = n] = F^{(n)}(W) - F^{(n+1)}(W), \quad n = 0, 1, \ldots \tag{6.9}$$

$$E[N(W)] = M(W) \tag{6.10}$$

$M(t)$ is called the *renewal function* and is discussed in Appendix B. The renewal function is obtained as the solution of the equation

$$M(W) = F(W) + \int_0^W M(W - t)f(t)dt \tag{6.11}$$

$$V[N(W)] = \sum_{n=1}^{\infty} (2n - 1)F^{(n)}(W) - [M(W)]^2 \tag{6.12}$$

*Warranty Costs*. The expected warranty cost to the manufacturer (assuming that all failures result in claims) is given by

**Table 6.1** $E[N(W)]$ for non-renewing FRW policy

| | $\beta = 1$ | $\beta = 2$ | $\beta = 4$ |
|---|---|---|---|
| $\mu = 2.0$ | 0.50 | 0.1843 | 0.1206 |
| $\mu = 2.5$ | 0.40 | 0.1308 | 0.0767 |

$$E[C_m(W)] = E[N(W)]E[C_i] = M(W)C_s \tag{6.13}$$

*Example 6.1* The manufacturer of a cell phone anticipates that a newly developed model will have an average lifetime (MTTF) of $\mu = 2.5$ years. Time to failure is expected to follow a Weibull distribution with shape parameter $\beta = 2$. The manufacturer is considering offering a one-year non-renewing free replacement warranty.

In an initial analysis of potential costs of the warranty, expected numbers of replacements required under warranty were calculated from (6.10) for the nominal values of the parameters as well as for $\mu = 2.0$ years and $\beta = 1$ (the exponential distribution) and $\beta = 4$.

The analysis requires evaluation of the renewal function at $W = 1$. This is discussed in [2], Chap. 3. For the exponential distribution, the renewal function is $M(t) = \lambda t$, where $\lambda = 1/\mu$. For the Weibull distribution, the parameter $\alpha$ must be chosen so that $\mu = 2$ and 2.5 for each $\beta$-value. These are obtained by solution of (A32). Finally, the renewal function for the Weibull distribution may be evaluated by table look-up or computer calculation. $E[N(W)]$ for various parameter combinations are shown in Table 6.1.

Note that these results indicate that the warranty would be very costly to the manufacturer. The extreme case is $\beta = 1$, $\mu = 2.0$, under which 50% of the items would fail under warranty. Even in the best case, $\beta = 4$ and $\mu = 2.5$, nearly 8% of the items would require replacement before the end of the warranty term. For decision making, more precise information regarding the parameter values is needed. The manufacturer may also consider decreasing the warranty period.

#### 6.4.1.2 Repairable Product

We confine our attention to the case where all failures over the warranty period are minimally repaired. This means that the item is restored basically to its condition immediately prior to the failure. This results because minimal repair usually involves simply replacing only failed components, so that after repair, the item is basically as it was at the time of failure since all non-replaced components have the same usage and age as before. As a result, warranty claims occur according to a non-homogeneous Poisson process with intensity function $\lambda(t) = h(t)$ (the hazard function associated with the failure distribution $F(t)$), as discussed in Sect. 3.7.2.

*Warranty Claims*. From (3.20) the probability function for the number of failures $N(W)$ is as follows:

**Table 6.2** Expected warranty costs for best and worst cases

| Time $t$ (in days) | Expected costs for | |
|---|---|---|
| | Worst case | Best case |
| 365 | $13.32c_r$ | $2.47c_r$ |
| 730 | $53.29c_r$ | $6.97c_r$ |
| 1,095 | $119.90c_r$ | $12.81c_r$ |

$$P[N(W) = n] = \frac{\{\Lambda(W)\}^n e^{-\Lambda(W)}}{n!}, \quad n = 0, 1, \ldots, \tag{6.14}$$

where

$$\Lambda(t) = \int_0^t h(t)dt \tag{6.15}$$

The first and second moments are given by

$$E[N(W)] = \Lambda(W) \tag{6.16}$$

$$V[N(W)] = \Lambda(W) \tag{6.17}$$

*Warranty Costs.* The expected warranty cost to the manufacturer is given by

$$E[C_m(W)] = \Lambda(W)c_r \tag{6.18}$$

*Example 6.2* The manufacturer of a copy machine provides a non-renewing FRW with a warranty period of six months. A large retailer of the machine plans to offer an extended warranty and is considering periods of one to three years. In studying potential costs, and analyst has concluded that the NHPP model with power law intensity is appropriate. The parameters are not known, but the analyst feels that the scale parameter is somewhere between 100 and 200 and the shape parameter between 1.5 and 2, indicating an aging effect. Within these ranges, the extreme cases are the worst (best) case, with smallest (largest) scale and largest (smallest) shape.

The MCF with power law intensity function (see Appendix B) for the worst case is

$$\Lambda_1(t; \alpha_1, \beta_1) = (t/\alpha_1)^{\beta_1} = (t/100)^{2.0}$$

and for the best case is

$$\Lambda_2(t; \alpha_2, \beta_2) = (t/\alpha_2)^{\beta_2} = (t/200)^{1.5}$$

The results for 1, 2 and 3 years (expressed in days) are given in Table 6.2.

Note that costs for the best and worst cases differ by from five to tenfold, indicating that more analysis may be required before a decision of the length of warranty and the price to charge is reached.

### 6.4.2 Renewing FRW Policy

We confine our attention to the case where failed items are replaced by new. There is no warranty claim if the item sold survives for a period $W$. If not, then the warranty ceases only when an item supplied as a replacement survives for a period $W$. As a result, the warranty period is $WP \geq W$.

*Warranty Claims*. Let $T_0$ denote the age at failure time for the unit sold and $T_i$ the age at failure the $i$th unit provided as a replacement under warranty. Then the number of warranty claims is $N(W) = n$ if and only if $T_i < W$ for $0 \leq i \leq (n-1)$ and $T_n > W$. Given that $N(W) = n$, the warranty period is a random variable given by

$$WP = \sum_{i=0}^{n-1} T_i + W \tag{6.19}$$

The probability function for $N(W)$ is given by

$$P\{N(W) = n\} = \bar{F}(W)[F(W)]^n, \quad n = 0, 1, 2, \ldots \tag{6.20}$$

and the expected number of failures is given by

$$E[N(W)] = \frac{F(W)}{\bar{F}(W)} \tag{6.21}$$

The expected warranty period is given by

$$E[WP] = W + \frac{F(W)}{\bar{F}(W)} \mu_W, \tag{6.22}$$

where

$$\mu_W = \int_0^W tf(t)dt \tag{6.23}$$

is the partial expectation of $T$.

*Warranty Costs*. The expected warranty cost is given by

$$E[C_m(W)] = E[N(W)]C_s = \left[\frac{F(W)}{\bar{F}(W)}\right] C_s \tag{6.24}$$

*Example 6.3* Suppose that the manufacturer in Example 6.1 offered a renewing FRW. To analyze this policy, we require the partial expectation $\mu_W$, given in (6.23). For the exponential distribution, the result is ([2], p. 174)

$$\mu_W = \mu[1 - (1 + W/\alpha)e^{-W/\mu}]$$

For the general Weibull ($\beta \neq 1$), calculation of the partial expectation requires evaluation of the incomplete gamma function (see [2], p. 176 for details). For the

**Table 6.3** Partial expectations, expected number of failure and expected length of warranty period for Example 6.3

| $\beta$ | $\mu$ | $\mu_W$ | $E[N(W)]$ | $E(WP)$ |
|---|---|---|---|---|
| 1 | 2.0 | 0.1804 | 0.6487 | 1.117 |
| | 2.5 | 0.1539 | 0.4918 | 1.076 |
| 2 | 2.0 | 0.1165 | 0.2169 | 1.025 |
| | 2.5 | 0.0777 | 0.1339 | 1.010 |
| 4 | 2.0 | 0.0330 | 0.0431 | 1.001 |
| | 2.5 | 0.0137 | 0.0174 | 1.000 |

parameter values used, the partial expectations with $W = 1$ are as indicated in the third column of Table 6.3.

From (6.21), the expected number of failures during the warranty period (and hence maximum number of claims) is easily calculated. The results for the combinations of parameter values used are as shown in the fourth column of Table 6.3. Thus the renewing FRW requires the manufacturer to supply more free replacements) than does the non-renewing FRW, as expected. The difference is about 25–30% if $\beta = 1$; 11–17% if $\beta = 2$; and 2–4% if $\beta = 4$.

The expected length of the warranty period is calculated from (6.22). The result is obtained by multiplying corresponding terms in the third and the fourth columns and adding the value of $W$. The results are shown in the last column of Table 6.3. The period of warranty coverage is extended only slightly.

### 6.4.3 Non-renewing PRW Policy

Under this policy, a rebate is given to the consumer rather than a repair or replacement in case of item failure. The customer is not required to purchase a replacement unit. Let $\alpha(T)$ be the rebate given if the age at failure is $T$. (Note that $\alpha(T) = 0$ for $T \geq W$). We confine our attention to the case in which the rebate is a linear function of age at failure, namely

$$\alpha(T) = \begin{cases} (1 - T/W)C_b, & 0 \leq T < W \\ 0, & T \geq W \end{cases} \tag{6.25}$$

*Warranty Claims*. For each item sold, there can either be a single warranty claim or no claim. The probability of a warranty claim is $F(W)$. The probability of no claim is $1 - F(W)$.

*Warranty Costs*. In this case, it is necessary to consider the cost from the manufacturer's and buyer's perspectives separately.

The warranty cost to the manufacturer, $C_m(W)$, is the rebate given. The expected value of this quantity is

$$E[C_m(W)] = \left[F(W) - \frac{\mu_W}{W}\right] C_b \tag{6.26}$$

**Table 6.4** Expected cost of warranty

| | $\beta = 1$ | $\beta = 2$ | $\beta = 4$ |
|---|---|---|---|
| $\mu = 2.0$ | $0.2131C_b$ | $0.0618C_b$ | $0.0083C_b$ |
| $\mu = 2.5$ | $0.1758C_b$ | $0.0404C_b$ | $0.0034C_b$ |

The cost per unit to the buyer is

$$\tilde{C}_b(W) = C_b - q(T) \tag{6.27}$$

As a result, the expected per unit cost to the buyer is

$$E[\tilde{C}_b(W)] = C_b\left[1 + \frac{\mu_W}{W} - F(W)\right] \tag{6.28}$$

Note that this includes the initial item purchased at full price.

*Example 6.4* We consider again the cell phone discussed in Examples 6.1 and 6.3. We assumed the Weibull distribution for time to failure, with $\beta = 1$, 2, and 4, and $\alpha$ values chosen so that $\mu = 2.0$ and 2.5 for each $\beta$. The length of the warranty period was $W = 1$. Values of $\mu_W$ for each combination of parameter values are given in Example 6.3.

It is instructive to compare the manufacturer's expected costs in the previous two cases with that of the rebate PRW Policy. For the rebate PRW policy, this expected cost is given in (6.26), in which $C_b$ is the selling price (cost to the buyer) of the item. For the parameter combinations used, the manufacturer's expected cost of warranty per unit sold for the rebate PRW are given in Table 6.4

The comparable manufacturer's cost for the nonrenewing FRW, given in (6.13), is $M(W)C_s$, where $C_s$ is the cost to the manufacturer of supplying a new item. The values of the renewal function are given in Example 6.1. Comparable values, which are not substantially higher, are given for the renewing FRW in Example 6.3.

The costs for the nonrenewing FRW and the rebate PRW (which is inherently nonrenewing) are identical if $\mathrm{a}C_b = \mathrm{b}C_s$, where a is obtained from Table 6.4 and b is the comparable factor in the corresponding table for the nonrenewing FRW. Otherwise, one or the other of these warranties would be less costly to the manufacturer. Which is the less costly depends on the relative magnitudes of the total cost of supplying a replacement item and the selling price of the item.

### 6.4.4 Renewing PRW Policy

The warranty claims and the warranty period *WP* are the same as for the renewing FRW policy discussed in Sect. 6.4.2.

*Warranty Costs*. For the analysis, we again need to consider costs from both the manufacturer's and buyer's perspectives. We first consider the cost to the buyer.

The cost to the buyer for item $I_{i+1}$, the replacement provided as a result of the $i$th failure under warranty, is given by

$$\tilde{C}_b(I_{i+1}, W) = C_b - \alpha(T_i) \tag{6.29}$$

This is a random variable with expectation[17]

$$E[\tilde{C}_b(I_{i+1}, W)] = C_b\left[\frac{\mu_W}{W} + (1 - F(W))\right] \tag{6.30}$$

Since the number of warranty claims over the warranty period (*WP*) is uncertain and the cost of each replacement is also uncertain, the total warranty cost to the buyer, $\tilde{C}_b(W)$, is also uncertain. The expected warranty cost to the buyer is given by

$$E[\tilde{C}_b(W)] = C_b\left[1 + \frac{\mu_W F(W)}{W\bar{F}(W)}\right] \tag{6.31}$$

Again, this includes the initial item purchased at full price.

The expected warranty cost to the manufacturer is the same as under the renewing FRW policy and is given by (6.24).

## 6.5 Life Cycle Cost Analysis per Unit Sale

Let $C_b(L, W)$ and $C_m(L, W)$ denote the life cycle cost to the buyer and the manufacturer, respectively. These costs are functions of $W$, the length of the warranty period, and of $L$, which we take to be greater than $W$. They are also functions of the other terms of the warranty (FRW, PRW, etc.), the distribution of time to failure of the item, and the way in which the life cycle is defined.

### *6.5.1 Non-renewing FRW Policy*

We consider a non-repairable product, so that failed items must be replaced by new ones. Under the non-renewing FRW, the first failure after expiration of the warranty results in a new purchase by the buyer and this comes with a new identical warranty. A typical plot is as indicated in Fig. 6.3. The time intervals, $Y$, between successive repeat purchases, with the first purchase occurring at $t = 0$, are the key units in the analysis. The $Y$'s are of the form $Y = W + Z$, where $Z$ is the remaining life of the item in use at the expiration of the warranty. Thus, after the

[17] The details of the derivations for the results presented in this section can be found in Sect. 5.3.1 of [2].

initial sale, the manufacturer incurs the cost of all replacements until time $W + Z$, receives income from a new sale at this time, and the cycle begins anew, ending when the total length of time reaches $L$. The buyer, on the other hand, during the initial warranty period has only the initial cost of purchasing the item, $C_b$, with the next cost, also in the amount $C_b$, occurring at time $W + Z$, and so on.

As a result, the cost analysis from the buyer's point of view is based on renewals associated with the variable $W + Z$, and cost models involve the distribution of this variable. For the manufacturer, the cost involves the distribution of time to failure of the item, with incomes equivalent to the cost to the buyer. Expressions for $C_b(L, W)$ and $C_m(L, W)$ for the FRW have been derived.[18]

### 6.5.2 Non-renewing PRW Policy

For this policy, warranty costs involve refunds to the consumer for claims occurring within the warranty period. Expressions for the expected life costs for the manufacturer and consumer have been derived.[19] Many of the results are quite complex, and we forego a detailed discussion.

## 6.6 Analysis of Life Cycle Cost over Product Life Cycle

Over the product life cycle, product sales (first and repeat purchases) occur over time in a dynamic manner. The sales rate $s(t)$ is given in (6.6), and the manufacturer must service the warranty claims resulting from each such sale.

Warranty claims occur over a period that is greater than $L$ and depend on the type of warranty. In the case of products sold with one-dimensional non-renewing warranty, this period is simply $L + W$. If the warranty is renewing, the period is longer than this.

The expected number of warranty claims per unit time changes dynamically and is a function of sales over time, product reliability, and other factors such as usage intensity, the usage environment, etc. This number is needed for planning of spares, repair facilities, and other service elements. The expected warranty cost per unit time also changes dynamically over time and is needed for determining the warranty reserves required to service warranty.

In this section we examine life cycle costs for non-renewing versions of both the FRW and PRW policies. We omit the derivations and present only the final results.[20]

---

[18] See [2], Sect. 4.5.1 for buyer's life cycle cost and Sect. 4.5.2 for seller's profit. Seller's cost is the difference between the two.

[19] See [2], Sect. 5.4.

[20] Details of the derivations can be found in Chap. 9 of [2].

## 6.6.1 Non-renewing FRW Policy

### 6.6.1.1 Non-repairable Product

Here the focus is on the demand for spares needed to replace items that fail that fail under warranty over the period $[0, L + W)$. The demand for spares in the interval $[t, t + \delta t)$ is due to failure of items sold in the period $[\psi, t)$, where $\psi$ is given by

$$\psi = \max\{0, t - W\}. \tag{6.32}$$

The expected demand rate for spares at time $t$, $\rho(t)$, is given by

$$\rho(t) = \int_{\psi}^{t} s(x)m(t - x)dx \tag{6.33}$$

for $0 \leq t \leq L + W$. $m(t)$ is the renewal density function associated with the failure distribution function $F(t)$ and is given by

$$m(t) = f(t) + \int_{0}^{t} m(t - x)f(x)dx \tag{6.34}$$

The expected total number of spares required to service the warranty over the product life cycle, ETS, is given by

$$ETS = \int_{0}^{L+W} \rho(t)dt \tag{6.35}$$

### 6.6.1.2 Repairable Product

Here the focus is on the demand for repairs to rectify failures under warranty over the period $[0, L + W)$. We confine our attention to the case where all failures under warranty are repaired through minimal repair. For each item sold, failures over the warranty period occur according to a non-stationary Poisson process with an intensity function $\lambda(t) = h(t)$, where $h(t)$ is the failure rate associated with the failure distribution function $F(t)$.

The expected repair rate at time $t$, $\rho_r(t)$, is given by

$$\rho_r(t) = \int_{\psi}^{t} s(x)r(t - x)dx \tag{6.36}$$

for $0 \leq t \leq L + W$, with $\psi$ given by (6.32). The total expected demand for repair over the warranty period, EDR, is given by

$$EDR = \int_{0}^{L+W} \rho_r(t)dt \tag{6.37}$$

### 6.6.2 Non-renewing PRW Policy

The rebate over the interval $[t, t + \delta t)$ is a result of failure of items that are sold in the interval $[t - \psi, t)$. $v(t)$, the expected refund rate (i.e., the amount refunded per unit time) at time $t$, is given by

$$v(t) = C_b \int_{\psi}^{t} s(x)\left(\frac{t-x}{W}\right)f(t-x)dx \tag{6.38}$$

for $0 \leq t \leq L + W$, with $\psi$ given by (6.32). The expected total reserve needed to service the warranty over the product life cycle, ETR, is given by

$$ETR = \int_{0}^{W+L} v(t)dt \tag{6.39}$$

## References


1. Bass FW (1969) A new product growth model for consumer durables. Manag Sci 15:215–227
2. Blischke WR, Murthy DNP (1994) Warranty cost analysis. Marcel Dekker, New York
3. Blischke WR, Murthy DNP (eds) (1996) Product Warranty Handbook. Marcel Dekker, New York
4. Chukova S, Arnold R, Wang D (2004) Warranty analysis: An approach to modeling imperfect repairs. Int J Prod Econ 89:57–68
5. Doyen L, Gaudoin O (2004) Classes of imperfect repair models based on reduction of failure intensity or virtual age. Reliab Eng Sys Saf 84:45–56
6. Henderson JM, Quandt RE (1958) Microeconomics theory. McGraw-Hill Inc., New York
7. Hill VL, Beall CW, Blischke WR (1991) A simulation model for warranty analysis. Int J Prod Econ 22:131–140
8. Jack N, Murthy DNP (2001) A servicing strategy for items sold under warranty. J Oper Res Soc 52:1284–1288
9. Jack N, van der Duyn Schouten F (2000) Optimal repair-replace strategies for a warranted product. Int J Prod Econ 67:95–100
10. Jiang X, Jardine AKS, Lugitigheid D (2006) On a conjecture of optimal repair-replacement strategies for warranted products. Math Comput Model 44:963–972

11. Kim CS, Djamaludin I, Murthy DNP (2001) Warranty cost analysis with heterogeneous usage intensity. Int Trans Oper Res 8:337–347
12. Kline MB (1984) Suitability of lognormal distribution for corrective maintenance repair times. Reliab Eng 9:65–80
13. Mahajan V, Wind Y (1986) Innovation diffusion models of new product acceptance. Ballinger Publishing Company Cambridge, Mass
14. Mahon BH, Bailey RJM (1975) A proposed improvement replacement policy for army vehicles. Oper Res Q 26:477–494
15. Murthy DNP, Djamaludin I (2001) Warranty and quality. In: Rahim MA, Ben-Daya M (eds) Integrated optimal modelling in PIQM: production planning, inventory, quality and maintenance. Kluwer Academic Publishers, New York
16. Murthy DNP, Nguyen DG (1988) An optimal repair cost limit policy for servicing warranty. Math Model 11:595–599
17. Nguyen DG, Murthy DNP (1986) An optimal policy for servicing warranty. J Oper Res 37:1081–1088
18. Nguyen DG, Murthy DNP (1989) Optimal replace-repair strategy for servicing items sold with warranty. Eur J Oper Res 39:206–212
19. Padmanabhan V (1996) Marketing and warranty. In: Blischke WR, Murthy DNP (eds) Product warranty handbook. Marcel Dekker, New York
20. Patankar JG, Mitra A (1996) Warranty and consumer behavior: warranty execution. In: Blischke WR, Murthy DNP (eds) Product warranty handbook. Marcel Dekker, New York
21. Sahin I, Polatogu H (1998) Quality, warranty and preventive maintenance. Kluwer Academic Publishers, Boston
22. Sheu SH, Yu SL (2005) Warranty strategy accounts for bathtub failure rate and random minimal repair cost. Comput Math Appl 49:1233–1242
23. Yun WY, Murthy DNP, Jack N (2008) Warranty servicing with imperfect repair. Int J Prod Econ 111:156–169
24. Zuo MJ, Liu B, Murthy DN (2000) Replacement–repair policy for multi-state deteriorating products under warranty. Eur J Oper Res 123:519–530

# Chapter 7
# Cost Models for Two-Dimensional Warranties

## 7.1 Introduction

Multi-dimensional warranties usually involve a time dimension similar to that in the case of one-dimensional warranties and one or more usage dimensions. There are several notions of usage. These are discussed in the next section. We confine our discussion to two-dimensional warranties, where the warranty is characterized by a rectangular region $\Omega \equiv [0, W) \times [0, U)$ and the warranty expires when the item reaches an age $W$ or usage reaches a level, $U$ whichever comes first.[1]

The manufacturer must service all claims under warranty. The cost analysis in this case is more involved than that of one-dimensional warranties, since failures are random points scattered over the two-dimensional warranty region as opposed to being random points along the time axis in the case of one-dimensional warranties. Three approaches have been proposed for modeling failures and warranty claims, and we discuss each of them. Costs depend not only on failures but also on several other factors. We discuss these and derive cost models for some of the simpler 2-D warranties.

The outline of the chapter is as follows. We begin with a brief discussion of the different notions of usage in Sects. 7.2. Section 7.3 deals with the system characterization for warranty cost analysis. This is similar to that given in Chap. 6 for the 1-D case, with some differences that we highlight. Modeling for warranty cost analysis is briefly discussed in Sect. 7.4. Modeling of failures is the critical element and three different approaches (Approach 1–3) to this are proposed and discussed in Sect. 7.5. The cost analysis of the FRW and PRW policies based on Approaches 1 and 3 are discussed in Sects. 7.6 and 7.7, respectively.

[1] Several other types of warranty regions have been proposed See, for example, [5, 6, 38].

W. R. Blischke et al., *Warranty Data Collection and Analysis*,
Springer Series in Reliability Engineering, DOI: 10.1007/978-0-85729-647-4_7,

## 7.2 Product Usage

In this section we discuss three distinct notions of usage. In the analysis, $X(t)$ denotes the usage up to time $t$.

### 7.2.1 Notions of Usage

Usage may be defined in a number of ways, as discussed in Sect. 3.6. We consider the following three types of usage which we label Types 1, 2, and 3.

*Type 1 Usage [Usage as Output Produced]* Many products are bought to produce some output and usage is the output. We use the term "output" in a very general sense as illustrated by the following examples:

- Automobile: $X(t)$ represents the distance traveled over the interval $[0, t)$.
- Photocopier: $X(t)$ represents the number of copies made over the interval $[0, t)$.
- Machine tool: $X(t)$ represents the number of components machined over the interval $[0, t)$.

In this case, one can define 'usage rate' as the output per unit time. In the case of an automobile, it could represent the distance traveled per year or per week, and so on.

*Type 2 Usage [Intermittent Usage-I]* In this case, $X(t)$ represents the total duration of usage over the interval $[0, t)$. Here $0 \leq X(t) \leq t$. Some examples are the following:

- Aircraft engine: $X(t)$ represents the number of hours flown by the engine over the interval $[0, t)$.
- Air-conditioner: $X(t)$ represents the number of hours in use over the interval $[0, t)$.

*Type 3 Usage [Intermittent usage-II]* Here $X(t)$ represents the number of times the product is used over the interval $[0, t)$. Some examples are the following:

- Landing gear of an aircraft: $X(t)$ represents the number of landings over the interval $[0, t)$.
- Large battery: $X(t)$ represents the number of times the battery has been recharged over the interval $[0, t)$.
- Washing machine: $X(t)$ represents the number of loads of washing done over the interval $[0, t)$.

### 7.2.2 Impact on Product Reliability

Usage impacts product reliability. The effect of intermittent usage I on product reliability is discussed in Sect. 3.6.4. The impact of other usage patterns on reliability will be discussed later in the chapter.

## 7.3 System Characterization for Warranty Cost Analysis

As in Sect. 6.2, the three different warranty costs of interest are the following:

- Warranty cost per unit sale
- Life cycle cost per unit sale (LCC-I)
- Cost over product life cycle (LCC-II)

In the case of 2-D warranty policies, warranty cost per unit has been studied extensively and the life cycle costs have received little or no attention.[2] As such, we confine our attention to warranty cost per unit sale.

### 7.3.1 Warranty Period (WP)

The two-dimensional warranty is characterized by the two parameters $W$ and $U$. The period $WP$ for which an item is covered under warranty can be less than or greater than $W$, and depends on a number of factors, including whether or not the warranty is renewing, the notion of usage, the usage rate, etc. We illustrate this for the three types of usage discussed in the previous section.

*Type 1 Usage* Fig. 7.1 shows the warranty period in the case of a non-renewing warranty for different usage rates (high and low). Note that the length of the warranty period is $WP < W$ for the high usage case and $WP = W$ for the low usage case.

Figure 7.2 shows the warranty period $WP$ for a renewing warranty. The duration depends on the number of failures occurring under warranty.

*Type 2 Usage* Here the usage is the duration that an item is used. As a result, the length of the warranty period is $WP < W$ for high intermittent usage and $WP = W$ for low usage, as indicated in Fig. 7.3. Note that the usage rate is zero when the item is not used, corresponding to the horizontal lines of the figure, and the slope of the trajectory is 45° whenever the item is in use.

*Type 3 Usage* Here the usage jumps by one every time the item is used, so that the trajectories are similar to those of Fig. 7.3, except that they are staircase functions. Again, the length of the warranty period is $WP < W$ for high intermittent usage and $WP = W$ for low usage.

---

2 One reason for this is that mathematical models for life cycle cost are very complex and not amenable to analysis.

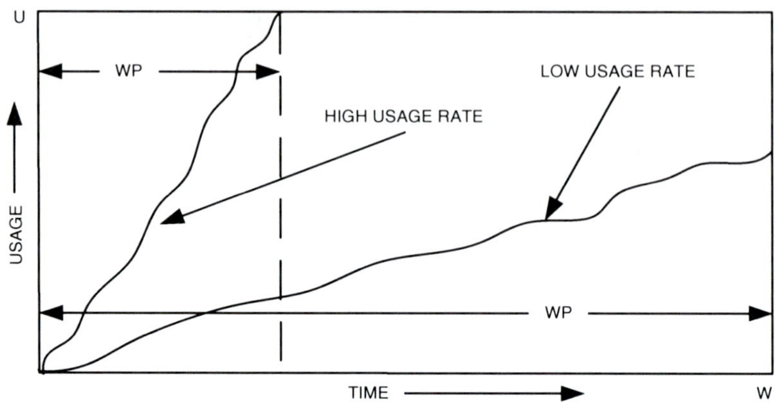


**Fig. 7.1** Warranty period (*WP*) for non-renewing warranty [two different usage rates]

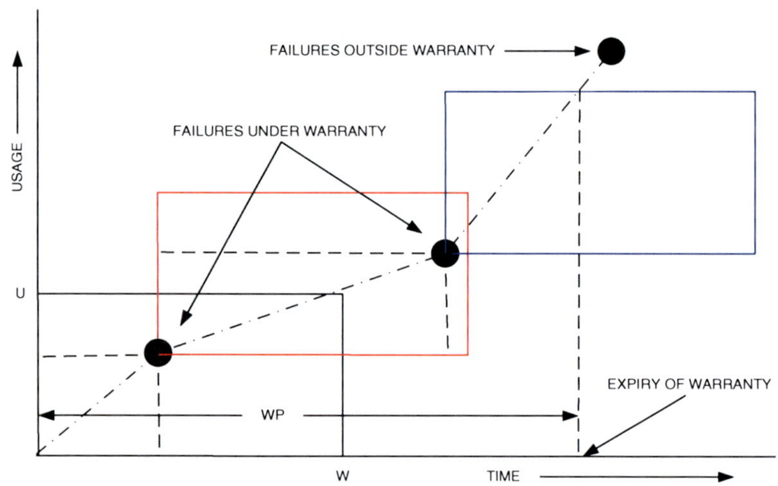


**Fig. 7.2** Warranty period (*WP*) for renewing warranty

## *7.3.2 Characterization for Cost per Item*

The system characterization for modeling warranty cost per unit sold is the same as that in Fig. 6.2. The modeling process requires characterization of the elements of this figure. For most of these, the result is the same as in Sect. 6.3. The two exceptions are (1) warranty servicing, which is discussed in the next section, and (2) product failures (and warranty claims), which is discussed in the subsequent section.

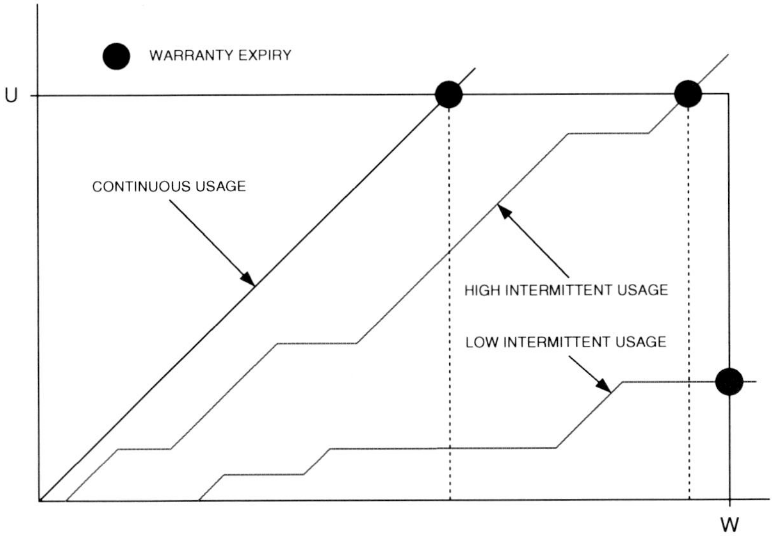


**Fig. 7.3** Expiry of warranty (Case 2)

## 7.4 Modeling for Warranty Cost Analysis

### *7.4.1 Warranty Servicing Strategy*

The two simplest strategies are (1) always repair or (2) always replace by new [8]. More complex strategies involve a choice between repair and replace based on average repair cost. Many such strategies have been proposed. These typically involve dividing the warranty region into several distinct regions. References [17] and [18] study two such strategies that involve rectangular regions. Reference [21] proposed a strategy which involves the complex shape indicated in Fig. 7.4. Here $\Gamma$ denotes the region enclosed by the curve. The servicing policy is as follows: replace by new at the first failure occurring in the region $\Gamma$ and minimally repair all other failures. The shape of the curve is selected to minimize the expected warranty servicing cost per unit sale.

We shall confine our attention to the simplified cases involving either replacement or minimal repair in the models discussed in later sections.

### *7.4.2 Simplifying Assumptions*

As discussed in Sect. 6.4, in order to simplify the building and analysis of the model, we make the following assumptions:

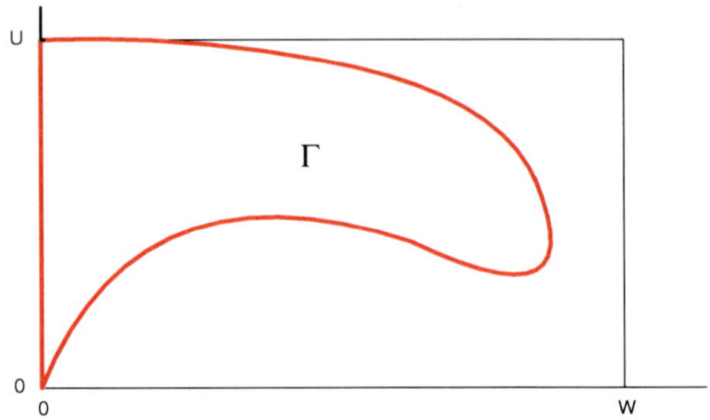


**Fig. 7.4** Warranty servicing strategy

1. All items are statistically similar. One can relax this assumption to include two types of items (conforming and non-conforming) to take into account quality variations in manufacturing.
2. Whenever a failure occurs, it results in an immediate claim. Relaxing this assumption involves modeling the delay time between failure and claim.
3. All claims are valid. This can be relaxed by assuming that a fraction of the claims are invalid, either because the item was used in a mode not covered by the warranty or because it was a bogus claim.
4. The time to rectify a failed item (either through repair or replacement) is sufficiently small in relation to the mean time between failures so that it can be considered to be approximately zero.
5. The manufacturer has the logistic support (spares and facilities) needed to carry out the rectification actions without any delays.

## 7.5 Modeling Failures and Claims [Type 1 Usage]

In the case of two-dimensional warranties, failures (and warranty claims) are random points in the two-dimensional plane, as indicated in Fig. 7.5. It is necessary to differentiate between first failures and subsequent failures; the latter depend on the warranty servicing strategy used.

As mentioned earlier, three different approaches have been used to model failures and claims under warranty. We briefly discuss each of these.

### *7.5.1 Approach 1 [1-D Models with Conditioning on Usage Rate]*

Here, the two-dimensional problem is effectively reduced to a one-dimensional problem by treating usage as a random function of age. In addition, it is assumed that the usage rate for a customer is constant over the warranty period but varies

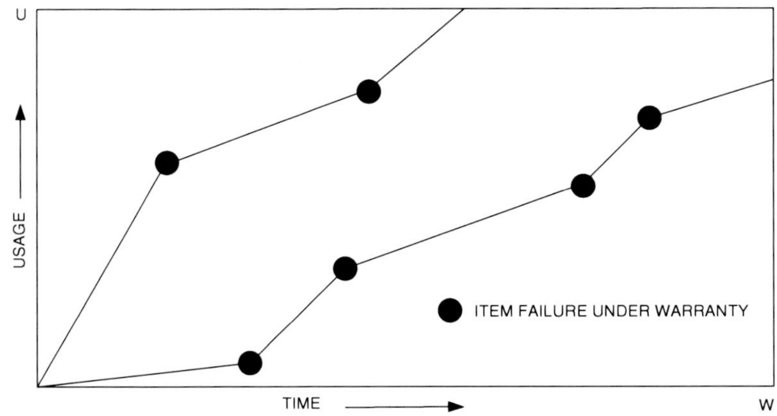


**Fig. 7.5** Typical time histories of failures over the over the warranty period

across the customer population as shown in Fig. 7.6. As a result, the usage rate $Z$ is a random variable. $Z$ can be modeled either as a discrete variable (for example, low, medium and high users) or as a continuous variable with a density function $g(z)$. Modeling of failures under warranty is done using 1-D models obtained by conditioning on the usage rate. The bulk of the literature deals with a linear relationship between usage and age.[3]

The warranty period (for a non-renewing warranty) depends on the usage rate and, conditional on $Z = z$, is given by $WP = \min\{W, (U/z)\}$. Models of failures under warranty are obtained by carrying out the un-conditioning, taking into account the dependence of the warranty period on the usage rate. This will be discussed later in the chapter.

#### 7.5.1.1 Approach 1-1

The time to first failure is modeled by a conditional hazard (failure rate) function $h(t|z) \geq 0$, that is a non-decreasing function in both $t$ and $z$. A simple such formulation is a polynomial relationship of the form[4]

$$h(t|z) = \theta_0 + \theta_1 t + \theta_2 z + \theta_3 zt, \tag{7.1}$$

where the parameters $(\theta_0, \theta_1, \theta_2, \theta_3)$ are all non-negative.

*First Failure*. The conditional distribution function of the time to first failure is given by

---

[3] See, for example, [5, 28, 11]. Reference [16] deals with motorcycle data. See [28, 41] for automobile warranty data analyses based on this approach.

[4] Special cases of these have been used in modeling of two-dimensional warranties [19, 30, 31, 34].

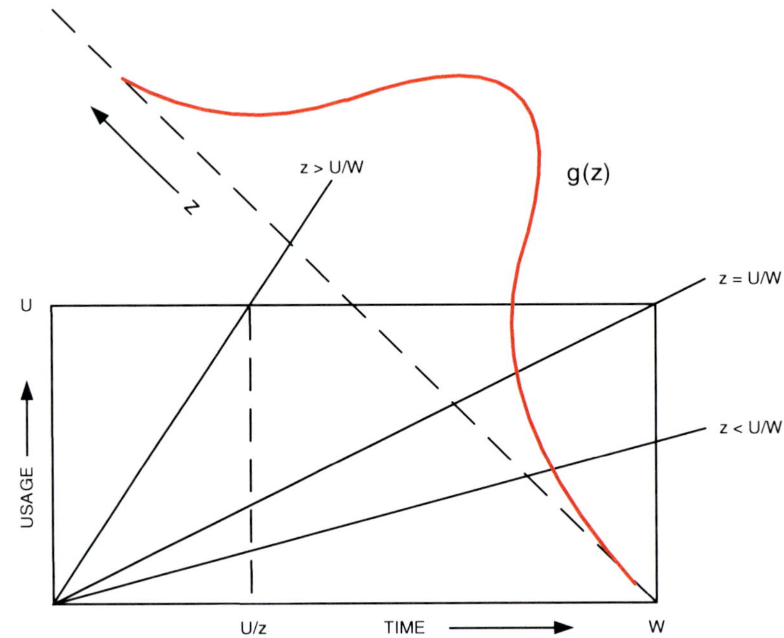


**Fig. 7.6** Typical time histories with different constant usage rates

$$F_z(t) \equiv F(t|z) = 1 - e^{-\int_0^t h(u|z)du} \tag{7.2}$$

*Subsequent Failures*. If failed units are replaced by new ones, then the counting process is the renewal process associated with $F(t|z)$. If failed units are repaired, then the counting process is characterized as a conditional intensity function $\lambda(t|z)$ that is a non-decreasing function in both $t$ and $z$. If failures are repaired minimally and the repair times are negligible, then $\lambda(t|z) = h(t|z)$, given by (7.1).

#### 7.5.1.2 Approach 1-2

Normally products are designed for some nominal usage rate $z_0$. As the usage rate increases [decreases], the rate of degradation increases [decreases] and this, in turn, accelerates [decelerates] the time to failure. As a result, the reliability decreases [increases] as the usage rate increases [decreases].[5]

[5] This approach uses the accelerated failure concept and has been used in the modeling of 2-D warranties by [2, 28].

*First Failure*. Let $\bar{F}_0(t)$ $(\equiv 1 - F_0(t))$ denote the base survivor function when the usage rate is the nominal value $z_0$. Conditional on the usage rate, the time to first failure is modeled by a survivor function

$$\bar{F}(t|z) = \bar{F}_0(t\tilde{z}^{\gamma}) \tag{7.3}$$

where $\tilde{z} = z/z_0$ and $\gamma > 1$.

*Subsequent Failures*. If failed units are replaced by new ones, then the counting process is the renewal process associated with $\bar{F}(t|z)$ as given by (7.3). If failures are repaired minimally and the time to repair is negligible, then failures over time follow a point process with intensity function that is given by the hazard function associated with the distribution function given by (7.3). As a result,

$$\lambda(t|z) = \tilde{z}^{\gamma}\lambda_0(t\tilde{z}^{\gamma-1}). \tag{7.4}$$

### 7.5.2 Approach 2 [1-D Composite Scale Models]

In this approach, the two scales—usage ($x$) and time ($t$)—are combined to provide a single composite scale ($v$).[6] In the simplest case, this scale is a linear combination of the form

$$v = at + bx \tag{7.5}$$

where $a$ and $b$ are parameters to be selected. References [24–26] discuss the selection of values for these parameters.[7]

A multiplicative combination [1] results in

$$v = x^{\alpha}t^{1-\alpha} \tag{7.6}$$

with $0 \leq \alpha \leq 1$.

### 7.5.3 Approach 3 [2-D Models]

Here one uses two-dimensional models involving bivariate distribution functions.[8] Let $(T, X)$ denote the age and usage at first failure.

---

[6] Reference [9] discusses alternative scales and their modeling. Three groups of models are proposed—(1) collapsible models, (2) generalized time transform (also referred to as accelerated failure time) models and, (3) hazard-based specifications. See also [24–27, 10].

[7] Reference [11] uses this approach in the context of the analysis of 2-D warranties.

[8] See Appendix A for a brief introduction to bivariate distribution functions.

#### 7.5.3.1 Approach 3-1

In this approach, the time to first failure is modeled by a bivariate failure distribution function $F(t,x)$ given by

$$F(t,x) = P\{T \le t, X \le x\} \tag{7.7}$$

$F(t,x)$ must be such that $E[X|T = t]$ is a non-decreasing function of $t$ in order to ensure that on the average usage increases with time.

The survivor function is given by

$$\bar{F}(t,x) = P\{T > t, X > x\} \tag{7.8}$$

The density function associated with $F(t,x)$ (provided the function is differentiable) is given by

$$f(t,x) = \frac{\partial^2 F(t,x)}{\partial t \partial x} \tag{7.9}$$

The hazard function associated with $F(t,x)$ is given by

$$h(t,x) = \frac{f(t,x)}{\bar{F}(t,x)} \tag{7.10}$$

with $h(t,x)\delta t \delta x$ defining the probability that the first system failure will occur in the rectangle $[t, t+\delta t) \times [x, x+\delta x)$, given that $T > t$ and $X > x$.

A variety of models have been proposed in the literature.[9]

If failed items are replaced by new, then the failures over the warranty region are modeled by a two-dimensional renewal process.[10] This has been used by several researchers in warranty cost analysis.[11]

*Example 7.1* Reference [29] proposed the following bivariate survivor function

$$\bar{F}(t,x) = \exp\{-(t/\alpha_1)^{\beta_1} - (x/\alpha_2)^{\beta_2} - \delta\psi(t,x)\} \tag{7.11}$$

with different forms of $\psi(t,x)$ defining families of survivor functions. We consider the case where

[9] For Weibull models, see [35]. Many non-Weibull models can also be used for modeling. For more on this, see [14, 21].

[10] See Appendix B for a brief introduction to two-dimensional renewal processes. Additional details can be found in [12, 13].

[11] References [22, 23, 33, 37, 40] have used this approach for 2-D warranty cost analysis.

Fig. 7.7 Plot of survivor function $\bar{F}(t,x)$

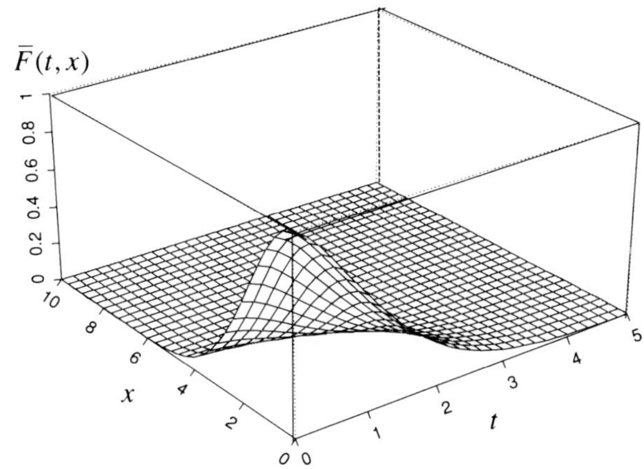


Fig. 7.8 Plot of hazard function $h(t,x)$.

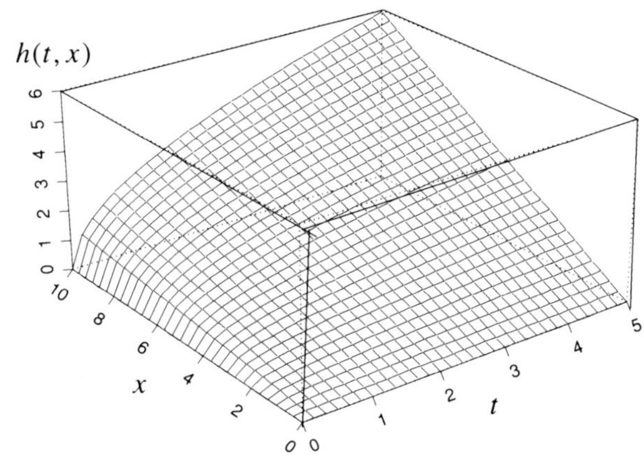


$$\psi(t,x) = [(t/\alpha_1)^{\beta_1/m} + (x/\alpha_2)^{\beta_2/m}]^m \tag{7.12}$$

with $\alpha_1, \alpha_2, \beta_1, \beta_2 > 0$, $\delta \geq 0$, and $0 < m \leq 1$. For the special case $m = 1$, the bivariate hazard function is given by

$$h(t,x) = (1+\delta)^2 \frac{\beta_1}{\alpha_1}\left(\frac{t}{\alpha_1}\right)^{\beta_1 - 1} \frac{\beta_2}{\alpha_2}\left(\frac{t}{\alpha_2}\right)^{\beta_2 - 1} \tag{7.13}$$

Let $\alpha_1 = 2$, $\alpha_2 = 3$, $\beta_1 = 1.5$, $\beta_2 = 2.0$, $\delta = 0.5$, $m = 1$. The units for age and usage are years and 10,000 km, respectively. The expected age at first system failure is given by $E(T_1) = \alpha_1 \Gamma(1/\beta_1 + 1) = 1.81$ (years). The expected usage at first failure is given by $E(X_1) = \alpha_2 \Gamma(1/\beta_2 + 1) = 2.66$ ($10^3$ km).

Figures 7.7 and 7.8 are plots of the survivor function $\bar{F}(t,x)$ and the hazard function $h(t,x)$. Note that $h(t,x)$ increases as $t$ (age) and $x$ (usage) increase, since $\beta_1$ and $\beta_2$ are greater than 1.

Finally, the concept of minimal repair is not fully developed and is a topic for new and further research.[12]

#### 7.5.3.2 Approach 3-2

This approach was proposed by [38]. They derive the bivariate density function $f(t,x)$by explicitly modeling usage as a function of time.[13]

#### 7.5.3.3 Approach 3-3

Here the joint distribution is derived in terms of the two marginal distributions using a copula function.[14] A special feature of the copula class is that the dependence structure is separated from the marginal effects, so the dependence relationship can studied without specifying the marginal distributions.

The bivariate distribution is given by

$$F(t,x) = C_\alpha\{F_T(t), F_X(x)\}, \tag{7.14}$$

where $\alpha$ denotes the association parameter, $F_T(t)$ and $F_X(x)$ are the two marginal distributions, and the copula function $C_\alpha(\cdot,\cdot)$ is itself a bivariate distribution function on $[0,1]\times[0,1]$.

A subclass of copulas are the Archimedean copulas, which index $C_\alpha(\cdot,\cdot)$ by means of a univariate function, resulting in a form that is more tractable for analysis.[15] The joint distribution is given by

---

[12] For further discussion, see [3, 4, 32].

[13] The logic of their derivation is as follows. Let $\psi(t)$ denote the cumulative usage at time $t$, and $X \equiv \psi(T)$, the cumulative usage at failure time $T$. The approach uses the following decomposition:

$$\begin{aligned} f(t,x) &= f_T(t) f_{\psi(T)|T}(x|t), && \text{since } X = \psi(t) \\ &= f_T(t) f_{\psi(t)|T}(x|t), && \text{since we condition on } T = t \\ &= f_{T|\psi(t)}(t|x) f_{\psi(t)}(x), && \text{by symmetry of the multiplicative law.} \end{aligned}$$

One first models cumulative usage to obtain $f_{\psi(t)}(x)$. Failure is then modeled through a hazard (failure rate) function of the form $r(t) = r_0(t) + \eta\psi(t)$, with $\eta > 0$ a specified constant. $r_0(t)$ is the base line hazard rate with no usage and the effect of usage is an increase in the hazard rate. $f_{T|\psi(t)}(t|x)$ is then obtained using the hazard rate and combining this with $f_{\psi(t)}(x)$ yields the bivariate density function $f(t,x)$.

[14] See page 10 of [36] for a definition of copula.

[15] See page 112 of [36] for more on Archimedean copulas.

$$F(t,x) = \phi_{\alpha}^{-1}\{\phi_{\alpha}[F_T(t)], \phi_{\alpha}[F_X(x)]\}, \tag{7.15}$$

where $\phi_{\alpha}(\cdot)$ is a convex function defined on $[0, 1]$ satisfying $\phi_{\alpha}(1) = 0$.

## 7.6 Warranty Cost Analysis per Unit [Approach 1]

We assume that the usage by time $t$, $X(t)$, is given by a linear function

$$X(t) = Zt \tag{7.16}$$

with $Z$ a non-negative random variable with density function $g(z)$. Define $\gamma$ as

$$\gamma = U/W \tag{7.17}$$

Note that conditional on $Z = z$, the warranty period $WP$is given by

$$WP = \begin{cases} W_z = U/z, & \text{if } z > \gamma \\ W, & \text{if } z \leq \gamma \end{cases} \tag{7.18}$$

As shown in Fig. 7.4. The conditional hazard function, given in (7.1), is a function of age ($t$), usage rate ($z$), and usage up to $t(zt)$. The conditional failure distribution $F_z(t)$ is given by (7.2).

We first derive the expected warranty costs by conditioning on $Z = z$ and then the expected warranty cost by un-conditioning.

### *7.6.1 Non-renewing FRW Policy*

We consider the following two cases: 1. Failed items replaced by new, and 2. Failed items repaired minimally.

#### 7.6.1.1 Replace by New

Since failed items are replaced by new ones, failures over the warranty period occur according to the renewal process associated with $F_z(t)$. As a result, the conditional expected warranty cost per unit is given by

$$E[C(W, U|Z = z)] = \begin{cases} C_s M_z(U/z), & \text{if } z > \gamma \\ C_s M_z(W), & \text{if } z \leq \gamma \end{cases} \tag{7.19}$$

where $C_s$ is the cost of providing a new replacement under warranty and $M_z(t)$ is given by

$$M_z(t) = F_z(t) + \int_0^t M_z(t - t')dF_z(t') \tag{7.20}$$

By un-conditioning, the expected warranty cost is found to be

$$E[C(W, U)] = C_s \left[ \int_0^\gamma M_z(W)g(z)dz + \int_\gamma^\infty M_z(U/z)g(z)dz \right]. \tag{7.21}$$

#### 7.6.1.2 Minimal Repair

Since failed items are repaired minimally and the repair times are assumed to be negligible, failures over the warranty period occur according to a NHPP process (see Appendix C) with conditional intensity function given by $\lambda(t|z) = h(t|z)$. Define

$$\Lambda_z(t) = \int_0^t \lambda(t|z)dt \tag{7.22}$$

Then the conditional expected warranty cost per unit is given by

$$E[C(W, U|Z = z)] = \begin{cases} c_r\Lambda_z(U/z), & \text{if } z > \gamma \\ c_r\Lambda_z(W), & \text{if } z \le \gamma \end{cases} \tag{7.23}$$

where $c_r$ is the average cost of a minimal repair. By un-conditioning, the expected warranty cost is found to be

$$E[C(W, U)] = c_r \left[ \int_0^\gamma \Lambda_z(W)g(z)dz + \int_\gamma^\infty \Lambda_z(U/z)g(z)dz \right] \tag{7.24}$$

### 7.6.2 Non-renewing PRW Policy

To determine specific policies of this type, it is necessary to specify the rebate/refund functions. Let $C_b$ denote the unit sale price. The linear rebate function is given by

$$\phi(t, x) = \begin{cases} C_b(1 - t/W)(1 - x/U), & (t, x) \in \Omega \\ 0, & \text{otherwise} \end{cases} \tag{7.25}$$

and the quadratic by

$$\phi(t,x) = \begin{cases} C_b(1 - t/W)^2(1 - x/U)^2, & (t,x) \in \Omega \\ 0, & \text{otherwise} \end{cases} \tag{7.26}$$

Conditional on $Z = z$, if a failure occurs under warranty with age at failure $t$, then the corresponding usage is given by $zt$. The refund is given by $\phi(t, tz)$. The conditional expected warranty cost in this case is

$$E[C(W, U|Z = z) = \int_0^{WP} \phi(t, tz)f(t|z)dt, \tag{7.27}$$

with $WP$ given by (7.18). In (7.18), $f(t|z)$ the density function associated with $F_z(t) = F(t|z)$.

The expected warranty cost, obtained by un-conditioning, is given by

$$E[C(W, U)] = \int_0^{\infty} E[C(W, U|Z = z)g(z)dz \tag{7.28}$$

## 7.7 Warranty Cost Analysis per Unit [Approach 2]

In Approach 2, discussed in Sect. 7.5.2, the 2-D data are reduced to 1-D by expressing the result as a combination of $x$ and $t$. A relatively simple such expression is the linear combination given in (7.6). References [25–27] use this approach with $a = \varepsilon$ and $b = 1 - \varepsilon$, with $0 < \varepsilon < 1$. $\varepsilon$ is chosen so that the coefficient of variation (discussed in Chap. 8) is minimized. The cost analysis is then done on the resulting 1-D data using the methods of Sect. 6.4.

## 7.8 Warranty Cost Analysis per Unit [Approach 3]

In this approach, the time to first failure is modeled by a bivariate distribution function $F(t, x)$. We omit the details of the derivation and present the final expressions. The details can be found in [5].

### *7.8.1 Non-renewing FRW Policy*

We confine our attention to the case where failed items are replaced by new. In this case, a two dimensional renewal process is used for modeling. The expected number of failures over $[0, t) \times [0, x)$ is given by $M(U, W)$, which is obtained as the solution of the two-dimensional integral equation

$$M(t,x) = F(t,x) + \int_0^t \int_0^x M(t-u, x-v) f(u,v) dv du. \quad (7.29)$$

The expected warranty cost is given by

$$E[C(W,U)] = C_s M(U,W) \quad (7.30)$$

In general, a computational approach must be used to obtain $M(t,x)$. A two-dimensional renewal equation solver can be found in [16].

*Example 7.2* The joint distribution of age and usage at failure of a non-repairable automobile component is given by the Beta-Stacy distribution with density function

$$f(t,x) = \frac{b t^{\alpha b - \theta_1 - \theta_2} (x/\eta)^{\theta_1 - 1} (t - x/\eta)^{\theta_2 - 1} \exp(-t/a)^b}{\Gamma(\alpha) B(\theta_1, \theta_2) a^{\eta b} \eta} \quad (7.31)$$

for $t > 0$, $0 < x < \eta t$, and all the model parameters greater than 0. The mean failure time and the mean usage at failure are given by

$$E[T] = \frac{a\Gamma(\alpha + 1/c)}{\Gamma(\alpha)} \quad \text{and} \quad E[X] = \frac{\theta_1 \eta}{\theta_1 + \theta_2} E[T] \quad (7.32)$$

Suppose that the parameter values are as follows: $\alpha = 2.0$, $b = 2.5$, $\theta_1 = 1.1$ and $\theta_2 = 1.5$. We consider the following three combinations of the remaining two parameters:

| Parameters | $a$ | $\eta$ | $E[T]$ (years) | $E[X]$ ($\times 10^4$ miles) |
|---|---|---|---|---|
| Set (a) | 2.65865 | 1.18180 | 4.0 | 2.0 |
| Set (b) | 1.99395 | 2.36371 | 3.0 | 3.0 |
| Set (c) | 1.3290 | 4.72742 | 2.0 | 4.0 |

Set (a) models light users, Set (b) medium users and Set (c) heavy users.

Table 7.1 gives the expected number of failures under warranty, $E[C(W, U)]/C_s$, for a range of $(W, U)$ combination where $C_s$ is the cost of supplying a new replacement unit. As can be seen, warranty costs increase as $W$ and/or $U$ increase, as expected. Here length of the warranty has a much greater impact on cost than does usage.

**Table 7.1** $E[C(W, U)]/C_S$ for non-renewing FRW policy

| U | W | | | | Model parameters |
|---|---|---|---|---|---|
| | 0.50 | 1.00 | 1.50 | 2.00 | |
| 0.50 | 0.0001 | 0.0013 | 0.0061 | 0.0165 | Set (a) |
| | 0.0002 | 0.0027 | 0.0109 | 0.0258 | Set (b) |
| | 0.0007 | 0.0080 | 0.0258 | 0.0461 | Set (c) |
| 1.00 | 0.0001 | 0.0020 | 0.0110 | 0.0316 | Set (a) |
| | 0.0003 | 0.0051 | 0.0219 | 0.0529 | Set (b) |
| | 0.0013 | 0.0165 | 0.0537 | 0.0969 | Set (c) |
| 1.50 | 0.0001 | 0.0020 | 0.0135 | 0.0425 | Set (a) |
| | 0.0003 | 0.0069 | 0.0316 | 0.0782 | Set (b) |
| | 0.0018 | 0.0245 | 0.0812 | 0.1478 | Set (c) |
| 2.00 | 0.0001 | 0.0020 | 0.0137 | 0.0479 | Set (a) |
| | 0.0003 | 0.0078 | 0.0394 | 0.1008 | Set (b) |
| | 0.0020 | 0.0316 | 0.1075 | 0.1975 | Set (c) |

**Table 7.2** $E[C(W, U)]/C_b$ for non-renewing PRW policy

| U | W | | | | Model parameters |
|---|---|---|---|---|---|
| | 0.50 | 1.00 | 1.50 | 2.00 | |
| 0.50 | 0.00001 | 0.00013 | 0.00064 | 0.00179 | Set (a) |
| | 0.00002 | 0.00027 | 0.00118 | 0.00303 | Set (b) |
| | 0.00007 | 0.00087 | 0.00317 | 0.00670 | Set (c) |
| 1.00 | 0.00001 | 0.00023 | 0.00119 | 0.00352 | Set (a) |
| | 0.00003 | 0.00054 | 0.00240 | 0.00627 | Set (b) |
| | 0.00013 | 0.00180 | 0.00665 | 0.01413 | Set (c) |
| 1.50 | 0.00001 | 0.00027 | 0.00157 | 0.00489 | Set (a) |
| | 0.00004 | 0.00075 | 0.00352 | 0.00936 | Set (b) |
| | 0.00019 | 0.00270 | 0.01012 | 0.02162 | Set (c) |
| 2.00 | 0.00001 | 0.00029 | 0.00178 | 0.00585 | Set (a) |
| | 0.00004 | 0.00090 | 0.00448 | 0.01220 | Set (b) |
| | 0.00023 | 0.00353 | 0.01347 | 0.02899 | Set (c) |

### 7.8.2 Non-renewing PRW Policy

The expected warranty cost to the manufacturer is given by

$$E[C(W,U)] = \int_{\Omega} \phi(t,x)dF(t,x) \tag{7.33}$$

where $\Omega = [0, W) \times [0, U)$, the warranty region.

**Table 7.3** $E[C(W, U)]/C_s$ for renewing PRW policy

| $C_b/C_s$ | $W$ | $U$ | | | |
|---|---|---|---|---|---|
| | | 0.50 | 1.00 | 1.50 | 2.00 |
| 1.1 | 0.50 | –0.00003 | 0.00006 | 0.00010 | 0.00011 |
| | 1.00 | 0.00005 | –0.00019 | 0.00070 | 0.00141 |
| | 1.50 | 0.00012 | 0.00055 | 0.00072 | 0.00367 |
| | 2.00 | 0.00007 | 0.00087 | 0.00317 | 0.00670 |
| 1.4 | 0.50 | –0.00041 | –0.00039 | –0.00034 | –0.00032 |
| | 1.00 | –0.00061 | –0.00659 | –0.00683 | –0.00592 |
| | 1.50 | –0.00052 | –0.00910 | –0.02673 | –0.02848 |
| | 2.00 | –0.00048 | –0.00807 | –0.03371 | –0.06436 |
| 1.8 | 0.50 | –0.00091 | –0.00098 | –0.00092 | –0.00089 |
| | 1.00 | –0.00149 | –0.01512 | –0.01687 | –0.01571 |
| | 1.50 | –0.00138 | –0.02198 | –0.06332 | –0.07134 |
| | 2.00 | –0.00132 | –0.02094 | –0.08334 | –0.15898 |

*Example 7.3* The failure distribution and parameters are the same as in Example 7.2. The refund function $\phi(t, x)$ is given by (7.25). Table 7.2 gives $E[C(W, U)]/C_b$ for a range of $(W, U)$ combinations obtained using (7.31) where $C_b$ is the sale price.

Again, the warranty costs increase as $U$ and/or $W$ increase. As for the FRW, the impact of $W$ is greater than that of $U$ over the range of values used. Note, however, that the costs for the PRW are less than those for the FRW, and significantly so for the larger values of $W$. In fact, warranty costs are quite low for nearly all parameter choices.

### 7.8.3 Renewing PRW Policy

This policy is offered by manufacturers to retain customers since giving a discount rather than a rebate encourages repeat purchases. The expected warranty cost for the renewing PRW is given by[16]

$$E[C_W(W, U)] = \frac{\int_\Omega [C_s - C_b + \phi(t, x)]dF(t, x)}{1 - F(\Omega)} \tag{7.34}$$

*Example 7.4* We again consider the Beta-Stacy distribution with the following parameter values:

[16] For details, see [39].

$$a = 0.6, \quad b = 1, \quad \alpha = 5, \quad \theta_1 = 18, \quad \theta_2 = 24, \quad \text{and} \quad \eta = 2.34.$$

Values of $E[C(W, U)]/C_s$ are given for various combinations of $C_b/C_s$, $W$ and $U$ in Table 7.3.

Here, unlike the non-renewing PRW policy, the expected warranty cost is not always positive or always increasing with $W$, $U$ and $C_b/C_s$. The negative values imply that the manufacturer is making a profit by offering warranty. This occurs because the replacement items are being sold at a reduced price, but one that, depending on the time and usage at failure, may be greater than the cost of replacement (i.e., $C_b\phi(t,x) > C_s$).

*Comment*: The expected warranty cost given by (7.33) assumes that all customers always execute the warranty claim on failure of an item. If this assumption does not hold, however, then, depending on the warranty execution function, the costs may become positive.

## References


1. Ahn CW, Chae KC, Clark GM (1998) Estimating parameters of the power law process with two measures of failure time. J Qual Technol 30:127–132
2. Baik J, Murthy DNP (2008) Reliability assessment based on two-dimensional warranty data. Int J Reliab Saf 2:190–208
3. Baik J, Murthy DNP, Jack N (2004) Two-dimensional failure modelling and minimal repair. Nav Res Logist 51:345–362
4. Baik J, Murthy DNP, Jack N (2006) Erratum: two-dimensional failure modelling with minimal repair. Nav Res Logis 53:115–116
5. Blischke WR, Murthy DNP (1994) Warranty cost analysis. Marcel Dekker Inc, New York
6. Blischke WR, Murthy DNP (eds) (1996) Product warranty handbook. Marcel Dekker, Inc., New York
7. Blischke WR, Murthy DNP (2000) Reliability. Wiley, New York
8. Chukova S, Johnston MR (2006) Two-dimensional warranty repair strategy based on minimal and complete repairs. Math Comput Model 44:1133–1143
9. Duchesne T, Lawless JF (2000) Alternate time scales and failure time models. Lifetime Data Anal 6:157–179
10. Finkelstein MS (2004) Alternative time scales for systems with random usage. IEEE Trans Reliab 50:261–264
11. Gertsbakh IB, Kordonsky HB (1998) Parallel time scales and two-dimensional manufacturer and individual customer warranties. IIE Trans 30:1181–1189
12. Hunter JJ (1974) Renewal theory in two dimensions: basic results. Adv Appl Probab 6:376–391
13. Hunter JJ (1996) Mathematical techniques for warranty analysis. In: Blischke WR, Murthy DNP (eds) Product warranty hand book. Marcel Dekker, New York
14. Hutchinson TP, Lai CD (1990) Continuous bivariate distributions: emphasising applications. Rumsby Scientific, Adelaide, Australia

15. Iskandar BP (1991) Two-dimensional renewal function solver. Research Report No, 4/91, Dept Mechan Eng, The University of Queensland, Brisbane, Australia
16. Iskandar BP, Blischke WR (2003) Reliability and warranty analysis of a motorcycle based on claims data. In: Blischke WR, Murthy DNP (eds) Case studies in reliability and maintenance. Wiley, New York
17. Iskandar BP, Murthy DNP (2003) Repair-replace strategies for two-dimensional warranty policies. Math Comput Model 38:1233–1241
18. Iskandar BP, Wilson RJ, Murthy DNP (1994) Two-dimensional combination warranty policies. Rech Oper 28:57–75
19. Iskandar BP, Murthy DNP, Jack N (2005) A new repair-replace strategy for items sold with a two-dimensional warranty. Comput Oper Res 32:669–682
20. Jack N, Iskandar B, Murthy DNP (2009) A repair-replace strategy based on usage rate for items sold with a two-dimensional warranty. Reliab Eng Sys Saf 94:611–617
21. Johnson NL, Kotz S (1972) Distributions in statistics: continuous multivariate distributions. Wiley, New York
22. Jung M, Bai DS (2007) Analysis of field data under two-dimensional warranties. Reliab Eng Sys Saf 92:135–143
23. Kim HG, Rao BM (2000) Expected warranty cost of a two-attribute free-replacement warranties based on a bi-variate exponential distribution. Comput Ind Eng 38:425–434
24. Kordonsky KB, Gertsbakh I (1993) Choice of best time scale for reliability analysis. Eur J Oper Res 65:235–246
25. Kordonsky KB, Gertsbakh I (1995) System state monitoring and lifetime scales–I. Reliab Eng Sys Saf 47:1–14
26. Kordonsky KB, Gertsbakh I (1995) System state monitoring and lifetime scales–II. Reliab Eng Sys Saf 49:145–154
27. Kordonsky KB, Gertsbakh I (1997) Multiple time scales and lifetime coefficient of variation: Engineering applications. Lifetime Data Anal 2:139–156
28. Lawless JF, Hu J, Cao J (1995) Methods for the estimation of failure distributions and rates from automobile warranty data. Lifetime Data Anal 1:227–240
29. Lu JC, Bhattacharyya CK (1990) Some new constructions of bivariate weibull models. Ann Inst Statisti Math 42:543–559
30. Moskowitz H, Chun YH (1994) A poisson regression model for two-attribute warranty policies. Nav Res Logist 41:355–376
31. Moskowitz H, Chun YH (1996) Two-dimensional free-replacement warranties. In: Blischke WR, Murthy DNP (eds) Product warranty handbook. Marcel Dekker, New York
32. Murthy DNP, Wilson RJ (1991) Modelling two-dimensional failure free warranties. Proceedings 5th symposium on applied stochastic models and data analysis: 481–492, Granada, Spain
33. Murthy DNP, Iskandar BP, Wilson RJ (1995) Two-dimensional failure free warranties: two-dimensional point process models. Oper Res 43:356–366
34. Murthy DNP, Xie M, Jiang R (2003) Weibull models. Wiley, New York
35. Murthy DNP, Baik J, Wilson RJ, Bulmer M (2006) Two-dimensional failure modelling. In: Pham H (ed) Handbook of engineering statistics. Springer-Verlag, London
36. Nelsen RB (2006) An introduction to copulas. Springer Verlag, New York
37. Pal S, Murthy GSR (2003) An application of gumbel's bivariate exponential distribution in estimation of warranty cost of motorcycles. Int J Qual Reliab Manag 20:488–502
38. Singpurwalla ND, Wilson SP (1998) Failure models indexed by two scales. Adv Appl Probab 30:1058–1072

39. Wilson RJ, Murthy DNP (1996) Two-dimensional pro-rata and combination policies. In: Blischke WR, Murthy DNP (eds) Product warranty hand book. Marcel Dekker, New York
40. Yang SC, Nachlas JA (2001) Bivariate reliability and availability modeling. IEEE Trans Reliab 50:26–35
41. Yang G, Zaghati Z (2002) Two-dimensional reliability modeling from warranty data. Proceedings annual reliability maintainability symposium, pp 272–278

# Chapter 8
# Preliminary Data Analysis

## 8.1 Introduction

We begin our discussion of the statistical analysis of data by looking briefly at some standard techniques for description and summarization of data. Prior to application of these techniques, however, it is important to inspect the data carefully to make sure that they are correct and appropriate for analysis in the context of the objectives of study.

The purposes of the initial inspection of the data are to

- Verify the source of the data
- Verify that the data include the variables specified
- Verify the units of measurement
- "Clean" the data by deleting or, if possible, correcting obviously incorrect results
- Identify outliers or otherwise unusual results
- Check for missing data
- Identify any other unusual data features

This activity is especially important when dealing with warranty claims data, which, as we have noted, are often very prone to error. For valid results, incorrect data must be removed from the analysis. (In doubtful cases, one might analyze the data with the suspect data included and well as excluded. If the results do not differ, use either analysis; if they do differ, judgment is required to determine which to accept.) Common sense is invaluable in cleaning data and preparing them for statistical analysis.

It is also necessary to determine exactly how the data were collected. The proper analysis of data depends crucially on an understanding of the data structure. For test and survey data, this is determined by the experimental design used in a laboratory or field study or the sampling plan used in a survey. For claims and

W. R. Blischke et al., *Warranty Data Collection and Analysis*,
Springer Series in Reliability Engineering, DOI: 10.1007/978-0-85729-647-4_8,

other supplementary warranty data, where the complete data set is to be analyzed, it is necessary to check that it is, in fact, complete and up to date.

Given an appropriate and complete data set, the objectives of a preliminary data analysis are to provide one or more of the following:

- A description of the key features of the data
- A summarization of the information content of the data in an easily understood format
- Graphical representations of the data
- Preparation of the data for detailed statistical analysis

This chapter will deal with many of the common tools used for these purposes. A number of examples will be included to illustrate the procedures. These will often involve data from case studies or mini-cases that will be analyzed in detail later in the book. Minitab will be used mainly in performing the analyses here and in the remainder of the book.[1]

Some of the material in this chapter is relatively elementary statistics and in intended as a review. As such, the coverage of these topics is brief. Additional details can be found in introductory statistics texts such as [12] and [10], and reliability books such as [2] and [9].

The contents of the chapter are as follows: In Sect. 8.2, we discuss several data issues in addition to those mentioned above, including scales of measurement, typical structures of failure data, and appropriate levels of analysis. Section 8.3 is concerned with summary statistics, including means, standard deviations, fractiles, etc. Graphical methods, including histograms, Pareto charts, and other common charts are discussed in Sect. 8.4. Probability plots of data and various distributions are given in Sect. 8.5. The use of graphical methods in data analysis is discussed in Sect. 8.6 and some comments on preliminary model selection are given in Sect. 8.7.

## 8.2 Data Related Issues

### *8.2.1 Large Data Sets, Data Mining and Preliminary Analysis*

Many companies compile modest to quite large data sets in tracking and servicing warranties. The data are typically stored in data banks and, depending on the use to which the information is to be put, analyzed to varying degrees by use of standard or specialized statistical methodology, as appropriate. In some instances, however, massive amounts of data are acquired, and special tools are needed to extract any useful information. This is typically true in the automobile industry, computers and

[1] Some other software packages, including Splus (http://www.insightful.com) and R-language (http://cran.r-project.org/) are also used in later Chapters.

other electronics industries, and any others having sales of millions of units and, even with relatively low claims rates, a large number of claims.

Data mining techniques have been developed to deal with massive data sets. We will not deal with data mining per se in this book, but will make note of some essential features that are of general interest in the context of analyzing claims data. A thorough treatment of data mining can be found in [3]. Very large data sets may be even more prone to error for a number of reasons, as discussed briefly below and in the references cited. A detailed discussion of "dirty" data, their causes, and the impact of this on data mining is given in [7].

An essential feature of data mining is a preliminary analysis called preprocessing. This is discussed in detail in [4] and [3], Chaps. 6–8. The objectives of preprocessing are to prepare the data for analysis by identifying and solving problems that may preclude any data analysis, understanding the nature of the data, and enabling the extraction of more meaningful knowledge. Data problems include

- Too much data (massive data sets, irrelevant data)
- Corrupt and/or noisy data
- Too little data (missing entries, missing variables, too few observations)
- Fractured data (multiple sources, incompatible data, data obtained at different levels)

Techniques for dealing with these problems include

- Data transformation (data filtering, ordering, editing, and modeling)
- Interactive techniques (data visualization, elimination, selection, identification of principal components, sampling)
- New information generation (time series analysis, data fusion, simulation, dimensional analysis, etc.)

For details, see [4].

Other approaches to preprocessing [3] include feature extraction, feature selection, and methods of discretization. Feature selection involves selection of a optimal (minimal) subset of data features to accomplish a data processing objective. A number of algorithms have been developed for this purpose. Feature extraction is concerned with the generation of new data features through transformation, reduction of dimensionality, and so forth, using principal component analysis, discriminant analysis, and other statistical methods. Discretization is used to reduce the number of distinct values in huge data sets to facilitate the analysis and enable the use of certain data mining algorithms that operate only in discrete spaces.

Some special problems are found in mining massive amounts of text data. For some approaches to statistical analysis and inference in this context, see [5] and [1].

In the remainder of this book, we shall, for the most part, assume that the data have been preprocessed or otherwise prepared for analysis, although a few techniques that are properly a part of that process (e.g., outlier detection, modeling of noise) will be briefly discussed in the chapters on basic analysis.

### 8.2.2 Scales of Measurement

As discussed in Chaps. 4 and 5, many types of data arise in dealing with warranty claims and the many related issues encountered in the use of warranty information. These data may be classified in several ways, for example:

- Discrete or continuous
- Complete or censored
- Univariate or multivariate
- Random or deterministic
- Quantitative or qualitative

Here we look at another approach to data classification, *scale of measurement*. Statistical techniques appropriate for analysis vary, and depend on the type of data. Scale of measurement is, in a sense, a refinement of some of the essential features of the above classifications and is important because it determines how numerical values are interpreted and what calculations can meaningfully be performed with the data.

There are four scales of measurement: nominal, ordinal, interval, and ratio. Their definitions and allowable calculations on each scale are as follows:

1. Nominal scale
   - Data are categorical
   - Examples—Item failed or operational; paint scratched, too thin, wrong color, blistered, imperfections
   - Allowable operations—counts only; no ranking or numerical operations
2. Ordinal scale
   - Data are categorical with a rank-order relationship
   - Examples—rating scales (severity of damage on a scale of 1 to 4; quality of sound of a speaker; seriousness of consequences of a failure)
   - Allowable operations—counts and ranking; no numerical operations
3. Interval scale
   - Data are numerical values on an equal-interval scale. (Note: on an interval scale, there is no true zero.)
   - Example—temperature
   - Allowable operations—ranking; addition and subtraction (and therefore averaging); multiplication and division are not meaningful
4. Ratio scale
   - Data are numerical values on an equal-interval scale with a uniquely defined zero
   - Examples—time to failure of an item, cost of repair, number of replacements under warranty
   - Allowable operations—all ordinary numerical and mathematical operations.

Note that in coding data, numerical values will usually be assigned to the observed values, regardless of the scale of measurement. Care should be taken in applying appropriate statistical methods and in interpretation of statistical results based on such data. Though often calculated, averages are completely meaningless for data that follow a nominal scale and of dubious value, at best, for data on an ordinal scale, though a median (defined in Sect. 8.3) may be acceptable for the latter.

### 8.2.3 Failure Data

As discussed in Chaps. 4 and 5, there are many types of data that may be relevant to warranty decisions. For many purposes, the most important of these are failure data. They are essential for estimating and predicting reliability and, along with various cost elements, for estimating and predicting the cost of warranty.

Failure data are always a part of the information collected in processing warranty claims. Some failure data are also collected during various stages of the product life cycle, particularly during design and development. All of these data can be useful in estimating and updating estimates of reliability, and ideally should be incorporated into a comprehensive model developed for this purpose.

In all of these situations involving failure data, the data may be complete or censored, or there may be some of each type. As discussed in Chap. 4, claims data on time to failure are inherently censored in that failure times are obtained only for items that fail during the warranty period (and then only for those for which claims are made). All other items are censored either at the service time if the warranty has not yet expired or at $W$, the length of the warranty.

Test data may be collected at the product level or at a number of levels below that, down to the part level. These data are often complete, though they may be censored as well, depending on the test design (e.g., if the design specified that testing be halted prior to the failure of all items on test.)

Another data structure sometimes encountered in claims data is grouping. Grouped data most frequently occur when items are tracked by lot and numbers of failures are reported periodically for the lot as a whole, without records of individual failure times.

As we shall see in the remainder of this and subsequent chapters, knowledge of the structure of the data is essential to a proper analysis. The techniques discussed in this chapter generally assume complete data, although several are appropriate for either censored or grouped data as well, or can be easily modified for this purpose.

### 8.2.4 Level of Analysis

As noted above, data may be collected at the product, part, and/or some intermediate level. This is true of claims data as well as test data. Correspondingly, the analysis may be done at various levels as well. This is particularly true with regard to reliability analysis. The appropriate level of analysis depends on the objectives of the study and the available data. As a rule of thumb, the appropriate level for analysis is the component level for design purposes, an intermediate level for monitoring and management of warranty servicing (which often involves replacement or repair of a module), and the product level for cost analysis.

## 8.3 Summary Statistics

In analyzing data, we begin with a *sample*, that is, a set of *observations* (measurements, responses, etc.), and perform various calculations and operations in order to focus and understand the information content of the sample data. The word "statistic" is used to refer to any quantity calculated from the data—averages, ranges, percentiles, and so forth. In this section, we will look at a number of statistics that are intended to describe the sample and summarize the sample information. These statistics also provide a foundation for *statistical inference*, which is the process of using the sample information to infer something about the *population* from which the sample was drawn.[2] Inferences are made concerning population parameters or other population characteristics. This is, in fact, the basic objective and primary thrust of the remainder of the book.

In this section, we will look at percentiles (fractiles) of a set of data, measures of center of a sample, measures of spread or dispersion, and measures of relationship between two or more variables in a data set. The measures will be applied to sets of warranty data obtained by various companies.

The procedures of this section are appropriate for complete data. They are not entirely inappropriate for censored data, but, if used in that context, must be interpreted with care. In particular, censoring will impose a conditionality on one form or another on the results. For example, one can calculate the mean time to failure of all items that failed while under warranty. This is a conditional mean, given that the lifetime is less than $W$. Inferences can be made only to the corresponding conditional population, and not directly to the population as a whole, i.e., not without some modification to take the conditionality constraint into account.

### 8.3.1 Notation

Here and in the ensuing chapters, we assume that we have a sample of size $n$ and denote the sample values $Y_1, Y_2, \ldots, Y_n$ if the values are considered to be random

[2] In a very real sense, probability and statistics are inverses of one another. Probability deals with models of randomness that can be used to make statements about the kinds of data that may occur. Statistics deals with the use of data to make statements about the model.

variables, and $y_1, y_2, \ldots, y_n$ if the values are observed values of the random variables. Note that $Y_i$ (or simply $Y$) is used as a generic variable. Other symbols ($U$, $V$, etc.) may be used in this way as well, except that in the context of warranty analysis $T$ will specifically indicate time and $X$ will be used to indicate usage. ($X$ may be used for other purposes as well.)

For some purposes, it will be convenient to order the observations from smallest to largest. The ordered set of observations will be denoted $y_{(1)}, y_{(2)}, \ldots, y_{(n)}$, with similar notations for $Y$, $T$, etc.

### 8.3.2 Fractiles

The *p-fractile*[3] of a continuous probability distribution is any value $y_p$ such that $F(y_p) = p$, where $0 \leq p \leq 1$. For a continuous CDF, $y_p$ is almost always uniquely determined.[4] In cases where it is not, the $p$-fractile can be taken to be any value in an interval, and there are several commonly used definitions for the term.

The $p$-fractile of a sample is defined as that value $y_p$ such that at least a proportion $p$ of the sample lies at or below $y_p$ and at least a proportion $1 - p$ lies at or above $y_p$. This value may also not be unique and there are a several alternative definitions that may be used. We define the $p$-fractile of a sample of observed values as follows: Let $k = [p(n+1)]$ and $d = p(n+1) - k$, where [x] denotes the integer part of x. If $k = 0$ or $k = n$ (corresponding to very small or very large values of $p$), the fractile is not defined. If $k = 1, \ldots, n-1$, then $y_p$ is given by

$$y_p = y_{(k)} + d(y_{(k+1)} - y_{(k)}) \tag{8.1}$$

If the CDF $F(y)$ is strictly increasing, then there is a unique value $y_p$ that satisfies $F(y_p) = p$, and the estimating equation for $y_p$ can be expressed as $y_p = F^{-1}(p)$, where $F^{-1}(.)$ denotes the inverse function.

Of particular interest in descriptive statistics are the .25, .50, and .75-fractiles, called the *quartiles*, and denoted $Q_1$, $Q_2$, and $Q_3$. Fractiles also have important applications in reliability, where the interest is in fractiles for small values of $p$. For example, if $t$ denotes the lifetime of an item, $t_{.01}$ is the time beyond which 99% of the lifetimes will lie.

We note that our definition agrees with that used in Minitab in calculating the sample quartiles.

*Example 8.1* Table F.2 gives kilometers driven and repair cost for $n = 32$ automotive warranty claims. The ordered values for repair cost, say $y$, are given in Table 8.1. Thus $y_{(1)} = 7.75$, and so forth. We calculate the quartiles and $y_{.05}$.

---

[3] Related terms are *percentile* and *quantile*.

[4] The exception occurs if the CDF is constant over some interval and increasing on either side of the interval.

**Table 8.1** Repair cost for automobile engines, ordered

| Repair cost (ordered) | | | | | | | |
|---|---|---|---|---|---|---|---|
| 7.75 | 11.70 | 24.60 | 26.35 | 27.58 | 27.78 | 29.91 | 34.68 |
| 42.71 | 42.96 | 48.05 | 60.35 | 77.22 | 77.24 | 77.57 | 78.42 |
| 98.90 | 127.20 | 149.36 | 253.50 | 388.30 | 432.89 | 556.93 | 658.36 |
| 831.61 | 918.53 | 1007.27 | 1101.90 | 1546.75 | 1638.73 | 1712.47 | 5037.28 |

For the first quartile, we have $k = [.25(33)] = 8$, and $d = .25$, so $Q_1 = \$34.68 + .25(42.71–34.68) = \$36.69$. Similarly, $Q_2 = \$88.66$, and $Q_3 = \$788.30$.

Note that these statistics on cost are of interest in and of themselves, even though the data, being claims data, are censored, as we have indicated. If one wished to estimate reliability in terms of kilometers driven from the data, the censoring would have to be taken into consideration.

We may, however, legitimately calculate descriptive statistics for the usage data, recognizing that we are describing a sample of kilometers driven until the first engine failure, given that a failure occurred during the warranty period. From the usage data of Table F.2, we find $Q_1 = 12.73$ km, $Q_2 = 17.25$ km, and $Q_3 = 27.23$ km. Again, these are interpreted as conditional values, not as descriptive of all drivers.

### 8.3.3 Measures of Center

The most common measures of the *center* of a sample (also called *measures of location*, or simply *averages*) are the sample mean and median. The sample median is the 0.50-fractile or $Q_2$. It is a natural measure of center since at least one-half of the data lie at or above it and at least one-half lie at or below $Q_2$. The sample mean of $Y$, denoted $\bar{y}$, is the simple arithmetic average given by

$$\bar{y} = \frac{1}{n}\sum_{i=1}^{n} y_i \tag{8.2}$$

For many statistical purposes, the sample mean is the preferred measure. It is the basis for many statistical inference procedures and is a "best" measure for these purposes for many types of populations. A problem is that the mean is sensitive to extreme values (*outliers*) in the data and can provide a somewhat distorted indication of center as a result of the outliers. In such cases, the median, which is not affected by extreme values, provides a more meaningful measure of the location of the center of the data.

Although both are measures of center, $\bar{y}$ and $Q_2$ measure this differently, and a comparison of the two provides additional information about the sample (and, by inference, about the population from which it was drawn). If the sample is perfectly symmetrical about its center, the mean and median are identical. If the two differ, this is an indication of *skewness*. If $Q_2 < \bar{y}$, the data are skewed to the right;

if $Q_2 > \bar{y}$, the data are skewed to the left. Failure data and the distributions used to model them (see Chap. 3) are often skewed right, which results from (usually) small numbers of exceptionally long-lived items.

Another approach to dealing with the distortion caused by outliers is to calculate a *trimmed mean*—remove a fixed proportion of both the smallest and largest values from the data and calculate the average of the remaining values.[5]

A few other measures are sometimes used. These include the mode, which is not of much in statistical inference, and various measures that can be defined as functions of fractiles, e.g., $(Q_3 - Q_1)/2$, $(y_{.90} - y_{.10})/2$, and so forth. In analyzing claims and other data, we will use the mean and median as measures of center, and occasionally look at the trimmed mean as well.

*Example 8.2* In Example 8.1, we looked at repair cost for the 32 warranty claims reported in Table F.2. The median was found to be $Q_2 = \$88.66$. For these data, the sample mean is $\bar{y} = \$536.09$. Note the very large difference between the mean and median, indicating skewness to the right. In fact, this is apparent in the data—there are several large and one very large value. The trimmed mean, eliminating the largest and smallest the observations (i.e., trimming about 5%), is \$371. This is still considerably larger than the median, indicating real skewness, beyond the influence of a few unusually large observations.

For the usage data, the sample conditional statistics are median = 17.25 km, mean = 18.32, and trimmed mean = 18.13. All of these are in fair agreement, indication that the conditional distribution of usage is not badly skewed, if at all.

### 8.3.4 Measures of Dispersion

A second descriptive measure commonly used in statistical analysis is a measure of *dispersion* (or *spread*) of the data. These measures reflect the *variability* in the data and are important in understanding the data and in properly interpreting many statistical results.

The most important measures of dispersion for most purposes are the sample variance and standard deviation. The *sample variance*, $s^2$, is given by

$$s^2 = \frac{1}{n-1}\sum_{i=1}^{n}(y_i - \bar{y})^2 = \frac{1}{n-1}\left\{\sum_{i-1}^{n} y_i^2 - \frac{1}{n}\left(\sum_{i=1}^{n} y_i\right)^2\right\} \tag{8.3}$$

The *sample standard deviation* is $s = \sqrt{s^2}$, and is the preferred measure for most purposes since it is in units of the original data.

---

[5] Minitab removes smallest and largest 5% (using the nearest integer to $.05n$). This usually removes the values causing the distortion and provides a more meaningful measure. Other (less drastic) methods of dealing with outliers will be discussed in Chap. 9.

The standard has a fairly direct interpretation as a measure of dispersion for samples and distributions that are symmetrical about their means or very nearly so. For the normal distribution (See Appendix A), the following statements are true with regard to the population mean and standard deviation:

- 68% of the values will lie within one standard deviation of the mean
- 95% of the values will lie within two standard deviations of the mean
- 99.7% of the values will lie within three standard deviations of the mean

These statements are also approximately true for unimodal, symmetrical distributions and samples generally, and may be used as guidelines for interpretation of dispersion. (*Note*: unless $n$ is quite large, the approximation may be quite crude for samples).

Another measure of variability sometimes used is the *interquartile range*, $I$, given by $I = Q_3 - Q_1$. An advantage of the interquartile range is that it is not affected by extreme values. A disadvantage is that it is not readily interpretable as is the standard deviation.

Finally, a useful measure of dispersion in certain application is the *coefficient of variation*, defined to be c.v. $= s/\bar{y}$. This measure is unit free and tends to remain relatively constant over measurements of different types (e.g., weights of individuals over different biological species, fuel consumption of engines of very different sizes).

*Example 8.3* For the repair cost data of Table F.2, the variance was found in Example 8.1 to be 941,463; the standard deviation is the square root of this value, or \$970.3. These large values reflect the significant amount of variability in the data. Both are influenced by the large outliers and the interpretation given above is not valid in this case because of these and the overall skewness of the data. In fact, 28 observations (86%) lie within one standard deviation of the mean, and 31 (97%) lie within both two and three standard deviations. For these data, the interquartile range is $I = \$788.30 - 34.68 = \$753.62$.

For the conditional usage data, the standard deviation is 9.74 km; the interquartile range is 14.52. For these data, 21 observations (66%) lie within one standard deviation of the mean, 31 (97%) lie within two standard deviations, and all 32 lie within three standard deviations, indicating an approximately symmetric distribution similar to the normal.

### 8.3.5 Measures of Relationship

When the data include two or more variables, measures regarding the relationship between the variables are of interest. Here we introduce two measures of *strength* of relationship for two variables, the ordinary correlation coefficient $r$, and a rank correlation coefficient, $r_s$.[6] Other measures of relationship and measures for more than two variables will be discussed in later chapters.

We assume a sample of bivariate data ($x_i$, $y_i$), $i = 1, \ldots, n$. The *sample correlation coefficient* is given by

$$r = \frac{\frac{1}{n-1}\sum_{i=1}^{n}(x_i - \bar{x})(y_i - \bar{y})}{s_x s_y}, \tag{8.4}$$

where $s_x$ and $s_y$ are the standard deviations of the two variables. The numerator of this expression, called the sample *covariance*, is itself used as a measure of relationship in certain applications.

$r$ is the sample equivalent of the population correlation coefficient $\rho$, a parameter of the biviariate normal distribution, and as such is a measure of the strength of *linear* relationship between the variables, with $\rho = 0$ indicating no linear relationship. In the case of the bivariate normal distribution, this is equivalent to independence of the variables. Note that the correlation coefficient is unit-free. In fact, $\rho$ and $r$ are in the interval $[-1,1]$, with the values $-1$ and $+1$ indicating that the variables are co-linear, with lines sloping downward and upward, respectively. The general interpretation is that values close to either extreme indicate a strong relationship and values close to zero indicate very little relationship between the variables.

An alternative measure of strength of relationship is *rank correlation*. Rank correlation coefficients are calculated by first separately ranking the two variables (giving tied observations the average rank) and then calculating a measure based on the ranks. The advantage of this is that a rank correlation is applicable to data down to the ordinal level and is not dependent on linearity. There are several such coefficients. The most straightforward of these is the Spearman rank correlation $r_s$, which is simply application of (8.4) to the ranks. Note that rank correlation can also be used to study trend in measurements taken sequentially through time. In this case, the measurements are ranked and these ranks and the order in which observations were taken are used in the calculation of $r_s$.

Another approach to the study of data relationships is *linear regression analysis*, in which the linear relationship between the variable is explicitly modeled and the data are used to determine values for the parameters of the model. The approach is applicable to nonlinear models as well. Regression analysis will be discussed Chaps. 10 and 13.

*Example 8.4* For the data of Table F.2, the correlation between repair cost and kilometers driven until first engine failure is found to be $r = 0.254$. The Spearman rank correlation is $r_s = 0.225$. The two are in fairly close agreement in this case and both are relatively small, indicating little relationship between the two variables.

---

[6] The subscript $s$ is for Charles Spearman, who devised the measure in 1904.

### *8.3.6 Descriptive Statistics with Minitab*

All of the measures discussed in this section are easily calculated using Minitab. The commands are found under Calc/Basic Statistics. The results of Examples 8.1–8.3 are found in the following Minitab output, obtained from "Descriptive Statistics:"

Descriptive Statistics: Repair Cost, km

| Variable | *N* | Mean | Median | TrMean | StDev | SE Mean |
|---|---|---|---|---|---|---|
| Repair Cost | 32 | 536 | 89 | 371 | 970 | 172 |
| km | 32 | 18.32 | 17.25 | 18.13 | 9.74 | 1.72 |

| Variable | Minimum | Maximum | Q1 | Q3 |
|---|---|---|---|---|
| Repair Cost | 8 | 5037 | 37 | 788 |
| km | 2.30 | 38.60 | 12.73 | 27.23 |

The value of "SE mean" is calculated as $s/\sqrt{n}$ and will be discussed in Chap. 9. The remaining output is self-explanatory.

Correlation coefficients are also easily obtained in Minitab, using Correlation, under Calc/Basic Statistics.

## 8.4 Basic Graphical Methods

### *8.4.1 Pareto Charts*

A *Pareto chart* is a plot of frequencies based on counts of observations falling into each class in a set of nominal data. This is a useful tool for dealing with categorical or qualitative data and is often used to represent certain failure data (e.g., frequency of occurrence of failure modes) and other nominal data in reliability. The chart is a bar chart of vertical bars representing frequencies, with the categories arranged in order of frequency of occurrence, from largest to smallest.

In the reliability context, the importance of a Pareto chart is it provides a means of easily identifying the most frequently occurring classes, which are usually the most important and which may require urgent attention. The chart "separates the vital few from the trivial many" (the "Pareto Principle"), which is especially useful in analyzing failure modes.

*Example 8.5* Data on 13 failure modes of a room air conditioner are given in Table F.1. A Pareto chart of the data is given in Fig. 8.1. It is easily seen that the most frequently occurring failures leading to warranty claims are "Grille door damaged" and that this along with "Front Panel/Grill Damaged," and "Vent Lever

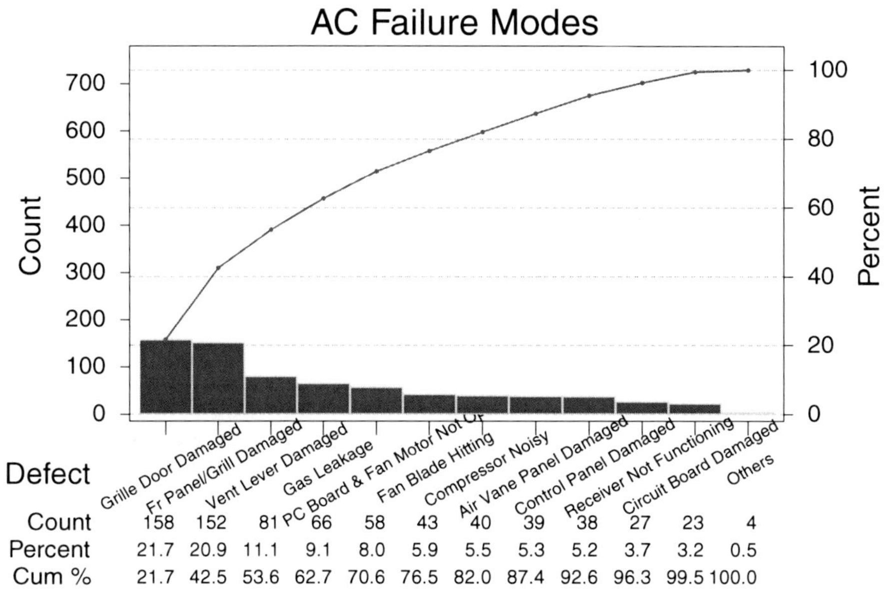


**Fig. 8.1** Pareto chart of home air conditioner failure modes

Damaged," jointly accounting for over one-half the claims. Likely causes of the problems may be faulty packaging, inadequate inspection, or improper handling in shipping or storage. Identification of the underlying cause could significantly decrease warranty costs.

*Note:* The chart given in Fig. 8.1 was produced by Minitab, using "Stat/Quality Tools." For this example, the "vital few" are easily identified directly from Table F.1. For the unaggregated data and for large data sets generally, however, this is not the case.

### *8.4.2 Histograms*

A *histogram* is a graphical representation of a frequency distribution for quantitative (ordinal, interval or ratio) data. It is formed by grouping the data according to some specified grouping algorithm, determining counts for the groups defined, and then plotting the frequencies in the form of a vertical bar chart. This provides a picture of the entire data set, resulting in a concise presentation of the data information in a form that displays the general shape of the distribution of values present in the data.

Groups are defined by specifying the upper and lower limits of each interval. There are a number of ways of doing this. For ordinal data, the natural groupings

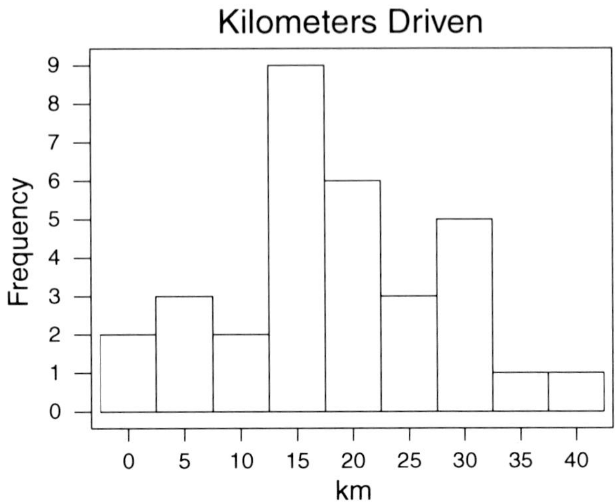


**Fig. 8.2** Histogram of kilometers driven

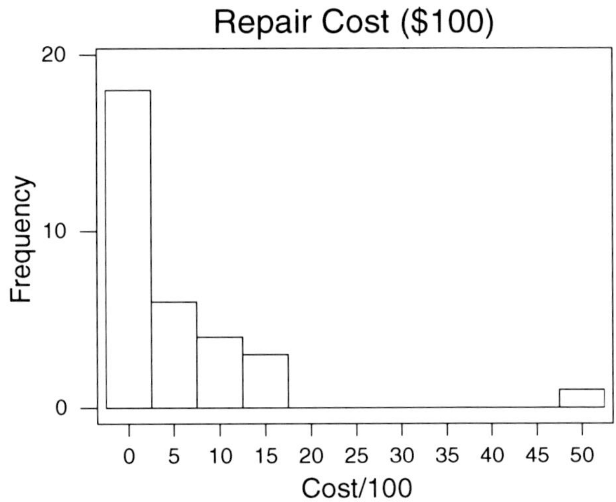


**Fig. 8.3** Histogram of repair cost

may be used or some groups be combined. For interval and ratio data, equal group widths are often used. In practice, it is easiest to use the groupings automatically specified in Minitab. If this does not provide a satisfactory picture, the user may specify the grouping desired.

*Example 8.6* Histograms of kilometers driven and repair costs under warranty for the data of Table F.2 are given in Figs. 8.2 and 8.3. Note that the histogram of km looks roughly symmetric (given the small number of values) and an extreme outlier is apparent in the histogram of repair costs. Even without the outlier, the distribution of repair cost is skewed, as noted previously.

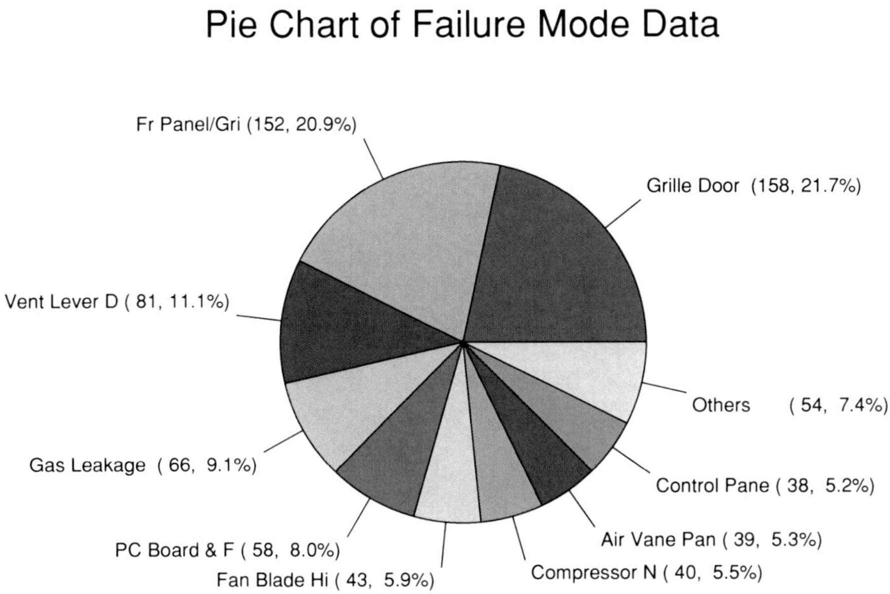


**Fig. 8.4** Pie chart of AC failure modes

The histogram of the repair cost data indicates that the majority of cases are clustered at the lower end of the scale, with most being below \$500. There is, however, a case in the class centered on \$5000. This high value for only a single case has a significant effect on the mean but little effect on the median, making the median a better indicator of central tendency for the repair costs data in this example, as noted in Example 8.2.

### *8.4.3 Pie Charts and Other Graphical Techniques*

Qualitative data may be represented graphically in a number of ways in addition to Pareto charts. A common alternative is a *pie chart*, which is a circle that is divided into sectors proportional to the frequencies in each group. Figure 8.4 is a pie chart of air conditioner failure mode data, labeled with modes, numbers in each class and corresponding percentages of failures. The pie chart provides the same information as the Pareto chart of Fig. 8.1.

Other alternatives to histograms have been developed as well. These include stem-and-leaf plots and box plots, which are available in Minitab under exploratory data analysis (EDA). For brief descriptions of the application of these to failure data see Chap. 3 of [2].

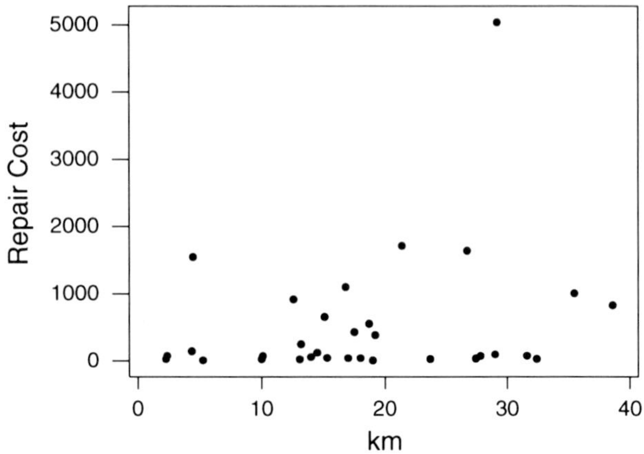


**Fig. 8.5** Plot of repair cost versus km driven

There are many other graphical methods of representing both qualitative and quantitative data. These are discussed in detail in [13] and [14–16].

### *8.4.4 Graphical Display of Data Relationships*

Regression analysis was mentioned briefly in Sect. 8.3.5 as a method of studying data relationships. The graphical approach to looking at the relationship between two variables is simply to plot one against the other. Regression analysis extends this concept by providing a linear or higher order model as a "fit" to the plotted data. Plots are easily obtained using many statistical packages, including Minitab. A plot of the km/repair cost data of Table F.2 is given in Fig. 8.5. Note that the apparent outlier is quite evident in the plot. The weak relationship between km driven and cost as measured by the correlation coefficient is reflected in the flat pattern of the data in Fig. 8.5. Three-dimensional plots are also easily obtained.

## 8.5 Probability Plots

The concept of a probability distribution (denoted $F(y)$) to express the uncertainty or variability in values of a random quantity such as time to failure, cost, etc., was discussed in Chap. 3. Many important probability distributions used in reliability theory and statistical analysis are listed in Appendix A. These may be thought of as possible models that underlie the data in applications such as those under consideration. One of our goals, as noted previously, is to reverse this process, using the data to determine, insofar as is possible, which model is the "correct" form for $F(y)$. In this section, we look at the sample CDF as well as a number of plots based on some common distributions $F(y)$. In both cases, plots will be given for complete and right-censored data.

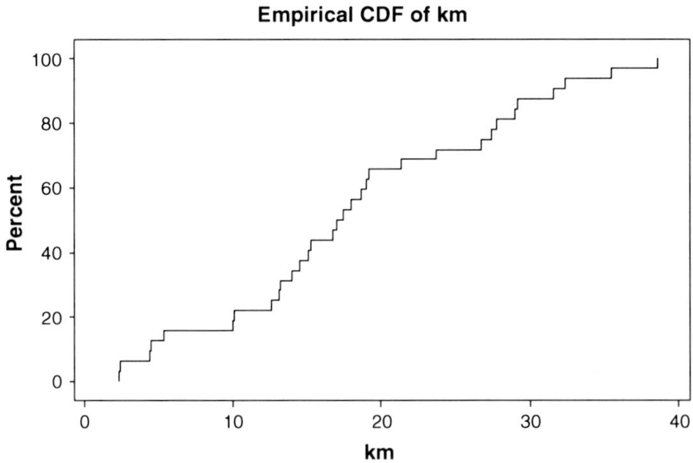


**Fig. 8.6** EDF for kilometers driven data of Table F.2

## 8.5.1 *Empirical Distribution Function*

One of the key tools for investigating the distribution underlying the data is the sample equivalent of $F(y)$, denoted $\hat{F}(y)$, and called the *empirical distribution function* (EDF). The EDF plots as a "step-function," with steps at data points. The form of the function depends on the type of population from which the sample was drawn. On the other hand, the procedure is nonparametric in the sense that no specific form is assumed in calculating the EDF.

The EDF and its calculation are discussed in Appendix C, Sect. C.2, for complete data, incomplete data and grouped data. We look at the first two of these.

Complete Data

For complete data, with ordered values $y_{(1)}, y_{(2)}, \ldots, y_{(n)}$, the EDF is given by

$$\hat{F}(y) = \begin{cases} 0 & y < y_{(1)} \\ \frac{1}{n+1} & y_{(i)} \le y < y_{(i+1)}, i = 1, \ldots, n-1 \\ 1 & y \ge y_{(n)} \end{cases} \tag{8.5}$$

*Notes:* (1) Minitab uses the divisor n rather than $(n + 1)$ in the middle expression of (8.5). Other choices have been used as well. The $(n + 1)$ divisor is now generally accepted as preferable [8]. (2) If there are tied observations in the array, the numerator of the middle expression is replaced by the number of tied values for each set of ties.

*Example 8.7* Minitab plots of the EDF for kilometers driven and repair costs for the data of Table F.2 are given in Figs. 8.6 and 8.7 (As noted above, these are in slight disagreement with equation (8.5)). As before, the result for km is interpreted as a conditional distribution. The pattern for km is roughly symmetrical, while that

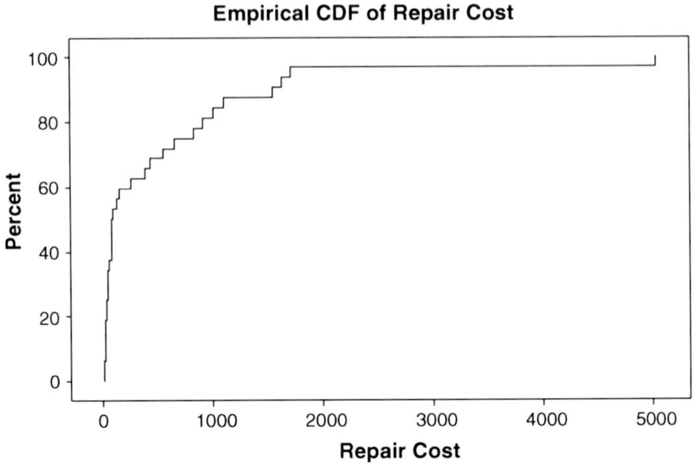


**Fig. 8.7** EDF for repair cost data of Table F.2

for cost is clearly skewed. This is consistent with the relationships of the mean and median and is more clearly seen in the histograms of Figs. 8.2 and 8.3.

*Right-Censored Data.* For censored data, we look only at censoring on the right since that is the most common censoring found in claims data and in many reliability applications. To calculate $\hat{F}(y)$, the observations are ordered, including both censored and uncensored values in the ordered array. Suppose that $m$ observations in the ordered array are uncensored. Denote these $y'_1, y'_2, \ldots, y'_m$. These are the locations of the steps in the plot of the EDF. To determine the heights of the steps, for $i = 1, \ldots, m$, form the counts $n_i$ = "number at risk" = number of observations greater than or equal to $y'_i$ in the original ordered array, and $d_i$ = number of values tied at $y'_i$(=1 if the value is unique), then calculate the "survival probabilities"

$$S_1 = 1 - \frac{d_1}{n_1} \quad \text{and} \quad S_i = \left(1 - \frac{d_i}{n_i}\right) S_{i-1}, \quad i = 2, \ldots, m \tag{8.6}$$

*Notes:* (1) This procedure for censored data may also be applied to grouped data. Since this is the sample version of F(y) it may be used to estimate the true CDF. In this context, the EDF is generally known as the Kaplan–Meier estimator. ([9], Sect. 3.7). (2) Again, there have been a number of alternatives considered with regard to the plotting positions. Equation (8.6) is used in Minitab. (3) Some versions of Minitab do not give the EDF for censored data, but do give values and plots of the survival function of (8.6). The EDF is easily calculated and plotted from these.

*Example 8.8* The data on battery life given in Table F.3 are right-censored. The ordered array, including both failures and censored observations (marked with a "*") is as follows: 64, 66, 131*, 162*, 163*, 164, 178, 185, 202*, 232*, 245*, 286*, 299, 302*, 315*, 319, 337*, 383, 385, 405, 482, 492, 506, 548, 589, 599, 619, 631, 639, 645, 656, 681, 722, 727, 728, 761, 765, 788, 801, 845*, 848, 852, 929, 948, 973, 977, 983*, 1084, 1100, 1100, 1259*, 1350, 1384*, 1421*.

The $y_i'$'s are the set of non-marked values in the array (64, 66, 164, 178, etc.). Here $n = 54$ and $m = 39$. The numbers at risk are $n_1 = 54$, $n_2 = 53$, $n_3 = 49$, $n_4 = 48$, etc. The $d_i$ are all 1 except for the case of the two values tied at 1100, for which we have $n_{38} = 6$ and $d_{38} = 2$. Thus $S_1 = 1–1/54 = 0.9815$, $S_2 = 0.9815(1 - 1/53) = 0.9630$, $S_3 = 0.9630(1 - 1/49) = 0.9433$, etc., with the EDF being the complement of these. The Minitab output given in Table 8.2 gives the complete set of values. (The final three columns will be discussed later.)

The survivor function is plotted in Fig. 8.8. Figure 8.9 shows the EDF, with connecting lines rather than as a step-function. In this graph, the EDF appears to more or less follow a sigmoid curve typical of the normal distribution. (See Appendix A, equation (A.26).)

The Minitab output includes as estimate of the MTTF with confidence interval along with other nonparametric estimates and details of the computation of the EDF. The results are given in Table 8.2.

### 8.5.2 Calculation of Probability Plots

In Example 8.8, it was noted that an EDF that plots as a sigmoid curve is indicative of the normal distribution. Nonnormal distributions give rise to data that plot in other specific characteristic shapes, which may or may not be easily recognizable. *Probability plots* have been developed to deal with this problem and to aid in the identification of distributions. The procedure for producing these plots involves transformation of the data (plotted on the horizontal scale) and/or the probability (vertical) scale so that the plot on the transformed scale is linear (within chance fluctuations). Equivalently, fractiles of the data may be plotted against fractiles of the theoretical scale. Such plots are referred to as "P–P plots." Plotting papers based on the transformed scales have been developed for a number of distributions and most statistical packages include options for probability plots.

Related plots of use in certain reliability applications are the hazard and cumulative hazard plots ([2], Sect. 11.2.3). The cumulative hazard function, given by $H(y) = -\log[1 - F(y)]$, is a measure of risk. Empirical hazard and cumulative hazard plots are plots of the sample (empirical) hazard functions, usually on a transformed scale such as that indicated above. Plotting papers and statistical program options are available for obtaining these plots as well. In most cases, the plots can be formed based on complete as well as censored data.

**Table 8.2** Minitab output for Kaplan–Meier estimation of the EDF

**Nonparametric estimates**

*Characteristics of variable (Mean)*

| Mean (MTTF) | Standard error | 95.0% Normal CI | |
|---|---|---|---|
| | | Lower | Upper |
| 726.3222 | 51.7679 | 624.8590 | 827.7855 |

*Characteristics of variable (Fractiles)*

| Median | IQR | Q1 | Q3 |
|---|---|---|---|
| 722.0000 | 456.0000 | 492.0000 | 948.0000 |

**Kaplan–Meier estimates**

| Time | Number at risk | Number failed | Survival probability | Standard error | 95% Normal CI | |
|---|---|---|---|---|---|---|
| | | | | | Lower | Upper |
| 64 | 54 | 1 | 0.9815 | 0.0183 | 0.9455 | 1.0000 |
| 66 | 53 | 1 | 0.9630 | 0.0257 | 0.9126 | 1.0000 |
| 164 | 49 | 1 | 0.9433 | 0.0318 | 0.8810 | 1.0000 |
| 178 | 48 | 1 | 0.9237 | 0.0367 | 0.8517 | 0.9956 |
| 185 | 47 | 1 | 0.9040 | 0.0409 | 0.8239 | 0.9841 |
| 299 | 42 | 1 | 0.8825 | 0.0452 | 0.7939 | 0.9711 |
| 319 | 39 | 1 | 0.8599 | 0.0494 | 0.7631 | 0.9566 |
| 383 | 37 | 1 | 0.8366 | 0.0532 | 0.7323 | 0.9410 |
| 385 | 36 | 1 | 0.8134 | 0.0566 | 0.7024 | 0.9243 |
| 405 | 35 | 1 | 0.7901 | 0.0596 | 0.6734 | 0.9069 |
| 482 | 34 | 1 | 0.7669 | 0.0622 | 0.6450 | 0.8888 |
| 492 | 33 | 1 | 0.7437 | 0.0645 | 0.6172 | 0.8701 |
| 506 | 32 | 1 | 0.7204 | 0.0665 | 0.5900 | 0.8508 |
| 548 | 31 | 1 | 0.6972 | 0.0683 | 0.5633 | 0.8311 |
| 589 | 30 | 1 | 0.6739 | 0.0699 | 0.5370 | 0.8109 |
| 599 | 29 | 1 | 0.6507 | 0.0712 | 0.5111 | 0.7903 |
| 619 | 28 | 1 | 0.6275 | 0.0724 | 0.4856 | 0.7693 |
| 631 | 27 | 1 | 0.6042 | 0.0733 | 0.4605 | 0.7480 |
| 639 | 26 | 1 | 0.5810 | 0.0741 | 0.4357 | 0.7262 |
| 645 | 25 | 1 | 0.5577 | 0.0747 | 0.4113 | 0.7042 |
| 656 | 24 | 1 | 0.5345 | 0.0751 | 0.3873 | 0.6817 |
| 681 | 23 | 1 | 0.5113 | 0.0754 | 0.3636 | 0.6590 |
| 722 | 22 | 1 | 0.4880 | 0.0754 | 0.3402 | 0.6359 |
| 727 | 21 | 1 | 0.4648 | 0.0753 | 0.3171 | 0.6124 |
| 738 | 20 | 1 | 0.4415 | 0.0751 | 0.2944 | 0.5887 |
| 761 | 19 | 1 | 0.4183 | 0.0746 | 0.2720 | 0.5646 |
| 765 | 18 | 1 | 0.3951 | 0.0740 | 0.2500 | 0.5401 |
| 788 | 17 | 1 | 0.3718 | 0.0732 | 0.2283 | 0.5153 |
| 801 | 16 | 1 | 0.3486 | 0.0722 | 0.2070 | 0.4902 |
| 848 | 14 | 1 | 0.3237 | 0.0712 | 0.1841 | 0.4633 |
| 852 | 13 | 1 | 0.2988 | 0.0700 | 0.1616 | 0.4359 |
| 929 | 12 | 1 | 0.2739 | 0.0684 | 0.1398 | 0.4080 |

(continued)

**Table 8.2** (continued)

| **Kaplan–Meier estimates** | | | | | | |
|---|---|---|---|---|---|---|
| Time | Number at risk | Number failed | Survival probability | Standard error | 95% Normal CI | |
| | | | | | Lower | Upper |
| 948 | 11 | 1 | 0.2490 | 0.0666 | 0.1185 | 0.3795 |
| 973 | 10 | 1 | 0.2241 | 0.0644 | 0.0978 | 0.3503 |
| 977 | 9 | 1 | 0.1992 | 0.0619 | 0.0779 | 0.3205 |
| 1084 | 7 | 1 | 0.1707 | 0.0592 | 0.0547 | 0.2868 |
| 1100 | 6 | 2 | 0.1138 | 0.0514 | 0.0131 | 0.2145 |
| 1350 | 3 | 1 | 0.0759 | 0.0462 | 0.0000 | 0.1664 |

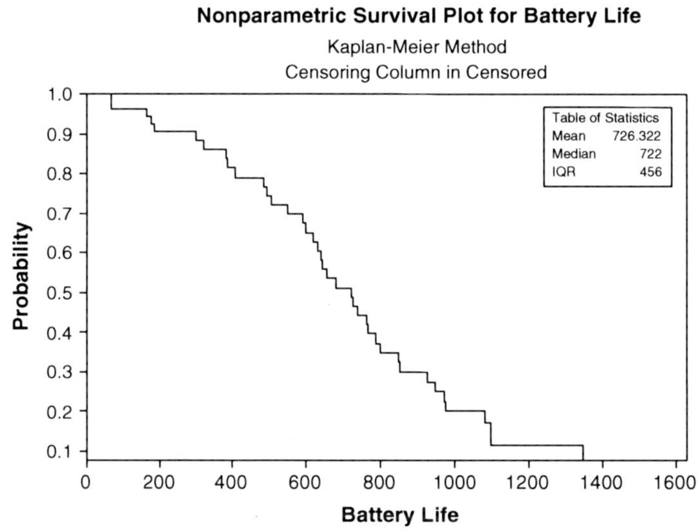


**Fig. 8.8** Battery life survival function

### *8.5.3 WPP Plots*

Weibull probability paper (WPP) plots are plots based on the Weibull distribution, with CDF

$$F(t) = 1 - e^{-(t/\alpha)^{\beta}} \tag{8.7}$$

for $t > 0$. (See Sect. 3.6.1 and Appendix A.) The distribution is commonly used to model failures. It was first used to model the breaking strength of materials [17] and is currently widely used in quality assurance and reliability applications of all kinds. The Weibull model is flexible, having several distinctive shapes and allowing for increasing, decreasing, and constant failure rates, and many generalizations of the model have been developed for modeling complex phenomena. The distribution is discussed in detail in ([6] Chap. 20, and [11]), which also

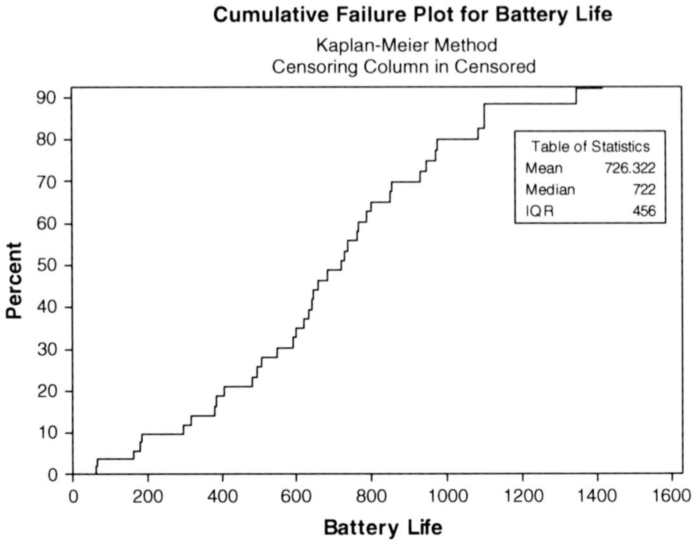


**Fig. 8.9** EDF of battery life

provides extensive coverage of a large number of extensions of the basic Weibull model and their use in reliability, as well as the use and interpretation of WPP plots in modeling data.

The basis of WPP plots is the transformation

$$y = \log[-\log(1 - F(t))] = \beta[\log(t) - \log(\alpha)] \tag{8.8}$$

The plot of $y$ against $x = \log(t)$ is linear. Weibull probability paper is scaled so that the plot of $F(t)$ versus $t$ is linear. This is also used to plot the EDF against the observed data values. If the data are a sample from a Weibull distribution, the plot will usually be approximately linear. The plots can also easily be produced using most statistical program packages. Minitab will provide WPP plots for both complete and censored data.

Many extensions of the Weibull distribution lead to distinctive nonlinear shapes when plotted on Weibull probability paper. These are discussed in Appendix C and in [11]. These sources also give useful guidelines for interpretation of the plots.

*Example 8.9* Table F.5 gives bond strength of an adhesive used in an audio component. Components were stored for varying lengths of time under uncontrolled (warehouse) conditions. A WPP plot of the strength data, obtained from Minitab using Stat → Reliability/Survival → Parametric distribution analysis—right censoring,[7] is given in Fig. 8.10. The data appear to follow a linear pattern reasonably well. The information to the right of the plot indicates that parameter values of

---

[7] The steps may vary with respect to the version of the Minitab software.

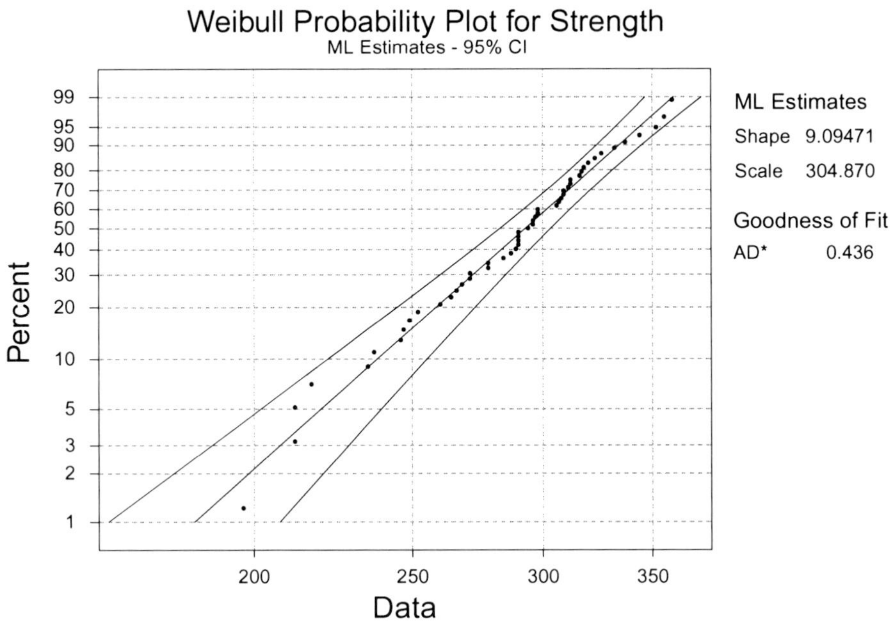


**Fig. 8.10** Bond strength of adhesive in warehoused audio parts

$\alpha = 304.87$ and $\beta = 9.095$, obtained from the data, were used in preparing the WPP plot. Note that this value of $\beta$ indicates a strongly increasing failure rate. (The remaining information in Fig. 8.10 will be discussed in Chaps. 9 and 10.)

The roughly linear pattern of the data on Weibull probability paper suggests that the Weibull model may be a reasonable choice for modeling failures (i.e., bond breakage) in this application. In fact, this is similar to the original application described by Weibull. Note, however, that this analysis has ignored the fact that the items were stored for varying lengths of time (in fact, ranging from 3 to 82 days). If strength and storage time are related, this could significantly affect the results. This question will also be addressed in Chap. 10.

*Example 8.10* The repair cost data of Table F.2 are not failure data and hence may not be expected to follow a Weibull distribution. On the other hand, the data are considerably skewed to the right, so the Weibull model is a possible choice. The WPP plot of the data is given in Fig. 8.11. Here the plot does not appear to be linear, so alternative models are indicated.

### *8.5.4 Other Probability Plots*

Probability plots for other distributions can be obtained by the method used for the Weibull distribution as long as the CDF can be expressed analytically and the

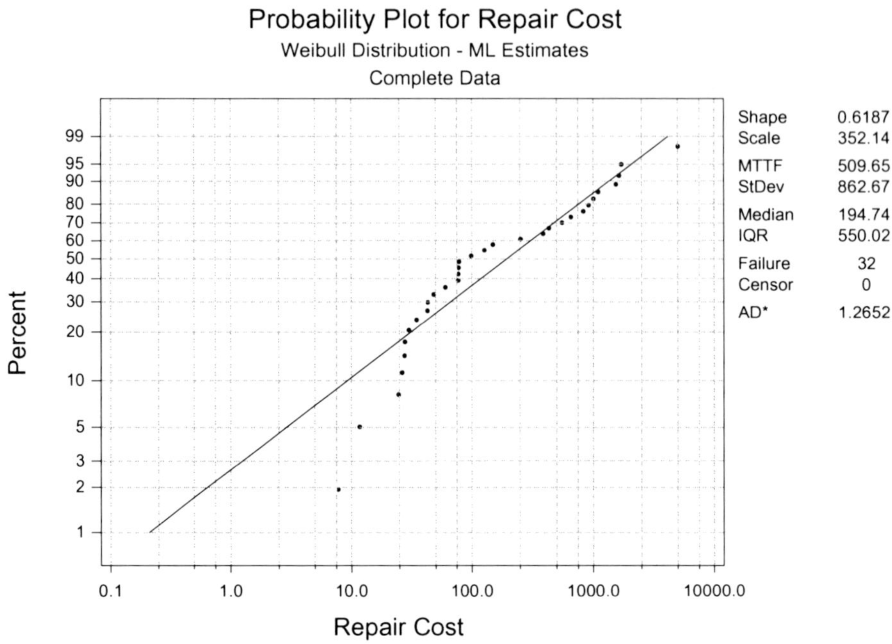


**Fig. 8.11** WPP plot of repair cost

resulting equation relating the scales can be solved. In other cases, numerical methods are used to devise the plotting papers.

The second distribution very frequently used in reliability applications is the exponential distribution. The exponential is the only distribution having constant failure rate. It is, in fact, a special case of the Weibull, resulting when $\beta = 1$. A linear plot is obtained for the exponential when $y = -\log(1 - F(t))$ is plotted against $t$.

Explicit solution of this type can also be found for a few other failure distributions—for example, the extreme value and Erlang distributions. Two other important distributions in reliability and life testing, the lognormal and gamma require numerical methods to obtain the plots. This is also true of the most important distribution in statistical inference, the normal distribution. For more on transformations for plotting, see ([2], Sect. 11.2) and ([9], Chap. 6).

Plotting papers are available for many of the distributions discussed above. In addition, statistical packages typically include many probability plot options. Minitab includes plots for the normal, lognormal, 3-parameter lognormal, gamma, 3-parameter gamma, 2-parameter exponential, smallest extreme value, Weibull, 3-parameter Weibull, largest extreme value, logistic, loglogistic, and 3-parameter loglogistic. (Formulas for most of these can be found in Appendix A.)

*Example 8.11* For purposes of illustration, we look at another choice of distribution for describing the repair cost data for the claims data of Table F.2, the

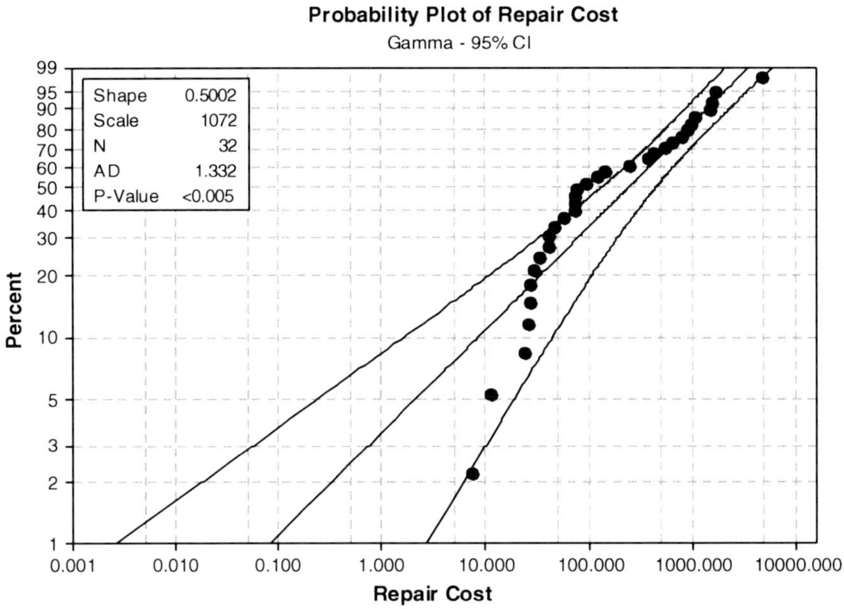


**Fig. 8.12** Gamma probability plot of repair data

gamma distribution. The probability plot is given in Fig. 8.12. The pattern on this chart is also highly nonlinear, indicating an unacceptable choice of model.

## 8.6 Use of Graphical Methods in Data Analysis

Graphs allow the display of patterns, relationships, and distributions of data that are difficult to discern simply by looking at the data. Selection of the right graph depends on the type of available data and the purposes of the analysis it.

Table 8.3 summarizes a number of commonly used graphs (given in Minitab and other statistical packages) and their uses in analysis of data. These tables can be used to select proper graphs for various purposes of analysis.

## 8.7 Preliminary Model Selection

We have noted in the preceding section and elsewhere that one of the essential first steps in analysis of failure data and in reliability analysis generally is selection of an appropriate failure distribution. In some instances, this can be done on the basis of theoretical considerations, engineering judgment, past experience, or other knowledge-based concepts. Often, however, the information available to the

**Table 8.3** Graphs and their uses

| Graphs | Objective of analysis | | | | | |
|---|---|---|---|---|---|---|
| | Assess relationships between pairs of variables | Assess distributions | Compare summaries or individual values of a variable | Assess distributions of counts | Plot a series of data over time | Assess relationships among three variables |
| Individual Value Plot | | | × | | | |
| Interval Plot | | | × | | | |
| Marginal Plot | × | | | | | |
| Matrix Plot | × | | | | | |
| Pie Chart | | | × | × | | |
| Probability Plot | | × | | | | |
| Scatter Plot | × | | | | × | |
| Stem-and-Leaf | | × | | | | |
| Time Series Plot | | | | | × | |
| 3D Scatter Plot | | | | | | × |
| 3D Surface Plot | | | | | | × |
| Area Graph | | | | | × | |
| Bar Chart | | | × | × | | |
| Box Plot | | × | × | | | |
| Contour Plot | | | | | | × |
| Dot Plot | | × | | | | |
| Empirical CDF | | × | | | | |
| Histogram | | × | | | | |

analyst is insufficient, and the choice must necessarily involve data as well. In these cases, statistical procedures for model selection are employed.

A relatively straightforward initial approach to the problem is to utilize the plotting methodology discussed above to select a tentative model. Prior to this analysis, however, it is important to clean the data, if necessary, and to look at the structure of the data. A key feature to be determined regarding data structure is the identification of multiple modes in the claims data. If these are present, it is useful to use Pareto charts to determine the most frequently occurring modes and their frequency of occurrence.

If sufficient data are available for each mode, the modes may be analyzed separately, which may simplify the analysis since tentative choices of failure distributions may be selected for each mode. If the data for separate modes is sparse, a single failure model may be sought, usually leading to a much more complex distribution. This will be discussed in Chap. 13.

The next step in the analysis is tentative model selection. Two approaches to this are:

1. Prepare WPP plots for each mode separately and/or data from combined modes (including all of the data) and examine the plots for recognizable patterns
2. Prepare plots for other distributions (e.g., selected from among those given in the Minitab list above) and examine the plots for approximately linear patterns

In analyzing data, typically both approaches may be used. Each may give insight into the data, and the whole will give a richer base from which to choose candidate failure distributions. In both cases, for warranty claims data, attention must be paid to the fact that the data are almost always incomplete. Proper plotting of the data requires censoring information as well. This may also be a part of the claims data or may be extracted from sales data, production data, or other relevant sources. If this information is not available, any plots of the failure data represent conditional distributions, and must be interpreted accordingly.

Tools for implementation of Approach (1) are given in Appendix C. The method is to examine the WPP plot(s) for patterns. A number of patterns that may occur are listed in Tables C.1. Some specific distributions that may give rise to the patterns are listed in Table C.2. In Approach (2), a number of plots will be generated and we look for linear or near-linear plots, as noted.

In this graphical analysis, it is important to keep in mind that the models selected are to be considered as *tentative*. What we are, in effect, modeling, is the particular set of data we happen to have gotten in the study. In order to accept any model selected by this procedure as the "true" model, validation at some level is necessary.[8] If one distribution stands out as apparently the only acceptable fit to the data, this may be taken as a self-validation. This rarely happens (except perhaps for small samples, which, in any case, do not provide much information), but

[8] As noted by a well-known statistician, Oscar Kempthorne, "No model is correct. But some are useful!".

even when it does, further validation may be needed. There are a number of steps and approaches that may be taken in the validation process. Some of these are:

- For large data sets, divide the sample into subsamples and analyze each separately. A commonly recommended approach is to divide the sample into thirds, using the first subsample to select a tentative model, the second to estimate the model, and the third for validation.
- Obtain an independent second sample and use it for validation.
- Develop a theory that leads to the selected model. If one is convinced that the theory is valid, accept the model.
- Develop a theory that leads to the selected model, then obtain additional data to test the theory.

Many other approaches to validation may be devised. Additional tools and techniques for validation will be discussed in the following two chapters as well as in later chapters of the book. Of particular relevance are estimation of parameters (graphical and analytical), tests of hypotheses, and goodness-of-fit tests.

In Example 8.9, we looked at the fit of the Weibull distribution to a set of data (from Table F.5) on bond strength. Visually, the fit of the Weibull model appears to be reasonably good.[9] Thus we may conclude that a relatively simple model is a good candidate for the distribution of bond strength. We conclude the chapter with a brief example in which this is not the case. Here we use the graphical approach to model selection for a set of data on battery life. This appears to be relatively simple application, but we find that none of the models used fits particularly well.

It should be noted that in both the bond strength and battery examples the analysis is not definitive, nor is any validation possible, since the sample is relatively small and no other data are available.

*Example 8.12* Probability plots for the battery lifetime data of Table F.3 are given in Fig. 8.13. The plots were generated by use of the Minitab "Distribution ID" option. All eleven plots listed as options are shown. All of the plots are generated using the maximum likelihood estimates of the parameters of the distribution. (This procedure is given in Appendix D and will be discussed is some detail in Chap. 9.) The first of the probability plots is the WPP plot. This plot is nearly linear, except for the first five points, indicating that the Weibull distribution itself may be a reasonable choice for the failure distribution. The first five points, in fact, appear to be out of the pattern of the remaining data in most of the plots, and may either be outliers or observations from a distribution different from that of the remaining data (possibly resulting from a different failure mode). None of the other shapes listed in Table C.1 appear to describe the pattern in the WPP plot.

Among the remaining distribution plots, perhaps the best fit (in the sense of most nearly linear) is the three-parameter lognormal. The "third parameter" referred to in the name of the distribution is a location parameter. A positive value

[9] In fact, the "goodness-of-fit" statistic is given as AD* = 0.436, which indicates a relatively good fit. This will be discussed further in Chap. 10.

for this parameter indicates that there is a lag period, before any failures occur, which seems unlikely. The other nearly linear patterns are for the three-parameter loglogistic and normal distributions. These, too, seem unlikely to be the true model. The reader may select other of the plots as providing a good fit, which demonstrates one of the principal difficulties with the purely graphical approach, namely its subjectivity. Note, however, that the value of the AD statistic is very large in all cases, which, as we shall see later, is indicative of a poor fit.

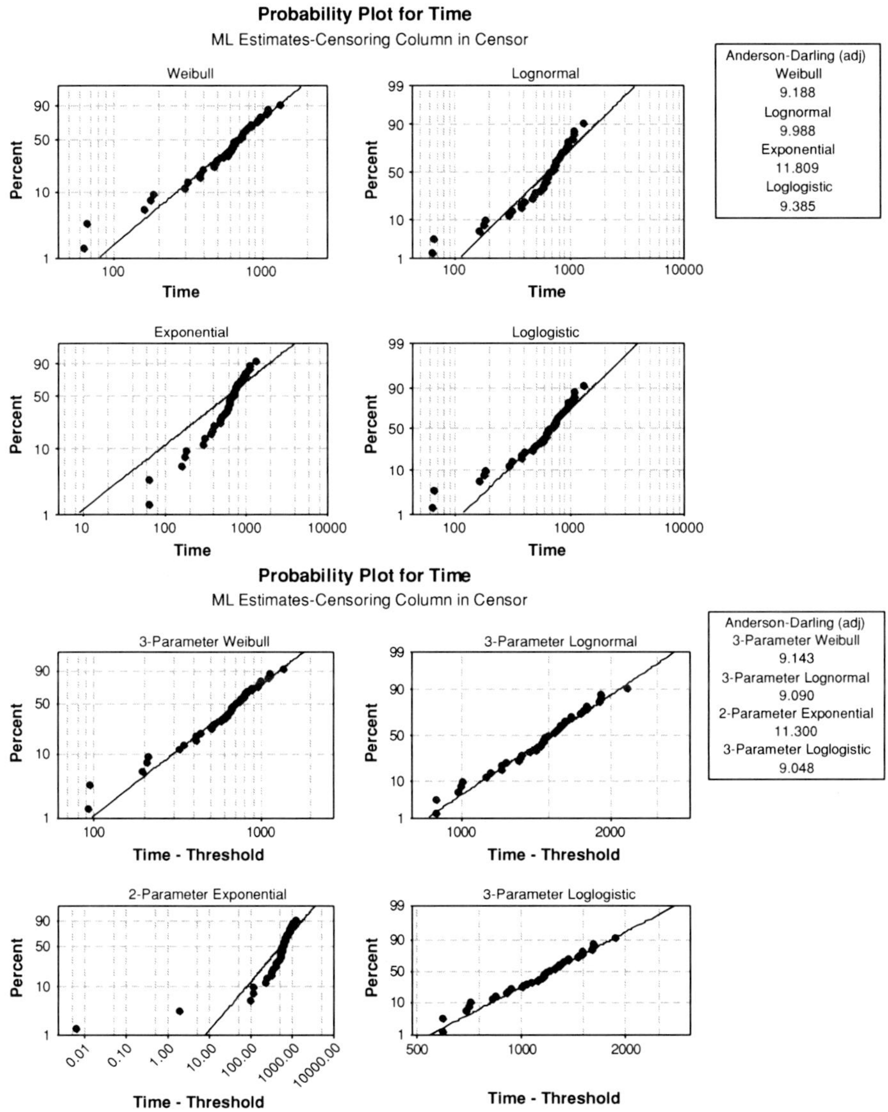


**Fig. 8.13** Minitab ID plots for time to failure of batteries (continued)

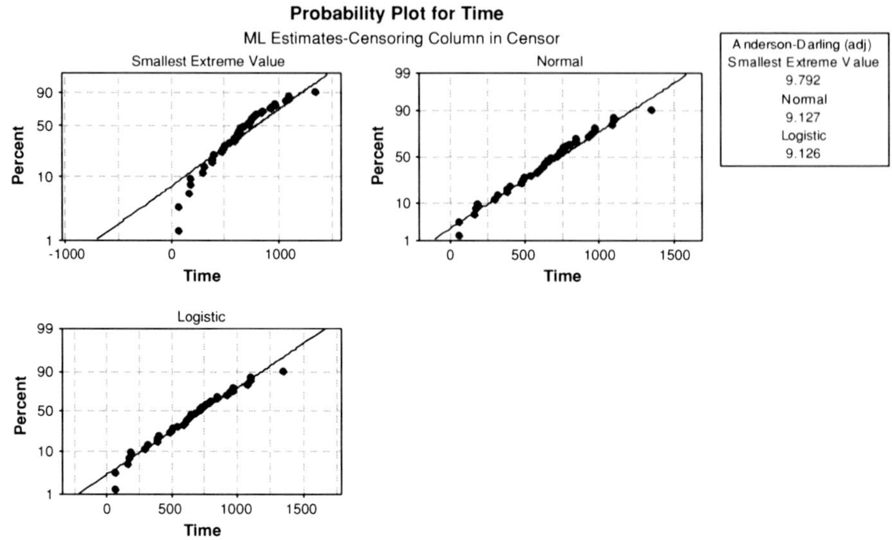


**Fig. 8.13** (continued)

As noted at the outset, we have not reached a definitive conclusion here with regard to the probability distribution for modeling time to failure. More data are needed (and, if possible, more information, e.g., failure modes) in order to make the selection. For purposes of further analysis, one may tentatively selects one of the distributions that fits reasonably well, at least visually, (say the Weibull) or, where possible, use nonparametric methods such as those discussed in Chap. 11 and elsewhere. Another alternative is to further analyze the data under each seemingly reasonable alternative distribution and compare the results. If the results do not differ in any material way, they may be accepted. If they do differ, further study is indicated.

## References

1. Berry M (2004) Survey of text mining: clustering, classification and retrieval. Springer, New York
2. Blischke WR, Murthy DNP (2000) Reliability. Wiley, New York
3. Cios KJ, Pedrycz W, Swiniarski RW, Kurgan LA (2007) Data mining: a knowledge discovery approach. Springer Science, New York
4. Famili A, Shen WM, Weber R, Simoudis E (1997) Data preprocessing and intelligent data analysis. Intell Data Anal 1:3–23
5. Jeske DR, Liu RY (2007) Mining and tracking massive text data: classification, construction of tracking statistics, and inference under misclassification. Technometrics 49:116–128

6. Johnson NL, Kotz S (1970) Continuous univariate distribution–1. Wiley Interscience, New York
7. Kim W, Choi BJ, Hong EK, Kim SO, Lee D (2003) A taxonomy of dirty data. Intell Data Anal 7:81–90
8. Makkonen L (2008) Bringing closure to the plotting position controversy. Commun Statist Theory and Methods 37:460–467
9. Meeker WQ, Escobar LA (1998) Statistical methods for reliability data. Wiley Interscience, New York
10. Moore DS, McCabe GP, Craig B (2007) Introduction to the practice of statistics. W H Freeman, New York
11. Murthy DNP, Xie M, Jiang R (2004) Weibull models. Wiley Interscience, New York
12. Ryan TP (2007) Modern engineering statistics. Wiley, New York
13. Schmid CF (1983) Statistical graphics. Wiley Interscience, New York
14. Tufte ER (1983) The visual display of quantitative information. Graphics Press, Cheshire, CT
15. Tufte ER (1989) Envisioning information. Graphics Press, Cheshire, CT
16. Tufte ER (1997) Visual explanations. Graphics Press, Cheshire, CT
17. Weibull W (1939) A Statistical theory of the strength of material. Igniörs Akademiens Handligar, Stockholm

# Chapter 9
# Basic Statistical Inference

## 9.1 Introduction

Effective warranty management requires proper data collection and data analysis. In Chaps. 4 and 5, we discussed warranty claims data and warranty supplementary data in detail. In Chap. 8, techniques for description and summarization of data were discussed. These provide a preliminary look at the data and its structure. The next step is data analysis to use the information from a sample to draw inferences concerning the population from which the sample was drawn. This requires an understanding of concepts and techniques of *inferential statistics*.

Inferences about populations typically are expressed in terms of population *parameters* (e.g., the shape and scale parameters of the Weibull distribution) or related population characteristics such as the mean, median, and so forth. This approach is called *parametric analysis* and depends crucially on, among other things, specific distributional assumption regarding the source of the data. For some inference problems, there exist alternate techniques that do not involve population parameters or require that the form of the distribution be known. These are variously called distribution-free or *nonparametric* procedures. In this chapter we deal mainly with parametric inference, but will look at a few nonparametric methods as well. We confine our attention to either complete data or data that are right- or interval-censored.[1]

It should be noted that the validity of any of the statistical inference procedures discussed here and in later chapters depends upon having random samples from the population(s) in question. While justified in the case of test and other experimental data and most of the other supplementary data discussed in Chap. 5, this assumption is rarely, if ever, satisfied when dealing with claims data. A consequence of this is that the results obtained in the analysis may be subject to considerable bias (e.g., if the sample is not really "representative" of the population as

[1] A more complete treatment of incomplete data is given in Chaps. 11–15.

W. R. Blischke et al., *Warranty Data Collection and Analysis*,
Springer Series in Reliability Engineering, DOI: 10.1007/978-0-85729-647-4_9,

a whole). The justification for use of the procedures in the case of claims data is to assume that the units actually observed can be considered to be a statistical sample of all the units that will eventually be manufactured and sold (Similar arguments are used to justify the analysis in other nonrandom situations, e.g., nonrandom survey data.). This may or may not be the case, and the data analysis and the user of the results (the decision maker) should be aware of the possibility of bias and use caution in interpreting the results.[2]

Other difficulties of dealing with claims and some related data are discussed in Chaps. 4 and 5. A key problem is that warranty claims do not necessarily represent true failures. Errors occur in both directions—claims made for items that have not, in fact, failed, and claims not made for failed items. This further confounds the problem of sample validity and should be considered in interpreting the results. This is discussed in Sects. 4.4.1, 4.7, and 8.2.3.

Given that the assumption of a proper random sample can be justified at some level, the methods of this chapter (as well as the chapters to follow) are applicable. The particular statistical procedure to be used depends on the objectives of the analysis, on the type of data available, and on the structure of the data, including the experimental design or survey design for test or sample survey data. For additional discussion, see [2], Sect. 3.5.2 and Chap. 10.

In this chapter, we look at a number of topics in statistical inference, including estimation of population parameters, tests of hypotheses, tolerance intervals, and nonparametric statistics. In applying the procedures, it is important to take into consideration the scale of measurement. This is discussed in Sect. 8.2.2. In particular, the parametric methods given below require interval or ratio data, while the nonparametric methods are also appropriate for interval data and, in some cases, for nominal data as well.

The outline of the chapter is as follows: Sect. 9.2 deals with basic notions of estimation, distributions to be considered, and properties of estimators. The maximum likelihood approach to estimation is discussed in Sect. 9.3. Sections 9.4 and 9.5 cover other methods of estimation and confidence intervals, respectively. Hypothesis testing is briefly reviewed in Sect. 9.6. Tolerance intervals, that is, intervals that cover a specified proportion of the population, are discussed in Sect. 9.7. Nonparametric methods, including methods for comparison of populations and rank correlation coefficients, are covered in Sect. 9.8.[3]

[2] In some cases, it is possible to use the data to check for evidence of bias. A better (but usually expensive) check is to randomly sample the entire population of sold items and compare key measures from that sample with those of the claims data.

[3] The chapter is intended to be a review of these topics, and, as such, treats most topics briefly, giving few details and no mathematical derivations. Details can be found in introductory statistical texts such as [13, 17]. Mathematical results and some techniques for their derivation are given in Appendix D; Details may be found in [6, 18].

## 9.2 Estimation

### 9.2.1 Basic Notions

In analyzing data, certain assumptions are made about the random mechanism that gave rise to the observed values. In most applications on interest here, these assumptions are expressed in terms of the underlying probability distribution used to model the population sampled, and its associated parameters. Here and in the next few sections, we discuss methods for *estimation* of the parameters, that is, using the data to assign numerical values to any parameters that are important for inference purposes and whose values are unknown.

We assume that the data consist of a random sample of size $n$ from the population in question and that we wish to estimate a parameter $\theta$ (which may be a vector). As before, the sample is denoted $Y_1, Y_2, \ldots, Y_n$ when considered random variables and $y_1, y_2, \ldots, y_n$ when considered a set of observed values of the random variables. An estimation procedure is a formula or equation (explicit or implicit) for estimating $\theta$. The result is denoted $\hat{\theta}$ (or $\theta^*$). To express $\hat{\theta}$ as a function of the sample, we write $\hat{\theta}(Y_1, Y_2, \ldots, Y_n)$ or $\hat{\theta}(y_1, y_2, \ldots, y_n)$, depending on whether we are considering $\hat{\theta}$ to be a random variable or a numerical value. To distinguish these, the first is called an *estimator* (or *point estimator*) and the second an *estimate* (or *point estimate*).

This notion will be extended to other, similar expressions. For example, the sample mean is often taken as an estimate or estimator of the population mean $\mu$. We write $\hat{\mu} = \bar{Y}$ for the estimator and $\hat{\mu} = \bar{y}$ to denote the estimate of the mean. Note that $\mu$ may be a parameter of the distribution (i.e., may appear explicitly in the formula for the CDF, as in the case of the normal distribution) or may be expressed as a function of the parameters of the distribution.

It is important to recognize that there is uncertainty in this process—different samples from the same population will lead to different numerical results for $\hat{\theta}$. This must be taken into account in selecting an appropriate methodology and in interpreting the results. One approach to this problem is *confidence interval estimation.* Confidence intervals are intervals of numbers determined in such a way that the probability that the interval contains the true value of $\theta$ is a specified quantity $\gamma$. Typically $\gamma$ is taken to be 0.95 or 0.99; the results are called 95 and 99% confidence intervals, respectively.

### 9.2.2 Distributions

In the following sections we will look at estimation of the parameters of a number of distributions of importance in reliability and warranty analysis. These are discussed in some detail in Chap. 3. Estimation results will be given for the following discrete distributions (listed here with reference to equations for CDF or density/probability distribution in Appendix A and parameters as indicated):

- Binomial distribution—Equation (A.12); parameter $\theta = (n, p)$
- Poisson distribution—Equation (A.18); parameter $\theta = \lambda$
- Hypergeometric distribution—Equation (A.16); parameter $\theta = (n, N, D)$

Continuous distributions to be considered include:

- Exponential distribution—Equation (A.21); parameter $\theta = \lambda$
- Gamma distribution—Equation (A.24); parameter $\theta = (\alpha, \beta)$
- Weibull distribution—Equation (A.30); parameter $\theta = (\alpha, \beta)$
- Normal distribution—Equation (A.26); parameter $\theta = (\mu, \sigma^2)$
- Lognormal distribution—Equation (A.44); parameter $\theta = (\mu, \sigma^2)$
- Inverse Gaussian distribution—Equation (A.43); parameter $\theta = (\mu, \lambda)$
- Three-parameter Weibull distribution—Equation (A.46); parameter $\theta = (\alpha, \beta, \tau)$
- Mixture of two exponential distributions—Equation (A.52) with $K = 2$ and CDF's given by Equation (A.21); parameter $\theta = (\lambda_1, \lambda_2.p_1)$

Estimation procedures for many of the remaining distributions listed in Appendix A can be found in the references cited.

### 9.2.3 Properties of Estimators

In any inference problem, there may be many seemingly logical ways to estimate the unknown parameters. For example, it was noted above that the sample mean may be used to estimate the population mean $\mu$. There are often a number of equally logical choices. In the case of the normal distribution, $\mu$ is also the median and mode of the distribution, so the corresponding sample quantities might be used to estimate $\mu$ as well. For the Weibull, $\mu$ is a function of $\alpha$ and $\beta$ (See (A.31)), and hence can be estimated as the corresponding function of any estimators $\hat{\alpha}$ and $\hat{\beta}$. To determine which procedures should be used in analyzing data, we look at the properties of the estimators. The objective is to select estimators that are optimal in some sense.

There are many ways of evaluating estimators. We look briefly at several of these. Estimators may be evaluated in terms of properties that hold for any sample size (sometimes called "small-sample properties") or in terms of asymptotic properties (as the sample size $n \to \infty$). The latter are particularly important in situations where the derivation of small sample properties is difficult or impossible due to the complexity of the mathematics.[4] Some key properties of optimal estimators are:

[4] Simulation studies can often be used to investigate alternative methods in this situation, but with obvious limitations.

- Unbiasedness—An estimator $\hat{\theta}$ is an *unbiased* estimator of $\theta$ if $E(\hat{\theta}) = \theta$ for all $n$ and all values of the parameters of the distribution, i.e., the average value of $\hat{\theta}$, averaging over all possible samples, is the true value of the parameter being estimated.
- Sufficiency—An estimator $\hat{\theta}$ is *sufficient* if it utilizes all of the sample information about the parameter $\theta$. The idea here is that it is sufficient to know the value of $\hat{\theta}$; if this is known, the individual sample values are no longer needed to estimate $\theta$ and no information is lost.
- Consistency—An estimator is *consistent* if for any $\varepsilon > 0$ and all $\theta$, $P(|\hat{\theta} - \theta| > \varepsilon) \to 0$ as $n \to \infty$. The interpretation of this is that the probability that the value of the estimator will differ from the true parameter value by any arbitrarily small amount converges to zero as the sample size becomes increasingly large. Consistency is often considered a minimal requirement of any estimator (otherwise no amount of data will be adequate).
- Efficiency—An unbiased estimator is *efficient* if no other unbiased estimator has smaller variance. The importance of this is that an *inefficient* estimator will require a larger sample size in order that its variance be equal to that of an efficient estimator, i.e., to achieve the same precision. To illustrate this, consider estimation of $\mu$ in a normal distribution. Since $\mu$ is both the mean and the median of the distribution, both the sample mean and sample median would appear to be logical choices. The median, however, is inefficient; the variance of the sample mean is 63% that of the sample median. Significantly more data is required to obtain the same precision using the sample median.
- Asymptotic properties—Consistency is by definition a large sample property of an estimator. Other large sample properties include asymptotic unbiasedness (bias $\to 0$ as $n \to \infty$) and asymptotic efficiency (the variance of the estimator approaches the Cramér-Rao bound (See Appendix D) as $n \to \infty$).

These measures of optimality are all reasonable, logical criteria. They cannot, however, be applied dogmatically or independently of one another, or illogical outcomes may result. For example, the first observation $Y_1$ is an unbiased estimator of $\mu$, but not one that would ever be used since it is not efficient or even consistent.

In the sections that follow, we discuss several estimation procedures, some of which are optimal, at least in some sense or in some applications (There is no procedure that satisfies all criteria of optimality for all distributions.) The most commonly applied estimation procedure is the method of maximum likelihood, which results in "best" estimators, at least in the asymptotic sense, for many distributions. It should be noted, however, that the use of asymptotic results does require large samples for validity. How large depends on the distribution and may depend on other factors as well, e.g., the parameters themselves. In many cases, $n$'s of 30–50 will do, and usually 100 or a few hundred will suffice, but there are pathological examples in which very large samples are required. For analysis of claims data, where large samples are the usual situation, asymptotic results are usually appropriate. The major difficulty in using asymptotic results where they are inappropriately applied is that probability statements, for example those associated with confidence interval estimation and tests of hypotheses, will be inaccurate and can differ substantially from the true probabilities.

## 9.3 Point Estimation: Method of Maximum Likelihood

### 9.3.1 Concept and Method

The *method of maximum likelihood* (ML) is a broadly applicable estimation procedure. In applying the method, we formulate the *likelihood function*, which is basically the joint distribution of the data, expressed as a function of the parameters of the distribution and the data $Y_1$, $Y_2$, ..., $Y_n$. The *maximum likelihood estimator* (MLE) of the parameters is the set of values that maximize this function. The idea is that the likelihood function in a sense represents the joint probability of the data and we are choosing parameter values that maximize the probability of obtaining the sample actually observed. (This is often called the *likelihood principle*.)

Maximization of the likelihood function is straightforward if the parameter space is unbounded and the distribution is differentiable with respect to $\theta$ (or the components of $\theta$ in the multi-parameter case). The likelihood equations are obtained by equating derivatives to zero and solving (numerically, if necessary) to obtain estimates of the parameters. In other cases, numerical methods of maximization are used.

The method is applicable to both complete and incomplete data and to grouped data as well. The type of data determines the form of the likelihood function and must be taken into consideration in its formulation. Given the appropriate formulation, the complexity of the data has no effect conceptually, but may significantly increase the complexity of the maximization problem. In the next sections, we look at ML estimators for the parameters of most of the distributions listed in Sect. 9.2.2 for complete data, and for a few distributions with incomplete data.

### 9.3.2 ML Estimators for Complete Data

ML estimators are given below for all of the distributions listed in Sect. 9.2.2 except the three-parameter Weibull and the mixed exponential, both of which involve some computational difficulties. A "*" will be used to indicate that the estimator or estimate is an MLE (e.g., $\theta^*$, $\alpha^*$, etc.). In most cases, variances or asymptotic variances of the estimators are listed as well. These may be estimated by substituting estimates for the parameters appearing in the formulas. Their use in inference will be discussed later in the chapter.

In the following list, the expressions are written in terms of the random variables involved—$Y_1, Y_2, \ldots, Y_n, X$, etc.—indicating that these are estimators (and hence also random variables). The estimates are numerical values calculated by substitution of the observed data values into these expressions. Except where indicated, we assume a random sample of size $n$. The likelihood function in this case is written in terms of the CDF, density, or probability distribution in question, as appropriate. In many instances, explicit expressions for the estimators are not possible, and a set of equations must be solved, usually by numerical methods, to calculate the estimates for a given set of data. The trial and error approach is often adequate. For some distributions, a better alternative is to use Minitab or some other program package to obtain the estimates. In Minitab, some graphical procedures (e.g., for probability plots) provide estimates, with ML and Least Squares (discussed later in the chapter) estimation as options.

The estimators for complete data are:

*Binomial Distribution* Let $X$ be the number of outcomes in $n$ trials having a specified characteristic (e.g., the number of defectives in a randomly selected sample of $n$ items). The MLE of $p$ is

$$p* = X/n \tag{9.1}$$

with variance

$$V(p*) = p(1-p)/n \tag{9.2}$$

*Poisson Distribution* The MLE of $\lambda$ is

$$\lambda* = \bar{Y} \tag{9.3}$$

with variance

$$V(\lambda*) = \lambda/n \tag{9.4}$$

*Hypergeometric Distribution* This distribution is encountered when sampling without replacement from a population of size $N$ in which $D$ items possess a specified characteristic (e.g., being defective). We assume a sample of size $n$ in which $X$ items are found to have the characteristic and wish to estimate the number $D$ (or proportion $D/N$) in the population. The MLE of $D$ is

$$D* = \left[\frac{X(N+1)}{n}\right] \tag{9.5}$$

where $[a]$ indicates the greatest integer less than or equal to $a$. The variance of the estimator is

$$V(D*) = \frac{(N+1)^2(N-n)}{n(N-1)}\left(\frac{D}{N}\right)\left(1-\frac{D}{N}\right) \tag{9.6}$$

*Exponential distribution* The MLE of $\lambda$ is

$$\lambda* = 1/\bar{Y} \tag{9.7}$$

The asymptotic variance of $\lambda*$ is

$$V_a(\lambda*) = \lambda^2/n \tag{9.8}$$

*Gamma distribution* The MLE of the shape parameter $\alpha$ is obtained as the solution of the equation

$$\frac{1}{n}\sum_{i=1}^{n}\log(Y_i) - \log(\bar{Y}) = \psi(\alpha*) - \log(\alpha*) \tag{9.9}$$

where $\psi(x)$ is the digamma function defined as

$$\psi(x) = \frac{\partial}{\partial x}\log[\Gamma(x)] \tag{9.10}$$

and $\Gamma(x)$ is the gamma function. The gamma and digamma functions are tabulated extensively [1]. Equation (9.9) must be solved numerically for $\alpha*$. The MLE of the scale parameter $\beta$ is then obtained as

$$\beta* = \bar{Y}/\alpha* \tag{9.11}$$

The asymptotic variances of the estimators are [9]

$$V_a(\alpha*) = \alpha/n[\alpha\psi'(\alpha) - 1] \tag{9.12}$$

and

$$V_a(\beta*) = \beta^2\psi'(\alpha)/n[\alpha\psi'(\alpha) - 1] \tag{9.13}$$

where $\psi'(x)$, the derivative of (9.10), is the trigamma function [1].

*Weibull distribution* The MLE of the shape parameter $\beta$ is obtained by solution of the equation

$$\frac{\sum_{i=1}^{n} Y_i^{\beta*}\log(Y_i)}{\sum_{i=1}^{n} Y_i^{\beta*}} - \frac{1}{\beta*} - \frac{1}{n}\sum_{i=1}^{n}\log(Y_i) = 0 \tag{9.14}$$

The MLE of the scale parameter $\alpha$ is obtained as

$$\alpha* = \left(\frac{1}{n}\sum_{i=1}^{n} Y_i^{\beta*}\right)^{1/\beta*} \tag{9.15}$$

The asymptotic variances of the MLE estimators are

$$V_a(\alpha*) = 1.1087\alpha^2/(n\beta^2) \tag{9.16}$$

and

$$V_a(\beta*) = 0.6079\beta^2/n \tag{9.17}$$

*Normal distribution* The MLE's are

$$\mu* = \bar{Y} \tag{9.18}$$

and

$$\sigma* = \left(\frac{1}{n}\sum_{i=1}^{n}(Y_i - \bar{Y})^2\right)^{1/2} \tag{9.19}$$

The variance of the MLE of $\mu$ is

$$V(\mu*) = \sigma^2/n \tag{9.20}$$

The variance of $\sigma^*$ is not needed since the distribution of $\sigma^*$ is a Chi-Square distribution, which is used in inference problems concerning $\sigma$ or $\sigma^2$.

*Lognormal distribution* Inferences about the lognormal parameters are dealt with by transforming to $X_i = \log(Y_i)$, which is normally distributed. Thus the data are transformed to the log scale, the parameters are estimated using (9.18) and (9.19), and results transformed back to the original scale.

*Inverse Gaussian distribution* The MLE's are

$$\mu* = \bar{Y} \tag{9.21}$$

and

$$\lambda* = n\left[\sum_{i=1}^{n}\left(\frac{1}{Y_i} - \frac{1}{\bar{Y}}\right)\right]^{-1} \tag{9.22}$$

The variance of $\mu^*$ is

$$V(\mu*) = \frac{\mu^3}{n\lambda} \tag{9.23}$$

The distribution of $\lambda^*$ is proportional to a Chi-Square distribution, which is used for inference problems regarding $\lambda$.

*Example 9.1* Failure data for air conditioner systems on three aircraft are given in Table F.10. We will look at the data for Aircraft #7909. This consists of $n = 29$ observations of time between failures of the AC unit. We assume that repair is good-as-new. It follows that the data may be considered to be realizations of independent and identically random variables. For purposes of illustration, both the Weibull and lognormal distributions will be considered in analyzing the data.

Estimates of the parameters of the Weibull distribution may be obtained by solving (9.14) for $\beta^*$ and substituting the result into (9.15) to calculate $\alpha^*$. Instead,

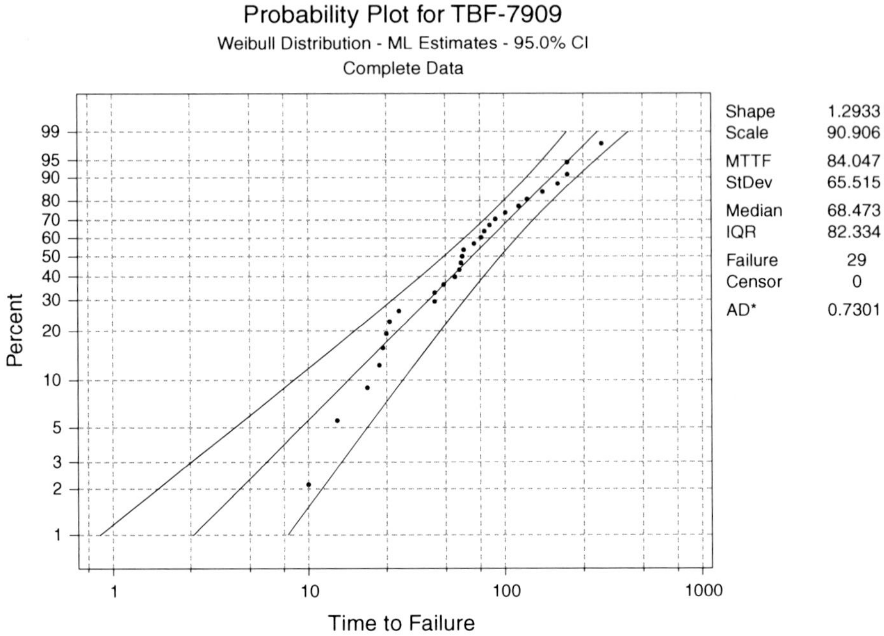


**Fig. 9.1** WPP plot, AC failure data, Aircraft 7909

we use the WPP plot from Minitab, which provides the output given in Fig. 9.1. From the output, the estimate of the shape parameter is found to be $\beta^* = 1.29$; the estimate of the scale parameter is $\alpha^* = 90.9$ (It is easily seen that these values satisfy (9.14) and (9.15).) Note that this value of the shape parameter corresponds to an increasing failure rate.

The MTTF for the Weibull distribution may be estimated by substitution of these results in the formula that expresses this in terms of the parameters of the distribution. This is given by (A.32) in Appendix A. The estimated MTTF is $\alpha^*\Gamma(1 + 1/\beta^*) = 84.05$,[5] a result also given in the Minitab output. The standard deviation, median and IQR given in the output may be calculated similarly (formulas not given here). The AD* statistic will be discussed in Chap. 10 under goodness-of-fit tests.

Note, incidentally, that the data in the WPP plot fall roughly along a straight line, with a few points at the lower end appearing either to be possible outliers or to indicate that the Weibull distribution is not an acceptable model. This will be discussed in more detail later.

This approach is also used to obtain estimates for the parameters of the lognormal distribution. The Minitab output is given in Fig. 9.2. From this, we obtain

[5] The gamma function may be evaluated by interpolation in the tables found in [1], or by use of the gamma command in Minitab Calculator.

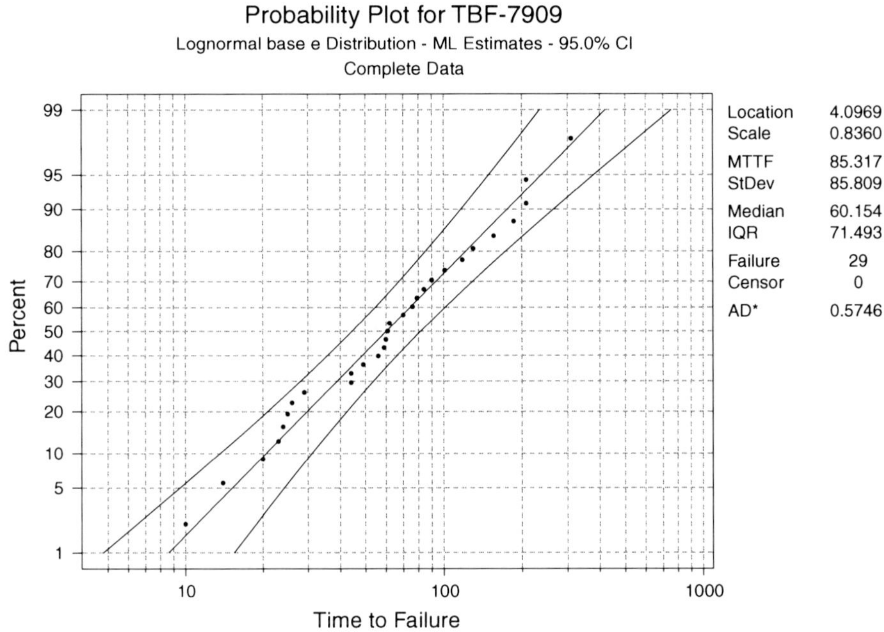


**Fig. 9.2** Lognormal plot, AC failure data, Aircraft 7909

the parameter estimates as $\mu^* = 4.0969$ and $\sigma^* = 0.8360$, which are the sample mean and standard deviation (with divisor $n$ rather than $n - 1$) of the data transformed to the log scale. The relationship between the parameters and the MTTF for this distribution, given in (A.45), is used to estimate this quantity. The result is $e^{\mu*+\sigma*^2/2} = e^{4.0969+.8360^2/2} = 85.32$, as shown in the Minitab output. The remaining quantities shown in the output are calculated similarly.

Note that the data appear to follow a roughly linear pattern in the lognormal plot as well (in fact, the linear fit appears to be a bit better here). How one might choose between the competing models is an important aspect of the analysis. This will be discussed in Chap. 10 as well as in later chapters.

### *9.3.3 ML Estimators for Incomplete Data*

In analysis of incomplete data, the structure of the data (i.e., the nature of the incompleteness) must be taken into account for a proper analysis. In this section, we look at ML estimation for several forms of censored data. A detailed discussion of these and a number of other types of censoring is given in Chap. 5. We also look briefly at ML estimation for grouped data. Additional results on estimation for incomplete data will be given in Part D of the book. The theory on which the results are based is given in Appendix D.

The likelihood function is given for various types of censoring in Appendix D. There are a large number of combinations of censoring schemes and the distributions listed above. We consider a few of these to illustrate the methodology. Estimators for others may be obtained by application of the ML method.[6] The results below include both singly and multiply censored data. Again, it is not always possible to express the estimators in closed form, and numerical methods may be required to solve the estimation equations. Minitab and other program packages include programs to handle censored data, and these may also provide ML estimates. The following are some ML estimators for censored data:

*Exponential Distribution, singly Type I censored data* Suppose $r$ items fail during the period of observation, with the remaining $n-r$ items censored at $T$. In the set of ordered observations $y_{(1)},\ldots,y_{(n)}$, the first $r$ are failure times and all of $y_{(r+1)},\ldots,y_{(n)}$ are all equal to $T$. The MLE of $\lambda$ is

$$\lambda* = \frac{r}{\sum_{i=1}^{n} y_i} = \frac{r}{\sum_{i=1}^{r} y_{(i)} + (n-r)T} \tag{9.24}$$

The MTTF is estimated as $\mu* = 1/\lambda*$. The asymptotic variances of $\lambda*$ and $\mu*$ are

$$V_a(\lambda*) = \frac{\lambda^2}{n(1-e^{-\lambda T})} \quad \text{and} \quad V_a(\mu*) = \frac{1}{n\lambda^2(1-e^{-\lambda T})} \tag{9.25}$$

*Exponential Distribution, Multiply Type I Censored Data* Suppose that $y_{(1)},\ldots,y_{(r)}$ are failure times as before and that $T_{r+1},\ldots,T_n$ are censoring times for the remaining items. The MLE of $\lambda$ is

$$\lambda* = \frac{r}{\sum_{i=1}^{r} y_{(i)} + \sum_{i=r+1}^{n} T_i} \tag{9.26}$$

The asymptotic variance of the MLE is

$$V_a(\lambda*) = \frac{\lambda^2}{n - \sum_{i=1}^{n} e^{-\lambda T_i}} \tag{9.27}$$

*Exponential Distribution, Type II Censored Data* Here $y_{(1)},\ldots,y_{(r)}$ are observed failure times, with $r$ fixed, and it is known only that the remaining lifetimes exceed $y_{(r)}$. The MLE of $\lambda$ is

$$\lambda* = \frac{r}{\sum_{i=1}^{r} y_{(i)} + (n-r)y_{(r)}} \tag{9.28}$$

[6] For additional results, see [8, 11, 15].

with variance

$$V(\lambda*) = \frac{(r-1)^2\lambda^2}{r^2(r-2)} \tag{9.29}$$

*Weibull Distribution, Type I Censoring* The data are as defined above. The MLE of $\beta$ is obtained as the solution of

$$\frac{\sum_{i=1}^{n} Y_i^{\beta*}\log(Y_i)}{\sum_{i=1}^{n} Y_i^{\beta*}} - \frac{1}{\beta*} - \frac{1}{r}\sum_{i=1}^{r}\log(Y_{(i)}) = 0 \tag{9.30}$$

The MLE of $\alpha$ is

$$\alpha* = \left(\frac{1}{r}\sum_{i=1}^{n} y_{(i)}^{\beta*}\right)^{1/\beta*} \tag{9.31}$$

*Weibull Distribution, Type II Censoring* The ML estimators are obtained from (9.14) and (9.15) with $n$ replaced by $r$.

*Example 9.2* Data on repair cost and kilometer driven for 32 vehicles that experienced engine problems under warranty out of a total of 329 vehicles are given in Table F.2. Some preliminary analyses of the data, with emphasis on repair cost, were included in Chap. 8. Under certain conditions, the repair cost data can be considered to be complete data. The usage data, however, are not. In fact, they are right-censored at 40,000 km, the length of the warranty, and no usage information is given for the remaining 297 vehicles. If we assume that all of these were driven 40 k km without engine failure, Minitab may be used to plot the resulting right-censored data and ML estimates obtained from the additional Minitab output. The result for the Weibull distribution is given in Fig. 9.3. For the actual observed lifetimes, the Weibull fit appears marginally acceptable. The estimated shape parameter is 1.0512, which suggests that the exponential distribution may be appropriate. A probability plot of this distribution is given in Fig. 9.4.[7] It is difficult to tell because of the way Minitab plots this distribution, but the fit to the exponential also appears to be reasonable. From the numerical output given with the Minitab plots, the estimated values of the MTTF are 341,350 km and 389,570 km, respectively. The results appear to be reasonable for vehicles of this type.

For the exponential, the ML estimate of $\lambda$ is $\lambda* = 1/389.57 = 0.002567$. Equation (9.29) may be used to estimate the variance of the estimator by substitution of the $\lambda*$ for $\lambda$. The result is $\hat{V}(\lambda*) = 31^2(.002567)^2/(32^2 \times 30) = 0.206 \times 10^{-6}$, giving an estimated standard deviation of the estimate as 0.000454.

[7] Note: In the Minitab parameterization of the exponential, the scale parameter is the inverse of that in the formula used in this book.

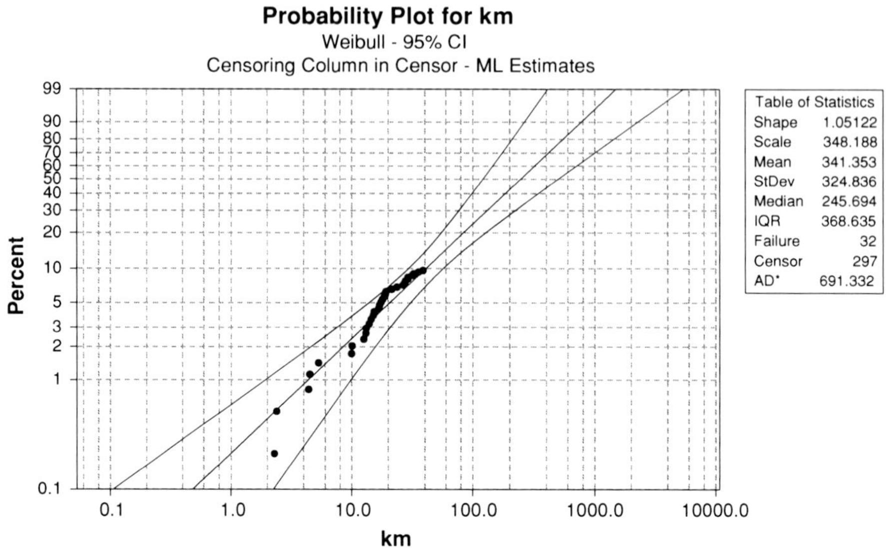


**Fig. 9.3** WPP plot for km data

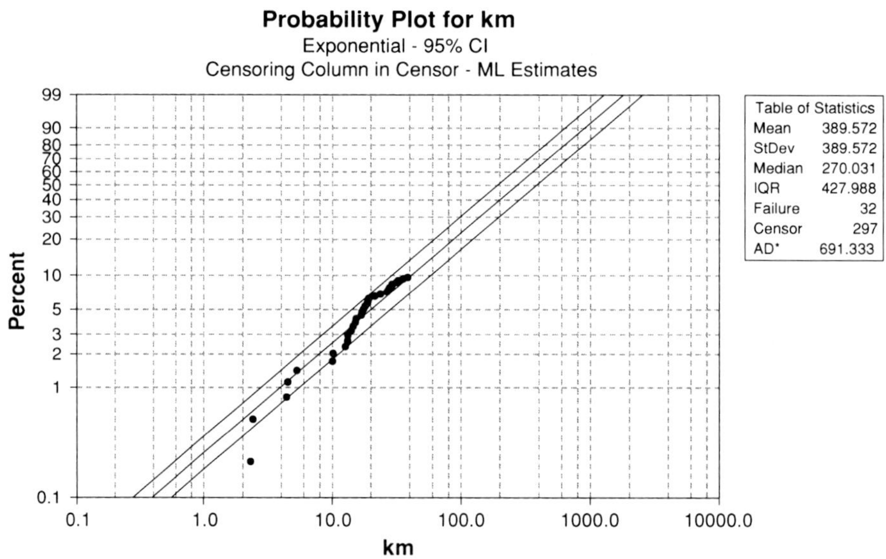


**Fig. 9.4** Exponential probability plot for km data

### 9.3.4 ML Estimation for Grouped Data

It is not uncommon to encounter grouped data in warranty applications. For grouped data, exact measurements are not recorded; it is only known that an item measurement falls into one of $k$ intervals defined by endpoints $y'_0, y'_1, \ldots, y'_k$. Flight hours for jet engines that had not failed are given in Table F.7 as grouped data, with groups 100 h in width. The final data set then becomes simply counts of numbers of observations falling into each group and the joint distribution of the sample (the likelihood function) is expressed as a multinomial distribution (A.20), as shown in Appendix D.

Once the likelihood function is obtained, derivation of ML estimators is conceptually straightforward, although operationally difficult because of the complexity of the ML equations for most distributions. Sources for solutions for specific distributions are given in [4].[8]

### 9.3.5 Properties of ML Estimators

These are important for at least two reasons: (1) In succeeding sections and in the remaining chapters of this and the next part of the book, we will discuss other aspects of statistical inference—confidence intervals, test of hypotheses, and a number of advanced techniques. These involve probability statements based on distributions of the appropriate statistics, often the MLE. (2) In analyzing data, it is desirable to employ methods that are optimal in some sense—"best" estimators, most powerful tests, and so forth. Knowledge of the properties of estimators is necessary for their evaluation in this context.

Under certain regularity conditions, maximum likelihood estimators are optimal in several senses, at least for large samples. The regularity conditions involve derivatives of the likelihood function (the first and second derivatives of the log of the likelihood function must exist), the information matrix (which must be a continuous function of the parameters), and constraints on nuisance parameters (parameters other than those being estimated).[9] For most of the distributions discussed above and elsewhere in the text, the regularity conditions are met. The key exceptions are distributions for which the range depends on a location or "shift" parameter, notably the two-parameter exponential and three-parameter Weibull distributions.

---

[8] For further information, see [15]. MLE's for grouped data from the three-parameter Weibull distribution (See (A.35)) are given by [5].

[9] For a thorough discussion, see [19].

Except for "non-regular" cases such as those cited, the MLE's are

- Consistent—the probability that the estimator differs from the true value of the parameter by any given amount tends to zero as $n \to \infty$
- Asymptotically unbiased—the bias of the estimator tends to zero as $n \to \infty$
- Asymptotically efficient—the covariance matrix of the estimator tends to the Cramér-Rao bound as $n \to \infty$ (See Appendix D.)
- Asymptotically normally distributed—the distribution of the estimator (properly standardized) tends to the normal distribution as $n \to \infty$

Note that all of these are asymptotic results (often called "large-sample results"). In practice, they are used in analysis of moderate to large data sets. The problem, of course, is what is meant by "large?" Common rules of thumb are "$n > 30$", "$n > 50$", and so forth. In truth, there is no such rule for which pathological examples for which the rule does not apply cannot be found. This is often not a problem in analysis of claims data, since large (sometimes very large) data sets are not uncommon.

The usefulness of these results is that procedures for obtaining approximate confidence intervals, tests of hypotheses and other statistical analyses can be based on the normal distribution. This is particularly convenient when exact results based on the actual distribution of the statistic in question (which should be used for inference wherever possible) cannot be obtained mathematically.

## 9.4 Other Methods of Estimation

There are many other methods of estimation as well. We look at a few of these that are also useful in certain cases in analysis of warranty claims data.

### *9.4.1 Method of Moments*

The method of moments is based on expressing the moments of a distribution (also called "population moments") in terms of the parameters of the distribution, equating sample moments to population moments, and solving the resulting equations for the unknown parameters. The population moments $\mu'_i$ (also called "moments about zero") are defined as

$$\mu'_i = E(Y^i), \quad i = 1, 2, \ldots \tag{9.32}$$

The corresponding sample moments $m'_i$ are

$$m'_i = \frac{1}{n}\sum_{j=1}^{n} y_j^i, \quad i = 1, 2, \ldots \tag{9.33}$$

If more convenient, the population and sample "central moments" or "moments about the mean" may be used instead (for $i > 1$). These are denoted $\mu_i$ and $m_i$, respectively, and are obtained by replacing $Y$ by $(Y - \mu)$ in (9.32) and $y_j$ by $(y_j - \bar{y})$ in (9.33), for $i = 2, 3, \ldots$, with $\mu_1 = \mu'_1$ and $m_1 = m'_1 = \bar{y}$ (Note that $\mu_1$ and $\mu_2$ are the mean and variance of the distribution.). Any other moments (e.g., factorial) or combinations of moments may be used as well. The conventional approach to estimating $k$ parameters is to use the first $k$ moments and apply the method as indicated.

As a simple illustration, we consider estimation of the two parameters, $\alpha$ and $\beta$, of the gamma distribution, given in (A.24). From (A.25), the mean and variance are $\mu = \alpha\beta$ and $\mu_2 = \alpha\beta^2$. From this, it follows easily that the moment estimates (indicated by a caret) are

$$\hat{\beta} = m_2/m_1 = s^2/\bar{y} \tag{9.34}$$

and

$$\hat{\alpha} = m_1^2/m_2 = \bar{y}^2/s^2 \tag{9.35}$$

The method of moments is a straightforward approach to estimation that has some intuitive appeal (sample moments ought to be "close" to the true population moments, especially for large samples). Furthermore, unless the moment equations are quite complex, the method is computationally relatively simple. As a result, one of the uses of moment estimates is as initial guesses in the numerical solution of more complex sets of equations, such as the ML equations (This option that is available in some Minitab routines.) The method of moments is also useful in and of itself in the case of large or very large samples, where estimation efficiency is not of concern.

Moment estimators are generally consistent and asymptotically unbiased. They are usually not efficient, however, except, of course when they are equivalent to MLE's or some other efficient estimators. Under fairly general conditions, they are also asymptotically normally distributed, with asymptotic variances that can be obtained by standard procedures.[10]

We conclude the section by looking at moment estimators for mixtures to two exponential distributions. This is a distribution that is important in some reliability applications since it may be used to represent time to failure of items having two different failure rates.

*Mixture of Two Exponential Distributions* The general form for the density function of a mixture of $K$ distributions is given in (A.53). Here $K = 2$, and the components of the mixture are the exponential distributions given in (A.22), with respective parameters $\lambda_1$ and $\lambda_2$, say, and mixing proportions $p_1 = p$ and $p_2 = 1 - p$. It is convenient to express the moment equations in terms of $p$ and $\mu_i = 1/\lambda_i$. The equations are

---

[10] For further information and additional applications, see [2, 20].

$$\bar{y} = \hat{p}\hat{\mu}_1 + (1 - \hat{p})\hat{\mu}_2 \tag{9.36}$$

$$\frac{1}{2}m_2' = \hat{p}\hat{\mu}_1^2 + (1 - \hat{p})\hat{\mu}_2^2 \tag{9.37}$$

$$\frac{1}{6}m_3' = \hat{p}\hat{\mu}_1^3 + (1 - \hat{p})\hat{\mu}_2^3 \tag{9.38}$$

Write $z_1$, $z_2$, $z_3$ for the left-hand sides of (9.36)–(9.38), respectively, and let $a = (z_3 - z_1 z_2)/(z_2 - z_1^2)$. The moment estimates are given by

$$\hat{\mu}_1, \hat{\mu}_2 = \frac{a \pm \sqrt{a^2 - 4(az_1 - z_2)}}{2} \tag{9.39}$$

where $\mu_1$ is taken to be the smaller value, and

$$\hat{p} = \frac{\hat{\mu}_2 - \bar{y}}{\hat{\mu}_2 - \hat{\mu}_1} \tag{9.40}$$

### 9.4.2 Least Squares Estimation

The method of least squares was developed in the context of linear models, regression analysis (see Chap. 10), and inference regarding the mean of a normal distribution. In the context of estimating $\mu$, the likelihood function is maximized by minimizing the expression $S = \sum_{i=1}^{n} (y_i - \mu)^2$. It is easily seen that the value of $\mu$ that minimizes $S$ is $\bar{y}$, i.e., the least squares estimate of $\mu$ is $\hat{\mu} = \bar{y}$, which, as we have seen, is the MLE.

More generally, $\mu$ may be a function (linear or nonlinear) of other variables and a number of parameters. The Least Squares Estimators (LSE) of the parameters of the model are obtained by minimization of the corresponding sum of squares. Many Minitab routines include an option for least squares estimation.

### 9.4.3 Bayes Estimation

The basic aim of *Bayesian statistical analysis* is to provide a framework and methodology for incorporating information from various sources into the analysis of a set of data. This information, called *prior* information, may be data from previous studies, subjective inputs, and so forth, and the idea is to use the current data to update the analysis and conclusions.

The key to the analysis is *Bayes Theorem* which expresses relationships between conditional probabilities (See Appendix A). In its simplest form, involving two possible events $A$ and $B$, the theorem states that

$$P(B|A) = \frac{P(A|B)P(B)}{P(A)} \tag{9.41}$$

In the context of warranty claims data, sources of information are very diverse, and may include design factors (e.g., reliability goals), part and item test data, subjective reliability evaluations, quality assurance data, information on similar products, estimates based on warranty claims data collected previously, and so forth. All of this information is incorporated into the analysis by formulating a *prior distribution*, interpreted as a probability distribution of the unknown parameters that we wish to estimate. Bayes' Theorem (in a more extensive form than given in (9.41)) is then applied to obtain the *posterior distribution*, and this is used to obtain a Bayes estimate of the parameter.

Bayesian analysis has important applications in reliability, particularly when dealing with highly reliable items where testing to demonstrate item reliability would be prohibitively expensive, but has not found widespread use in analysis of claims data. For brief discussions of Bayesian analysis, see [2], Chap. 8, and [20], Chap. 11. For a thorough treatment of Bayesian reliability analysis, see [12].

### *9.4.4 Graphical Methods*

In Sect. 8.5, we discussed graphical methods for data description. Probability plots were presented for various distributions for both complete and incomplete data. These plots can also be used to obtain parameter estimates. We note at the outset that, as in the case of moment estimators, the graphical estimators are often not optimal and should only be used in a preliminary analysis of the data or as preliminary guesses in numerical equation solution.

There are a number of ways of obtaining parameter estimates from probability plots. The following are a few possibilities[11]:

- To estimate $k$ parameters, select $k$ fractiles (See Sect. 8.3.2) or sets of fractiles of the assumed or known distribution, express these as functions of the parameters, equate sample fractiles (read from the probability plot) to population fractiles, and solve the resulting equations (This is a variation on the method of moments.). As an illustration, in the case of a normal distribution, the 0.5-fractile (the median) may be used as an estimate of $\mu$, and, for example, (0.84-fractile – 0.16-fractile)/2 as an estimate of $\sigma$. (Another estimate of $\sigma$ is (0.975-fractile – 0.025-fractile)/3.92.
- Use ruler and other instruments to deduce parameter values based on the known shape of the probability plot.

[11] There are many other approaches, depending on the data and the assumed distribution. For additional discussion and examples of applications, see [2, 14].

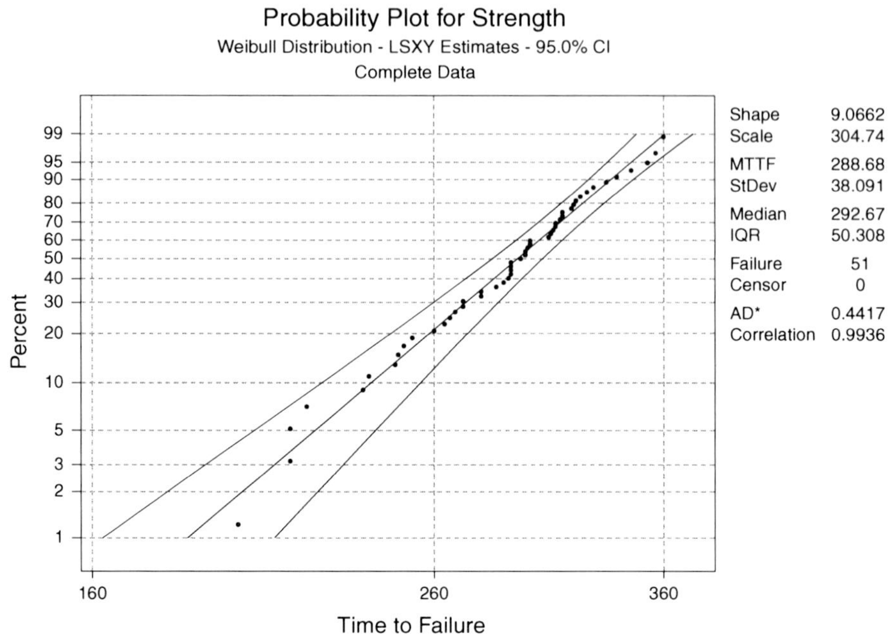


**Fig. 9.5** WPP plot for bond strength, least squares estimates

- Plot the data on a linearized scale and use the relationship between the line and the distribution parameters to calculate estimates, e.g., by means of least squares regression analysis (See Sect. 10.4).
- Easiest approach: Employ the options in Minitab and other programs to obtain the estimates (Minitab provides options for calculating least squares estimates and MLE's).

We illustrate the use of graphical methods by considering the Weibull distribution. In this case, the most straightforward approach is to use the linearizing transformation discussed in Sect. 8.5 and Appendix C, fit a line to the data on the linear scale, and use the relationship between the parameters of the distribution and the slope and intercept of the line to estimate the parameters. The transformation to obtain the linear model for the Weibull is given in (8.8). This is easily estimated by least squares. It follows immediately that the slope of the estimated line is the LSE $\hat{\beta}$ of the shape parameter $\beta$. $\alpha$ is estimated by $\hat{\alpha} = e^{-a/\hat{\beta}}$, where $a$ is the intercept of the fitted line.

*Example 9.3* We consider the bond strength data of Table F.5. The WPP plot for the data is given in Fig. 8.10, which also includes the MLE's of $\alpha$ and $\beta$ as part of the Minitab output. The results are $\alpha^* = 304.9$ and $\beta^* = 9.09$. For comparison, the Minitab results using least squares estimation are given in Fig. 9.5. The estimates are $\hat{\alpha} = 9.066$ and $\hat{\beta} = 304.74$, in very close agreement with the MLE's.

An advantage of the graphical approach is its simplicity and (usually) computational ease. In addition, probability plots are easily gotten for both complete and censored data. The difficulties with the graphical approach are:

- The process is often subjective, with no formal mathematical structure
- For purely graphical procedures, it is not possible to determine the properties of the estimators.
- Except for estimators obtained by analytical methods, it is not possible to determine standard errors of the estimators and hence confidence interval estimation (see the next section) is not possible.

## 9.5 Confidence Interval Estimation

### 9.5.1 Basic Concept

In the previous sections, we have been concerned with point estimation—using a single point (which may be multi-dimensional) to estimate an unknown parameter. This in itself does not convey an important additional piece of information, the uncertainty in the estimator or estimate. For MLE's (and some other estimators), we have also given variances (asymptotic or exact) of the estimators. The square root of this variance, called the *standard error* of the estimator, is a measure of this uncertainty. A *confidence interval* (or *confidence interval estimator*) takes the uncertainty in account by providing as an estimator an interval of numbers (or, more generally, a region), along with a measure of the "confidence" or likelihood that the true value of the parameter is contained within the interval or region.

For a single unknown parameter $\theta$ and a sample of size $n$, a confidence interval is defined as follows:

*Definition* A confidence interval with confidence coefficient $\gamma$ for a parameter $\theta$ is an interval defined by two limits, $L_1(Y_1, Y_2, \ldots, Y_n)$ and $L_2(Y_1, Y_2, \ldots, Y_n)$, having the property that

$$P\{L_1(Y_1, Y_2, \ldots, Y_n) < \theta < L_2(Y_1, Y_2, \ldots, Y_n)\} = \gamma \tag{9.42}$$

If $\theta$ is a $k$-dimensional parameter, one can form confidence intervals for each parameter separately, or can form a joint confidence region that simultaneously contains all parameter values with confidence $\gamma$. In this section, we restrict attention to confidence intervals for individual parameters (or, in the multi-parameter case, confidence intervals for individual parameters).

The confidence interval given in (9.42) is two-sided confidence interval. In some applications, a one-side confidence interval is appropriate. This may be either a lower confidence bound, formed by omitting $L_2$ in (9.42), or an upper confidence bound, formed by omitting $L_1$. The interpretation is that we are $100\gamma\%$

confident that $\theta$ is at least (at most) as large as the bound. This is particularly important in certain reliability applications, where we wish to be confident that the reliability exceeds a specified value (e.g., a design goal).

Confidence intervals are often based on best point estimators such as the MLE. To construct such intervals, knowledge of the distribution of the MLE is required. Other approaches to the construction of confidence intervals have been developed.[12]

In cases where the exact distribution of the MLE (or other appropriate statistic) cannot be determined, approximate confidence intervals may be used. The most common approximation is based on the asymptotic distribution of the estimator (properly normalized). In the case of the MLE and other estimators of this type,[13] this is a normal distribution with variance obtained as specified in Appendix D. Note that the asymptotic confidence intervals are approximate in the sense that the confidence coefficient $\gamma$ will not be achieved exactly. In most cases, the approximation will be quite good for modest to large sample sizes, as discussed previously.

### 9.5.2 Confidence Intervals for the Parameters of Selected Distributions

In this section, we list confidence intervals for the parameters of some of the distributions previously discussed. We assume complete data, with samples of size $n$. Exact intervals will be given where possible. In other cases, large sample results based on the normal approximation will be given. The calculations require fractiles of various distributions. These are tabulated in Appendix E.

Below we give confidence intervals for the mean of a normal distribution with known and unknown variance, the variance of a normal distribution, the proportion $p$ in a binomial distribution, and the parameters of the Poisson, exponential, gamma, Weibull, and lognormal distributions. Two-sided confidence intervals are given in the list below. One-sided intervals can be obtained as indicated in the previous section. These require using either the $\gamma$-fractile or the $(1 - \gamma)$-fractile instead of those indicated in the expressions given below.

The results are:

*Normal distribution, confidence interval for $\mu$, $\sigma$ known* The confidence interval for $\mu$ is

[12] For further information and a number of examples, see [19], Chap. 20, [20], Chap. 8, and [2], Chaps. 5 and 8.

[13] The estimators in this class are known as Best Asymptotically Normal (BAN) estimators. See [19] for details.

$$\bar{y} \pm z_{1-(1-\gamma)/2} \frac{\sigma}{\sqrt{n}} \tag{9.43}$$

where $z_p$ is the $p$-fractile of the standard normal distribution, given in Table E.1. This result is key to the computation of asymptotic confidence intervals. These are obtained by replacing $\bar{y}$ in (9.43) by the MLE of the parameter in question and replacing $\sigma/\sqrt{n}$ by an estimate of the standard error of the MLE, gotten by substituting parameter estimates into the expression for the standard error.

*Normal distribution, confidence interval for $\mu$, $\sigma$ unknown* The result is

$$\bar{y} \pm t_{n-1,1-(1-\gamma)/2} \frac{s}{\sqrt{n}} \tag{9.44}$$

where $t_{n-1,p}$ is the $p$-fractile of the Student-t distribution with $(n - 1)$ degrees of freedom (*df*), given in Table E.2, and $s$ is the square root of the sample variance with divisor $(n - 1)$.

*Normal distribution, confidence interval for $\sigma$* The interval is

$$\left( \frac{(n-1)s^2}{\chi^2_{n-1,1-(1-\gamma)/2}}, \frac{(n-1)s^2}{\chi^2_{n-1,(1-\gamma)/2}} \right) \tag{9.45}$$

where $s^2$ is the sample variance with divisor $(n - 1)$, and $\chi^2_{n-1,p}$ is the $p$-fractile of the Chi-Square distribution with $(n-1)$ *df*, given in Table E.3. To obtain a confidence interval for the standard deviation $\sigma$, calculate the interval of (9.45) and take square roots of the resulting values.

*Binomial distribution, confidence interval for $p$* If $n$ is large enough so that $np*$ and $n(1 - p*)$ are both at least 5, the normal approximation may be used. The result is

$$p* \pm z_{1-(1-\gamma)/2} \sqrt{\frac{p*(1-p*)}{n}} \tag{9.46}$$

where $p*$ is given in (9.1). If the conditions on $n$ are not satisfied, exact results based on the binomial must be used [3].

*Poisson distribution, confidence interval for $\lambda$* The point estimator of $\lambda$, given in (9.3), is the sample mean. The confidence interval for $\lambda$ is

$$\left( \frac{1}{2n} \chi^2_{2n\bar{y},(1-\gamma)/2}, \frac{1}{2n} \chi^2_{2n\bar{y}+2,1-(1-\gamma)/2} \right) \tag{9.47}$$

For large $n$, asymptotic confidence intervals based on the normal distribution may be used instead of (9.47). See [2], p. 151.

*Exponential distribution, confidence interval for $\lambda$, complete data* The MLE $\lambda*$ is given in (9.7). The confidence interval for $\lambda$ is

$$\left( \frac{\chi^2_{2n,(1-\gamma)/2}}{2n\bar{y}}, \frac{\chi^2_{2n,1-(1-\gamma)/2}}{2n\bar{y}} \right) \tag{9.48}$$

To obtain a confidence interval for the MTTF $\mu = 1/\lambda$, use the reciprocals of the values in (9.48).

*Exponential distribution, confidence interval for $\lambda$, censored data* For Type I censored data, use (9.48) with $n\bar{y}$ replaced by the denominator of (9.26) and with $2r$ *df* instead of $2n$, where $r$ is the number of uncensored observations. For Type II censored data, the normal approximation may be used; see (9.28) and (9.29) for the estimator and its asymptotic variance.

*Weibull distribution, confidence intervals for $\alpha$ and $\beta$* The MLE's are calculated as the solution of (9.14) for $\beta^*$ and as given in (9.15) for $\alpha^*$. Asymptotic variances are estimated by substituting $\alpha^*$ for $\alpha$ and $\beta^*$ for $\beta$ in (9.16) and (9.17). The asymptotic confidence intervals are then obtained by use of the normal approximation as indicated below. The results are[14]

$$\alpha^* \pm 1.0529 z_{1-(1-\gamma)/2} \frac{\alpha^*}{\beta^* \sqrt{n}} \tag{9.49}$$

and

$$\beta^* \pm z_{1-(1-\gamma)/2} \beta^* \sqrt{0.6079/n} \tag{9.50}$$

*Gamma distribution, confidence intervals for $\alpha$ and $\beta$* The MLE's are obtained from (9.9) and (9.11), with variances given in (9.12) and (9.13). Asymptotic confidence intervals are obtained by use of the normal approximation as illustrated above for the Weibull distribution.

*Lognormal distribution, confidence intervals for $\mu$ and $\sigma^2$* As in the case of point estimation, the confidence intervals are obtained by transforming the data to the log scale, using the appropriate results for the normal distribution to obtain the intervals, and then transforming back to the original scale of measurement.

Note: Additional results, including confidence intervals based on incomplete data, are listed in [2], Chap. 5 (complete data) and Chap. 8 (censored data), and in [11]. Exact results are available in a few cases (e.g., $\lambda$ in the Inverse Gaussian). In other cases, asymptotic results, based on the MLE's and calculated as indicated previously, may be used if the sample sizes are sufficiently large. These may be used for incomplete data as well, employing the results of Sect. 9.3.3.

*Example 9.4* An initial analysis of data on bond strength of an adhesive was given in Example 8.9. A Minitab WPP plot showed a near-linear pattern and provided MLE's for the Weibull parameters as $\alpha = 304.9$ and $\beta^* = 9.095$. The results were based on a sample of size $n = 51$. Asymptotic confidence intervals for $\alpha$ and $\beta$, based on the normal approximation, are obtained from (9.49) and (9.50).

[14] For other approaches and many additional results, see [11], Chap. 4.

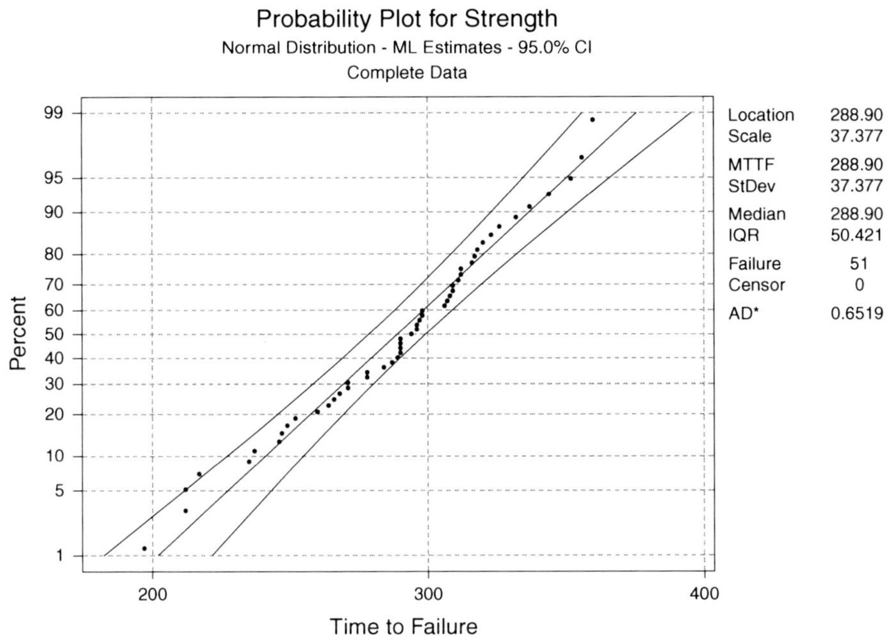


**Fig. 9.6** Normal probability plot for bond strength

Estimates of the variances for the approximating normal distributions, obtained from (9.16) and (9.17), are given by $\hat{V}_a(\alpha*) = 1.1087(304.9)^2/(51(9.095)^2) = 24.432$ and $\hat{V}_a(\beta*) = 0.6079(9.095)^2/51 = 0.9860$. The estimated standard errors are 4.943 and 0.9930, respectively. To calculate 95% confidence intervals, we use $z_{.975} = 1.960$. The resulting intervals are $304.9 \pm 1.96\ (4.943)$ or (295.2, 314.6) for $\alpha$, and $9.095 \pm 1.96\ (0.993)$ or (7.15, 11.04) for $\beta$. We are 95% confident that the shape parameter $\beta$ is between 7.15 and 11.04, indicating a strongly increasing failure rate.

*Normal Distribution* For values of the shape parameter $\beta$ greater than 3, which the confidence interval for $\beta$ strongly indicates is the case here, the shape of the Weibull distribution is quite close to that of a normal distribution. A normal probability plot of the data is given in Fig. 9.6. Again the data fall in a near-linear pattern, indicating that the normal distribution may also be a reasonable choice.

For purposes of illustration, we calculate confidence intervals for the normal parameters. These require fractiles of the Student-t and Chi-Square distributions. The procedures are based on the use of the sample variance with divisor $(n-1)$ rather than $n$. The corresponding standard deviation in this example is $s = 37.75$; the variance is 1425.1. The sample mean is $\bar{y} = 288.90$. A 95% confidence interval for $\mu$ requires the fractile $t_{50,0.975} = 2.009$, obtained from Table E.2. From (9.44), the interval is $288.9 \pm 2.009(37.75)(51)^{-1/2} = 288.9 \pm 10.6$, or (278.2, 299.5). Note that the results for $\mu$ are in close agreement with those for the Weibull distribution, for which the estimated MTTF is given in Fig. 9.5 as 288.68.

We calculate a 90% confidence interval for $\sigma^2$. The required values from Table E.3 are $\chi^2_{50,0.05} = 34.76$ and $\chi^2_{50,0.95} = 67.50$. From (9.45), the confidence interval is (50(1425.1)/67.50, 50(1425.1)/34.76) = (1055.6, 2049.9). From this, a 90% confidence interval for $\sigma$ is obtained as (32.49, 45.28).

*One-Sided Interval* Since bond strength is a measure of reliability, it may be desirable to calculate a lower one-sided confidence interval instead of the two-sided interval given above. For this we need the fractile $t_{50,\ 0.05} = -t_{50,0.95} = -1.676$. The lower confidence bound is $288.9-1.676(37.75)(51)^{-1/2} = 280.0$. We are 95% confident that the bond strength is at least 280 pounds.

## 9.6 Hypothesis Testing

Hypothesis testing is an important area of statistical inference wherein various hypotheses (usually two) are formulated concerning a population, and one of these is selected on the basis of sample information. The hypotheses are usually expressed in terms of relevant population parameters, but may be in terms of other population characteristics as well (e.g., a population fractile or a function of the parameters such as a coefficient of variation). The statistical procedures employed in hypothesis testing require that a random sample be drawn from the population of interest. As noted, this is, strictly speaking, ordinarily not the case when dealing with claims data, and certain assumptions are necessary to justify use of the procedures.

Here we briefly review basic concepts of hypothesis testing, look at the relationship between this and confidence intervals, and list a few commonly used tests of importance in reliability and analysis of warranty data.

### *9.6.1 Basic Notions*

In the classical approach to hypothesis testing, two hypotheses, the *null hypothesis* $H_0$, and the *alternate hypothesis* $H_1$ are formulated. By definition, $H_0$ is the hypothesis that is tested; $H_1$ is the hypothesis that is accepted if $H_0$ is rejected. In essence, $H_0$ is tested by determining statistically whether or not the data support the hypothesis. If the data are found to be unlikely to occur under $H_0$, the hypothesis is rejected and $H_1$ is accepted. The definition of "unlikely" requires specification of the probability of occurrence of an event and, in parametric analysis, of a probability model underlying the data. If the data do not lead to rejection of $H_0$, the statistical conclusion is that we fail to reject $H_0$ rather than that we accept it. In some cases (e.g., when immediate action is required), the analyst or manager may proceed as though $H_0$ were, in fact, true; in others, failing to reject may mean that no decision will be made until further data are available.

The following is a common set of hypotheses concerning a population mean:

$$\mathrm{H}_0 : \mu = \mu_0 \quad \mathrm{H}_1 : \mu \neq \mu_0$$

where $\mu_0$ is a specified numerical value. Here $\mathrm{H}_0$ is a simple hypothesis; it consists of a single number. $\mathrm{H}_1$ is a *composite hypothesis*, consisting of more than one value. $\mathrm{H}_1$ is a *two-sided* alternative; the corresponding test will be a two-tailed test.

In some situations, *one-sided* alternatives are desired. In reliability, for example, one may wish to determine whether or not the MTTF is at least as great as some specified value (e.g., a target value). In this case, the appropriate hypotheses are

$$\mathrm{H}_0 : \mu \leq \mu_0 \quad \mathrm{H}_1 : \mu > \mu_0$$

The corresponding test in this case will be a one-tailed test (Here equality is made part of $\mathrm{H}_0$ because, formally, that is what is tested.) A typical example of hypotheses in which the inequalities are in the opposite direction would be a test of whether or not the proportion of defectives in a population is less than a specified value.

Note that in this illustration, the engineer or analyst will ordinarily wish to conclude that the alternate hypothesis is true, since this will indicate that a design objective has been met. The basic principle here, true in hypothesis testing in general, is that *the burden of proof is put on $H_1$*.

Because of the way in which the null and alternate hypotheses are structured, one or the other must be true. Correspondingly, there are two possible errors that may occur:

- Type I Error: Reject $\mathrm{H}_0$ when it is true
- Type II Error: Fail to reject $\mathrm{H}_0$ when it is false

The probability of a Type I error, called the *level of significance* of the test and denoted $\alpha$, is under the control of the data analyst. Typical values of $\alpha = P\{\text{Reject } \mathrm{H}_0 | \mathrm{H}_0 \text{ true}\}$ used in data analysis are 0.10, 0.05, and 0.01[15] (called "testing at the 10% level," etc.). The probability of a Type II error, denoted $\beta$, depends on $\alpha$, the underlying probability distribution, and the distance between the hypothesized value and the true value. $(1 - \beta)$ is called the *power* of the test. "Best" tests are those with highest power (i.e., highest probability of rejecting $\mathrm{H}_0$ when it is false).

In performing tests, appropriate test statistics are employed. These are typically functions of "best" estimators (because these often lead to most powerful tests). Procedures for testing hypotheses about parameters of several important distributions are given below. Many are based on the MLE's for the parameters in

---

[15] Note: most statistical program packages do not require the specification of a value for $\alpha$ as an input. Instead, the analysis is performed and the output includes a "$p$-value," which is the probability of obtaining the observed value of the test statistic, given that $H_0$ is true. If $p \leq \alpha$, $H_0$ is rejected; if $p > \alpha$, it is not.

question. In cases where the distribution of the MLE is not known, tests based on the asymptotic distribution of an estimator are used.

### 9.6.2 Relationship between Hypothesis Testing and Confidence Interval Estimation

Hypothesis testing is related to confidence interval estimation as follows: a confidence interval with confidence coefficient $\gamma$ is the set of all null-hypothesized values of the parameter that would not be rejected when testing at level of significance $\alpha = 1 - \gamma$. As a result, $H_0$ may be tested by calculating the corresponding confidence interval and noting whether or not it contains the null hypothesized value.

In implementation of this notion, it is necessary to identify the proper confidence interval corresponding to the test situation. For a two-tailed test with level of significance $\alpha$, a two-sided confidence interval with confidence coefficient $\gamma = 1 - \alpha$ is employed. For an upper-tail test, a *lower* one-sided confidence bound is used. For a lower-tail test, an *upper* one-sided confidence bound is used.

### 9.6.3 Tests of Hypotheses for Parameters of Selected Distributions

Test procedures may be described in two ways:

1. Specification of an appropriate test statistic and defining a range of values of the test statistic that lead to rejection of $H_0$ (the *rejection region*) and a range of values that do not lead to its rejection (the *acceptance region*)
2. Specification of an appropriate confidence interval to be used in testing as indicated in the previous section

Below we list procedures for several commonly occurring inference situations.

*Testing hypotheses about the mean of a normal distribution, $\sigma$ known* The test statistic is

$$z = \frac{\bar{y} - \mu_0}{\sigma/\sqrt{n}} \tag{9.51}$$

If $\mu_0$ is the true mean of the population and the assumption of normality is valid, the random variable corresponding to $z$ follows a standard normal distribution ($\mu = 0$ and $\sigma = 1$). Percentiles $z_p$ of this distribution are given in Table E.1. Rejection regions are as follows:

- $H_0$: $\mu = \mu_0$ versus $H_1$: $\mu \neq \mu_0$—Reject at significance level $\alpha$ if $|z| > z_{1-\alpha/2}$
- $H_0$: $\mu < \mu_0$ versus $H_1$: $\mu > \mu_0$—Reject at significance level $\alpha$ if $|z| > z_{1-\alpha}$
- $H_0$: $\mu > \mu_0$ versus $H_1$: $\mu < \mu_0$—Reject at significance level $\alpha$ if $|z| < z_{\alpha}$

Note: A version of the statistic of (9.51) is used for asymptotic tests regarding a parameter $\theta$. For these, $\bar{y}$ is replaced by the MLE (or any other asymptotically normal estimator) of $\theta$, $\mu_0$ by the null hypothesized value of $\theta$, and $\sigma/\sqrt{n}$ be the estimated asymptotic standard error of the estimator of $\theta$.

*Testing hypotheses about the mean of a normal distribution, $\sigma$ unknown* The possible sets of hypotheses are as indicated above. The test statistic is

$$t = \frac{\bar{y} - \mu_0}{s/\sqrt{n}} \tag{9.52}$$

Under $H_0$ and normality, the statistic has a Student-t distribution with $(n - 1)$ *df*. The rejection regions are as indicated for the z-test, except that fractiles of the normal distribution are replaced by the corresponding fractiles of the Student-t distribution with $(n - 1)$ *df*.[16]

*Testing hypotheses about the variance of a normal distribution, unknown $\mu$* The hypotheses may be one- or two-sided, analogous to those for the mean given above. Note that in many applications (e.g., in quality assurance and reliability) the alternate $H_1$: $\sigma^2 < \sigma_0^2$ is the appropriate choice, since the objective is to demonstrate that the variability does not exceed a required or target value.

The test statistic is a Chi-Square statistic given by

$$\chi^2 = \frac{(n - 1)s^2}{\sigma^2} \tag{9.53}$$

Under $H_0$ and normality, this statistic is distributed as Chi-Square with $(n - 1)$ *df* (see (A.58)). The acceptance regions are analogous to those given above for $\mu$, except that percentiles of the Chi-Square distribution, given in Table E.3, are used.

To test a hypothesis about the standard deviation, translate to a hypothesis about the variance. For samples from non-normal distributions, a test based on asymptotic normality may be used and large samples are required.

*Tests hypotheses about p, binomial distribution* As in confidence interval estimation, if $n$ is large enough so that $n\hat{p}$ and $n(1 - \hat{p})$ are both at least 5, asymptotic normality of the MLE may be used. The test statistic is

$$z = \frac{\hat{p} - p_0}{\sqrt{p_0(1 - p_0)/n}} \tag{9.54}$$

The rejection regions are the same as those indicated for the $z$-tests of hypotheses about a population mean. For small values of $n$, the exact binomial is used to determine the probability of occurrence of the observed proportion under $H_0$.

[16] *Note*: The version of the z-statistic used for asymptotic tests may be replaced by the corresponding modification of the t-test for small sample sizes, say $n < 100$. In either case, the test is approximate in the sense that the level of significance is only approximately achieved. How good the approximation may be depends, among other things, on the true distribution of the test statistic and the sample size.

*Testing hypotheses about λ, exponential distribution, complete or censored data* For complete or type I censored data, exact confidence intervals for $\lambda$ (and hence for $\mu = 1/\lambda$) can be obtained. Calculate as appropriate confidence interval for $\lambda$ as indicated in Sect. 9.5.2, and use the procedure of Sect. 9.6.2 to test the corresponding hypothesis.

For Type II censored data, use the asymptotic normality of the estimator and the results of (9.28) and (9.29) to obtain the appropriate test. For details and alternative procedures, see [11], Sect. 3.2 and [8], Sect. 9.3.

*Testing hypotheses about the parameters of the Weibull distribution* For testing hypotheses about $\alpha$ and $\beta$, if large samples of complete data are available, the most straightforward approach is to use asymptotic normality of the MLE's and the asymptotic variances given in (9.16) and (9.17). For incomplete data, see [11], Chap. 4.

*Testing hypotheses about the parameters of the lognormal distribution* Transform to the log scale and perform the analysis on the results using the procedures for the normal distribution.

Other distributions To test hypotheses about the parameters of other distributions discussed in Sects. 9.3–9.5, calculate confidence intervals (exact or asymptotic) and use the approach of Sect. 9.6.2.[17]

*Example 9.5* Suppose that bond strength data in Table F.5 are a sample from a normal distribution, that the mean bond strength of bonds in items in storage is nominally $\mu = 300$ pounds, and that the analyst wishes to determine whether or not there is evidence that the mean is, in fact, different from 300. The appropriate hypotheses are $H_0$: $\mu = 300$ versus $H_1$: $\mu \neq 300$. Test at $\alpha = 0.05$. We assume that $\sigma$ is not known. The sample mean and variance are $\bar{y} = 288.9$ and $s^2 = 1425.1$. The value of the test statistic is $t = (288.9 - 300)/\sqrt{1425.1/51} = 2.100$. The appropriate tabulated value is $t_{50,0.025} = 2.009$. Since the calculated value exceeds this, the null hypothesis is rejected at the 5% level and we conclude that there is evidence that $\mu \neq 300$.

Note that the test could have been done by referring to the two-sided 95% confidence interval. From Example 9.4, the result was (278.2, 299.5). Since this does not include the null hypothesized value of 300, $H_0$ is rejected.

### *9.6.4 Comparing the Means of Two Populations*

An important inference problem in many applications is the comparison of two population means. Examples are average output of a standard and a new production process, comparison of the average cost of repairs in two regions, and so

[17] Many additional results may be found in [6, 8, 11, 15, 19, 20], and numerous other texts on theoretical and applied statistics and related field.

forth. The methodology of the previous several sections may be applied here as well. We consider two of the more important cases, the normal and exponential distributions. Results for other situations may be found in the references cited in the previous section.

We assume that independent random samples are available from two populations and that we wish to compare population means. The sample sizes are $n_1$ and $n_2$, the sample means are $\bar{y}_1$ and $\bar{y}_2$, and the sample variances are $s_1^2$ and $s_2^2$. The null and alternate hypotheses are

$$\mathrm{H}_0 : \mu_1 = \mu_2 \text{ vs. } \mathrm{H}_1 : \mu_1 \neq \mu_2$$

or either of the one-sided alternatives, as appropriate.

*Comparison of the means of two normal distributions, unknown variances,* $\sigma_1^2 = \sigma_2^2$ Denote the common variance $\sigma^2$. The estimate of $\sigma^2$ used in the analysis is the pooled estimate

$$s_p^2 = \frac{(n_1 - 1)s_1^2 + (n_2 - 1)s_2^2}{n_1 + n_2 - 2} \tag{9.55}$$

The test statistic is

$$t = \frac{\bar{y}_1 - \bar{y}_2}{s_p\sqrt{\frac{1}{n_1} + \frac{1}{n_2}}} \tag{9.56}$$

Under $\mathrm{H}_0$, this has a Student-t distribution with $(n_1 + n_2 - 2)$ *df*, and the test is carried out as indicated for the $t$-test in the previous section.

*Other problems involving comparisons of normal means* Related problems and appropriate procedures are:

- Comparison of two means, known variances—Replace the denominator of (9.56) by $\sqrt{(\sigma_1^2/n_1) + (\sigma_2^2/n_2)}$, where the $\sigma^2$'s are the variances of the respective populations. The result is a $z$-test of the form given in the previous section.
- Comparison of two means, unequal population variances, large samples—use (9.56) with denominator replaced by $\sqrt{(s_1^2/n_1) + (s_2^2/n_2)}$. For large $n$'s, this is approximately a $z$-statistic and Table E.1 is used for the test.
- Comparison of two means, unequal population variances, small samples—calculate the $t$-statistic as for large samples (using separate variances). This has a distribution that may be approximated by a Student-t with *df* given by the nearest integer to

$$df = \frac{\left[(s_1^2/n_1) + (s_2^2/n_2)\right]^2}{\frac{(s_1^2/n_1)^2}{n_1} + \frac{(s_2^2/n_2)^2}{n_2}} \tag{9.57}$$

- Comparison of more than two populations—nonparametric approach discussed in Sect. 9.8 and comparison of normal means in Chap. 10.

*Comparison of the means of exponential distributions* We assume independent samples from exponential distributions with parameters $\lambda_1$ and $\lambda_2$. Since the means are the reciprocals of the $\lambda_i$, comparison of the means is equivalent to comparison of the $\lambda$'s. For complete data, the test statistic is

$$U = \lambda_1^*/\lambda_2^* \tag{9.58}$$

where $\lambda_i^*$ is the MLE of $\lambda_i$. Under $H_0$, $U$ is distributed as $F$ with $2n_1$, $2n_2$ *df.*[18] Fractiles $F_{df_1,\, df_2,\, p}$ of the F-distribution are given in Table E.4. The acceptance region for the two-tailed test is $\left(1/F_{2n_1,\, 2n_2,\, 1-\alpha/2}, F_{2n_2 2n_1,\, 1-\alpha/2}\right)$. For one-tailed tests, the acceptance region is modified as before.

Comparison of two means with Type II censored data proceeds as above, with *df* modified to account for censoring. For Type I censoring and for comparing more than two exponential means, a likelihood ratio test is used. This and related problems are discussed in detail in [11], Chap. 3.

*Example 9.6* Data on tensile strength of silicon carbide fibers are given in Table F.14 for fibers of various lengths. We compare the mean tensile strength of 12.7 mm fibers with that of 25.4 mm fibers (denoted Groups 1 and 2, respectively). It is anticipated that shorter length fibers will be stronger and the objective is to determine if there is evidence that this is the case. Probability plots suggest that the normal distribution provides a much better model for the data than does the exponential. We assume normality with equal population variances.

The appropriate hypotheses are $H_0$: $\mu_1 < \mu_2$ versus $H_1$: $\mu_1 > \mu_2$. We test at the 5% level. Minitab output provides the following information: $n_1 = 50$, $n_2 = 64$, $\bar{y}_1 = 3.091$, $\bar{y}_2 = 2.850$, $s_1 = 0.635$, and $s_2 = 0.668$. From the last two, we find the pooled standard deviation to be $s_p = 0.654$. From the output or (9.56), we obtain $t = 1.95$, with 112 *df.* Minitab also provides a $p$-value of 0.027. Since this is less than 0.05, the null hypothesis is rejected and we conclude that there is evidence that average tensile strength of 12.7 mm fibers is greater than that of 25 mm fibers.

## 9.7 Tolerance Intervals

A *tolerance interval* is essentially a confidence interval for a population fractile. In particular, it is an interval that contains a specified proportion $P$ of a population with confidence $\gamma$. As in the case of confidence intervals, tolerance intervals (also called *tolerance limits*) may be one-sided or two-sided.

Tolerance intervals have important applications in reliability. Two-sided intervals, for example, may be used to determine a region within which a specified

[18] In general, an $F$-statistic is the ratio of two Chi-Square variables. *df* are those associated with the numerator and denominator of this ratio.

proportion of manufactured parts will lie with regard to an important characteristic. Lower one-sided tolerance limits on lifetimes of items are useful in many applications since they provide a value beyond which a specified proportion of items will survive with confidence $\gamma$.

We look briefly at tolerance intervals for the normal and exponential distributions and at nonparametric tolerance intervals. Few results are available for other distributions.

*Normal tolerance intervals* Normal tolerance limits are based on $\bar{y}$ and $s$ and are of the form $\bar{y} \pm Ks$ (for two-sided intervals). The factor $K$ is obtained from Table E.5. $K$ depends on the sample size $n$, the coverage $P$, and the confidence coefficient $\gamma$. Lower and upper tolerance limits are of the form $\bar{y} \mp K's$ and $\bar{y} \pm K's + K'$s, respectively, with $K'$ obtained from Table E.6 for a limited set of combinations of $n$, $\gamma$, and $P$. More extensive tables may be found in [16].

*Tolerance limits for the exponential distribution* Tolerance intervals for the exponential employ the Chi-Square distribution. The two-sided tolerance interval with coverage $P$ and confidence $\gamma$ is

$$\left(\frac{2n\bar{y}\log[2/(1+P)]}{\chi^2_{2n,\,1-(1-\gamma)/2}},\ \frac{2n\bar{y}\log[2/(1-P)]}{\chi^2_{2n,\,(1-\gamma)/2}}\right) \tag{9.59}$$

In many reliability applications, a lower tolerance limit is desired. The result is

$$L = \frac{2n\bar{y}\log(1/P)]}{\chi^2_{2n,\,\gamma}} \tag{9.60}$$

The limit is such that we are $100\gamma\%$ confident that a proportion $P$ of items will have a lifetime of at least $L$.

*Nonparametric tolerance intervals* Here we look at nonparametric tolerance intervals. Additional nonparametric results are given in the next section. The nonparametric limits are based on the ordered observations $y_{(1)}, y_{(2)}, \ldots, y_{(n)}$. The two-sided interval consists of two order statistics, $y_{(r)}$ and $y_{(q)}$, having the property that at least a proportion $P$ lies between the two values with confidence $\gamma$. The results hold for any continuous distribution. Moderate to large sample sizes are required.

Values of $(r, s)$, where $q = n - s + 1$, are given in Table E.7 for $\gamma = 0.9$ and $0.95$, $P = 0.75$, $0.90$, and $0.95$, and selected $n$'s ranging from 50 to 1,000.[19]

*Example 9.7* For the bond strength data of Table F.5, we calculate 95% tolerance intervals having coverage 0.90, for all three procedures discussed above. For these data, $n = 51$, $\bar{y} = 288.9$, and $s = 37.75$.

For the normal distribution, the factor needed for calculation of the tolerance interval is found by interpolation in Table E.5 to be $K = 1.992$. The tolerance interval is $288.9 \pm 1.992(37.75)$, giving the interval (213.7, 364.1). Given that

[19] More extensive tables as well as factors for one-sided nonparametric tolerance intervals are given in [2, 16, 18], who initially published the results.

bond strengths are normally distributed, we are 95% confident that 90% of the bond strengths will lie within this interval.

Under the assumption of an exponential distribution for bond strengths, the tolerance interval is[20]

$$\left(\frac{2(51)(288.9)\log(1.053)}{\chi^2_{102,\,0.975}}, \frac{2(51)(288.9)\log(20)}{\chi^2_{102,\,0.025}}\right) = \left(\frac{1521.8}{131.84}, \frac{88277.6}{75.956}\right)$$
$$= (11.54,\ 1162.4).$$

For the nonparametric tolerance interval, we use the tabulated value for $n = 50$. The value is $(r, s) = (1, 1)$, indicating that the interval is formed by the smallest and largest observations in the data. The resulting interval is (197, 360). This agrees reasonably well with the result for the normal distribution. Both of these, however, differ markedly from the result for the exponential distribution. This is not surprising, since the Weibull distribution that appeared to be a reasonable fit to the data is much more nearly normal than exponential.

## 9.8 Nonparametric Methods

In Sect. 9.6, tests of hypotheses about means were discussed. The hypotheses involved individual population means as well as comparisons of the means of two populations. Tests were carried out under specified distributional assumptions. In this section, we look at nonparametric methods that are basically alternatives to these tests. The nonparametric procedures do not, in fact, test hypotheses about means. Instead, they typically do one of the following:

1. Test a null hypothesis about the median
2. Test a null hypothesis about a median difference using paired samples from two populations
3. Compare the medians of two or more populations
4. Compare two or more distributions

The advantage of the nonparametric approach is that it does not depend on a specific distribution for validity[21] and hence is much more broadly applicable. The disadvantage is that nonparametric tests are not as powerful as the parametric alternatives when the required distributional assumptions hold. In the following, we look briefly at several nonparametric tests that may be used in one or more of the items in the list above. Also included will be a more thorough treatment of rank correlation, a procedure introduced in Chap. 8.

---

20 *Note*: Since Table E.3 gives values only to 100 *df*, the required fractiles of the Chi-Square distribution were obtained by use of Minitab.

21 A key difficulty is that parametric tests applied to data from distributions very different from that assumed may result in error rates quite different from the nominal.

### 9.8.1 Sign Test

The Sign Test may be used for Items 1 and 2 in the above list. The null and alternate hypotheses in this case are expressed in terms of population medians rather than means.

To test a hypothesis about a single population median, calculate the differences between each observation and the null hypothesized median. Let $n_1$ be the number of positive differences that result and $n_2$ be the number of negatives (differences of zero are ignored). Under $H_0$, we expect half the differences to carry each sign. The test of $H_0$ is therefore a binomial test of $H_0$: $p = ½$, given in Sect. 9.6.3 (with one-sided or two-sided alternatives as appropriate). If the sample size is adequate, the test statistic is (9.54), with $p_0 = 0.5$, and $\hat{p} = n_1/(n_1 + n_2)$.

The Sign Test may be used to compare two population medians if samples of paired observations from the two populations are available. This occurs, for example, is measurements are obtained from $n$ individuals at two different times. Signs of the differences between the two measurements are recorded (eliminating ties) and the test is a binomial test of $H_0$: $p = ½$, as above.

The test is found in most computer packages. The Minitab command for the Sign Test is Stat/Nonparametrics/1-Sample Sign.

### 9.8.2 Wilcoxon Signed Rank Test

The Signed Rank Test [21] can be used as an alternative to the t-test in both the situations in which the Sign Test is used. It is a more powerful alternative, but requires special tables unless the sample is of at least moderate size. The test is done as follows: Calculate differences as for the Sign Test (either differences between observations and a hypothesized median, or differences between paired observations), then rank the absolute values of the differences. The test statistic is the sum of the ranks associated with positive differences. The Minitab command is Stat/Nonparametrics/1-Sample Wilcoxon.

*Example 9.8* The bond strength data of Table F.5 were used in Example 9.5 to test the null hypothesis $H_0$: $\mu = 300$ versus the alternative that $\mu \neq 300$. The test was a $t$-test based on the assumption of normality. $H_0$ was rejected at the 5% level. The Sign Test may be used to test the same hypothesis concerning the median. Minitab gives the values $n_1 = 20$, $n_2 = 31$, median $= 294$, and a $p$-value of 0.1614, which does not lead to rejection of $H_0$.[22] The data appear to be normal, so the *t-test* used in Example 9.5 would be the better test since the Sign Test is not as powerful.

---

[22] Note: Minitab calculates the exact binomial probability of obtaining a result this or more discrepant from that expected under $H_0$, which may differ somewhat from that obtained using the normal approximation unless $n$ is quite large.

For the Signed Rank Test, the sum of positive ranks is 485. For testing $H_0$ against the stated alternative, Minitab gives a $p$-value of 0.096, also not significant at the 5% level.

### 9.8.3 Rank Sum Tests

Several nonparametric procedures involve ranking observations in two or more groups and then summing the ranks in various ways to form a test statistic. The resulting tests, called Rank Sum Tests, do not compare means or medians, but test whether or not the samples are from the same population. The tests require independent observations and samples, and assume continuous distributions. Procedures for three types of experimental data are listed and briefly discusses below. All of these tests are available in Minitab and other statistical packages, where the tests statistics and $p$-values are given on use of any of the procedures.

#### 9.8.3.1 Mann–Whitney Test

This test, also called the Mann–Whitney $U$, the Wilcoxon Rank Sum Test, etc., was first given by [21] for samples of equal size and later extended to unequal sample sizes by Mann and Whitney. The test statistic is denoted $U$, and the test compares two populations.[23]

In applications, the test is most easily performed using a package such as Minitab (Stat/Nonparametrics/Mann–Whitney). It is also easy to calculate the value of the test statistic by hand. Assume samples of sizes $n_1 < n_2$. Rank all $n = n_1 + n_2$ observations in a single array, assigning average ranks to ties, if any. Calculate $R_1$ = sum of the ranks in Sample 1. The value of $U$ is calculated as

$$U = R_1 - \frac{n_1(n_1 + 1)}{2} \tag{9.61}$$

If both sample sizes exceed 10, the normal approximation may be used. The test statistic is $z = (U - \mu_U)/\sigma_U$, where $\mu_U = n_1 n_2/2$ and $\sigma_U = [n_1 n_2(n_1 + n_2 + 1)/2]^{1/2}$.[24] For smaller samples, tables of critical values are needed.[25]

---

[23] The test is sometimes described as a comparison of the medians of two populations, which is not strictly a correct interpretation.

[24] Note: The formula for the standard deviation assumes no ties. If there are only a few, this can safely be ignored. Otherwise, the formula requires an adjustment given in most texts on nonparametrics. Computer packages automatically perform the adjustment.

[25] See texts on nonparametric statistics for the tables. Additional details regarding the Mann–Whitney U may be found in the Wikipedia article by that title and in [7].

#### 9.8.3.2 Kruskal–Wallis Test

An extension of the Mann–Whitney $U$ to more than two populations was given by [10]. The data in samples from $k$ populations are ranked in a combined array and rank sums are calculated for each sample. The analysis done is basically an Analysis of Variance (ANOVA) for a one-way classification (to be discussed in Chap. 10), using the ranks instead of the original data. The test statistic is a Chi-Square with $(k - 1)$ *df*. Minitab includes a program for performing the test. For additional details, see [7].

#### 9.8.3.3 Friedman Test

The Friedman Test is a further extension of the Mann–Whitney to the Randomized Complete Block Design (See Chap. 10). It is also an application of ANOVA to ranks, with ranks determined separately for each block (a feature of the structure of the experimental design.) For sufficiently large experiments, the test statistic is a Chi-Square. For smaller experiments, exact tables are needed. Details may be found in [7] Minitab includes a program for performing the test.

*Example 9.9* In Example 9.6, a $t$-test of $H_0$: $\mu_1 < \mu_2$ versus $H_1$: $\mu_1 > \mu_2$ (where $\mu_1$ is the mean tensile strength of 12.7 mm fibers and $\mu_2$ that of 25.4 mm fibers) led to rejection of $H_0$ at $\alpha = 0.05$; the $p$-value was 0.027. The Mann–Whitney Test compares the two populations as a whole. Minitab output describes this as a one-tailed test of the medians and gives a value of the test statistic of 3186 and $p = 0.038$ (adjusted for ties). The results of the tests are comparable.

If a comparison the distribution of strengths of fibers of all four lengths is desired, the appropriate test is the Kruskal–Wallis. For this data set, Minitab gives a value of 109.5 for the test statistic (adjusted for ties). This is approximately a Chi-Square with 3 *df*; The $p$-value for the test is 0.000 and the hypothesis that these are samples from the same population is rejected.

### 9.8.4 Rank Correlation

The rank correlation coefficient was introduced in Sect. 8.3.5 as a nonparametric measure of the strength of relationship between two variables $x$ and $y$. For a sample of size $n$, each variable is ranked separately (with ties, as usual, being given the average rank), resulting in pairs of ranks $(x_1, y_1)$, ..., $(x_n, y_n)$. The Spearman rank correlation coefficient $r_s$ is then calculated as indicated in (8.4). If there are no ties, a simpler calculation is

$$r_s = 1 - \frac{6\sum (x_i - y_i)^2}{n(n^2 - 1)} \tag{9.62}$$

For $n > 20$, a $t$-test can be used to test whether or not the true correlation is different from zero. The test statistic is

$$t = \frac{r_s}{\sqrt{(1 - r_s^2)/(n - 2)}} \tag{9.63}$$

with $(n - 2)$ *df*. For additional details, see [7].

*Example 9.10* In Example 8.4, the rank correlation between kilometer driven and cost of repairs under warranty was found to be $r_s = 0.225$, with $n = 32$. The $t$-test of $H_0$: $\rho_s = 0$, where $\rho_s$ denotes the population rank correlation coefficient, results is $t = 0.225/[(1-.2252)/(30)]^{1/2} = 1.265$. Minitab provides a $p$-value of 0.216. $H_0$ is not rejected at any $\alpha$ less than 0.20. We conclude that there is no evidence of a relationship between km and repair cost.

## References

1. Abramowitz M, Stegun IA (1964) Handbook of mathematical functions with formulas, graphs and mathematical tables. U.S. Government Printing Office, Washington
2. Blischke WR, Murthy DNP (2000) Reliability. Wiley, New York
3. Dixon WJ, Massey FJ Jr (1969) Introduction to Statistical Analysis, 3rd edn. McGraw-Hill, New York
4. Govindarajulu Z (1964) A supplement to Mendenhall's bibliography on life testing and related topics. J Am Statist Assoc 59:1231–1291
5. Hirose H, Lai TL (1997) Inference from grouped data in three-parameter Weibull models with applications to breakdown-voltage experiments. Technometrics 39:199–210
6. Hogg RV, Craig A, McKean JW (2004) Introduction to mathematical statistics, 6th Edition edn. Prentice Hall, New York
7. Hollander M, Wolfe DA (1999) Nonparametric statistical methods, 2nd Edition edn. Wiley, New York
8. Høyland A, Raussand M (1994) System reliability theory. Wiley Interscience, New York
9. Johnson NL, Kotz S (1970) Distribution in statistics: continuous univariate distributions–I. Wiley, New York
10. Kruskal W, Wallis A (1952) Use of ranks in one-criterion variance analysis. J Am Statist Assoc 47:583–621
11. Lawless JF (1982) Statistical models and methods for lifetime data. Wiley, New York
12. Martz HF, Waller RA (1982) Bayesian reliability analysis. Wiley, New York
13. Moore DS, McCabe GP, Craig B (2007) Introduction to the practice of statistics. W H Freeman, New York
14. Murthy DNP, Xie M, Jiang R (2004) Weibull models. Wiley, New York
15. Nelson W (1982) Applied life data analysis. Wiley, New York
16. Owen DB (1962) Handbook of statistical tables. Addison-Wesley, Reading, MA
17. Ryan TP (2007) Modern engineering statistics. John Wiley, New York
18. Somerville P (1958) Tables for obtaining non-parametric tolerance limits. Ann of Math Statist 29:599–601
19. Stuart A, Ord JK (1991) Kendall's advanced theory of statistics, vol. 2, 5th Edition edn. Oxford University Press, New York
20. Wackerly D, Mendenhall W, Scheaffer RL (2007) Mathematical statistics with applications, Duxbury, New York 2007
21. Wilcoxon F (1945) Individual comparisons by ranking methods. Biometrics 1:80–83

# Chapter 10
# Additional Statistical Techniques

## 10.1 Introduction

The previous two chapters dealt with data description and summarization and several topics in basic statistical inference—estimation, hypothesis testing, tolerance intervals, and so forth. In this chapter, we continue the review of topics in inference, with brief description of the methodology, discussion of their usefulness in analysis of warranty data, and worked examples.

Topics covered include:

- Tests for outliers
- Goodness-of-fit tests—Chi-Square, Kolmogoroff–Smirnov, Anderson–Darling
- $t$- and $F$-tests for comparing two or more normal populations
- Techniques for comparing exponential and Weibull populations
- Linear regression and correlation analysis
- Additional topics in estimation—estimation of functions of parameters, including reliability
- Tests of assumptions

In this chapter, we will deal primarily with complete samples, the exception being goodness-of-fit tests, where some results for censored samples are discussed. Extensions to other analyses of incomplete data and to other, more complex data sets will be given in the chapters that follow.

The chapter outline is as follows: Methods for detecting and dealing with outliers are discussed in Sect. 10.2. In Sect. 10.3, we deal with goodness-of-fit tests and their use and interpretation in various situations. In Sect. 10.4, the use of these tests in model selection is briefly discussed. Tests for comparing the means of two or more normal populations are discussed in Sect. 10.5 for two commonly used experimental designs. Basic regression and correlation analysis based on linear models are discussed in Sect. 10.6. Estimation of functions of parameters generally is discussed in Sect. 10.7, with application to estimation of the coefficient of

W. R. Blischke et al., *Warranty Data Collection and Analysis*,
Springer Series in Reliability Engineering, DOI: 10.1007/978-0-85729-647-4_10,

variation, estimation of a warranty cost model, and estimation of a reliability function. Section 10.8 is concerned with tests of some of the various assumptions made in the statistical analyses.

## 10.2 Tests for Outliers

Extreme observations, relatively large or relatively small values that appear to be far removed from the bulk of the data, are occasionally encountered in data analysis. In truly messy data, not uncommon in dealing with warranty claims, unusual observations may occur with disconcerting frequency. The extreme observations, called outliers, may occur for a number of reasons, including at least the following:

- Errors in measurement
- Errors in recording, transcription, etc.
- Valid measurements of items from a distribution having very long tails
- Valid measurements of items from a different population

Whatever the cause, outliers can significantly affect the outcome of tests and other statistical procedures.

There are two key issues in dealing with outliers: (1) determining whether suspicious results are outliers; (2) deciding on a course of action if it is determined that one or more observations are indeed outliers. We look briefly at both issues.

### *10.2.1 Graphical Methods for Detection of Outliers*

Graphical techniques discussed in Sect. 8.4 are useful in detecting possible outliers. Visual inspection of frequency distributions, histograms and other charts may easily identify outliers, particularly in extreme cases. This is true, for example, of the histogram of warranty-related repair cost data (Table F.2) given in Fig. 8.3, where the value in excess of \$5,000 clearly stands out from the remainder of the data.

Another frequently used graphical device for identifying outlying observations is the *boxplot* (originally called a "box-and-whisker plot"). This is a plot in which the middle 50% of the data. i.e., all observations between the quartiles $Q_1$ and $Q_3$, is represented by a rectangle (the "box") and the remaining data are indicated by lines outside the box (the "whiskers") or by points beyond the lines (the outliers). The length of the lines is calculated as 1.5 times the *Interquartile Range* (IQR), where IQR $= Q_3 - Q_1$. Observations between 1.5(IQR) and 3(IQR) are designated "mild" outliers; those beyond 3(IQR) are "extreme" outliers. The box also shows a line for the median, $Q_2$.

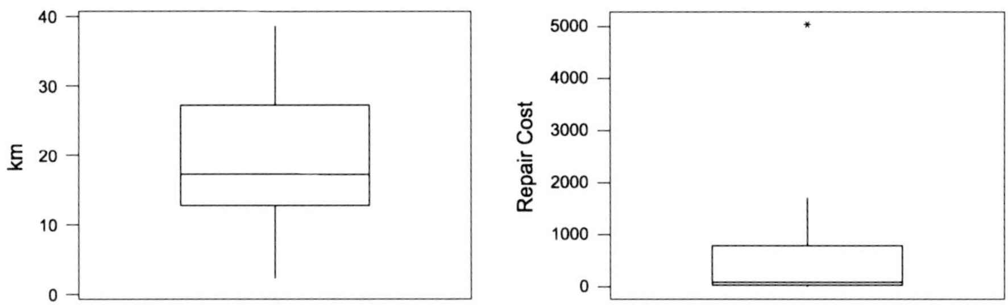


**Fig.10.1** Boxplots of kilometers driven and repair cost under warranty

**Fig. 10.2** Boxplot of repair cost after removal of the maximum value

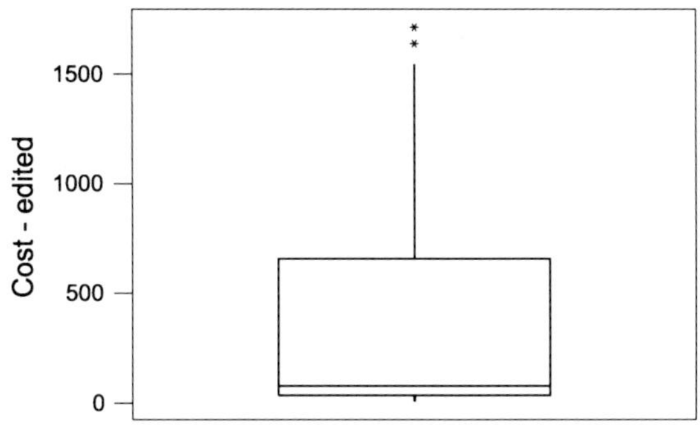


*Example 10.1* Figure 10.1 shows boxplots of the usage and repair costs for the data of Table F.2. For kilometers driven, the boxplot looks quite reasonable, though the median is below the center of the box. The boxplot for repair costs, however, shows a very skewed distribution, with the median near the lower edge of the box, the lower whisker so small that it does not appear in the plot, and the extreme outlier indicated by a "*." (Note: Minitab does not distinguish between mild and extreme outliers in its notation.)

For a better picture of the data, the extreme observation may be removed. A boxplot of the remaining data is given in Fig. 10.2. Here the skewness of the distribution remains obvious and two additional possible outliers are identified.

Other graphical procedures may also be useful in displaying outliers. These include stem-and-leaf plots, dotplots, and others, as listed in Minitab and other statistical program packages.

### *10.2.2 Outlier Tests for the Normal Distribution*

There are a number of statistical approaches to the detection of outliers. Procedures have been developed for this purpose for the normal, exponential, and other distributions. We look briefly at the most commonly used technique for detecting outliers in normal data, the *Grubb's Test* [5]. Ordinarily, a test for normality should be conducted prior to application of the test for outliers.

The test proceeds as follows: Calculate the sample mean and standard deviation using all of the sample data. The test statistic for a two-tailed test (i.e., to detect either the max or the min as an outlier) is

$$G = \frac{\max_{i=1,\ldots,n} |y_i - \bar{y}|}{s} \tag{10.1}$$

The rejection region for the test at significance level $\alpha$ is

$$G > \frac{(n-1)t_{n-2,\,\alpha/2n}}{\sqrt{n(n-2+t^2_{n-2,\,\alpha/2n})}} \tag{10.2}$$

For a one-tailed test, the numerator of $G$ is replaced by either $(\bar{y} - y_{(1)})$ or $(y_{(n)} - \bar{y})$, as appropriate, and $\alpha/2n$ is replaced by $\alpha/n$ in obtaining the tabulated $t$ in (10.2). If an observation is found to be an outlier, it is removed from the data and the test is repeated on the remaining data. The process is continued until no further outliers are found.

For further information on outliers, see [1]. and the Wikipedia articles on Outliers, Grubb's Test, Chauvenet's Criterion, and Pierce's Criterion.

### *10.2.3 Dealing with Outliers in Data Analysis*

When outliers have been detected, visually or through testing, the action to be taken in analyzing the data is often a difficult decision. On the one hand, it is clearly desirable to avoid tainting the results by inclusion of erroneous data. On the other hand, data that are legitimate should not be excluded even if they appear to be extreme values. The following are a few considerations when dealing with outliers:

- The source of any data that appear to be suspicious should be investigated to assure that the data have been collected and recorded properly.
- Any observations, suspect or not, should be removed if found to be erroneous (e.g., due to faulty measuring devices, fault procedure, recording errors, etc.).
- When sample sizes are large, observations that appear to be extreme occur with increasing frequency. These should not be removed unless there is reason to believe that they are erroneous.
- If the normal distribution model is found to be acceptable, observations found to be extreme by Grubb's test may be removed.
- If the normal distribution does not appear to be tenable, nonnormal models should be investigated and suspicious data eliminated only in the context of a tenable model.

- Robust statistical techniques should be employed instead of the usual normal inference procedures. In this context, nonparametric methods are often not sensitive to outliers, e.g., ranks are not much affected, if at all.
- Finally, it is sometimes instructive to perform analysis with the outlier(s) included as well as with them excluded to assess the effect of the outliers on the results.

## 10.3 Goodness-of-Fit Tests

The primary objective of a goodness-of-fit test is to investigate how well a specified probability distribution "fits" a set of data. This notion was introduced in Chap. 8, where we looked at graphical methods and judged the fit visually, for example by noting whether or not the a linear pattern was obtained in a probability plot, whether an EDF appeared to have a certain shape, and so forth. Here we extend the analysis by applying statistical tests to judge the adequacy of the fit. This provides a rigorous framework for the analysis. It also provides a method of comparing fits of a data set to many different distributions, which is an important component of model building.

In this section, we will look briefly at the Chi-Square, Anderson–Darling, and Kolmogorov–Smirnov Tests. The first of these is most appropriate for fitting discrete distributions. The latter two are used for fitting continuous distributions and are commonly available in statistical program packages. We will illustrate the tests by use of Minitab.

The basic theory underlying goodness-of-fit tests [18] assumes that the distribution whose fit is being tested is completely specified. In practice, the much more common situation is that the form of the distribution (e.g., Weibull) is specified, but the parameters of the distribution are estimated from the data. This significantly complicates the theory and requires that the rejection regions of the tests be modified. This is discussed in [2], Sect. 11.3.4 and in the references cited.

Modifications of the rejection region when parameters are estimated have been devised for complete data and a number of standard distributions. Further modification of the critical values of the tests is required in the case of incomplete data. These may depend on the specific distribution, the type and amount of censoring, and the sample size. Studies of this problem have not been done and further research in this area is needed.

Note, however that statistical packages typically give values of one or more of the goodness-of-fit test statistics as part of the output of graphical and other procedures for complete and incomplete data and specified or estimated parameters. Caution should be used in interpreting these statistics. In particular, comparison of calculated values to the standard critical values based on completely specified distributions and complete data may not be meaningful unless the sample size is quite large. The values of goodness-of-fit statistics in these cases may be useful as comparative measures in model fitting, but not as tests of individual distributions.

## 10.3.1 Chi-Square Tests

### 10.3.1.1 Fitting Discrete Distributions

The classic Chi-Square goodness-of-fit test was first used for fitting discrete distributions. It is assumed that a sample of size $n$ is observed, with each observation falling into one of $k$ classes. $O_i$ is the number of observations falling into class $i$ and $E_i$ the number expected under the hypothesized distribution for class $i$, $i = 1, 2, \ldots, k$. $E_i$ is calculated as $np_i$, where $p_i$ is the probability that an observation falls into class $i$ under the hypothesized distribution. The test statistic is

$$\chi^2 = \sum_{i-1}^{k} \frac{(O_i - E_i)^2}{E_i} = \sum_{i=1}^{k} \frac{O_i^2}{E_i} - n \tag{10.3}$$

Under $H_0$, the test statistic is distribution as Chi-Square with $(k - 1)$ *df* if the distribution is completely specified under the null hypothesis. If parameters are estimated from the data, degrees of freedom are reduced by the number of parameters estimated. Since large deviations of observed from expected lead to rejection, the test is an upper-tail test. Critical values are given in Table E.3

For fitting discrete distributions, the classes are determined by the possible values of the variable. If the number of possible values is finite, classes would initially be defined by associating a class with each of the possible values. For example, for the binomial distribution with parameters $p$ and $m$, say, $m + 1$ classes would be defined, corresponding to the values 0, 1, …, $m$. The usual procedure is then to check expected values; if any are less than 1,[1] the corresponding classes are combined with adjacent classes. This is continued until all classes have $E_i$'s of at least 1. For distributions with an infinite number of possible values, such as the Poisson, the natural classes are used as well, with classes in the tails and elsewhere if necessary combined as indicated.

*Example 10.2* In a test of accuracy of data transcription from warranty claims forms, numbers of errors found per 500 entries were recorded for $n = 36$ sets of 500 entries. If errors occur randomly and independently, the distribution of counts should be a Poisson (given by (A.18)). Counts of errors ranged from 0 to 5, with a mean of $\bar{y} = 1.4722$. The $E_i$ were calculated using this as an estimate of $\lambda$. The results were as shown in Table 10.1.

By our rule, the last two groups are combined. The calculated value of the test statistic is $\chi^2 = 7^2/8.26 + \cdots + 1/2.24 - 36 = 1.848$, with $5 - 1 - 1 = 3df$.

[1] Traditionally, this number is taken to be 5. We prefer a less conservative approach.

**Table 10.1** Observed and expected frequencies, Poisson distribution

| Number of errors | 0 | 1 | 2 | 3 | 4 | 5 |
|---|---|---|---|---|---|---|
| Frequency | 7 | 14 | 8 | 6 | 0 | 1 |
| Expected frequency | 8.26 | 12.16 | 8.95 | 4.39 | 1.62 | 0.62 |

From Table E.3, the tabulated value for testing at the 5% level is 7.81. The hypothesis of a Poisson distribution is not rejected.

#### 10.3.1.2 Fitting Continuous Distributions

Using Chi-Square for testing goodness-of-fit to continuous distributions is appropriate for large samples. The advantage of the test is that it is relatively easy to apply and is broadly applicable. Its disadvantage is that it is not very powerful, and alternatives, such as those given in the following sections, are preferred.

To apply the Chi-Square test to data from continuous distribution, the data are first grouped into $k$ intervals as is done in forming a frequency distribution or preparing a histogram. The intervals are determined by limits $L_0 = -\infty < L_1 < \cdots < L_k = \infty$, and $p_i$ is the probability that an observation falls in the interval $(L_{i-1}, L_i)$. The $p_i$ are given by

$$p_i = \int_{L_{i-1}}^{L_i} dF_0(x) \tag{10.4}$$

where $F_0(.)$ is the null-hypothesized CDF. As before, the resulting $\chi^2$ has *df* equal to $k - 1$ minus the number of parameters estimated from the data.

### 10.3.2 Kolmogorov–Smirnov Test

We assume that the null-hypothesized distribution $F_0(.)$ is completely specified. In this case, the Kolmogorov–Smirnov (K–S) test is nonparametric in that the distribution of the test statistic does not depend on $F_0$. The test statistic, denoted $D_n$, is the maximum difference between the EDF and $F_0$. This is calculated as $D_n = \max\{D_n^-, D_n^+\}$, with

$$D_n^- = \max_{i=1,\ldots,n}\left[\frac{i}{n} - F(y_{(i)})\right], \quad D_n^+ = \max_{i=1,\ldots,n}\left[F(y_{(i)}) - \frac{i-1}{n}\right], \tag{10.5}$$

where the $y_{(i)}$ are the order statistics. $H_0$ is rejected at level $\alpha$ if $D_n$ exceeds the critical value $d_\alpha(n^{1/2} + 0.11n^{-1/2} + 0.12)^{-1}$, where $d_\alpha = 1.224, 1.358$, and $1.628$, for $\alpha = 0.10$, 0.05, and 0.01, respectively. The K–S Test is available in some statistical program packages.

### 10.3.3 Anderson–Darling Test

The Anderson–Darling (A–D) test is also nonparametric if $F_0$ is completely specified. The test statistic is

$$A^2 = \frac{-1}{n}\sum_{i=1}^{n}(2i-1)\log\{(F_0(y_{(i)}))[1-F_0(y_{(n-i+1)})]\} - n \tag{10.6}$$

The critical values of $A^2$ do not depend on $n$. For $\alpha = 0.10, 0.05,$ and $0.01$, they are $a_a = 1.933, 2.492,$ and $3.857$.

### 10.3.4 K–S and A–D Tests for Selected Distributions, Parameters Estimated

The results above are applicable to any distributions for which the parameters are known or specified. The test statistics can be calculated as indicated using estimates of the parameter instead. For this purpose, the MLE's are used and it is necessary to adjust the critical values. The required adjustment depends on the hypothesized distribution (and hence the procedure is no longer nonparametric in the sense described above) and on the sample size. The exact distributions of the test statistics is not known and approximate critical values have been obtained by numerical methods. This has been done for only a few distributions. The results are given below.

Values of the K–S and/or the A–D statistic are given in the output of many program packages. These are not interpretable in the usual sense if critical values are not known. Nonetheless, the statistics are useful in model building in that the values of the test statistics can be used to compare fits of various candidate distributions.

#### 10.3.4.1 Tests for the Normal and Lognormal Distributions, Parameters Estimated

Modifications of the K–S and A–D critical values when parameters are estimated are as follows[2]:

- K–S Test: the critical values (for $n \geq 10$) are given by $d_a(n^{1/2} + 0.83n^{-1/2} - 0.01)^{-1}$, where $d_a = 0.819, 0.895,$ and $1.035$, for $\alpha = 0.10, 0.05,$ and $0.01$, respectively.

[2] Factors for calculating critical values for other choices of $\alpha$ can be found in [2] and the references cited.

- A–D Test: the critical values are $a_\alpha(1 + 4n^{-1} - 25n^{-2})^{-1}$, where $a_\alpha =$ 0.656, 0.787, and 1.092 for $\alpha = 0.10, 0.05$, and 0.01, respectively.

There are many other tests for normality. Critical values and additional discussion of several of these are given in [3] and [10].

To test for goodness-of-fit to the lognormal distribution, transform the data to the log scale and use the test for normality.

#### 10.3.4.2 Test for the Exponential Distribution, Parameter Estimated

Modifications of the K–S and A–D critical values when the parameter $\lambda$ is estimated are as follows:

- K–S Test: the critical values are given by $0.2n^{-1} + d_a(n^{1/2} + 0.50n^{-1/2} + 0.26)^{-1}$, where $d_a = 0.990, 1.094$, and 1.308 for $\alpha = 0.10, 0.05$, and 0.01, respectively.
- A–D Test: the critical values are $a_\alpha(1 + 0.60n^{-1})^{-1}$, where $a_\alpha = 1.078$, 1.341, and 1.957 for $\alpha = 0.10, 0.05$, and 0.01, respectively.

An alternate approach is to assume a Weibull distribution and use the methods of Chap. 9 to test the hypothesis that the value of the shape parameter is one.

#### 10.3.4.3 Test for the Weibull and Extreme Value Distributions, Parameters Estimated

We consider the extreme value distribution of (A.33). To estimate the parameters $\mu$ and $\sigma$, use the fact that if $Y$ has this distribution, then $X = e^Y$ has a Weibull distribution with parameters $\alpha = e^\mu$ and $\beta = 1/\sigma$. Calculate MLE's $\alpha^*$ and $\beta^*$ using the data on the X-scale, transform back to the Y-scale, and calculate the test statistics using as $F_0$ the CDF of (A.33) with estimated parameter values $\mu^* = \log(\alpha^*)$ and $\sigma^* = 1/\beta^*$. For testing goodness-of-fit, large samples ($n \geq 100$) are needed for the K–S Test. The critical values for the two tests are:

- K–S Test: the critical values are given by $d_a n^{-1/2}$, where $d_a = 0.803, 0.874$, and 1.007 for $\alpha = 0.10, 0.05$, and 0.01, respectively.
- A–D Test: the critical values (good for any $n$) are $a_\alpha(1 + 0.20n^{-1/2})^{-1}$, where $a_\alpha = 0.637, 0.757$, and 1.038 for $\alpha = 0.10, 0.05$, and 0.01, respectively.

For fitting the Weibull distribution, calculate the MLE's $\alpha^*$ and $\beta^*$, then transform to the log scale and fit to the extreme value distribution as indicated above.

*Example 10.3* Data on fiber strength (Table F.14) are partially analyzed in Examples 9.6 and 9.9. For further parametric analysis, distributional assumptions are required. Figure 10.3 is a Minitab four-way probability plot of strength of fibers of four lengths.

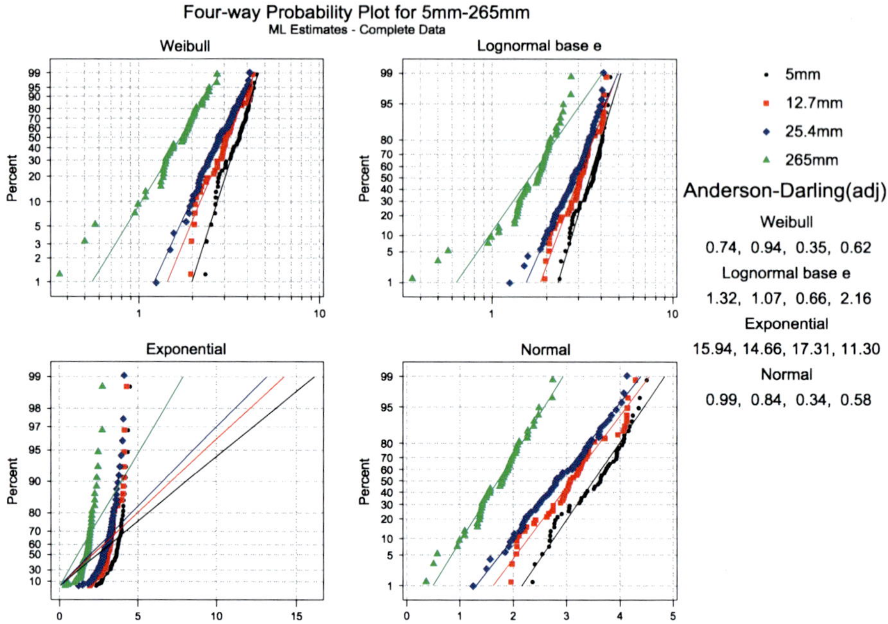


**Fig. 10.3** A–D Tests for fiber data

Included in the output are values of the A–D statistic obtained in fitting four distributions, the Weibull, lognormal, exponential and normal, using MLE's of the parameters in each case. It is clear from the figure that the lognormal and exponential distributions give very poor fits and can be eliminated as possible models. Comparison of the $A^2$-values given to the right of the graph indicates that the Weibull provides a somewhat better fit for the first group (5 mm) and the normal for the remaining three. To test the significance of the fits, we calculate the critical values (the sample sizes are $n = 64$ for 25.4 mm fibers and $n = 50$ for the rest).

The critical values for testing at the 5% level are:

- Weibull: 0.739 for 25.4 mm and 0.736 for the rest
- Normal: 0.745 for 25.4 mm and 0.736 for the rest

Testing at the 5% level, both the normal and Weibull distributions are rejected for the first two groups, but not for the last two. Testing at the 1% level, neither distribution would be rejected for any of the groups.

## 10.4 Model Selection

An initial look at model selection was provided in Sect. 8.7, where the notion of the use of plots in preliminary selection of probability models was discussed. For complete data, a number of basic distributions (exponential, normal, Weibull, etc.)

can be fit using Minitab. The user has the option of choosing either a least squares fit or maximum likelihood. In either case, the value of the A–D statistic is given as part of the output. For the least squares fit,[3] the correlation between the observed and fitted values is also given. Similar results may be obtained in SAS, R, and other program packages. A *p*-value or adjusted p-value (adjusted for estimating parameters) associated with the A–D test statistic may be given as well.

As in hypothesis testing generally, these results may be used to reject models that provide poor fits to the data, leaving those with relatively larger p-values as candidate models. The continuing process of model selection is then based on additional data, either additional samples or a subsample reserved for this purpose in the analysis of existing data. In any case, in the early stages of the process it is usually advisable to use a relatively small level of significance (1% or less) in selecting candidate models, so that few models that might conceivably fit are rejected.

In summary, the goodness-of-fit tests of Sect. 10.3 are useful in model building in that:

1. If the distribution being tested for fit is completely specified, the A–D and K–S procedures are nonparametric in that the critical values do not depend on the specified distribution
2. If the when parameters are unknown and estimated from the data, good approximations of the critical values are available for selected standard distributions
3. The goodness-of-fit tests may be useful in tentative selection of distributions in other situations as discussed below.

In the remainder of this section, we look at the third item.

Most computer packages also include goodness-of-fit tests for the simpler distributions with incomplete data and provide values of the test statistic in those cases as well, though omitting the *p*-values.[4] For incomplete data with parameters estimated, neither exact nor approximate critical values are known. Thus, as noted above and in Sect. 8.7, results of goodness-of-fit tests on such data may be used only to compare fits of various candidate models, and to select one or more as "better" than others, or occasionally as "best."

Often the data are from more complex populations, e.g., mixtures and other multi-modal distributions, competing risk, multiplicative models, generalized distributions, and so forth.[5] In these cases, most packages do not offer a method (graphical or analytical) for fitting.

---

[3] See Sect. 10.6 for a discussion of least squares estimation.

[4] Minitab uses the A–D statistic for this purpose.

[5] See Chaps. 3, 12, 13, 14, and Appendix for discussion and use of these and other complex models.

An approach to model selection when dealing with complex models that are not included in standard packages, is to prepare a list of candidate models, use least squares to estimate the parameters, calculate the value of the sum of squares of deviations in each case, and compare the results. This may require the use of other (non-statistical) program packages or the preparation of programs specifically for this purpose.

Some caution must be used in comparing results using this approach, particularly if the sample sizes are relatively small. In general, models with more parameters tend to result in a smaller sum of squares than those with fewer parameters. This may be accounted for by adjusting the value of the sum of squares, which is usually done by dividing by $n - 1 - k$, where $k$ is the number of parameters estimated. This adjustment is particularly important if the sample size is small.

A detailed description of this approach based on Weibull models, along with instructive worked examples, is given by [16].[6] The analysis there (in Chap. 16) uses up to a total 21 models, including the basic two- and three-parameter Weibull distributions of (A.30) and (A.46), mixtures of Weibulls, competing risk formulations, sectional formulations based on the Weibull, and so forth. Data are plotted and shapes identified, and parameters are estimated and models compared by the least squares method. The approach is illustrated by application to samples of small, medium and large sizes, including the data of Tables F.16 and F.17.

A similar procedure based on Weibull models may be employed by use of the tables and formulas given in Appendix C. Select a list of candidate models from those given, or use those given in Table C.2. Use appropriate computer routines to obtain WPP plots (or other probability plots, as appropriate) of the data. For the WPP plots, use judgment to determine the shape of the plot and use Tables C.1 and C.2 to determine candidate models. Calculate the sums of squares for all of the models tested and compare the results.

In conclusion, the following is a list of recommendations that the practitioner may find useful in model selection:

- Use graphical procedures as indicated in Chap. 8 in an initial analysis.
- For complete data, proceed as indicated in the first two paragraphs above.
- If none of the simpler models provide an adequate fit, proceed to the more complex models such as those in [16]. Other complex models may be used as well. Use least squares to determine estimates and compare sums of squares or adjusted sums of squares.
- When $p$-values are not provided, use tests only comparatively to pick candidate models.
- Follow the procedures as specified above and in Chap. 8.

[6] Many results on graphical procedures for fitting based on these models are also given in this source.

- Do not accept the results as definitive unless a comprehensive validation study has been done.

Note that the last recommended procedure is only feasible for large samples. If limited data are available, there will be very limited information, in any case, regarding the true model. This is discussed further in Chap. 12.

## 10.5 Comparing Two or More Population Means

Methods for comparison of the means of two normal populations and of two exponential populations are given in Chap. 9. Comparison of two or more population means is important in many experimental contexts—comparing reliabilities of several models of a product, comparing error rates of three shifts of workers, sales of marketing personnel over many districts, and so forth.

In this section, we look at Analysis of Variance (ANOVA), which is an extension of the t-test for comparison of two population means given in Chap. 9 to the case of more than two normal populations.[7] The analysis assumes equal population variances as well as normality, but is fairly robust to violations of these assumptions.[8] If the assumptions are severely violated, nonparametric methods such as those given in Chap. 9 should be used instead.

ANOVA is a well-known, broadly applicable tool for comparing population means. Here we give a brief review of the basic methodology. The structure of the analysis depends upon how the data were collected, i.e., on the structure of the experimental or test design. Design of experiments (DOE) deals with many facets of the experimental or data collection process, including[9]:

- Treatment design—single or multiple factors, complete or incomplete treatment sets, etc.
- Randomization—assignment of treatments to experimental units
- Replication—amount and method
- Models—representations of the data and structure, assumptions

Here we limit attention to two simple experimental designs, the Completely Randomized Design (CRD), and the Randomized Complete Block Design (RCBD). These are also called one-way and two-way classifications, respectively. As is apparent from their names, both of these designs assume specific

[7] For comparison of means of several exponential or Weibull populations, see [2].

[8] In the sense that actual error rates are reasonably close to the nominal. In addition, the fact that sample means are approximately normally distributed by the Central Limit Theorem is sometimes used to justify the use of ANOVA.

[9] These concepts, along with some basic designs, are discussed in [2], Chap. 10. Further discussion and additional designs may be found in texts on DOE such as [11, 15, 21].

randomization patterns. As noted previously, true randomization is not a feature of claims data and it is necessary to assume that the data obtained are random in the sense that they represent all of the items on which claims may have been made.

## *10.5.1 The Completely Randomized Design*

The CRD and its analysis are a direct extension of the two-sample problem discussed in Sect. 9.6.4. Here we assume that random samples have been drawn from $k \geq 2$ normal populations (also called *treatments* or *treatment groups*). Sample sizes may be unequal, and are denoted $n_1, \ldots, n_k$, with $\sum_{i=1}^{k} n_i = n$. ANOVA is used to test the null hypothesis $H_0 : \mu_1 = \mu_2 = \cdots = \mu_k$ against the alternative $H_1 : \mu_i \neq \mu_j$ for some *i, j* pair.

### 10.5.1.1 Model and Assumptions

The data obtained are denoted $Y_{ij}, i = 1, \ldots, k; j = 1, \ldots, n_i$. A model for the data expressing the observations in terms of the structure of the CRD is given by

$$Y_{ij} = \mu + \tau_i + \varepsilon_{ij} \tag{10.7}$$

This is a linear model that expresses the observation as an overall mean $\mu$ plus a "treatment effect" $\tau_i$ plus and an "error" term $\varepsilon_{ij}$. Here $\tau_i = \mu - \mu_i$ and $\varepsilon_{ij}$ represents a random deviation of the observed value from its expectation.

We assume that $\varepsilon_{ij}$ is normally distributed with mean zero and variance $\sigma^2$ for all *i, j*. The treatment effect $\tau_i$ is considered to be a random variable if (i) the treatment groups used in the experiment are selected at random from a larger set of treatments, and (ii) we wish to draw inferences to this larger set. In this case, $\tau_i$ is assumed to be normally distributed with mean zero and variance $\sigma_\tau^2$ for all *i*. Otherwise, $\tau_i$ is called a "fixed" (i.e., nonrandom) effect, with $\sum_{i=1}^{k} \tau_i = 0$. The models are called, respectively, the random effects and fixed effects models. We will focus primarily on the fixed effects model. Many of the test procedures are the same for the two models (which is not necessarily true for more complex designs), but the interpretation is different in the two cases. (See the references cited in footnote 9.)

In the fixed effects case, the hypotheses stated above are expressed in terms of the model parameters as $H_0 : \tau_1 = \cdots = \tau_k = 0$ versus $H_1 : \tau_i \neq 0$ for some *i*. For the random effects case the hypotheses are expressed in terms of $\sigma_\tau^2$.

### 10.5.1.2 ANOVA

The notion of Analysis of Variance is to look at the total variability in the data and analyze it, i.e., break it into its component parts. The "parts" or "sources of

**Table 10.2** Analysis of Variance for the Completely Randomized Design

| Source | *df* | SS | MS | F |
|---|---|---|---|---|
| Treatments | $k - 1$ | SST | SST/$(k - 1)$ | MST/MSE |
| Error | $n - k$ | SSE | SSE/$(n - k)$ | |
| Total | $n - 1$ | SSTot | | |

variability" are identified by the structure of the design, and in fact, are associated with the terms of the model. For the CRD, the identifiable sources of variability are "Total" and "Treatments." Any remaining variability is called "Error" or "Experimental Error." The variance associated with this source is a measure of the inherent variability in the data, which includes inherent variation in the population sampled, measurement error, variability associated with other sources not recognized or accounted for, and so forth. It may also be thought of as the variability among experimental units treated alike.

In calculating variances associated with the identified sources, we calculate (in ANOVA terminology) a Sum of Squares (*SS*) for each, determine *df* for each source, and finally calculate a sample variance, called a Mean Square (*MS*), for each as $MS = SS/df$. For the CRD, the resulting ANOVA is given in Table 10.2.

Sums of Squares are calculated based on the observed values $y_{ij}$ as follows:

$$SSTot = \sum_{i=1}^{k}\sum_{j=1}^{n_i} y_{ij}^2 - \frac{T^2}{n} \tag{10.8}$$

$$SST = \sum_{i=1}^{k} \frac{T_i^2}{n_i} - \frac{T^2}{n}, \tag{10.9}$$

and

$$SSE = SSTot - SST, \tag{10.10}$$

where $T_i$ is the total for Treatment $i$ and $T$ is the grand total.

In ANOVA, $H_0$ is tested by means of an $F$-test,[10] as indicated in Table 10.1. The $F$-statistic has $k - 1, n - 1 df$; the test is an upper tail test, with critical values given in Table E.4. Formulas (10.8, 10.9, 10.10) are used for hand calculation. Minitab[11] provides two programs for the calculations, depending on how the data are set up. Minitab also provides a "$p$-value" as a part of the output. The interpretation of this is as follows: Prior to the analysis, select a level of significance $\alpha$ for the test. If the $p$-value is less than $\alpha$, reject $H_0$.

[10] When $k = 2$, the $F$-test is equivalent to the $t$-test of Sect. 9.6.4.

[11] Similar programs are included in all statistical packages.

```
One-way ANOVA: Strength versus Length

Analysis of Variance for Strength

Source     DF        SS       MS      F      P
Length      3    88.001   29.334  78.61  0.000
Error     210    78.362    0.373
Total     213   166.363

S = 0.6109    R-Sq = 52.90%    R-Sq(adj) = 52.22%

                                   Individual 95% CIs For Mean Based on
                                   Pooled StDev
Level        N    Mean    StDev    ----+---------+---------+---------+-----
12.7mm      50  3.0908   0.6353                          (--*-)
25.4mm      64  2.8495   0.6683                      (-*--)
265.0mm     50  1.7128   0.5318    (--*-)
5.0mm       50  3.5004   0.5818                                  (-*--)
                                   ----+---------+---------+---------+-----
                                     1.80      2.40      3.00      3.60

Pooled StDev = 0.6109
```

**Fig. 10.4** Minitab output of Anova and confidence intervals for means of fiber data

#### 10.5.1.3 Multiple Comparisons

The overall test addresses the question of whether or not all of the population means are equal. If $H_0$ is rejected, the experimenter or analyst will typically wish to know which means differ from which others. There are a number of procedures—called multiple comparison tests—that may be used to investigate this [2], Sect. 10.4.4. Several options are available in Minitab and other program packages, and the reader is referred to these for use of this methodology.

*Example 10.4* In Example we looked at the comparison of means of two treatments from the set of four given in Table F.14. In Example 9.9, the Kruskal–Wallis nonparametric test was used to compare all four treatments. The result was a $p$-value of 0.000. Here we use ANOVA to compare the four treatment means and include a multiple comparison procedure as well. Minitab output displaying the results of the ANOVA is shown in Fig. 10.4.

The $F$-value of 78.61 is significant at any level; the $p$-value given is 0.000. We conclude that treatment means differ. The output includes individual confidence intervals for the means. These give a good indication for this data set of which population means may differ. Minitab output for a statistical test of this is given in Fig. 10.5. The results are confidence intervals for differences between population means, calculated in such a way that the overall Type I error rate (over all six comparisons) is 5%. The procedure used is known as the Tukey Test, and is relatively conservative.[12] The interpretation of the results is that if the confidence interval for a given pair or treatments includes the value zero, the treatment means are considered not different. Here all fiber lengths are found to be different from each other with regard to average strength except for 25 mm versus 12.5 mm.

[12] See [15] Chap. 3, for this and other multiple comparison procedures.

```
Tukey's pairwise comparisons

Family error rate = 0.0500   Individual error rate = 0.0103
Intervals for (column level mean) - (row level mean)

              12.5mm      25mm       265mm
 25mm        -0.0571
              0.5397

 265mm        1.0618     0.8383
              1.6942     1.4351

 5mm         -0.7258    -0.9493    -2.1038
             -0.0934    -0.3525    -1.4714
```

**Fig. 10.5** Minitab output for Tukey's multiple comparison test

## *10.5.2 Analysis of Other Experimental Designs*

In general, ANOVA for other experimental designs follows the same principles. Sources of variability are identified, *df* are determined for each, and so forth, usually leading to F-tests of hypotheses about treatment means. Linear models such as that of (10.7) may be written for each design. These reflect the structure of the design and may lead to tests of structural elements other than treatments. Multiple comparison procedures may be used in these analyses as well.

Treatments may, in fact, be more complex than a simple classification such as that considered above in the CRD. In many experiments, more than one treatment factor (e.g., temperature and humidity as well as length of fiber) may be of interest. Experiments in which two or more levels of each of several factors are varied are called *factorial* experiments. In these, a treatment is a combination of one level of each factor, and treatment *df* and sums of squares are further broken down to additional sources called main effects and interactions. The models, analysis, and tests are somewhat more complex. These, along with many other aspects of DOE, are discussed in the references cited. See especially [21] for more complex experiments, including fractional factorials, incomplete block designs, and so forth.

### 10.5.2.1 ANOVA for the RCBD

We conclude the section with a discussion of the Randomized Complete Block Design. In the basic RCBD, experimental units are first grouped into $r$ groups of size $k$. The groups are called *blocks* or *replicates* and treatments are randomly assigned to experimental units in each block. The model for this design is

$$Y_{ij} = \mu + \tau_i + \rho_j + \varepsilon_{ij}, \tag{10.11}$$

**Table 10.3** ANOVA for the RCBD

| Source | *df* | SS | MS | F |
|---|---|---|---|---|
| Blocks | $r - 1$ | SSB | SSB/$(r - 1)$ | |
| Treatments | $k - 1$ | SST | SST/$(k - 1)$ | MST/MSE |
| Error | $(r - 1)(k - 1)$ | SSE | SSE/$(r - 1)(k - 1)$ | |
| Total | $rk - 1$ | SSTot | | |

where $\mu$, $\tau_i$, and $\varepsilon_{ij}$ are as before and $\rho_j$ is the replicate effect, which may be fixed or random.

The ANOVA breakdown for this design is given in Table 10.3. Sources of variability include the additional factor, Blocks. Sums of Squares are calculated using the principles exhibited in the formulas for the CRD. Minitab and other packages include programs, usually titled "Two-way Classification," for completing the computations.

The overall test comparing treatment means is the usual $F$-test. The nonparametric test for this experiment is the Friedman Test, discussed in Sect. 9.8.3. As noted, multiple comparison procedures may be applied here as well. Note that no test is given for Block effects. See [11] for comments regarding such tests.

## 10.6 Basic Linear Regression

### *10.6.1 Concept, Model and Assumptions*

In Chap. 8, we looked at the correlation coefficient as a measure of strength of linear relationship between two variables. Here we look at a related concept, *regression analysis*, which explicitly expresses, and uses data to estimate, this relationship. We assume a sample of size $n$ of bivariate observations $(x_i, y_i)$, where $x_i$ is a *predictor* variable (also called an *explanatory* or *independent* variable) and $y_i$ is the *response* (or *dependent*) variable, $i = 1, 2, \ldots, n$. The predictor variable may be deterministic (with values selected by the experimenter) or random. In the latter case, the analysis is a conditional analysis, given the values of the predictor. The response is assumed to be a normally distributed random variable. Examples are length of storage as a predictor of bond strength, ambient temperature as a predictor of number of breakdowns, and fiber strength as a function of length of fibers.

In regression analysis, we look at the relationship between the predictor variable $x$ and the expected values of the response variable $Y$. In linear regression, this is modeled as

$$Y_i = \alpha + \beta x_i + \varepsilon_i. \tag{10.12}$$

The error term $\varepsilon_i$ is assumed to be normally distributed with mean zero and variance $\sigma^2$ for all $x$.[13] As a result, $Y_i$ is normally distributed with mean $E(Y_i) = \alpha + \beta x_i$ and variance $\sigma^2$. Thus $Y_i$ is, on average, linearly related to $x_i$ with intercept $\alpha$ and slope $\beta$ (also called the *regression coefficient*), and $\sigma^2$ is a measure of the variability about the line.

A generalization of this set-up is the *multiple linear regression* model, in which there are $k$ predictors, with $E(Y_i)$ expressed as a multiple linear function of these. This model will also be developed in Chap. 13. Many other regression models are used in data analysis. These include polynomial regression (which may be analyzed by means of a multiple linear regression model), exponential models (some of which can be linearized by appropriate transformation of variables and some of which are intrinsically nonlinear), and so forth. For a thorough treatment of regression analysis, see [4].

## 10.6.2 Inference

We restrict attention to the simple linear regression of (10.12). Inference problems include:

- Estimation of the model parameters $\alpha$, $\beta$, and $\sigma^2$
- Test of relationship, formulated as a test of $H_0 : \beta = 0$ versus $H_1 : \beta \neq 0$
- Point and interval estimation of $E(Y_i)$ at a specified $x$-value
- Prediction of an individual $Y$-value at a specified $x$-value, with confidence interval

The method of least squares is used to estimate $\alpha$ and $\beta$.[14] The estimates, denoted $a$ and $b$, respectively, are

$$b = \frac{\sum_{i=1}^{n} x_i y_i - \frac{1}{n}\left(\sum_{i=1}^{n} x_i\right)\left(\sum_{i=1}^{n} y_i\right)}{\sum_{i=1}^{n} x_i^2 - \frac{1}{n}\left(\sum_{i=1}^{n} x_i\right)^2} \tag{10.13}$$

and

$$a = \bar{y} - b\bar{x}. \tag{10.14}$$

Denote the numerator and denominator of the RHS of (10.13) $S_{xy}$ and $S_{xx}$, respectively, and let $S_{yy}$ denote the sum of squares of $y$'s corresponding to $S_{xx}$. To estimate $\sigma^2$, the unbiased estimator is used. The estimate is

[13] Alternative assumptions will be discussed in Chap. 13.

[14] Under the stated assumptions, these are also the MLEs of $\alpha$ and $\beta$.

$$s^2 = \frac{S_{yy} - b^2 S_{xx}}{n-2}. \tag{10.15}$$

Under the stated conditions $b$ is normally distributed with mean $\beta$ and variance $V(b) = \sigma^2/S_{xx}$, which is estimated by

$$s_b^2 = s^2/S_{xx} \tag{10.16}$$

The test statistic for testing $H_0 : \beta = 0$ is

$$t = \frac{b}{s_b}. \tag{10.17}$$

Under $H_0$, this statistic has a Student-$t$ distribution with $(n - 2)$ *df*, and the $t$-test is performed in the usual way.[15] The test may be modified to test values other than $\beta = 0$ by subtracting the hypothesized value in the numerator of (10.15), and one-tailed alternatives may be tested by modifying the rejection region in the usual way. In addition, regression coefficients from different samples may be compared by modification of the procedure given for normal means in Sect. 9.6.4.

The basic test of $H_0 : \beta = 0$ may also equivalently be done by means of an $F$-test in an ANOVA. This is part of the output of the Minitab and other regression programs and will be discussed in more detail in Chap. 13.

A point on the sample regression line, $\hat{y} = a + bx$ is used for both estimating $E(Y)$ and predicting a single $Y$-value at a specified $x$-value. The variances used in calculating a confidence interval, however, differ in these two cases. The two-sided confidence interval for a predicted mean at a specified value $x_0$ is

$$\hat{y} \pm t_{n-2,1-(1-\gamma)/2}\, s\sqrt{\frac{1}{n} + \frac{(x_0 - \bar{x})^2}{S_{xx}}}, \tag{10.18}$$

where $\bar{x}$ is the sample mean and $\gamma$ is the confidence coefficient. For predicting a single observation, the corresponding confidence interval is

$$\hat{y} \pm t_{n-2,1-(1-\gamma)/2}\, s\sqrt{1 + \frac{1}{n} + \frac{(x_0 - \bar{x})^2}{S_{xx}}}. \tag{10.19}$$

### *10.6.3 Relation to Correlation Analysis*

As noted previously, the sample correlation coefficient $r$ is a measure of strength of linear relationship. In the above notation, $r$ is calculated as

[15] An equivalent test is an $F$-test with $1, n - 2$ *df* which is given in an ANOVA associated with this analysis. This will also be discussed in Chap. 13.

```
Regression Analysis: Repair Cost versus km

The regression equation is
Repair Cost = 73 + 25.3 km

Predictor    Coef  SE Coef     T      P
Constant     73.2    363.8  0.20  0.842
km          25.26    17.59  1.44  0.161

S = 954.089   R-Sq = 6.4%   R-Sq(adj) = 3.3%

Analysis of Variance

Source          DF        SS       MS     F      P
Regression       1   1876707  1876707  2.06  0.161
Residual Error  30  27308583   910286
Total           31  29185290

Unusual Observations

            Repair
Obs    km     Cost  Fit  SE Fit  Residual  St Resid
  2  29.2     5037  811     255      4226     4.60R

R denotes an observation with a large standardized residual.

Predicted Values for New Observations

New
Obs  Fit  SE Fit       95% CI          95% PI
  1  705     206  (285, 1125)  (-1288, 2698)

Values of Predictors for New Observations

New
Obs    km
  1  25.0
```

**Fig. 10.6** Minitab Regression Output for Data of Table F.2

$$r = \frac{S_{xy}}{\sqrt{S_{xx}S_{yy}}} \tag{10.20}$$

The statistic for testing $H_0 : \rho = 0$, where $\rho$ is the true correlation in the population is a $t$-test with $(n - 1)$ $df$, and is identical to that for testing $H_0 : \beta = 0$.

The square of the correlation coefficient, usually written $R^2$, can also be calculated as the ratio of the Regression SS to Total SS in the ANOVA mentioned above. The interpretation of $R^2$ is that it is the proportion of the total variability in $Y$ that is "explained" by the regression relationship.[16]

*Example 10.5* In Example 8.4, we looked at the correlation between repair cost and kilometers driven for a small set of warranty claims data. The data are given in Table F.2. The calculated correlation coefficient was $r = 0.254$. Here we look at a regression analysis of these data, with km driven as the predictor of cost. The Minitab output for this analysis is given in Fig. 10.6.

From the output we see that $a = 73.2, b = 25.26$, and $s^2 = 910,286$ (gotten as the MS for Error in the ANOVA). The entry for kilometers in the "SE Coef" column in Fig. 10.6, 17.59 is $s_b$. The calculated $t$ for testing $H_0 : \beta = 0$

[16] In computer output, $R^2$ is usually expressed as a percent and interpreted as the percent of explained variability.

is $t = 25.26/17.59 = 1.44$, with 30 *df* and, from the output, the *p*-value is 0.161. $H_0$ is not rejected at any level of significance less that 0.161, and we conclude that there is no evidence of a linear relationship between warranty cost for repair and kilometers driven. Note that the *F*-test in the ANOVA results in the same *p*-value since the two tests are equivalent. Both are also equivalent to the test of $H_0 : \rho = 0$, and we conclude that the two variables are uncorrelated.

Also included in the output are predicted values of the mean and a single *y*-value at $x = 25$ (i.e., 25,000 km driven at the time of the claim). Both predictions are $\hat{y} = 705$ ("Fit" in the output). The confidence interval for the mean is given as "CI" in the table, and is quite large; that for a predicted value ("PI" in the table) is so large as to be useless. These are a result of the fact that there is little, if any, relationship between the variables. This is also evident from the fact that the value of $R^2$ given in the output is 6.4%.

## 10.7 Estimation of Functions of Parameters

In analyzing warranty claims and related data, we often encounter situations in which we wish to estimate various quantities that are expressed as functions of the parameters of the underlying distributions of lifetimes or other characteristics for which data are available. Typical applications involve cost models and reliability functions. In some cases, it is possible to reparameterize the distribution and apply methods such as maximum likelihood directly. When that is not possible, a common approach is to estimate a function $\tau(\theta)$ (where $\theta$ may be a vector) by $\tau(\theta^*)$, where $\theta^*$ is an estimator of $\theta$, usually the MLE.

To extend the results to other inference problems, such as calculating confidence intervals or testing hypotheses, the distribution or asymptotic distribution of $\tau(\theta^*)$ must be determined. Some results that are useful in this regard are given in Appendix D. Included are a general approach for obtaining the asymptotic mean and variance of an estimator of a function $\tau(\theta)$ and specific results for sums, products and ratios of random variables. Under certain conditions, such estimators are asymptotically normal and the variance results can be used to form asymptotic confidence intervals and tests. This is done based on the variance formulas and using fractiles of the normal distribution. Here we look briefly at two examples to illustrate the approach, estimation of the coefficient of variation, and estimation of a warranty cost model. The method will be applied to estimation of system reliability in the next section.

### *10.7.1 Estimation of the Coefficient of Variation*

The coefficient of variation (CV), defined as $C = \sigma/\mu$ (often expressed as a per cent) was suggested in Sect. 8.3.4 as a measure of dispersion that is unit-free and is a useful measure for comparing relative variability in populations that may otherwise be very different. $C$ is estimated by $C^* = \sigma^*/\mu^*$, where $\mu^*$ and $\sigma^*$, the MLE's, are, in the normal case, the sample mean and the sample standard deviation, with the latter being calculated as the root of the sample variance with divisor $n$. We assume samples from normal populations, in which case the sample mean and standard are independent, so the covariance between the two is zero.

To determine the approximate variance of $C^*$, we use the approximation of (D.19). We have two variables, $\mu^*$ and $\sigma^*$, with $\tau(\mu^*, \sigma^*) = \sigma^*/\mu^*$. The required variances are $V(\mu^*) = \sigma^2/n$ and[17]

$$V(\sigma^*) = \sigma^2\left\{1 - \frac{1}{n} - \frac{2\Gamma^2(n/2)}{n\Gamma^2((n-1)/2)}\right\}, \tag{10.21}$$

where $\Gamma(.)$ is the gamma function. The required derivatives are

$$\frac{\partial\tau}{\partial\mu} = \frac{-\sigma}{\mu^2}, \quad \text{and} \quad \frac{\partial\tau}{\partial\sigma} = \frac{1}{\mu} \tag{10.22}$$

From (D.19) and the normality assumption,

$$V(C^*) \approx \frac{\sigma^2}{\mu^4}V(\mu^*) + \frac{1}{\mu^2}V(\sigma^*) = \frac{\sigma^4}{n\mu^4} + \frac{\sigma^2}{\mu^2}\left\{1 - \frac{1}{n} - \frac{2\Gamma^2(n/2)}{n\Gamma^2((n-1)/2)}\right\} \tag{10.23}$$

As usual, $V(C^*)$ is estimated by substitution of $\mu^*$ and $\sigma^*$ into the RHS of (10.23).

*Example 10.6* Data on strength of bonds between components in an audio subsystem (given in Table F.5) were partially analyzed in Examples 8.10 and 9.4. For these data, the MLEs are $\mu^* = 288.9$ and $\sigma^* = 37.38$, so $C^* = 0.1294$ or 12.9%. From (10.23), the estimated asymptotic variance of $C^*$ is

$$\hat{V}_a(C^*) = \frac{1}{51}(0.1294)^4 + (0.1294)^2\left\{1 - \frac{1}{51} - \frac{2}{51}\left(\frac{\Gamma(25.5)}{\Gamma(25)}\right)^2\right\} = 0.0001687 \tag{10.24}$$

The estimated standard deviation of $C^*$ is 0.0130, indicating that $C$ is known relatively precisely.

### 10.7.2 Estimation of a Warranty Cost Model

As a second example of the derivation of asymptotic results regarding a function of random variables, we look at estimation of the cost to the manufacturer, $C_m(W)$, of

[17] From [2], p. 147.

the nonrenewing PRW for which the buyer is given a pro-rata rebate on failure of an item under warranty. The cost model for this warranty is given in (6.26). For purposes of illustration, we assume that the time to failure of the item is exponentially distributed with parameter $\lambda$.

Estimation of the cost model requires estimation of the CDF and of the partial expectation $\mu_W$. For the exponential distribution, the latter is given by

$$\mu_W = \frac{1}{\lambda}\left(1 - e^{-\lambda W} - \lambda W e^{-\lambda W}\right). \tag{10.25}$$

The CDF is $F(y) = 1 - \mathrm{e}^{-\lambda y}, y > 0$. $\lambda$ is estimated by $\lambda^* = 1/\bar{Y}$, with asymptotic variance given by (9.8) as $V_a(\lambda^*) = \lambda^2/n$. From (6.26), the cost model assuming exponentially distributed lifetimes becomes

$$E(C_m(W)) = C_b\left\{1 - \frac{1 - e^{-\lambda W}}{\lambda W}\right\}. \tag{10.26}$$

Thus we may take $\tau(\lambda)$ to be $(1 - e^{-\lambda W})/\lambda W$, so

$$\frac{\partial \tau}{\partial \lambda} = \frac{(1 + \lambda W)e^{-\lambda W} - 1}{\lambda^2 W}. \tag{10.27}$$

As before, all of these are estimated by substitution of the MLE $\lambda^*$ for $\lambda$. The asymptotic variance of $C^*$ is obtained from (D.11) as

$$V_a(C^*) = C_b^2\left(\frac{(1 + \lambda W)e^{-\lambda W} - 1}{\lambda^2 W}\right)^2\left(\frac{\lambda^2}{n}\right) \tag{10.28}$$

### *10.7.3 Estimation of Reliability*

The reliability of a product may be estimated at many stages—during product development, based on test data on components, prototypes, etc.; during production, based on sampling and testing of finished products, and post-production, based on further sampling and testing, warranty claims or follow-up studies.

As discussed in Chap. 3, the reliability assessments at the various stages are dealing with different measures of reliability, including inherent reliability, reliability at sale, and field reliability. The last is of primary practical importance, for both the consumer and the manufacturer. For the consumer, this is the experienced product characteristic. For the manufacturer, it is the reliability that drives warranty costs, consumer complaints and company reputation.

Here we look at data-based assessment of reliability. The data may be test data at various levels and from various sources or warranty data of the types discussed in Chaps. 4 and 5, including warranty claims data. In connection with the latter, we note again that claims data present special difficulties when used to estimate field reliability because of warranty execution and reporting problems. It is important to recognize that as a result of these, claims rates almost always underestimate the actual field failure rate and hence overestimate the actual field reliability.

In the remainder of this section, we look at estimation of reliability based on various types of data, ranging from simple pass-fail (binomial) data to time-to-failure data from several life distributions, to estimation based on system models. Except for the binomial case, symbols such as the following will be used to indicate estimators or estimates: $R$, $R^*$ (to indicate MLE's), $\hat{R}$ (as a generic symbol), $R(t)$ (to indicate dependence on time), etc. We assume a random sample of size $n, y_1, y_2, \ldots, y_n$, which may be complete or incomplete.

#### 10.7.3.1 Binomial Data

The simplest type of failure data is simply a count of the number of failures observed in a sample of $n$ items. The items may be parts, components, systems, etc. Failure may be inability of the item to perform its function, inability to produce a specified output, etc. Under the assumption that the data are a random sample of independent items, the distribution of number of failed items $X$ is a binomial distribution (A.12) with parameters $n$ and $p$, where $p$ is the proportion of failed items and is the parameter we wish to estimate. The solution is given in (9.1), with variance given in (9.2) and confidence interval given in (9.46). Note that in most reliability applications, we are interested in verifying that the reliability is at least a specified value, in which a lower one-sided confidence interval is appropriate.

#### 10.7.3.2 Exponential Failure Data

Here are assume a random sample of $n$ items from a population in which time to failure is exponentially distributed (A.21) with parameter $\lambda$. The data may by failure data on parts, systems, etc., and may be complete or right-censored. The reliability function is

$$R(t) = P(T > t) = e^{-\lambda t} \tag{10.29}$$

For complete data, the MLE of $\lambda$ is given in (9.7), from which we obtain the MLE of $R(t)$ as

$$R^*(t) = e^{-t/\bar{y}}, \tag{10.30}$$

where $\bar{y}$ is the sample mean. A lower one-sided confidence limit $R_L(t)$ with confidence $\gamma$ is given by

$$R_L(t) = e^{-[t\chi^2_{\gamma,2n}]/2n\bar{y}}, \tag{10.31}$$

where $\chi^2_{\gamma,\, 2n}$ is the $\gamma$-fractile of the chi-square distribution with $2n$ *df*.

For right-censored data, with Type II censoring, the procedure is the same, except that $n\bar{y}$ is replaced by the sum of all observations and censoring times. For more on this and other types of censoring, see [10].

### 10.7.3.3 Weibull Failure Data

We consider the two-parameter Weibull distribution of (A.30), and assume complete data. The reliability function is

$$R(t) = R(t; \alpha, \beta) = e^{-(t/\alpha)^\beta}. \tag{10.32}$$

The MLE is $R^* = R(t; \alpha^*, \beta^*)$, where $\alpha^*$ and $\beta^*$ are the MLE's of the scale parameter $\alpha$ and shape parameter $\beta$, given in (9.14) and (9.15), respectively.

The asymptotic variance of $R^*$ [2, p. 266], is:

$$V_a(R^*) = V_a(\alpha^*)\left(\frac{\partial R}{\partial \alpha}\right)^2 + V_a(\beta^*)\left(\frac{\partial R}{\partial \beta}\right)^2 + 2Cov_a(\alpha^*, \beta^*)\left(\frac{\partial R}{\partial \alpha}\right)\left(\frac{\partial R}{\partial \beta}\right) \tag{10.33}$$

The asymptotic variances of $\alpha^*$ and $\beta^*$ that appear in this expression are given in (9.16) and (9.17). The asymptotic covariance is

$$Cov_a(\alpha^*, \beta^*) = \frac{0.254}{n\alpha}. \tag{10.34}$$

The required derivatives are

$$\frac{\partial R}{\partial \alpha} = \frac{\beta t^\beta}{\alpha^{\beta+1}} e^{-(t/\alpha)^\beta}, \quad \frac{\partial R}{\partial \beta} = \left(\frac{t}{\alpha}\right)^\beta \log(\alpha/t) e^{(t/\alpha)^\beta} \tag{10.35}$$

As before, these quantities are estimated by replacing the parameters by their MLE's. The results may be used to obtain asymptotic confidence intervals for reliability at part, component, or other levels for which appropriate failure data are available.

*Example 10.7* In example 10.3 we looked at the fiber strength data of Table F.14, and concluded that the Weibull distribution was a reasonable fit for the various samples. Here we look at 265 mm fibers. A WPP plot of the data is given in Fig. 10.7. Note from the figure that the A–D statistic is 0.6259. As noted in Example 10.3, this value is not significant at the 5% level. The Weibull fit appears to be good except for the three smallest data points, which can be considered possible outliers. For illustration, we use the entire sample in the reliability analysis.

Suppose that in a certain application fiber strength of 0.075 pounds is required for fibers of this length. The reliability of the fiber is then

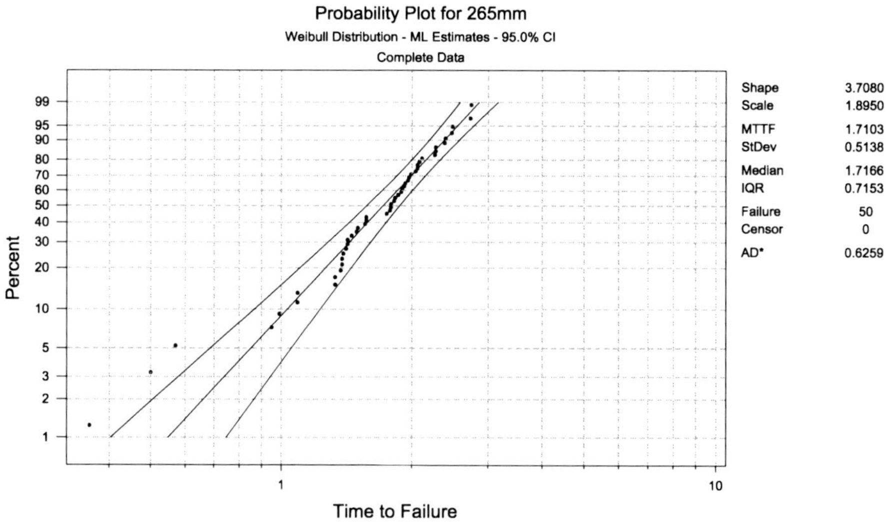


**Fig. 10.7** Weibull plot of 265 mm fiber strength data

$$R(0.75) = e^{-(0.75/\alpha)^{\beta}}$$

From Fig. 10.7, we find the MLE's of the Weibull parameters to be $\alpha^* = 1.895$ and $\beta^* = 3.708$, so that the MLE of $R$ is $R^*(0.75) = 0.9683$. From (9.16), (9.17), and (10.34), the estimated asymptotic variances and covariance are found to be

$$\hat{V}_a(\alpha^*) = \frac{1.1087(1.895)^2}{50(3.708)^2} = 0.00579, \quad \hat{V}_a(\beta^*) = \frac{0.6079(3.708)^2}{50} = 0.0167$$

and

$$\hat{Cov}_a(\alpha^*, \beta^*) = \frac{0.254}{50(1.895)} = 0.00268$$

The estimates of the derivatives are obtained from (10.35) as

$$\frac{\hat{\partial R}}{\partial \alpha} = 0.0629, \quad \frac{\hat{\partial R}}{\partial \beta} = 0.0289$$

From (10.33), the estimated asymptotic variance of $R^*(0.75)$ is calculated as

$$\begin{aligned}\hat{V}_a(R^*) &= 0.00579(0.0629)^2 + 0.167(0.0289)^2 + 2(0.00268)(0.0629)(0.0289)\\ &= 0.000172.\end{aligned}$$

Thus the estimated standard deviation of $R^*$ is 0.0,131. A 95% confidence interval for $R$ is roughly $0.97 \pm 0.03$ or (0.94, 1.00). We conclude that the reliability is not known very precisely and that the part is apparently not very reliable.

#### 10.7.3.4 Other Distributions

For reliability estimation for other failure distributions, including the gamma, lognormal, and extreme value, see [2]. For additional discussion, see [8] and [10].

#### 10.7.3.5 Estimating System Reliability from Component Reliabilities

We assume $k$ components, operating independently. The basis of estimating system reliability in this way is modeling of the system, e.g., by means of a block diagram. We consider two relatively simple systems, series and parallel structures. This is discussed in Sect. 3.8.3, with models for series systems given in (3.26) and for parallel systems in (3.27). More complex systems can usually be modeled as combination of these two structures.[18]

*Series systems* For $k$ independent components connected in series, the system reliability is

$$R = R_1 R_2 \ldots R_k, \tag{10.36}$$

where $R_i$ is the reliability of the $i$th component. If the $R_i$'s are estimated from independent sets of data, with estimates denoted $\hat{R}_i$, the estimate of $R$ is calculated as the product of the $\hat{R}_i$. If the estimators of the component reliabilities are unbiased, so is the estimator of $R$, since under independence the expectation of the product is the product of the expectations. (See Appendix D.) The variance of the estimator is somewhat more complex. For $k = 2$, the result is

$$V(\hat{R}) = [E(\hat{R}_1)]^2 V(\hat{R}_2) + [E(\hat{R}_2)]^2 V(\hat{R}_1) + V(\hat{R}_1)V(\hat{R}_2) \tag{10.37}$$

Note that is the estimators are unbiased, the expectation terms in (10.37) are simply the true values of the reliabilities. For larger values of $k$, the result becomes increasingly complex.[19] The variance is estimated as before and may be used to obtain asymptotic confidence intervals in the usual way.

*Parallel systems* For $k$ components operating independently in a parallel configuration, with reliabilities denoted as above, the system reliability is given by

$$R = 1 - (1 - R_1)(1 - R_2)\ldots(1 - R_k) \tag{10.38}$$

This is estimated in the usual way. Since for any random variable $X$, $V(1 - X) = V(X)$, the variance of the estimator may be obtained by simple modification of (10.37). This may be used in inference problems regarding system reliability as before.

---

[18] For details regarding the modeling of more complex systems, see [2], Chap. 7, [6] and [8].

[19] See [2], Appendix A for $k = 3$. The result extends to higher values in an obvious way.

*Confidence intervals for system reliability* The above results may be used to obtain confidence intervals (exact or asymptotic) for reliability in the usual way. As the size of the system increases and becomes more complex, the confidence interval calculations become increasingly difficult, and computer algorithms are needed to complete the analysis. There are a number of commercial products available for this purpose.

An alternative is to use approximations or bounds. One such, based on reducing the problem to binomial system data, is given by [17].[20]

#### 10.7.3.6 Other Inference Problems in Reliability

We list and briefly discuss several other approaches to reliability analysis and the role of data in applications of these.

*Bayesian Analysis* Bayesian statistical analysis provides a unique approach to estimation of system reliability based on component reliabilities. This involves a two-stage analysis, beginning at the sub-system level (or a multi-stage analysis beginning at a lower level) and the use of these results in the formulation of a prior distribution at the system level. Once data are obtained at the system level, the Bayesian methodology is applied to estimation of system reliability, thus providing a well-structured framework for incorporating information at all levels into to final result A thorough discussion of the method along with a worked example is given in [14].[21]

*Estimation in Stress-strength models* In stress-strength modeling, failure is characterized by means of two variables, the strength of the item $X$, and the stress $Y$ to which the item is subjected ([2], Sect. 6.5). The item fails if $Y > X$, and the reliability of the item is calculated as $P(X > Y)$. $X$ and $Y$ are assumed to be random variables with specified probability distributions. In experimental studies, measurements of $X$ and $Y$ are taken on failure of the test item and the analysis proceeds by estimating the parameters of distributions of $X$ and $Y$ and using these to estimate $P(X > Y)$. Details and examples for the normal and exponential distributions are given in [2], Sect. 8.7.1.[22]

*Estimation of the parameters of intensity functions* Estimation of intensity functions is key to analysis of data in situations where system failures occur according to a point process (See Appendix B). Here we assume that failures occur independently and the objective is to estimate the parameters of an intensity function $\Lambda(t, \theta)$. Approaches to the problem for the following two data structures

---

[20] See [2], Chap. 8, for a discussion of this result.

[21] For additional details, extensions of these results, and more thorough discussions of the approach, see [13] and [2], Sect. 8.6.4. For application to lifetime data, see [7]. For more on the Bayesian approach to reliability analysis generally, see [12].

[22] Additional models of this type are given in [9], Chap. 6. For a nonparametric approach, see [19].

are given in [2]: (i) actual failure times observed for all items, and (ii) failure time grouped into intervals. This is discussed in Chap. 12.

## 10.8 Tests of Assumptions

In this and the preceding two chapters and well as in the remaining chapters on data analysis techniques, many assumption are made concerning the data, including assumptions regarding the populations from which the data were obtained, how they were obtained, the type and structure of data, and so forth. In fact, the selection of a proper method of analysis depends crucially on these assumptions. An incorrect methodology may lead to incorrect or, in extreme cases, misleading conclusions and interpretations of the results. In analysis of data, it is important to consider the validity of the assumptions and to be aware of the potential consequences of their violation.

Typical assumptions made in data analyses include the following:

- In parametric data analysis generally, it is typically assumed that observations (scalar or vector) are independent, identically distributed (IID) random variables drawn from a specified distribution (exponential, normal, Weibull, etc.), though the parameters (or other, related characteristics) of the distribution may vary for different subsets of the data.
- In nonparametric data analysis, the same is true except that the form of the distribution is not specified.
- In ANOVA, the assumptions are independent observations from normal distributions with the same population variances, with means given by a model that reflects the structure of the experiment and lack of interaction between experimental factors unless specifically included in the model.
- In regression analysis the data on two or more variables are assumed to follow (on the average) a specific model that expresses the relationship between the variables. This may be linear, multiple linear, or nonlinear, and the individual data elements are assumed to be independent and normal, with equal variances.

In the remainder of this section, we look briefly as tests of some of these assumptions.

### *10.8.1 Tests of Independence*

Violation of the assumption of independence is of particular concern in analysis of time series data, that is, data on the same item or process collected sequentially in time, but it may occur in other data as well. In general, independence is assured by selecting the individuals to be included in the sample by use of a random process.

The time series type of dependence is not often a problem in dealing with warranty data since most of the data (claims or supplementary) can reasonable be assumed to have been generated by some random process—either actual randomization as in a properly designed experiment, or the random occurrence of failures. Violations of this may result when more than one failure occurs on the same item or due to the fact that customers are not actually chosen randomly, but these are unlikely to introduce any significant amount of dependence in any of the data we ordinarily encounter.

A simple nonparametric test of randomness or independence is the *runs test* ([20], Sect. 15.9). In this, the signs of the deviations of the observations from the estimated model (called *residuals*) are listed sequentially in the order in which the data were collected. The test statistic is a count of the number of runs of like symbols (positive or negative deviations). The number of runs expected under independence is calculated; significantly fewer than this expected number leads to the conclusion of non-independence. (See the reference cited for details.)

A second approach, developed in the context of time series analysis, is to look at *autocorrelations*, that is, the correlation between successive observations (ordered by time), the correlation between observations two apart, etc. (called autocorrelations of lag 1, lag 2, etc.). Details of the calculation and use of autocorrelations can be found in texts on time series as well as in the documentation of program packages such as Minitab, SAS, etc.

### *10.8.2 Tests of Distributional Assumptions*

We discuss tests of distributions very briefly since the two approaches most used for this are

1. Graphical methods of assessing fits (WPP and other probability plots; other aids to fitting)—discussed in detail in Chap. 8 and Appendix C
2. Goodness-of-fit tests (chi-square, A–D, K–S)—discussed in detail in Sect. 10.3 above and the references cited

Note that for more complex models (e.g., regression, ANOVA), the tests are to be performed on the residuals (see the previous section). This leads to additional complications with the statistical goodness-of-fit tests, since the residuals are not independent, even if the data are, nor are exact or approximate critical values known for this application. Nonetheless, the procedures are useful in model building as discussed previously.

If there is evidence of violation of distributional assumptions, the usual advice is to look at transformations of the data to a scale on which the assumption is more tenable. Common transformations are logarithmic, square root or other power transformations, and so forth.[23] If a scale on which the assumption (normal or exponential, etc.) appears to hold, at least approximately, cannot be found, caution

[23] See [11] and [15] for discussions of this in the context of ANOVA.

in interpretation of the results should be exercised since any probabilistic statements made depend on this assumption for validity. This is particularly true in confidence interval estimation where the true confidence may be very different from the nominal (e.g., 75% instead of 95%) and in hypothesis testing, where the level of significance may be similarly affected.

## *10.8.3 Tests of Assumption in ANOVA*

Tests of independence and normality were discussed previously. Here we look briefly at tests of homogeneity of variance and tests of the model based on residuals. We restrict attention to the Completely Randomized Design. The model for this design is given in (10.7).

### 10.8.3.1 Tests of Homogeneity of Variance

There are several tests for the hypothesis that $k$ population variances are equal. Two commonly used tests are Bartlett's test, which is sensitive to the assumption of normality, and Levene's test, which is distribution-free. Both are available in Minitab, which also provides confidence intervals for the population variances (based on normality).

The test statistic for Bartlett's test is

$$\chi^2 = \frac{1}{c}\left[\left(\sum_{i=1}^{k}(n_i - 1)\right)\log(s^2) - \sum_{i=1}^{k}[(n_i - 1)\log(s_i^2)]\right], \tag{10.39}$$

where

$$c = 1 + \frac{1}{3(k-1)}\left(\sum_{i=1}^{k}\frac{1}{n_i - 1} - \frac{1}{n-k}\right) \tag{10.40}$$

and $n = \Sigma n_i$. Under $H_0$, the test statistic is distributed as $\chi^2$ with $(k - 1)$ *df*. The test is an upper-tail test.

If normality is not a tenable assumption, the Levene test may be used. This is based on the absolute deviations of the individual observations from their respective sample medians. An ANOVA is performed on these quantities. A one-side test at the 1% level is suggested.[24]

*Example 10.8* In Example 10.4, ANOVA was used to compare the mean strengths of fibers of 4 lengths. From the Minitab output given in Fig. 10.4, it is seen that the sample variances in these four groups vary from 0.2828 to 0.4466, with df ranging from 49 to 63. The Bartlett and Levene tests for these data are given in the Minitab

[24] For more on these tests and some alternatives, see [11], Sect. 2.5 and [21], Sect. 3.14.

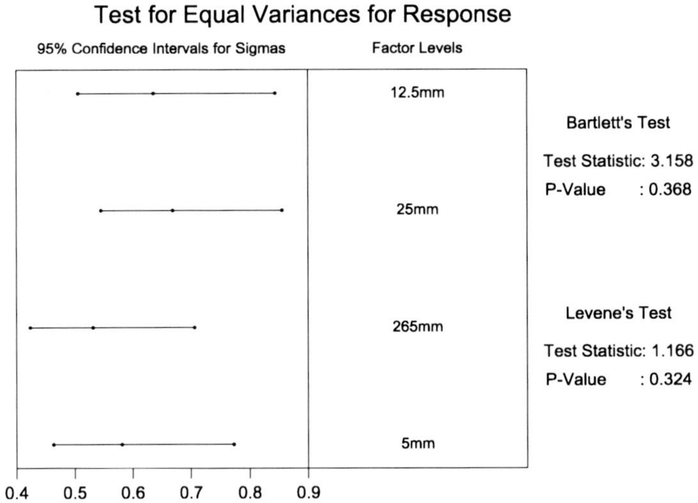


**Fig. 10.8** Test of homogeneity of variance

output shown in Fig. 10.8. Neither test leads to rejection of the null hypothesis that the population variances are equal.

#### 10.8.3.2 Other Tests of Assumptions in ANOVA

Other tests and plots based on residuals can be informative with regard to the model and assumptions. The residuals (denoted $e_i$) in ANOVA are obtained by estimating the model elements ($\mu$ and the $\tau$'s), and then calculating the residuals as

$$e_{ij} = y_{ij} - \hat{\mu} - \hat{\tau}_i = y_{ij} - \bar{y}_i \tag{10.41}$$

Residual plots are also widely used in regression analysis and we defer the discussion on their interpretation until the next section. For additional discussion on this in the context of designed experiments and ANOVA, see [21], Sect. 1.9 and [2], Chap. 10.

### *10.8.4 Tests of Assumption in Regression Analysis*

Regression analysis is also based primarily on linear (or intrinsically linear[25]) models. In regression analysis, most tests are based on residuals. For simple linear regression, with model (10.12), the residuals are given by

[25] An intrinsically linear model is one that can be transformed to linearity. A simple example is the polynomial model $y_i = \beta_0 + \beta_i x_i + \beta_2 x_i^2 + \cdots + \beta_k x_i^k$, where each power of $x$ can be treated as a separate variable and the model becomes a multiple linear model.

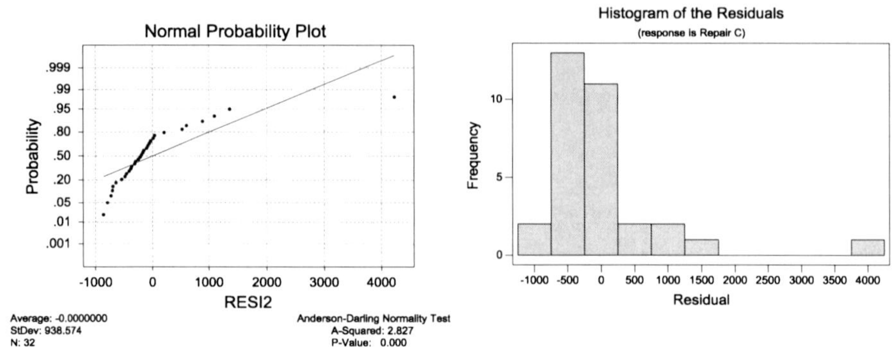


**Fig. 10.9** Normal probability plot and histogram of residuals in km-repair cost regression

$$e_i = y_i - a - bx_i \tag{10.42}$$

As in ANOVA, these are simply the difference between the observed values and the estimated model (the regression line in this case). The residuals may be used to test the normality assumption as indicated above. They may also be used to test equality of variances and adequacy of the linear model.

Plots of residuals are often used to obtain graphical evidence regarding the adequacy of assumptions. Residuals may be plotted against $y_i$, against fitted values, and against any other variables of interest. The normal pattern that indicates no departure from assumed conditions is a horizontal pattern with random scatter about the center of the plot. Patterns other than this may indicate anomalies such as skewed distributions, unequal variance about the line (or plane or hyperplane in higher dimensions), on an incorrect model. Possible problems are nonlinearity, outliers in the data, interactions among predictors in multiple regression, improper choice of predictors, etc.[26] In the following example, we look at a partial residual analysis for the repair cost data of Table F.2.

*Example 10.9* In Example 10.5, we looked the relationship between km driven and the repair costs for 32 items that had failed under warranty. A regression analysis was done and it was concluded that there is apparently no linear relationship between the two variables. Plots of residuals are given in Figs. 10.9 and 10.10.

It is apparent from Fig. 10.9 that the assumption of normality is invalid and that there is at least one outlier. Both the histogram and the normal probability plot of the residuals indicate that the data are highly skewed. In addition, the value of the A–D statistic is given as 2.827, with an associated $p$-value of 0.000, which, in spite of the failure of assumptions of the test to hold, supports the conclusion on non-normality.

---

[26] Residual plots and their interpretation are discussed in detail in [4] and most other texts on regression analysis.

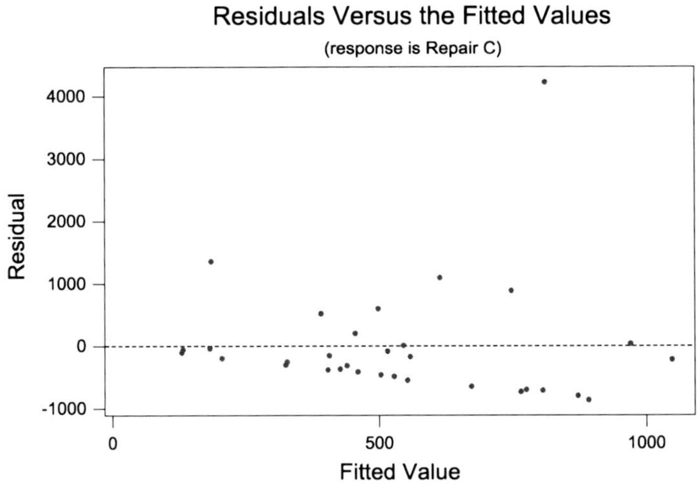


**Fig. 10.10** Plot of residuals versus fitted values

We note that the skewness of repair cost itself was apparent in the histogram of the data in Fig. 8.3. This does not, however, necessarily lead to skewness of the residuals. Under some circumstances if there is a strong linear relationship between the predictor variable and response variables, this could account for the skewness in the analysis where the predictor was ignored.

The plot of residuals versus fitted values supports the conclusions already stated. The extreme outlier is apparent in the plot, and it appears that, in fact, there may be several less extreme outliers as well. Even without the outlier(s), the regression assumptions appear not to be valid.

## References

1. Barnett V, Lewis T (1994) Outliers in statistical data. Wiley, New York
2. Blischke WR, Murthy DNP (2000) Reliability. Wiley, New York
3. D'Agostino RB, Stephens MA (1986) Goodness-of-fit techniques. Dekker, New York
4. Draper NR, Smith H (1998) Applied regression analysis, 3rd edn. Wiley-Interscience, New York
5. Grubb FE (1969) Simple criteria for detecting outlying observations. Technometrics 1:1–24
6. Høyland A, Rausand M (1994) System reliability theory. Wiley-Interscience, New York
7. Hulting FL, Robinson JA (1994) The reliability of a series system of repairable subsystems. Nav Res Logist Q 41:483–506
8. Kececioglu D (1994) Reliability engineering handbook, vol 2. Prentice-Hall, Englewood Cliffs, NJ
9. Kapur KC, Lamberson LR (1977) Reliability in engineering design. Wiley, New York
10. Lawless JF (1982) Statistical models and methods for lifetime data. Wiley, New York
11. Lorenzen TJ, Anderson VL (1993) Design of experiments. Marcel Dekker Inc., New York
12. Martz HF, Waller RA (1982) Bayesian reliability analysis. Wiley, New York

13. Martz HF, Waller RA (1990) Bayesian reliability analysis of complex series-parallel systems of binomial subsystems and components. Technometrics 32:407–416
14. Martz HF, Waller RA, Fickas ET (1988) Bayesian reliability analysis of series systems of binomial subsystem s and components. Technometrics 30:143–159
15. Montgomery DC (2005) Design and analysis of experiments, 6th edn. Wiley, New York
16. Murthy DNP, Xie M, Jiang R (2004) Weibull models. Wiley, New York
17. Spencer FW, Easterling RG (1986) Lower confidence bounds on system reliability using component data. Comm Statist Theory and Methods 18:4211–4227
18. Stuart A, Ord JK (1991) Kendall's advanced theory of statistics, 5th edn, vol 2. Oxford University Press, New York
19. Ury HK (1972) On distribution free confidence bounds for $P(Y > X)$. Technometrics 14:577–581
20. Wackerly D, Mendenhall W, Scheaffer RL (2007) Mathematical statistics with applications. Duxbury, New York
21. Wu CFJ, Hamada M (2000) Experiments. Wiley Interscience, New York

# Part IV
# Warranty Data Analysis

# Chapter 11
# Nonparametric Approach to the Analysis of 1-D Warranty Data

## 11.1 Introduction

Data analysis begins with the use of graphical and analytical approaches in order to gain insights and draw inferences without making any assumptions regarding the mathematical formulation that is appropriate for modeling the data. *Nonparametric methods* play an important role in this approach to the analysis of data. They provide an intermediate step toward building more structured models that allow for more precise inferences with a degree of assurance that the model assumptions are valid. As such, nonparametric methods are also referred to as distribution-free methods. This is in contrast to parametric methods (discussed in Chap. 12), which begin with a probabilistic model and then carry out the analyses as appropriate to that model.

In either case, the analysis depends on the data and the data structure. If the data are limited to just warranty claims data (discussed in Chap. 4), then we have only failure data. Supplementary data (such as censored data, etc., as discussed in Chap. 5) provide additional information. We consider the three different data structures that are discussed in Sect. 5.12. Other issues such as repair or replace, the type of warranty, etc., are also important. As a result there can be many different scenarios for each of the three data structures, as indicated in Sect. 5.13.

This chapter deals with the nonparametric approach to the analysis of 1-D warranties for these scenarios. The outline of the chapter is as follows: In Sect. 11.2, we discuss the nonparametric approach and indicate the quantities for which this approach can be used to draw inferences using estimates based on data of various types. The next three sections (Sects. 11.3–11.5) deal with the nonparametric approach for deriving the estimators. We give the form of the estimators and illustrate their use by means of examples based on data sets from Appendix F. We conclude with some comments and suggestions for practitioners in Sect. 11.6.

W. R. Blischke et al., *Warranty Data Collection and Analysis*,
Springer Series in Reliability Engineering, DOI: 10.1007/978-0-85729-647-4_11,

## 11.2 Non-parametric Approach to Data Analysis

The nonparametric approach allows the user to analyze data without assuming an underlying distribution. That is, the approach does not require that the form of the sampled population be known. The nonparametric approach has a number of advantages as well as some inherent disadvantages. The ability to analyze data without assuming an underlying life distribution avoids many potential errors that may occur because of incorrect assumptions regarding the distribution. On the other hand, the use of nonparametric procedures with data that can be handled with a parametric procedure is inefficient in the sense that it results in loss of information. In particular, the confidence bounds associated with the nonparametric approach are usually wider than those calculated via the parametric approach, and predictions outside the range of the observations are usually not possible. It is recommended that any set of warranty data first be subjected to a nonparametric analysis before moving on to parametric analyses based on the assumption of a specific underlying distribution.

In Chap. 9, we looked at nonparametric methods based on medians, ranks and similar measures. Procedures included nonparametric tolerance intervals, confidence intervals based on medians, and comparisons of populations. Here we look at the nonparametric approach to inferences regarding quantities such as the distribution function $F(t)$, density function $f(t)$, reliability function $R(t)$, hazard function $h(t)$, cumulative hazard function $H(t)$, renewal function $M(t)$, mean cumulative function (MCF) $\mu(t)$, as well as warranty claim rates (WCR) at the product, component or some intermediate level.[1] These quantities may be used to estimate and forecast future warranty costs. For example, estimates of $M(t)$ are needed for estimation of warranty costs for the non-renewing FRW policy where failed items are replaced by new, and of $\mu(t)$ when failed items are rectified through minimal repair.

The nonparametric approach has been used in some of the earlier chapters without detailed discussion on the properties of the estimators.[2] In this chapter, we focus our attention on alternate forms of estimators for some of the quantities mentioned above and the statistical properties of the resulting estimators (e.g., expected value, variance, asymptotic properties, etc.).

## 11.3 Analysis of Structure 1 Data

Structure 1 data consists of detailed information on each item and is most appropriate for estimating the failure distribution, reliability, hazard functions and associated functions such as the renewal function and MCF. In Sect. 5.13 we

[1] These quantities are discussed in Chap. 3, Appendix A, and C. The objectives or goals for analysis are discussed in Sect. 1.6.1.

[2] The EDF discussed in Sect. 8.5.1 is based on the nonparametric approach.

discussed four different scenarios (Scenarios 1.1–1.4). In Scenarios 1.1 and 1.2, the data are simply the failures times, with failed items replaced by new ones in the former case and rectified through minimal repair in the latter. Scenarios 1.3 and 1.4 are the data of Scenarios 1.1 and 1.2, respectively, with the addition of censored times for unfailed items (including original items and unfailed replacements of items that failed under warranty.)

In this section we look at the nonparametric approach for estimation of $F(t)$, $M(t)$, $\mu(t)$, etc., for these four scenarios.

### 11.3.1 Data Scenario 1.1

The nonparametric estimation of the CDF, also known as the EDF, is discussed in Sect. 8.5.1 and in Appendix C (Sect. C.2.1). The resulting estimate is denoted $\hat{F}(t)$. The estimate of the reliability function is given by $\hat{R}(t) = 1 - \hat{F}(t)$. The estimate of the cumulative hazard function is given by $\hat{H}(t) = -\log \hat{R}(t)$. Example 8.7 provides an illustration of the method.

We focus our attention on the estimation of the renewal function and look at three different methods.

*Method 1* An approach to nonparametric estimation of the renewal function is given in [5, 6] under the assumption that the inter-event times have positive mean $\mu$ and finite variance $\sigma^2$. If $\hat{\mu}$ and $\hat{\sigma}$ are estimates of $\mu$ and $\sigma$, respectively, an estimate of $M(t)$ is given by[3]

$$\hat{M}_1(t) = \frac{t}{\hat{\mu}} + \frac{\hat{\sigma}^2}{2\hat{\mu}^2} - \frac{1}{2} \tag{11.1}$$

for $F(t)$ nonarithmetic.[4] For this method, the asymptotic variance of $N(t)$, denoted $V_1(t)$, is given by[5]

$$\hat{V}_1(t) = \frac{\hat{\sigma}^2 t}{\hat{\mu}^3} + \frac{5\hat{\sigma}^4}{4\hat{\mu}^4} - \frac{2\hat{\mu}_3}{3\hat{\mu}^3} + \frac{1}{12}, \tag{11.2}$$

where $\hat{\mu}_3$ is the sample third moment about the mean, which is assumed to be finite. This implies that the estimates of $M_1(t)$ and $V_1(t)$ are obtained by use of estimates of the moments.

[3] Some properties of this estimator are discussed in [2], p. 549.

[4] A distribution function is said to be arithmetic if its support is on $\{0, \pm d, \pm 2d, \ldots\}$ for some constant $d$ and otherwise is said to be non-arithmetic, [6].

[5] See [7] and ([2] p. 554) for details.

*Method 2* Let $\{i_1, i_2, \ldots, i_k\}$ be a subset of size $k$ of the set $\{1, 2, \ldots, n\}$ and $\sum_c$ be the sum over all $\binom{n}{k}$ distinct subsets of $\{i_1, i_2, \ldots, i_k\}$. Then an unbiased estimator of $P(S_k = T_{i_1} + T_{i_2} + \cdots + T_{i_k} \le t) = F^{(k)}(t)$ is defined [1, 2] as

$$\hat{F}^{(k)}(t) = \binom{n}{k}^{-1} \sum\nolimits_c I(T_{i_1} + T_{i_2} + \cdots + T_{i_k} \le t) \tag{11.3}$$

Here $I(.)$ is the indicator function of a set. Let $m = m(n)$ be a positive integer depending on $n$ such that $m \le n$ and $m \to \infty$ as $n \to \infty$. References [1, 2] defines another nonparametric estimator of the renewal function as

$$\hat{M}_2(t) = \sum_{k=1}^{m} \hat{F}^{(k)}(t) \tag{11.4}$$

For this method, the variance of $N(t)$, denoted $V_2(t)$, can be estimated by use of (B.10). Asymptotic properties and an interval estimator of $M_2(t)$ can be found in [5]. Numerical comparisons between the above two estimators and their use in warranty analysis are discussed in [5, 6].

*Method 3* Reference [3] proposed nonparametric estimation of the renewal function through the nonparametric estimate of the cumulative probability distribution $F(t)$ of the lifetime variable $T$. They present the necessary formulas for nonparametric estimation of $M(t)$ by (11.4) and suggest that $M(t)$ could be estimated by substituting the step function $\hat{F}(t)$, the PL estimate[6] of $F(t)$, in those formulas. For this method, the variance of $N(t)$ can be estimated using (B.10).

*Comparison of the Methods* Reference [6] investigated the relative performance of the estimators $\hat{M}_1(t)$ and $\hat{M}_2(t)$ by Monte Carlo simulation. He found in the simulation study that $\hat{M}_1(t)$ does well compared with $\hat{M}_2(t)$ and the variance of $\hat{M}_1(t)$ is smaller than that of $\hat{M}_2(t)$. However, $\hat{M}_1(t)$ does not perform well when $t$ is small relative to the mean, as is the case in Example 11.1 below.

For large $t$, $\hat{M}_1(t)$ is the preferred estimator because of ease of computation and its performance, as shown by the simulation studies reported by [6]. $\hat{M}_1(t)$, however, is not a consistent estimator. For small $t$, $\hat{M}_2(t)$ is the preferred estimator. This estimator is consistent for all $t$, but the amount of computation needed increases at a very rapid rate as $t$ and/or $m$ [used in (11.4)] increase.

*Example 11.1* We consider the data of Table F.2. Here the independent variable is distance (in KM) as opposed to time. We assume that the data can be viewed as

---

[6] Discussed in Sect. 11.3.1.

corresponding to Scenario 1.1. For this data set, $n = 32$, $\bar{t} = 18.3219$, $s^2 = 94.8624$ and $m_3 = (1/n)\sum_{i=1}^{n}(t_i - \bar{t})^3 = 185.0126$. Thus, from (11.1) and (11.2),

$$\hat{M}_1(t) = \frac{t}{\bar{t}} + \frac{s^2}{2\bar{t}^2} - \frac{1}{2} = \frac{t}{18.3219} + \frac{94.8624}{2(18.3219)^2} - \frac{1}{2} = 0.0546t - 0.3587$$

and

$$\hat{V}_1(t) = \frac{s^2 t}{\bar{t}^3} + \frac{5s^4}{4\bar{t}^4} - \frac{2m_3}{3\bar{t}^3} + \frac{1}{12} = 0.0154t + 0.1631.$$

For example, if $t = 10$, we obtain $\hat{M}_1(10) = 0.1871$ and $\hat{V}_1(10) = 0.3171$.

To illustrate Method 2, we need to estimate $F^{(k)}(.)$. This requires various counts, using the data given in Table F.2. For example, $\hat{F}^{(1)}(10)$ is the proportion of observations less than or equal to 10. We find $\hat{F}^{(1)}(10) = \binom{32}{1}^{-1} \times 6 = 0.1875$. $\hat{F}^{(2)}(10)$ is the number of pairs with sums less than or equal to 10, over $\binom{32}{2}$. We find $\hat{F}^{(2)}(10) = 10/496 = 0.0202$. Similarly, $\hat{F}^{(3)}(10) = 3/4960 = 0.0006$. Higher-order terms with respect to $k$ given in (11.3) are zero. It follows that

$$\hat{M}_2(10) = \sum_{k=1}^{3} \hat{F}^{(k)}(10) = 0.1875 + 0.0202 + 0.0006 = 0.2083,$$

and the corresponding nonparametric estimate of the variance of the estimate [obtained from (B.10)] is

$$\hat{V}_2(10) = \sum_{k=1}^{3} (2k-1)\hat{F}^{(k)}(10) - [\hat{M}_2(10)]^2 = 0.3663$$

Note that $\hat{M}_2(10)$ is different[7] from $\hat{M}_1(10)$, but not markedly so. The same is true of the estimated variances of the estimates.

These estimates can be used to estimate expected warranty costs of the engine. The engine is sold with a non-renewing FRW policy with warranty 10 K Km and failed items are replaced by new. Let $C_s$ denotes the average cost to the manufacturer of servicing a warranty claim. An estimate of this is \$536 (the average warranty cost from F.2). From (Table 6.13), the expected warranty cost to the manufacturer, is given by $E_1[C_m(W = 10)] = \hat{M}_1(10)C_s = 0.1871 \times 536 =$ \$100.29 or $E_2[C_m(W = 10)] = \hat{M}_2(10)C_s = 0.2083 \times 536 =$ \$111.65, depending on whether one uses $\hat{M}_1(10)$ or $\hat{M}_2(10)$.

[7] Comparisons between these estimates are given in [5, 6].

### 11.3.2 Data Scenario 1.2

Scenario 1.2 data consists of the ages of all items that failed in the warranty period, with failed items being repaired minimally. For this Scenario, the MCF can be estimated by simply modifying the method for Scenario 1.4 data discussed in Sect. 11.3.4.

### 11.3.3 Data Scenario 1.3

The data consist of the ages of items that failed under warranty and the censoring times for items that replaced failed ones and did not fail within the warranty period.

Reference [9] derived the nonparametric estimator of the survival function for censored data which is known as the *product-limit* (PL) *estimator*. This estimator is also widely known as the *Kaplan–Meier estimator* of the survival function.[8] ([10] p. 71) discusses nonparametric estimation of the survival function for both complete and censored data. The PL estimate of survival function can be used to estimate the cumulative hazard function for censored data. A nonparametric estimator of the cumulative hazard function was derived by [1 and 14–16]. This estimator is asymptotically equivalent, as the sample size increases, to the product limit estimator. Reference [21] describe nonparametric confidence interval estimation of survival probabilities for censored data. Example 8.9 provides an illustration of the procedure.

### 11.3.4 Data Scenario 1.4

In this scenario, the data consists of failure times for items that failed under warranty (as in Scenario 1.2), and supplementary data consisting of censored times for all other items. All failures are rectified through minimal repair and hence the focus is on the estimation of the MCF. The procedure is as follows[9]:

1. Order the unique recurrence times $t_{ij}$ among all of the $I$ customers. Let $n$ denote the number of unique times. These ordered unique times are denoted $t_{(1)} < t_{(2)} < \cdots < t_{(n)}$.
2. Compute $n_{it_{(k)}}$, the total number of recurrences for customer $i$ at $t_{(k)}$.
3. Let $\delta_{it_{(k)}} = 1$ if item $i$ is still being observed at time $t_{(k)}$ and $\delta_{it_{(k)}} = 0$ otherwise.
4. Compute

---

[8] Estimation of the survival function for censored data also given in Sect. 8.5.1.

[9] This is based on a method proposed by [17, 19, 20]. Additional details can be found in ([12], p. 397).

$$\hat{\mu}(t_{(j)}) = \sum_{k=1}^{j} \left[ \frac{\sum_{i=1}^{I} \delta_{it_{(k)}} n_{it_{(k)}}}{\sum_{i=1}^{I} \delta_{it_{(k)}}} \right] = \sum_{k=1}^{j} \left[ \frac{n^*_{\cdot t_{(k)}}}{\delta_{\cdot t_{(k)}}} \right] = \sum_{k=1}^{j} \bar{n}_{t_{(k)}} \tag{11.5}$$

for $j = 1,2,\ldots,n$, where $n^*_{\cdot t_{(k)}} = \sum_{i=1}^{I} \delta_{it_{(k)}} n_{it_{(k)}}$, $\delta_{\cdot t_{(k)}} = \sum_{i=1}^{I} \delta_{it_{(k)}}$, and $\bar{n}_{t_{(k)}} = n^*_{\cdot t_{(k)}} / \delta_{\cdot t_{(k)}}$.

Note that $n^*_{\cdot t_{(k)}}$ is the total number of item recurrences at time $t_{(k)}$, $\delta_{\cdot t_{(k)}}$ is the size of the risk set[10] at $t_{(k)}$, and $\bar{n}_{t_{(k)}}$ is the average number of recurrences per item at $t_{(k)}$. As in the case of nonparametric estimation of a CDF, the estimator $\hat{\mu}(t)$ is a step function, with jumps at recurrence times, but constant between recurrence times. The estimator of the variance of $\hat{\mu}(t_{(j)})$ is[11]

$$V\left[\hat{\mu}(t_{(j)})\right] = \sum_{i=1}^{I} \left\{ \sum_{k=1}^{j} \frac{\delta_{it_{(k)}}}{\delta_{\cdot t_{(k)}}} \left[ n_{it_{(k)}} - \bar{n}_{t_{(k)}} \right] \right\}^2 \tag{11.6}$$

*Example 11.2* We consider Data Set 10 given in Table F.11 (valve seat replacement data for diesel engines) and assume that failed items are minimally repaired, so that the data set corresponds to Scenario 1.4. The data include 46 instances of valve replacement. We estimate the MCF to investigate whether the replacement rate increases or decreases with age.

Figure 11.1 is an event plot of the data, showing the observation period and the reported repair times, created by SPLIDA[12] [13]. The event plot can also be created by MINITAB.[13] The plot shows when the failures occurred for each system. Each line extends to the last day of observation. Note that service times range from just under 400 days (System 409) to nearly 700 days (System 251).

Estimates of the means $\mu(t_{(j)})$ for the valve seat as a function of engine age along with point-wise approximate 95% confidence intervals are given in Table 11.1 and shown graphically in Fig. 11.2. For example, $\hat{\mu}(92)$ is 0.122, indicating that the mean (cumulative) number of repairs per valve seat at age 92 days is 0.122. Equivalently, the mean number of repairs per hundred valve seats at age 92 days is 12.2 or 12.2%. We are 95% confident that the true mean value of the cumulative function at age 92 days is between 0.0536 and 0.2773.

As can be seen from Table 11.1 and Fig. 11.2, the estimate increases sharply between 620 and 650 days, but it is important to recognize that this part of the estimate is based on only a small number of systems, as indicated by the wide confidence intervals in that region. The slightly concave up pattern of the plot of MCF versus time (age) indicates that the time between repairs is slightly

---

[10] The set contains counts of the items working and of the number uncensored just prior to $t_{(k)}$.

[11] Detailed derivations can be found in [11, 18], and ([12], p. 398).

[12] SPLIDA (S-Plus Life Data Analysis), see, www.public.iastate.edu/~splida.

[13] In MINITAB, use Stat → Reliability/Survival → Nonparametric Growth Curve.

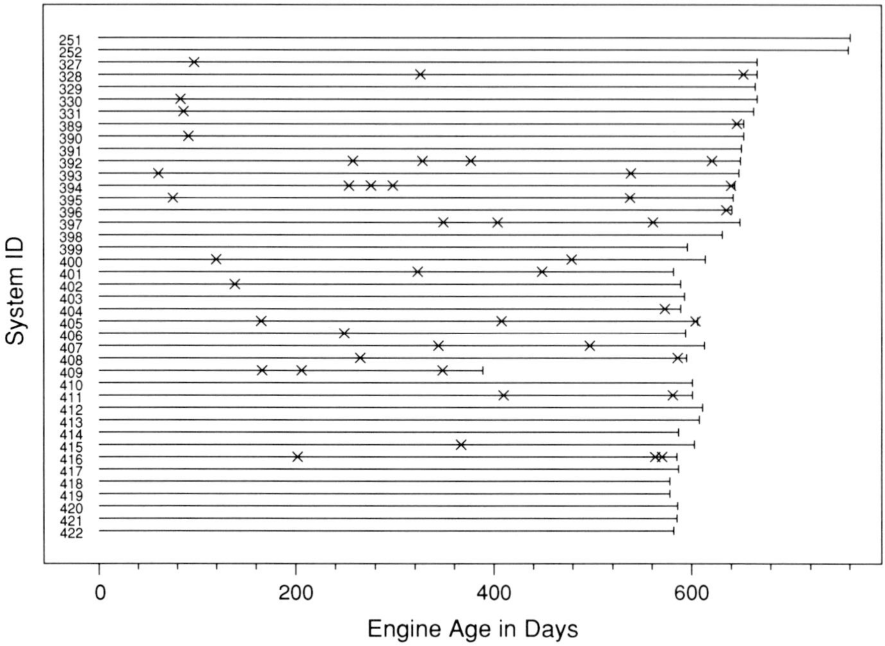


**Fig. 11.1** Event plot for Data Set 10

**Table 11.1** Estimates of MCF, its Standard Errors and 95% Confidence Interval

| Age in Days $(t_{(j)})$ | $\hat{\mu}(t_{(j)})$ | $S\hat{E}[\hat{\mu}(t_{(j)})]$ | 95% Confidence Interval | |
|---|---|---|---|---|
| | | | Lower | Upper |
| 61 | 0.0244 | 0.0241 | 0.0035 | 0.1690 |
| 76 | 0.0488 | 0.0336 | 0.0126 | 0.1885 |
| 84 | 0.0732 | 0.0407 | 0.0246 | 0.2175 |
| 87 | 0.0976 | 0.0463 | 0.0385 | 0.2475 |
| 92 | 0.1220 | 0.0511 | 0.0536 | 0.2773 |
| | | 35 rows omitted | | |
| 604 | 1.0606 | 0.1852 | 0.7532 | 1.4935 |
| 621 | 1.1194 | 0.2076 | 0.7782 | 1.6102 |
| 635 | 1.1819 | 0.2070 | 0.8385 | 1.6661 |
| 640 | 1.2444 | 0.2264 | 0.8711 | 1.7777 |
| 646 | 1.3210 | 0.2286 | 0.9413 | 1.8550 |
| 653 | 1.6070 | 0.3515 | 1.0468 | 2.4670 |

decreasing over time, that is, the system reliability is deteriorating. This issue is investigated further in Chap. 12.

These results can be used to estimate repair costs, a key factor in the cost of warranty. Let $c_r$ denote the average cost incurred by the manufacturer for repairing

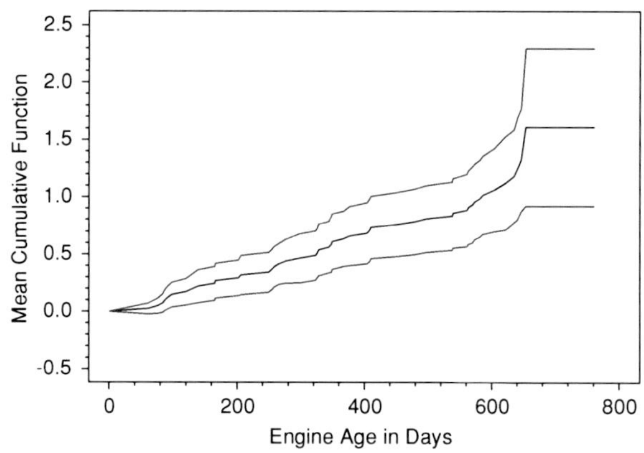


**Fig. 11.2** MCF for Data Set 10 with 95% confidence intervals

a claim for the valve seat. An estimate of the expected repair cost at age 640 days is $E[C_m(W = 640)] = \hat{\mu}(640)c_r = 1.2444c_r$. We conclude that a two-year (or 730 day) warranty would be very costly to the manufacturer.

## 11.4 Analysis of Structure 2 Data

Structure 2 data consist of counts of failures for each item over different time intervals. Nonparametric estimates and plots of the hazard function using the age clock are appropriate for data of this type. Here we discuss analysis of data under Scenarios 2.1 and 2.3.

### *11.4.1 Data Scenario 2.1*

The methods proposed for Scenario 2.3 below are applicable for this scenario if the censored observations are ignored and only the uncensored complete observations are considered. In this case, MINITAB may be used to calculate and graph the Turnbull survival plot, which is an actuarial survival plot, and the corresponding hazard plot.[14] References [22, 23] developed an iterative algorithm for computing a nonparametric maximum likelihood estimate of the cumulative distribution function for the data. Minitab provides the Turnbull estimate of the interval probabilities, along with standard errors for these probabilities. The actuarial model is an alternative nonparametric analysis that displays information for groupings of failure times. The Kaplan–Meier method assumes that the

[14] See MINITAB for details and use Stat → Reliability/Survival → Distribution Analysis (Arbitrary Censoring) → Nonparametric Distribution Analysis.

suspensions in an interval occur at the end of that interval, after the failures have occurred. Minitab's actuarial method assumes that the suspensions occur in the middle of the interval, which has the effect of reducing the number of available units in the interval, with the number available for analysis becoming $N'_j = N_j - \tilde{n}_j/2$.

## 11.4.2 Data Scenario 2.3

Here we give a method for estimation of the hazard function based on computing the number of failures in time interval $[(j-1)\Delta, j\Delta)$, denoted $n_{\cdot j}$, and the number of items operating in that time interval, denoted $N_j, j = 1, 2, \ldots, J$, where $J$ is the total number of intervals. The method can be used for multiply-censored data that are grouped.[15] The equation for the estimator of the hazard function is:

$$\hat{h}_j = \frac{\sum_{i=1}^{I} n_{ij}}{N_j} = \frac{n_{\cdot j}}{N_j}, \quad j = 1, 2, \ldots, J \tag{11.7}$$

The number of items at risk just prior to interval $j$, $N_j$, is calculated as $N_j = N - \sum_{k=0}^{j-1} n_{\cdot k} - \sum_{k=0}^{j-1} \tilde{n}_k$, where $N$ denotes the total number of items and $\tilde{n}_j$ the number of censored items in interval $j, j = 1, 2, \ldots, J$. $\hat{h}_j$ is an estimate of $h(t)$ for $t = j\Delta$, where $\Delta$ denotes the interval for discretization of time. The estimates of $H(t)$, $F(t)$ and $R(t)$ associated with $\hat{h}_j$ are easily obtained based on their interrelationships.

*Example 11.3* If the data of Data Set 10 of Table F.11 (Structure 1 Data) were in Structure 2 format with $J = 8$ intervals and $\Delta = 100$ days, then the data would be as given in Table 11.2. In this table, the resulting data 5 out of the 41 systems are shown, with "F" and "C" used to represent, respectively, the failure and censored data.

Using the data from Table 11.2, we create Table 11.3, where estimates of the hazard and reliability functions are also given. A plot of $\hat{H}(t)$ can be created from these results to assess whether the distribution of Valve Seat failure data has a constant, decreasing, or increasing hazard function.[16] Since $\hat{H}(t) = -\ln[1 - \hat{F}(t)]$, plots involving $\hat{H}(t)$ are used in exactly the same way[17] as plots of $\hat{F}(t)$. The estimates of Table 11.3 are indicative of an increasing hazard function.[18]

[15] See Sect. 5.12.2 for a graphical view of Structure 2 data.

[16] See Chap. 8 for more on plotting procedures.

[17] See Sect. 8.5 and Appendix C.

[18] This will be investigated further in Chap. 12 by fitting parametric distributions to the data.

**Table 11.2** Valve Seat failure data formatted in Structure 2

| System ID | Age (days) in interval ($\Delta = 100$ days, $J = 8$) | | | | | | | |
|---|---|---|---|---|---|---|---|---|
| | 0–100 | 100–200 | 200–300 | 300–400 | 400–500 | 500–600 | 600–700 | 700–800 |
| 251 | | | | | | | | C |
| 252 | | | | | | | | C |
| 327 | F | | | | | | C | |
| 328 | | | | F | | | F,F,C | |
| | | | | 36 systems omitted | | | | |
| 415 | | | | F | | | | |
| Total | 6F | 5F | 8F | 8F, 1C | 6F | 8F, 15C | 7F, 23C | 2C |

**Table 11.3** Estimates of hazard and reliability functions for the valve seat data of Structure 2

| $j$ | Start–End | $n_{\cdot j}$ | $\tilde{n}_j$ | $N_j$ | $\hat{h}_j \equiv \hat{h}(t)$ $[t = j\Delta]$ | $\hat{H}_j = \hat{H}(t)$ | $\hat{R}(t)$ |
|---|---|---|---|---|---|---|---|
| 1 | 0–100 | 6 | 0 | 89 | 0.0674 | 0.0674 | 0.9348 |
| 2 | 100–200 | 5 | 0 | 83 | 0.0602 | 0.1277 | 0.8801 |
| 3 | 200–300 | 8 | 0 | 78 | 0.1026 | 0.2302 | 0.7944 |
| 4 | 300–400 | 8 | 1 | 70 | 0.1143 | 0.3445 | 0.7086 |
| 5 | 400–500 | 6 | 0 | 61 | 0.0984 | 0.4429 | 0.6422 |
| 6 | 500–600 | 8 | 15 | 55 | 0.1455 | 0.5883 | 0.5553 |
| 7 | 600–700 | 7 | 23 | 32 | 0.2188 | 0.8071 | 0.4462 |
| 8 | 700–800 | 0 | 2 | 2 | 0.0000 | 0.8071 | 0.4462 |

The function survfit,[19] given in R and S-plus software, provides Kaplan–Meier and Fleming–Harrington estimates of the reliability function.[20] Figure 11.3 shows these estimates for the valve seat data of Structure 2. The Kaplan–Meier and Fleming–Harrington estimates are shown, respectively, by solid and dotted lines. The dashed lines are the 95% confidence intervals for the reliability function based on the Kaplan–Meier estimate. The $\hat{R}(t)$ values given in Table 11.3 coincide with the Fleming–Harrington estimate shown in Fig. 11.3. In this example, the Fleming–Harrington estimator is slightly higher than the KM estimate at every time point. We can see from Table 11.3 and Fig. 11.3 that about 55.53% of the valve seats are estimated to survive until 600 days.

[19] The survfit function computes an estimate of a survival curve for censored data using either the Kaplan–Meier or the Fleming and Harrington method and computes the predicted survivor function for a Cox proportional hazards model.

[20] For more information, see [4], S-plus (www.insightful.com), and R-language (http://cran.r-project.org).

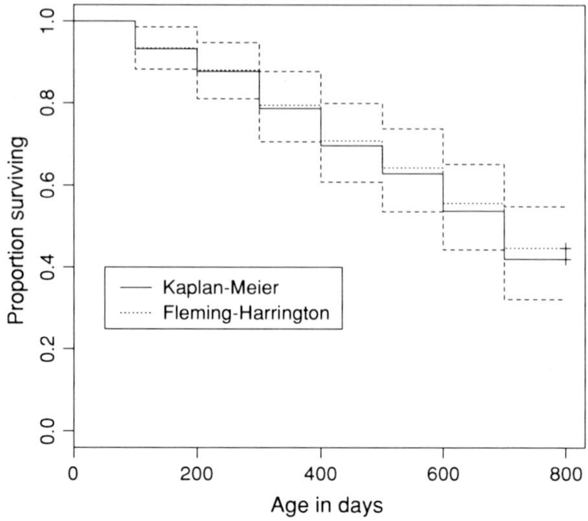


**Fig. 11.3** Plots of reliability functions for the data given in Table 11.2

## 11.5 Analysis of Structure 3 Data

Structure 3 data are the aggregated number of failures, for a group of items, over discrete time intervals. In the following 3 subsections we discuss analysis of Scenario 3.3 data by use of the failure distribution function, MOP–MIS (Month of Production–Month in Service) diagrams, and warranty claim rates (WCR).

### *11.5.1 Estimation of the CDF Using Scenario 3.3 Data*

The data consist of failure times aggregated over discrete time intervals. We consider a month as the unit, so that $\Delta = 1$ month.[21] Here we use the data to estimate the distribution, reliability and hazard functions. We consider a non-parametric estimator of the CDF, the standard error of the estimator, and confidence intervals for the CDF. The estimated CDF can be used to estimate the $R(t)$, $H(t)$ and other quantities of interest.

For this structure,[22] suppose that in the *jth* month $S_j$ units are sold. Based on the age clock, these items have age zero when put into use. Let $n_{jt}$ denote the number of units that failed in the $t$-th age interval $(t - 1, t]$, $j = 1, 2, \ldots, J$; $t = 1, 2, \ldots, \min(K, W)$. Here $K$ denotes the number of observed months ($K \geq J$) and $W$ denotes the warranty period. Then the size of the risk set at the beginning of interval $t$ is

---

[21] If necessary, the unit "month" can be easily substituted with "week", "day" and so on [8].

[22] See Sect. 5.12.3 for more details of Structure 3 data.

$$N_t = N - \sum_{i=0}^{t-1} n_{\cdot i} - \sum_{i=0}^{t-1} \tilde{n}_i \tag{11.8}$$

where

$$n_{\cdot t} = \begin{cases} \sum_{j=1}^{\min(J,K-t+1)} n_{jt}, & \text{if } t = 1, 2, \ldots, \min(W, K) \\ 0, & \text{otherwise} \end{cases}$$

is the total number of units that failed in interval $(t - 1, t]$,

$$\tilde{n}_t = \begin{cases} S_{\min(J,K-t+1)} - \sum_{i=1}^{t} n_{\min(J,K-t+1),i}, & \text{if } t = K - J + 1, \ldots, \min(W-1, K-1) \\ \sum_{j=1}^{\min(J,K-t+1)} \left( S_j - \sum_{i=1}^{t} n_{ji} \right), & \text{if } t = \min(W, K) \\ 0, & \text{otherwise} \end{cases}$$

is the number of units censored at $t$, and $N = \sum_{j=1}^{J} S_j$ is the total number of units sold during $J$ months. The nonparametric estimator of $F(t)$ is obtained using the Kaplan–Meier estimator. The result is

$$\begin{aligned} \hat{F}(t) &= 1 - \prod_{i=1}^{t} \left[ 1 - \frac{n_{\cdot i}}{N_i} \right] \\ &= 1 - \prod_{i=1}^{t} [1 - \hat{p}_i] = 1 - \hat{R}(t), \quad t = 1, 2, \ldots, \min(K, W), \end{aligned} \tag{11.9}$$

where $\hat{p}_i = n_{\cdot i}/N_i$ is an estimator of the sample proportion of failed items, and $\hat{R}(t) = \prod_{i=1}^{t} [1 - \hat{p}_i]$ is an estimator of the reliability function.

Reference ([12], p. 54) discuss estimation methods for the variance and pointwise normal-approximation confidence intervals[23] for $F(t)$. By using the logit transformation, they show that two-sided approximate $100(1 - \alpha)\%$ confidence intervals for $F(t)$ can be calculated as

$$\left[ \frac{\hat{F}(t)}{\hat{F}(t) + (1 - \hat{F}(t)) \times w}, \frac{\hat{F}(t)}{\hat{F}(t) + (1 - \hat{F}(t))/w} \right] \tag{11.10}$$

where $w = \exp\left\{ z_{(1-\alpha/2)} \hat{se}_{\hat{F}(t)} / [\hat{F}(t)(1 - \hat{F}(t))] \right\}$ and

[23] Chapter 9 gives general definition and derivation of confidence intervals.

**Table 11.4** Nonparametric estimates of $F(t)$ for Data Set 11

| $t$ | $n_{.t}$ | $\tilde{n}_t$ | $N(t)$ | $\hat{p}$ | $1-\hat{p}$ | $\hat{R}(t)$ | $\hat{F}(t)$ |
|---|---|---|---|---|---|---|---|
| 1 | 9 | 631 | 104005 | 0.000087 | 0.999913 | 0.999913 | 0.000087 |
| 2 | 22 | 1506 | 103365 | 0.000213 | 0.999787 | 0.999701 | 0.000299 |
| 3 | 64 | 3229 | 101837 | 0.000628 | 0.999372 | 0.999072 | 0.000928 |
| 4 | 80 | 3774 | 98544 | 0.000812 | 0.999188 | 0.998261 | 0.001739 |
| 5 | 61 | 3677 | 94690 | 0.000644 | 0.999356 | 0.997618 | 0.002382 |
| 6 | 47 | 3695 | 90952 | 0.000517 | 0.999483 | 0.997103 | 0.002897 |
| 7 | 28 | 3689 | 87210 | 0.000321 | 0.999679 | 0.996783 | 0.003217 |
| 8 | 36 | 3656 | 83493 | 0.000431 | 0.999569 | 0.996353 | 0.003647 |
| 9 | 22 | 3351 | 79801 | 0.000276 | 0.999724 | 0.996078 | 0.003922 |
| 10 | 31 | 3242 | 76428 | 0.000406 | 0.999594 | 0.995674 | 0.004326 |
| 11 | 32 | 3873 | 73155 | 0.000437 | 0.999563 | 0.995239 | 0.004761 |
| 12 | 25 | 4298 | 69250 | 0.000361 | 0.999639 | 0.994879 | 0.005121 |
| 13 | 13 | 4270 | 64927 | 0.000200 | 0.999800 | 0.994680 | 0.005320 |
| 14 | 7 | 3600 | 60644 | 0.000115 | 0.999885 | 0.994565 | 0.005435 |
| 15 | 4 | 3091 | 57037 | 0.000070 | 0.999930 | 0.994495 | 0.005505 |
| 16 | 4 | 3117 | 53942 | 0.000074 | 0.999926 | 0.994422 | 0.005578 |
| 17 | 8 | 3208 | 50821 | 0.000157 | 0.999843 | 0.994265 | 0.005735 |
| 18 | 5 | 47600 | 47605 | 0.000105 | 0.999895 | 0.994161 | 0.005839 |

$$\hat{se}_{\hat{F}(t)} = \sqrt{V(\hat{F}(t))} = \hat{R}(t)\sqrt{\sum_{j=1}^{t}\frac{\hat{p}_j}{n_j(1-\hat{p}_j)}} \tag{11.11}$$

*Example 11.4* Consider Data Set 11 given in Table F.12 (aggregated warranty claims and censored data for a specific automobile component). Table 11.4 shows the numerical computations of $F(t)$ by (11.9) based on the values of $n_{.t}$, $\tilde{n}_t$ and $N_t$ for $t = 1, 2, \ldots, 18$.

The estimates of $p$ in Table 11.4 indicate that the age-based proportion of failing increases rapidly in the early life (ages 1–4) of the component. From ages 6 to 12 it becomes approximately constant and it decreases from ages 13 to 18.

The estimates of $F(t)$ along with approximate 95% point-wise confidence intervals for $F(t)$ are shown in Fig. 11.4. Figure 11.4 shows a smoothly increasing relationship for the early ages (1–13 months) compared with that of ages greater than 13 months.

In Fig. 11.4, the final value of the CDF at age 18 is $F(t = 18) = 0.0058$, indicating a claims rate of 0.58%, which is a very low rate for an automobile component. This is evidence of a reliable product. From Fig. 11.4, we are 95% confident that the probability of failing of the component within the warranty period (age 18 month) is between 0.0053 and 0.0064.

An estimate of the expected warranty cost per unit to the manufacturer is $C_s$ times the proportion of the units expected to fail within $W$. Thus the nonparametric

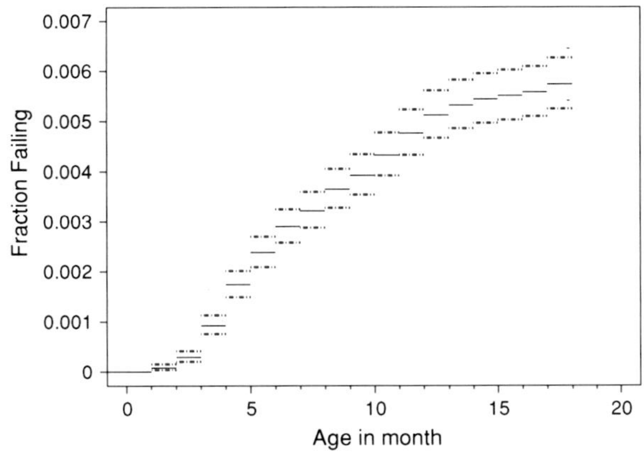


**Fig. 11.4** Nonparametric estimates of $F(t)$ for Data Set 11 with approximate 95% pointwise confidence intervals

estimate of the expected warranty cost to the manufacturer (with an 18 month warranty) is $E[C_m(W = 18)] = \hat{F}(18)C_s = 0.0058C_s$.

### *11.5.2 MOP–MIS Diagram*

The MOP-MIS (Month of Production–Month in Service) diagram[24] is used to relate production dates, failures and age at failure. It may also be used to evaluate and compare the quality of products from warranty claims by each MOP and MIS. In this analysis, we are looking at monthly data, but the period can be shorter—for example, a week or a day.[25] To construct MOP–MIS diagrams, we use the following notations.

$M_i$: number of units produced in period $i$, $i = 1, 2, \ldots, I$

$S_{ij}$: number of units produced in period $i$ and sold in period $j$, $j = 1, 2, \ldots, J$ $(i \le j)$,$(M_i = \sum_{j \ge i} S_{ij}$, if all products produced in period $i$ are sold)

$n_{ijk}$: number of warranty claims in period $k = 1, 2, \ldots, K$ for units produced in period $i$ and sold in period $j$ $(i \le j \le k)$

Figure 11.5 shows the time ordering sequence between these items. Note that $I$ represents the total number of production periods, $J$ is the total number of sale periods and $K$ is the total number of claim periods with $I \le J \le K$.

$n_{i \cdot k}$: number of warranty claims for units produced in period $i$ that fail in period $k$ $(i \le k)\left(n_{i \cdot k} = \sum_{j=i}^{\min(J,k)} n_{ijk},\ i \le k\right)$.

$n_t = n_{\cdot\cdot t}$: Total number of warranty claims in period $t$ $(t \ge 1)\left(n_t = n_{\cdot\cdot t} = \sum_{i=1}^{t} n_{i \cdot t}\right)$

[24] Chapter 15 presents a graphical representation of MOP–MIS plot.

[25] In some cases it can be a shift, if a company operates more than one shift per day.

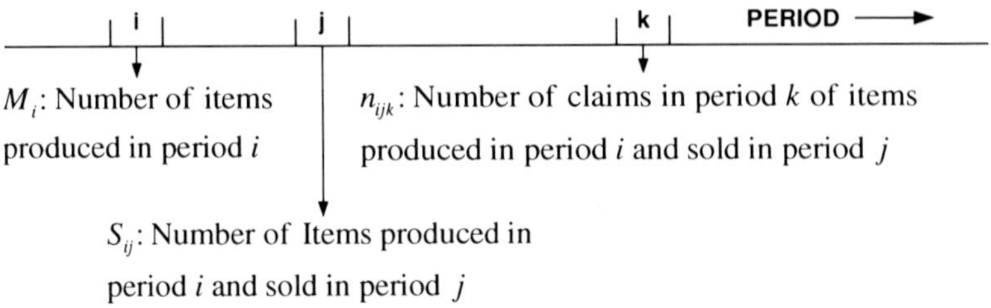


**Fig. 11.5** Production, sales and warranty claims over time

**Table 11.5** Monthly sales $S_{ij}$ for different MOP

| MOP ($i$) | $M_i$ | Sale period ($j$) | | |
|---|---|---|---|---|
| | | 1 | 2 | 3 |
| 1 | $M_1 = 500$ | $S_{11} = 250$ | $S_{12} = 150$ | $S_{13} = 50$ |
| 2 | $M_2 = 550$ | | $S_{22} = 300$ | $S_{23} = 200$ |
| 3 | $M_3 = 525$ | | | $S_{33} = 300$ |

$\bar{S}(t)$: Number of units under warranty in period$t$ $(t \geq 1)$
MOP: Month of Production (indexed by subscript $i$)
MOS: Months of Sale (indexed by subscript $j$)
MIS: Months in Service (duration for which the item is in use—indexed by $t = k - j$)
$n_{it}$: Number of items from MOP $i$ that fail at age $t$

To illustrate the MOP–MIS diagram, we will consider the following two tables with $I = J = K = 3$, $W = 3$, and arbitrary values of $M_i$, $S_{ij}$, and $n_{ijk}$. The first, Table 11.5, is sales data obtained from retailers.

The second, Table 11.6, gives warranty claims data obtained from service agents.

From these two tables, one can create a table for the number of failures in age group ($t$) for each MOP. The result is given in Table 11.7. This table would be useful for analysis of a non-renewing warranty with warranty duration equal to 3 periods. The formats of all three tables are standard for MOP–MIS analysis.[26]

From Tables 11.5 and 11.6, we can compute the number of items under warranty for each time period, $\bar{S}(t)$. In the case of non-renewing 1-D warranty policies, it is not difficult to compute $\bar{S}(t)$. For the 1-D FRW policy, this quantity is given by $\bar{S}(t) = \sum_{i=1}^{t} \sum_{j=i}^{t} S_{ij}$, and for the 1-D PRW policy (with refund) it is given by $\bar{S}(t) = \sum_{i=1}^{t} \left[ \sum_{j=i}^{t} \left\{ S_{ij} - \sum_{k=j}^{t} n_{ijk} \right\} \right]$.

[26] The MOP–MIS tables may be of different forms; see Table 11.10 for another form.

**Table 11.6** Monthly failures $n_{ijk}$ for different MOP

| MOP ($i$) | MOS ($j$) | $S_{ij}$ | Failure period ($k$) | | |
|---|---|---|---|---|---|
| | | | 1 | 2 | 3 |
| 1 | 1 | $S_{11} = 250$ | $n_{111} = 1$ | $n_{112} = 2$ | $n_{113} = 2$ |
| | 2 | $S_{12} = 150$ | | $n_{122} = 2$ | $n_{123} = 1$ |
| | 3 | $S_{13} = 50$ | | | $n_{133} = 1$ |
| 2 | 2 | $S_{22} = 300$ | | $n_{222} = 2$ | $n_{223} = 3$ |
| | 3 | $S_{23} = 200$ | | | $n_{233} = 1$ |
| 3 | 3 | $S_{33} = 300$ | | | $n_{333} = 2$ |
| Total failures in period $k$ | | | $n_1 = 1$ | $n_2 = 6$ | $n_3 = 10$ |

**Table 11.7** Failures $n_{it}$ indexed by MOP ($i$) and months in use ($t$)

| MOP ($i$) | $\{n_{it}\}$ in MIS ($t$) under warranty | | |
|---|---|---|---|
| | 1 | 2 | 3 |
| 1 | $n_{11}(= n_{111} + n_{122} + n_{133})$ $= 1 + 2 + 1 = 4$ | $n_{12}(= n_{112} + n_{123})$ $= 2 + 1 = 3$ | $n_{13}(= n_{113}) = 2$ |
| 2 | $n_{21}(= n_{222} + n_{233}) = 2 + 1 = 3$ | $n_{22}(= n_{223}) = 3$ | |
| 3 | $n_{31}(= n_{333}) = 2$ | | |
| Total failures at MIS $t$ | 9 | 6 | 2 |

The number of items under warranty at the beginning of age group $t$ (MIS $t$) for the MOP $i$ denoted $\bar{S}(i,\, t)$, is given by

$$\bar{S}(i,\, t) = \begin{cases} \sum\limits_{j=i}^{\min(J,K-t+1)} S_{ij}, & \text{if } t = 1 \\ \sum\limits_{j=i}^{\min(J,K-t+1)} \left\{ S_{ij} - \sum\limits_{k=j}^{j+t-2} n_{ijk} \right\}, & \text{if } t > 1 \end{cases} \tag{11.12}$$

For example, using (11.12) and Table 11.6, we obtain

$$\bar{S}(1,\, 1) = \sum_{j=1}^{\min(3,3-1+1)} S_{1j} = S_{11} + S_{12} + S_{13} = 250 + 150 + 50 = 450,$$

and

$$\begin{aligned} \bar{S}(1, 2) &= \sum_{j=1}^{2} \left\{ S_{1j} - \sum_{k=j}^{j+2-2} n_{1jk} \right\} = (S_{11} - n_{111}) + (S_{12} - n_{122}) \\ &= (250 - 1) + (150 - 2) = 397. \end{aligned}$$

Similarly, $\bar{S}(1, 3) = 147$, $\bar{S}(2, 2) = 500$, $\bar{S}(2, 3) = 298$ and $\bar{S}(3, 3) = 300$.

From Tables 11.6 and 11.7, one can calculate the number of warranty claims *WC* and the warranty claims rate *WCR*, and then generate MOP–MIS plots or tables.

### 11.5.3 Warranty Claims (WCs) and Warranty Claim Rates (WCRs)

There are two different ways of defining the number of warranty claims. The first is the number of warranty claims as a function of time. This is given by

$$WC_1(t) = n_t = \sum_{i=1}^{t} \sum_{j=i}^{t} n_{ijt}, \quad t = 1, 2, \ldots$$

Note that here the claims are not differentiated by month in service or month of production. The second way of defining this is

$$WC_2(i,t) = n_{it} = \sum_{j=i}^{\min(J,K-t+1)} n_{i,j,j+t-1}, \quad i = 1, 2, \ldots, I; \; t = 1, 2, \ldots \min(K, W),$$

which characterizes claims as a function of MOP $i$ and MIS $t$.

The calculated values of $WC_2(i, t)$ are given in Tables 11.7 and 11.8. From these tables, we see, for example, that

$$WC_2(1,1) = n_{11} = \sum_{j=1}^{3} n_{1,j,j+1-1} = n_{111} + n_{122} + n_{133} = 1 + 2 + 1 = 4,$$

and

$$WC_2(1,2) = n_{12} = \sum_{j=1}^{2} n_{1,j,j+2-1} = n_{112} + n_{123} = 2 + 1 = 3.$$

The WC analysis does not take into account the number of units still under warranty. This depends on sales as well as the terms of the warranty policy. The warranty claims rate (WCR) analysis looks at claims as a fraction of the units still under warranty. The WCR can be defined in three ways. The first is

$$WCR_1(t) = \frac{WC_1(t)}{\bar{S}(t)} = \frac{n_t}{\bar{S}(t)}, \quad t = 1, 2, \ldots \tag{11.13}$$

Note that $WCR_1$ is the ratio of the total count of claims in period $t$ and the count of items under warranty prior to that period.

**Table 11.8** Estimation of $\mathrm{WCR}_3(i, t)$ for each $i$ and $t$

| MOP ($i$) | MIS ($t$) | $\bar{S}(i,t)$ | $\mathrm{WC}_2(i, t)$ or $n_{it}$ | $WCR_3(i,t)$ |
|---|---|---|---|---|
| 1 | 1 | 450 | 4 | 0.0089 |
| | 2 | 397 | 3 | 0.0076 |
| | 3 | 147 | 2 | 0.0136 |
| 2 | 2 | 500 | 3 | 0.0060 |
| | 3 | 298 | 3 | 0.0101 |
| 3 | 3 | 300 | 2 | 0.0067 |

The second definition is

$$WCR_2(i) = \frac{\sum_{t=1}^{\min(K,W)} n_{it}}{M_i}, \quad i = 1, 2, \ldots, I \tag{11.14}$$

This gives the count of claims over claims reported for a particular production month $i$ divided by the total number of items produced in that month.

The third definition is

$$WCR_3(i,t) = \frac{WC_2(i,t)}{\bar{S}(i,t)} = \frac{n_{it}}{\bar{S}(i,t)}, \quad i = 1, 2, \ldots, I;\ t = i, i+1, \ldots, \min(W, K) \tag{11.15}$$

$WCR_3(i, t)$ indicates the age-based or MIS-based claim rates for each production month. The values of $WCR_3(i, t)$ for $i = 1, 2, 3$ and $t = 1, 2, 3$, obtained from (11.15), are given in Table 11.8. This serves to illustrate the procedure for calculating $WCR_3(i, t)$ based on Structure 3 data.

*Example 11.5* This example considers warranty claims data for an automobile component. Data relating to components manufactured over a 12 month period ($I = 12$) were collected. The warranty period was 18 months ($W = 18$). Monthly sales and failure count data for the production month September are given in Table F.13. The components produced in this month were sold over a 20 month period ($J = 20$) and claims under warranty were collected during a 21 month period ($K = 21$). Detailed data for all production months are not given.

The warranty claims rates $WCR_2(i)$[27] for each of the 12 months ($i = 1, 2,\ldots,12$) of production (January–December), are shown in Fig. 11.6. As can be seen, the warranty claims rate is initially decreasing and there is a significant increase in the month of June ($i = 6$) and another increase toward the ends of the year. This suggests that there are some problems with the June and possibly December months of production.

[27] The detailed data that are required to calculate $\mathrm{WCR}_2(i)$ for all production months are not given in the book.

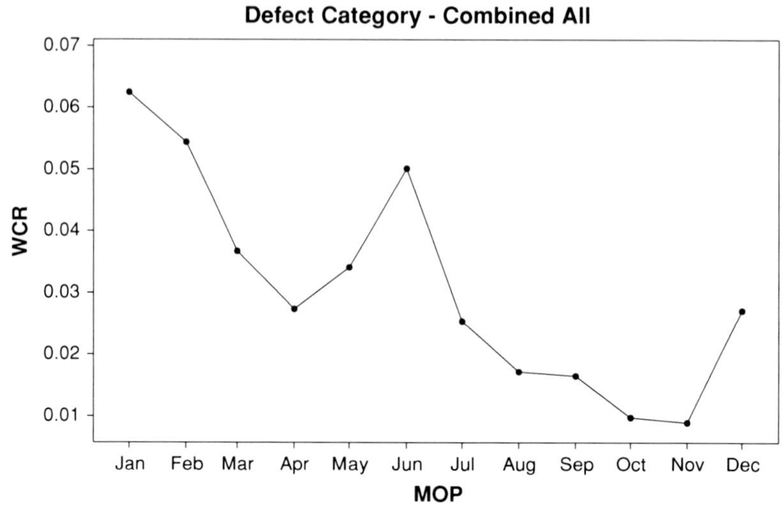


**Fig. 11.6** $WCR_2(i)$, $i =$ Jan, Feb, …, Dec, as a function of MOP

**Table 11.9** Calculated values of $\mathrm{WCR}_3(i,\ t)$ for a particular MOP ($i = 9$) of an automobile component

| MIS ($t$) | $\mathrm{WC}_2(i,\ t)$ | $\bar{S}(9, t)$ | $WCR_3(9, t)$ |
|---|---|---|---|
| 1 | 6 | 8938 | 0.0007 |
| 2 | 8 | 8932 | 0.0009 |
| 3 | 3 | 8917 | 0.0003 |
| 4 | 5 | 8902 | 0.0006 |
| 5 | 6 | 8877 | 0.0007 |
| 6 | 7 | 8863 | 0.0008 |
| 7 | 9 | 8829 | 0.0010 |
| 8 | 12 | 8769 | 0.0014 |
| 9 | 8 | 8679 | 0.0009 |
| 10 | 10 | 8608 | 0.0012 |
| 11 | 10 | 8582 | 0.0012 |
| 12 | 13 | 8523 | 0.0015 |
| 13 | 8 | 8422 | 0.0010 |
| 14 | 8 | 8343 | 0.0010 |
| 15 | 9 | 8241 | 0.0011 |
| 16 | 6 | 8064 | 0.0007 |
| 17 | 9 | 7830 | 0.0011 |
| 18 | 10 | 7387 | 0.0014 |
| Total | 147 | | 0.0173 |

The monthly sales data ($S_{ij}$) and counts of failures as a function of MIS and MOS for the month of September are given in Table F.13. For the MOP data, the number of items under warranty at the beginning of time period $t$, $\bar{S}(i,\ t)$, the

**Table 11.10** MOP–MIS table of $WCR_3(i, t)$ of automobile component for all MOP

| | MIS-1 | MIS-2 | MIS-3 | MIS-4 | MIS-5 | MIS-6 | MIS-7 | MIS-8 | MIS-9 | MIS-10 | MIS-11 | MIS-12 | MIS-13 | MIS-14 | MIS-15 | MIS-16 | MIS-17 | MIS-18 |
|---|---|---|---|---|---|---|---|---|---|---|---|---|---|---|---|---|---|---|
| MOP-1 | 0.0017 | 0.0006 | 0.0013 | 0.0021 | 0.0008 | 0.0019 | 0.0034 | 0.0036 | 0.0038 | 0.0025 | 0.0045 | 0.0064 | 0.0059 | 0.0034 | 0.0074 | 0.0050 | 0.0043 | 0.0064 |
| MOP-2 | 0.0011 | 0.0012 | 0.0013 | 0.0012 | 0.0015 | 0.0015 | 0.0037 | 0.0017 | 0.0037 | 0.0023 | 0.0033 | 0.0049 | 0.0049 | 0.0044 | 0.0050 | 0.0051 | 0.0050 | 0.0042 |
| MOP-3 | 0.0009 | 0.0003 | 0.0009 | 0.0017 | 0.0010 | 0.0024 | 0.0017 | 0.0016 | 0.0014 | 0.0016 | 0.0037 | 0.0028 | 0.0027 | 0.0025 | 0.0023 | 0.0031 | 0.0029 | 0.0042 |
| MOP-4 | 0.0008 | 0.0006 | 0.0009 | 0.0002 | 0.0009 | 0.0011 | 0.0009 | 0.0009 | 0.0013 | 0.0023 | 0.0015 | 0.0023 | 0.0021 | 0.0020 | 0.0026 | 0.0024 | 0.0036 | 0.0016 |
| MOP-5 | 0.0012 | 0.0007 | 0.0009 | 0.0022 | 0.0015 | 0.0010 | 0.0009 | 0.0022 | 0.0015 | 0.0013 | 0.0020 | 0.0023 | 0.0028 | 0.0023 | 0.0017 | 0.0031 | 0.0043 | 0.0035 |
| MOP-6 | 0.0009 | 0.0008 | 0.0005 | 0.0016 | 0.0014 | 0.0021 | 0.0011 | 0.0030 | 0.0030 | 0.0035 | 0.0042 | 0.0053 | 0.0045 | 0.0029 | 0.0057 | 0.0050 | 0.0040 | 0.0035 |
| MOP-7 | 0.0012 | 0.0006 | 0.0006 | 0.0008 | 0.0011 | 0.0016 | 0.0008 | 0.0014 | 0.0020 | 0.0019 | 0.0022 | 0.0015 | 0.0013 | 0.0016 | 0.0017 | 0.0017 | 0.0017 | 0.0029 |
| MOP-8 | 0.0004 | 0.0006 | 0.0006 | 0.0004 | 0.0004 | 0.0010 | 0.0003 | 0.0006 | 0.0009 | 0.0007 | 0.0012 | 0.0013 | 0.0010 | 0.0013 | 0.0017 | 0.0015 | 0.0017 | 0.0026 |
| MOP-9 | 0.0007 | 0.0009 | 0.0003 | 0.0006 | 0.0007 | 0.0008 | 0.0010 | 0.0014 | 0.0009 | 0.0012 | 0.0012 | 0.0015 | 0.0010 | 0.0010 | 0.0011 | 0.0007 | 0.0011 | 0.0014 |
| MOP-10 | 0.0007 | 0.0007 | 0.0000 | 0.0005 | 0.0002 | 0.0002 | 0.0011 | 0.0007 | 0.0002 | 0.0019 | 0.0005 | 0.0007 | 0.0005 | 0.0002 | 0.0010 | 0.0005 | 0.0003 | 0.0003 |
| MOP-11 | 0.0005 | 0.0005 | 0.0005 | 0.0003 | 0.0005 | 0.0005 | 0.0003 | 0.0006 | 0.0003 | 0.0009 | 0.0006 | 0.0008 | 0.0002 | 0.0010 | 0.0005 | 0.0005 | 0.0012 | 0.0003 |
| MOP-12 | 0.0011 | 0.0010 | 0.0010 | 0.0010 | 0.0013 | 0.0008 | 0.0024 | 0.0013 | 0.0016 | 0.0023 | 0.0021 | 0.0034 | 0.0037 | 0.0024 | 0.0027 | 0.0013 | 0.0005 | |

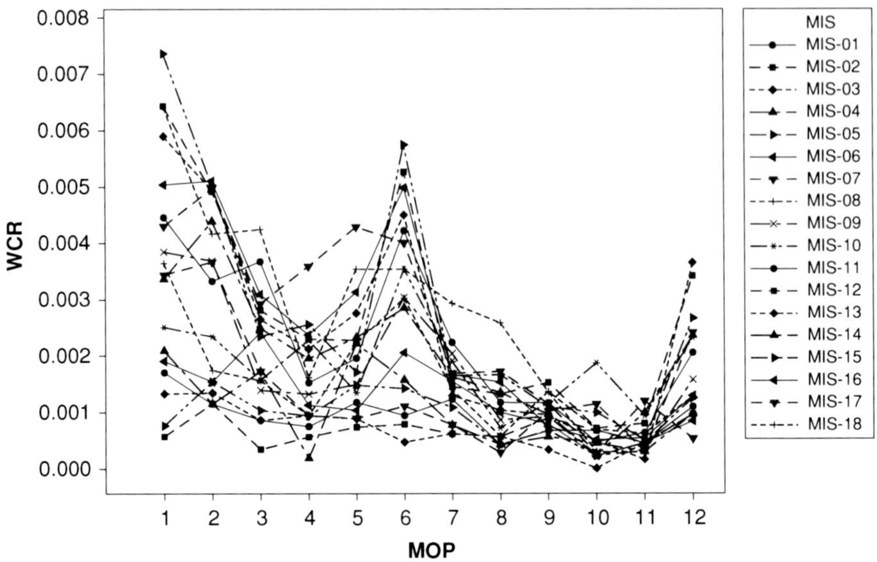


**Fig. 11.7** MOP–MIS Chart of $WCR_3(i, t)$

number of warranty claims, $WC_2(i, t)$, and the warranty claims rates $WCR_3(i, t)$ can be calculated. These values are given in Table 11.9.

The estimates of $WCR_3(i, t)$ for the other eleven MOP can be calculated similarly. The results are given in Table 11.10.

Based on Table 11.10, the MOP–MIS plot of $WCR_3(i, t)$ for all MOP ($i = 1, 2, \ldots, 12$) and MIS ($t = 1, 2, \ldots, 18$) is shown in Fig. 11.7. This figure is useful in determining if the failure rates are related to month in service and/or month of production. The figure indicates that the warranty claims rates are initially decreasing with respect to month of production and that there is a significant increase for the 6th month (June). In MOP June, the high claims rates are for 10, 12, 13, 15, 16, 17, and 18 MIS.

Figure 11.8 shows the estimates of WCR for each MIS separately. This figure indicates that the production period July–November (MOP 7–MOP 11) is the best in the sense that the claim rates are low and stable for all MIS in this period. For MIS from 1 to 10, the claim rates are low and approximately constant in all MOP. The variation in claim rates in different MOP increases as MIS increases. This example will be discussed further in Chap. 16.

Numerical computation of $WCR_3(i, t)$ requires writing a program for executing (11.15). This can done using any one of the many programming languages or using Microsoft Excel. Figures 11.7 and 11.8 were generated after importing the estimated $WCR_3(i, t)$ from Table 11.10 in a MINITAB Worksheet and choosing Graph → Scatterplot → With Connect and Group.

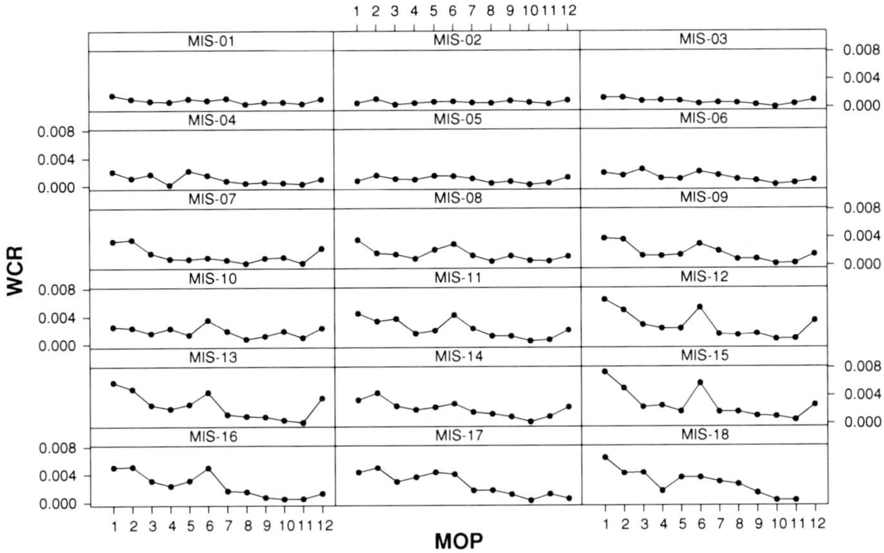


**Fig. 11.8** MOP–MIS Chart of $WCR_3(i, t)$ for each MIS separately

## 11.6 Conclusion

This chapter discussed nonparametric approaches to the analysis of warranty data under different data scenarios and data structures. The focus was on obtaining nonparametric estimates of the distribution function $F(t)$, reliability function $R(t)$, cumulative hazard function $H(t)$, renewal function $M(t)$, mean cumulative function $\mu(t)$, the warranty claims rate (WCR), etc., based on the various types of data. Some of these estimates will be used in further analyses in Chaps. 17 and 18.

Structure 1, Scenario 3 or Scenario 4 data are needed for reliability analysis. When a Structure 1 data set is very large, it may be useful to convert the data into Structure 3 data, so that it is more manageable. Data of this form are also better suited to analysis of the continuous improvement process, as discussed in Chap. 15.

## References

1. Aalen O (1976) Nonparametric inference in connection with multiple decrement models. Scand J Statist 3:15–27
2. Blischke WR, Murthy DNP (1994) Warranty cost analysis. Marcel Dekker, New York
3. Blischke WR, Scheuer EM (1975) Calculation of the cost of warranty policies as a function of estimated life distributions. Nav Res Logist Q 22:681–696
4. Fleming T, Harrington DP (1984) Nonparametric estimation of the survival distribution in censored data. Commun Statist 13(20):2469–2486

5. Frees EW (1986) Nonparametric renewal function estimation. Ann Statist 14:1366–1378
6. Frees EW (1986) Warranty analysis and renewal function estimation. Nav Res Logist Q 33:361–372
7. Frees EW (1986) Estimating the cost of a warranty. J Bus Econ Statist 6:79–86
8. Kalbfleisch JD, Lawless JF, Robinson JA (1991) Methods for the analysis and prediction of warranty claims. Technometrics 33:273–285
9. Kaplan EL, Meier P (1958) Nonparametric estimation from incomplete observations. J Am Statist Assoc 53:457–481
10. Lawless JF (1982) Statistical models and methods for lifetime data. Wiley, New York
11. Lawless JF, Nadeau C (1995) Some simple robust methods for the analysis of recurrent events. Technometrics 37:158–168
12. Meeker WQ, Escobar LA (1998) Statistical methods for reliability data. Wiley, New York
13. Meeker WQ, Escobar LA (2005) SPLIDA (S-Plus Life Data Analysis). www.pulic.iastate.edu/~splida
14. Nelson W (1969) Hazard plotting for incomplete failure data. J Qual Technol 1:27–52
15. Nelson W (1972) Theory and application of hazard plotting for censored survival data. Technometrics 14:945–966
16. Nelson W (1982) Applied life data analysis. Wiley, New York
17. Nelson W (1988) Graphical analysis of system repair data. J Qual Technol 20:24–35
18. Nelson W (1995) Confidence limits for recurrence data—applied to cost or number of product repairs. Technometrics 37:147–157
19. Nelson W (1998) An application of graphical analysis of repair data. Qual Reliab Eng Int 14:49–52
20. Nelson W (2008) Repair data set of: how to graph, analyze, and compare. Encyclopedia of statistics in quality and reliability. Wiley, New York
21. Thomas DR, Grunkemeier GL (1975) Confidence interval estimation of survival probabilities for censored data. J Am Statist Assoc 70:865–871
22. Turnbull BW (1974) Nonparametric estimation of a survivorship function with doubly censored data. J Am Statist Assoc 69:169–173
23. Turnbull BW (1976) The empirical distribution function with arbitrary grouped, censored, and truncated data. J Royal Statist Soc 38:290–295

# Chapter 12
# Parametric Approach to the Analysis of 1-D Warranty Data

## 12.1 Introduction

This chapter deals with parametric approaches to the analysis of one-dimensional warranty data, complementing the nonparametric approach discussed in Chap. 11. The parametric approach involves the use of models. Warranty data play a critical role in model selection, estimation of model parameters and validation of the model. Once these tasks are completed, the model may be used for prediction and other inferences. We consider the various data structures and scenarios discussed in Sects. 5.12 and 5.13, discuss the issues involved, and illustrate the methodology by application of the parametric analyses to data sets from Appendix F. Selected results are contrasted with comparable nonparametric results of Chap. 11 for the same data sets in order to compare the two approaches.

The outline of the chapter is as follows. In Sect. 12.2, we discuss the parametric approach and its advantages in the analysis of warranty data. Sects. 12.3–12.5 deal with parametric approaches for analysis of data of the three structures discussed in Sect. 5.12. For each data structure, we look at the various scenarios discussed in Sect. 5.13. Sect. 12.6 discusses the prediction of warranty claims and costs and we conclude with some comments and suggestions for practitioners in Sect. 12.7.

## 12.2 Parametric Approach to Data Analysis

### *12.2.1 Basic Concepts*

The parametric approach to data analysis is concerned with the construction, estimation, and interpretation of mathematical models as applied to empirical data. This involves the following three steps:

W. R. Blischke et al., *Warranty Data Collection and Analysis*, Springer Series in Reliability Engineering, DOI: 10.1007/978-0-85729-647-4_12, 

- Step 1: Model selection
- Step 2: Estimation of model parameters
- Step 3: Model validation

*Model selection* As discussed in Sect. 3.5, there are two basic approaches to selecting a model—(1) Physics-based modeling, where the model is based on a physical theory, and (2) Data-dependent modeling, where the model is developed solely on the basis of the available data.[1] In this chapter we focus our attention on the second approach. We consider several of the data scenarios discussed in Chap. 5.

The models involve either statistical parametric life distributions or the density or hazard functions associated with them. There are a number of parametric models that can be used successfully in modeling warranty data. In this chapter, we confine our attention to models involving a number of distributions discussed in Chap. 9 and/or Appendix A. More complex models (e.g., those involving other modifications of standard distribution or two or more distributions) are discussed in Chap. 13.

*Estimation of model parameters* As discussed in Chap. 3, the model will ordinarily involve one or more parameters whose values are unknown. In Chap. 9, several methods for using sample data to estimate unknown parameters were discussed. Here, we confine ourselves to the method of maximum likelihood. The method involves deriving an expression for the likelihood function,[2] the form of which depends on the type of data available. In Chap. 5 we looked at three data structures and several scenarios for each. Below we look at the likelihood function and the resulting estimators for each of these.

*Model validation* Validation is the process of determining the degree to which a selected model (along with the assigned or estimated parameter values) is an accurate representation of the real-world problem of interest. A poor fit of model (either graphical or analytical) may occur for two reasons: (1) the model is incorrect, or (2) the model is correct, but the parameter values specified or estimated may differ from the true values by too great an amount. Several approaches can be used for model validation. A straightforward approach to validating the model involves a *goodness-of-fit* test. Some of the commonly used statistical tests for validating model are the Chi-Square test, the Kolmogorov–Smirnov (KS) test and the Anderson–Darling (AD) test, all of which are discussed in Chap. 10.[3] Some of the steps that may be taken in the validation process are discussed in Sect. 8.7.

*Data and Notation* We consider the three data structures discussed in Sect. 5.12. Structure 1 deals with detailed data, Structure 2 with aggregated data for each

[1] See Sects. 8.7 and 10.4.

[2] See Appendix D.

[3] See also ([23], Chap. 5).

customer, and Structure 3 with data aggregated over all customers. We discuss the issues involved in parametric modeling under the data scenarios of Sect. 5.12 (which depend on data collection intervals and warranty servicing) and the three structures given in Sect. 5.13. The results are illustrated by examples using data sets from Appendix F. The notation given in Sects. 5.12 and 5.13 will be used in this chapter.

### 12.2.2 Akaike Information Criterion (AIC)

Akaike [1, 2] proposed an information theoretic interpretation of the likelihood function and proposed it as a model selection criterion, which is known as *Akaike Information Criterion* (AIC). Here we give a brief review of the argument that leads to AIC. Let g be the density function of the true model and $f(\theta)$ the density function that specifies the model with parameter vector $\theta$. The model $g$ is considered to be unknown. A pseudo-distance between $g$ and $f(\theta)$ is given by the Kullback–Leibler information quantity [15] as

$$I(g;f(\theta)) = E_g\left[\log\frac{g}{f(\theta)}\right] = E_g[\log g] - E_g[\log f(\theta)] \tag{12.1}$$

It is known that $I(g;f(\theta)) \geq 0$ with equality if and only if $g = f(\theta)$. It follows that $I(g;f(\theta)) \neq I(f(\theta);g)$ in general. The first term of the right-hand side of (12.1) is constant for all estimators of $f(\theta)$. Hence the second term, $E_g[\log f(\theta)]$, becomes the quantity on interest. Akaike [1] derived an asymptotically unbiased estimator of $E_g[\log f(\theta)]$and defined AIC as

$$\text{AIC} = -2(\text{maximum loglikelihood}) + 2(\text{number of model parameters}) \tag{12.2}$$

A model for which the AIC is the minimum is considered to be the best approximating model among a set of alternative models.[4]

### 12.2.3 Comparison with the Nonparametric Approach

Some of the advantages of using parametric models in the analysis of warranty data are as follows:

- Parametric models can be expressed as a function of a small number of parameters, as opposed to an entire curve.

[4] The detailed derivation, properties and applications of AIC are given in [1, 2, 7, 24, 26] and references cited therein.

- Parametric models provide smooth estimates of $F(t)$, $R(t)$, $H(t)$, etc., rather than step functions.
- The models can be used to draw inferences regarding quantities such as the MTTF, $R(t)$, B10-life (fractiles), etc., at product, component or some intermediate level. Properties of these quantities can be investigated either exactly or by means of asymptotic results, such as variance, confidence intervals, and so forth.
- A parametric model can be used to extrapolate with respect to the lifetime of a variable to the lower or upper tail of a distribution. This is useful in forecasting warranty claims and costs for a given warranty period.
- There are a variety of software packages dealing with parametric methods, and relatively fewer that deal with nonparametric procedures.

Although there are advantages and disadvantages of nonparametric versus parametric methods, the decision regarding which method is more appropriate depend very much on individual circumstances. In practice, it is often useful to apply both the parametric and nonparametric methods to a data set, if both are applicable. A more detailed discussion on the parametric approach can be found in most textbooks on statistical inference (e.g., [10, 14, 17, 19]).

## 12.3 Analysis of Structure 1 Data

Structure 1 data are most appropriate for estimating the failure distribution, reliability, hazard functions and associated functions such as the renewal function and MCF. In Sect. 5.13 we discussed four different scenarios (Scenarios 1.1–1.4) for this structure, resulting from the different warranty servicing strategies (replace or repair) and whether the data are only claims data or claims plus supplementary data. In this section, we look at the parametric approach to modeling for Scenarios 1.1–1.4. The analyses for Scenarios 1.1 and 1.3 are done by use of failure distribution functions and Scenarios 1.2 and 1.4 by means of the MCF.

### *12.3.1 Data Scenario 1.1*

Scenario 1.1 data consist only of failure times of items that failed under warranty. We assume a parametric model $f(t;\, \theta)$ for the failure time $T$. The models to be used will be selected from those listed in Sect. 9.2.2 and Appendix A. A suitable model for the data is selected from a set of competing models based on a model selection criterion, and the method of maximum likelihood is used to estimate the parameters of the model.

Under Scenario 1.1 data, the likelihood function for data from a single customer $i$ $(i = 1, 2, \ldots, I)$ is given by

**Table 12.1** Estimates of AD* and AIC for the eleven distributions for AC failure data, Aircraft 7909

| Distribution | AD* | AIC |
|---|---|---|
| Weibull (A.30) | 0.808 | 315.676 |
| Lognormal (A.44) | 0.653 | 313.532 |
| Exponential (A.21) | 1.312 | 316.654 |
| Loglogistic | 0.636 | 314.818 |
| 3-Parameter Weibull (A.46) | 0.706 | 313.334 |
| 3-Parameter Lognormal | 0.658 | 315.510 |
| 2-Parameter Exponential | 0.702 | 311.334 |
| 3-Parameter Loglogistic | 0.623 | 316.530 |
| Smallest Extreme Value (A.33) | 2.959 | 348.100 |
| Normal (A.26) | 1.991 | 332.358 |
| Logistic | 1.773 | 329.402 |

$$L_i(\theta) = \prod_{j=1}^{J} f\left(t_{ij} - t_{i,j-1}; \theta\right) \tag{12.3}$$

where $t_{ij}$ denotes the calendar time for the $j$th failure experienced by customer $i$, and $t_{i0}$ denotes the sale time for customer $i$ ($i = 1, 2, \ldots., I, j = 1, 2, \ldots, J$). Since failures for separate customers are independent, the complete likelihood (including all customers) is simply the product of the individual customer likelihoods, i.e.,

$$L(\theta) = \prod_{i=1}^{I} L_i(\theta) = \prod_{i=1}^{I}\prod_{j=1}^{J} f\left(t_{ij} - t_{i,j-1}; \theta\right) = \prod_{i=1}^{n_1} f(t_i; \theta), \tag{12.4}$$

where $n_1$ denotes the total number of failed items, with ages at failure given by $t_i$, $(1 \le i \le n_1)$. Inferences concerning the model parameter $\theta$ are made based on the likelihood function of (12.4).

*Example 12.1* In Example 9.1, Weibull and lognormal parametric models are employed in the analysis of failure data for air conditioner systems (Table F.10) on Aircraft #7909. It is reasonable to assume that the data were collected under Scenario 1.1. For purposes of illustration, a variety of distributions, given in Table 12.1, will be used in analysis of the data.

Likelihood function (12.2) is used for finding the MLEs of the parameters of each of the distributions. We employ the adjusted AD test statistic (AD*) and Akaike Information Criterion (AIC) as the model selection criteria. Table 12.1 shows the calculated adjusted AD test statistic and the AIC for each of the eleven distributions. This table indicates that the smallest extreme value, normal, logistic and exponential distributions can be eliminated as possible models. The three-parameter loglogistic distribution has the smallest AD*-value and the two-parameter exponential distribution has the smallest AIC, and hence are selected by the respective criteria as the "best" distributions. However, the values of AD* for the loglogistic, lognormal and 3-paramter lognormal are nearly as small as the AD* for the 3-parameter loglogistic and the AIC for the three-parameter Weibull

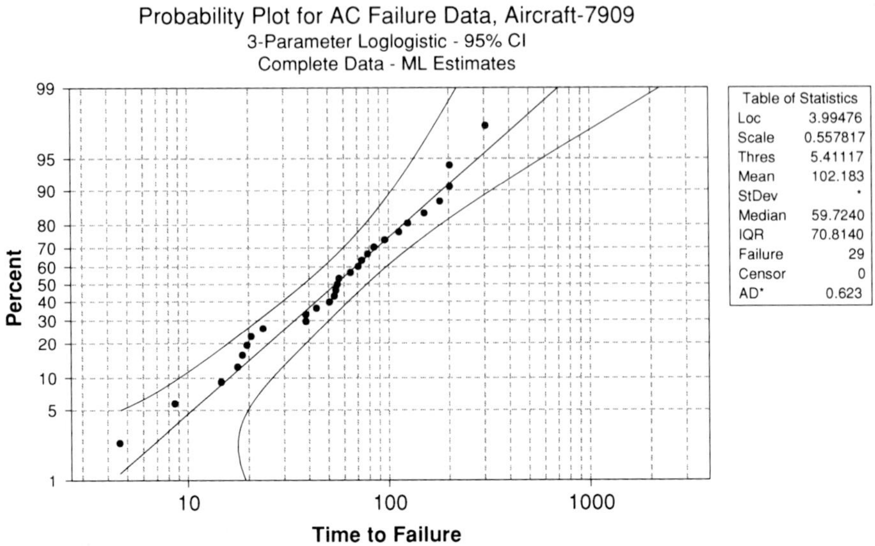


**Fig. 12.1** Three-parameter loglogistic probability plot for AC failure data for aircraft-7909

and lognormal are nearly as small as the AIC for the two-parameter exponential. Therefore, loglogistic, lognormal, 3-paramter lognormal and 3-parameter loglogistic distributions may be selected as a list of possible distributions for the data.

Minitab probability plots for the 3-parameter loglogistic and 2-parameter exponential distributions are given in Figs. 12.1 and 12.2. These figures display the transformed probability plots for the two distributions along with the ML estimates of parameters, mean, median, etc. The smallest data point in Fig. 12.2 appears to be a possible outlier for the 2-parameter exponential distribution. Aside from that, both distributions appear to provide reasonable fits.

The maximum likelihood estimates of the parameters for 3-parameter loglogistic are: location parameter $\hat{\mu} = 3.99476$, scale parameter $\hat{\sigma} = 0.557817$ and threshold parameter $\hat{\lambda} = 5.41117$. For 2-parameter exponential, the parameter estimates are: scale parameter $\hat{\mu} = 73.6172$, and threshold parameter $\hat{\lambda} = 9.9$. The 3-parameter loglogistic distribution estimate of the mean time to failure (102.183) is considerably higher than that of the 2-parameter exponential distribution (83.5172). There are not enough data to perform a validity test here, but it is definitely necessary before one can believe that the "true" model has been found.

### *12.3.2 Data Scenario 1.2*

Scenario 1.2 data include ages of all repairable items that failed under warranty and for which a claim was made, up to and including the last such failure.

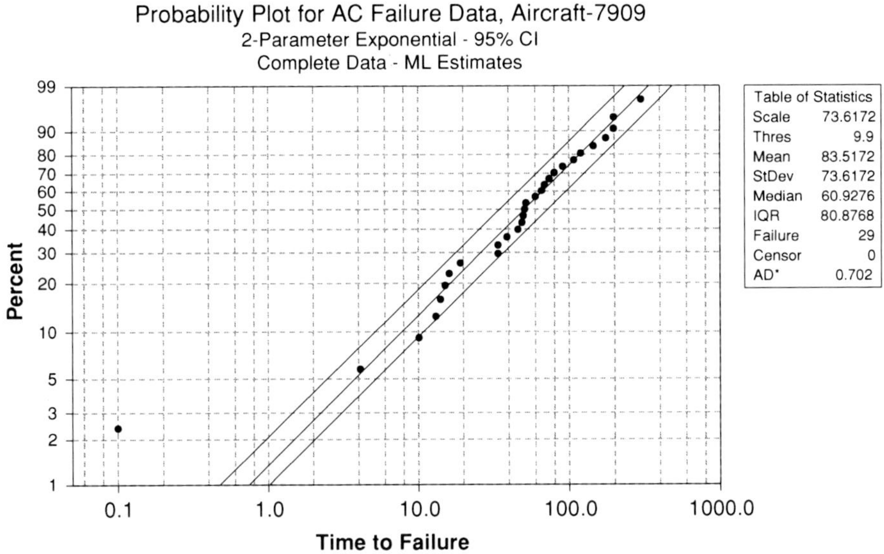


**Fig. 12.2** Two-parameter exponential probability plot for AC failure data for aircraft-7909

For this scenario, we assume that warranty servicing involves minimal repair and that repair times are negligible. Under these assumptions, failures for each customer can be viewed as occurring according to a counting process characterized by a mean cumulative function (MCF). The most commonly used counting process models for the failure process of a repairable system are a renewal process (RP), including the homogeneous Poisson process (HPP), a non-homogeneous Poisson process (NHPP), and superimposed versions of these processes (see Appendix B). The HPP differs from the NHPP only in that the rate of occurrence of failures is constant over time in the former and variable in the latter. An HPP is equivalent to a renewal process with exponential inter-failure times.

Here we derive the parametric estimate of the MCF of an NHPP model, and use the result to investigate whether the repair rate increases or decreases with the lifetime of the item. We also include an example in which the parametric estimate of the MCF is compared with the nonparametric estimate given in Chap. 11.

To specify an NHPP model, we use the intensity function (also known as the rate of occurrence of failures or ROCOF), $\lambda(t) = \lambda(t;\theta)$, where $\theta$ is a vector unknown of parameters. The corresponding cumulative intensity function[5] (or mean cumulative number of recurrences over (0, t]) is $\Lambda(t) = \Lambda(t;\theta)$.

*Single customer* Suppose a single customer $i$ is observed for the time period $(t_{i0},\ t_{a_i}]$, with $t_{i0} = 0$ the time at which the sale occurs and $t_{a_i}$ the time of the last failure for customer $i$ $(i = 1, 2, \ldots, I)$. The observed period $(t_{i0},\ t_{a_i}]$ is divided into

[5] Intensity function and cumulative intensity function are defined in Appendix B.

non-overlapping small time intervals in which either zero or one failure occurs. The likelihood function for customer $i$ can be expressed as a function of the probabilities of these outcomes [20, 25].

If we observe customer $i$ for the time interval $(0,\ t_{a_i}]$, with $n_i$ failures occurring at exact times $t_{i1}, t_{i2}, \cdots, t_{in_i}$ rather than occurring in intervals, then, using a limiting argument, it follows that the likelihood in terms of the density approximation is

$$L_i(\theta) = \prod_{j=1}^{n_i} \lambda(t_{ij};\theta)\exp[-\Lambda(t_{a_i};\theta)], \quad i = 1, 2, \ldots, I \tag{12.5}$$

*Comment:* Another possible scheme for a repairable system is to observe the system until the $n_i$th failure. In this case, likelihood (12.3) still applies, but with $t_{a_i}$ replaced by $t_{n_i}$.

*Multiple customers* In warranty data, we may observe multiple failures for multiple customers. We assume that all $I$ customers are identical and thus have the same intensity function $\lambda(t)$.[6] The NHPP complete likelihood function (based on all customers) is

$$L(\theta) = \prod_{i=1}^{I} L_i(\theta) = \prod_{i=1}^{I}\left[\left\{\prod_{j=1}^{n_i}\lambda(t_{ij};\theta)\right\}\exp\{-\Lambda(t_{a_i};\theta)\}\right] \tag{12.6}$$

In (12.6), $\sum_{i=1}^{I} n_i$ is the total number of repaired items over all customers.

To specify an NHPP model, we consider two choices of $\lambda(t;\ \theta)$:

1. Weibull or power law process intensity function, namely

$$\lambda_1(t;\theta) = \lambda_1(t;\alpha,\beta) = \frac{\beta}{\alpha}\left(\frac{t}{\alpha}\right)^{\beta-1}, \quad \alpha > 0,\ \beta > 0,\ t \geq 0 \tag{12.7}$$

   The corresponding value of $\Lambda(t;\theta)$ is $\Lambda_1(t;\alpha,\beta) = (t/\alpha)^{\beta}$.

2. The log linear intensity function,[7] given by

$$\lambda_2(t;\theta) = \lambda_2(t;\gamma_0,\gamma_1) = \exp(\gamma_0 + \gamma_1 t), \quad -\infty < \gamma_0, \gamma_1 < \infty,\ t \geq 0 \tag{12.8}$$

   The corresponding value of $\Lambda(t;\theta)$ is $\Lambda_2(t;\gamma_0,\gamma_1) = [\exp(\gamma_0)][\exp(\gamma_1 t) - 1]/\gamma_1$.

---

[6] The assumption that all customers (or systems) have the same intensity function is a strong assumption and might be inappropriate in some applications. If all systems are different, then each system can be modeled by its own intensity function with parameter $\theta_i$ [3].

[7] Intensity function (12.6) is also known as the exponential law or the Cox-Lewis intensity function.

*Comments*

- If $\beta = 1$ [$\gamma_1 = 0$], the power-law [log linear] NHPP model becomes the HPP model.
- If $\beta < 1$ [$\gamma_1 < 0$], the power-law [log linear] intensity is a strictly decreasing function of age. The result is known as a happy system since it improves with age.
- If $\beta > 1$ [$\gamma_1 > 0$], the power-law [log linear] intensity is an increasing function of age. The system is known as a sad system in that it degrades with age.

*Likelihood and ML estimation* With $\lambda_1(t;\alpha,\beta)$ and exact recurrence times, it follows from (12.5) that the single-customer (or system) likelihood can be expressed as

$$L(\alpha,\beta) = \left(\frac{\beta}{\alpha^{\beta}}\right)^{n} \prod_{i=1}^{n} t_i^{\beta-1} \exp[-\Lambda_1(t_a;\alpha,\beta)] \tag{12.9}$$

where $t_1, t_2, \ldots, t_n$ are the *exact* times of the $n$ failures. For this model, the closed-form MLEs are available. They are given by

$$\hat{\beta} = \frac{n}{n\log(t_a) - \sum_{i=1}^{n}\log(t_i)} \quad \text{and} \quad \hat{\alpha} = \frac{t_a}{n^{1/\hat{\beta}}} \tag{12.10}$$

With $\lambda_2(t;\gamma_0,\gamma_1)$ as specified in (12.8), we can express the single-customer likelihood, using (12.5), as

$$L(\gamma_0,\gamma_1) = \exp\left(n\gamma_0 + \gamma_1\sum_{i=1}^{n} t_i\right)\exp[-\Lambda_2(t_a;\gamma_0,\gamma_1)] \tag{12.11}$$

For this model, the MLEs of $\gamma_0$ and $\gamma_1$ are obtained by solving

$$\sum_{i=1}^{n} t_i + \frac{n}{\hat{\gamma}_1} - \frac{nt_a}{1-\exp(-\hat{\gamma}_1 t_a)} = 0 \quad \text{and} \quad \hat{\gamma}_0 = \log\left[\frac{n\hat{\gamma}_1}{\exp(\hat{\gamma}_1 t_a) - 1}\right] \tag{12.12}$$

Using $\lambda_1(t;\alpha,\beta)$ in (12.6), the likelihood for multiple failures for multiple customers becomes

$$L(\alpha,\beta) = \left(\frac{\beta}{\alpha^{\beta}}\right)^{N} \prod_{i=1}^{I}\prod_{j=1}^{n_i}\left\{t_{ij}^{\beta-1}\right\}\exp\left[-\sum_{i=1}^{I}(t_{a_i}/\alpha)^{\beta}\right] \tag{12.13}$$

where $N = \sum_{i=1}^{I} n_i$. For this model, the MLEs of $\alpha$ and $\beta$ are obtained by solving the equations

$$\hat{\alpha} = \left(\frac{\sum_{i=1}^{I} t_{a_i}^{\hat{\beta}}}{N}\right) \quad \text{and} \quad \hat{\beta} = \frac{N}{\hat{\alpha}^{-\hat{\beta}}\sum_{i=1}^{I} t_{a_i}^{\hat{\beta}}\log(t_{a_i}) - \sum_{i=1}^{I}\sum_{j=1}^{n_i}\log(t_{ij})} \tag{12.14}$$

Similarly, the MLEs of $\gamma_0$ and $\gamma_1$ for multiple-customer data can be obtained by using $\lambda_2(t; \gamma_0, \gamma_1)$ in (12.6).

*Tests for trend in interfailure times* In applying an NHPP model, it is necessary to test for the existence of trends in the failure data. Several statistical tests for trend in inter-failure times are available [16, 20]. We look at two such procedures.[8]

*Military Handbook Test* The Military handbook test is constructed for the null hypothesis of an HPP with no trend versus the alternative of an NHPP with monotone trend. The test statistic for more than one system (or customer) is

$$\chi^2_{\mathrm{MHB}} = 2\sum_{i=1}^{I}\sum_{j=1}^{n_i} \log\left[t_{a_i}/t_{ij}\right] \tag{12.15}$$

which is chi-square distributed with $2\sum_{i=1}^{I} n_i$ degrees of freedom under the null hypothesis of an HPP model. It is a powerful tool for testing NPP versus the NHPP power law model, i.e., the NHPP with intensity function $\lambda_1(t; \alpha, \beta)$ given in (12.7).

*Laplace Test* The Laplace test is constructed for the null hypothesis of an HPP versus the alternative hypothesis of an NHPP with monotonic intensity function. The test statistic for $I$ independent systems is

$$Z_{\mathrm{LP}} = \frac{\sum_{i=1}^{I}\sum_{j=1}^{n_j} t_{ij}/t_{a_i} - 1/2\sum_{i=1}^{I} n_i}{\sqrt{\frac{1}{12}\sum_{i=1}^{I} n_i}}. \tag{12.16}$$

Under the null hypothesis that the underlying process is an HPP, the asymptotic distribution of $Z_{\mathrm{LP}}$ is a standard normal distribution. The test is a powerful tool for testing HPP versus NHPP with a log linear intensity function $\lambda_2(t; \gamma_0, \gamma_1)$.

For the case of a single system or customer, (12.15) and (12.16) may be used as well, with $I = 1$.

*Example 12.2* In this example, we analyze the photocopier service data of Table F.17 (Data Set 16). Each row of the table lists a part that was replaced, giving the number of copies made at the time of replacement, the age of the machine in days, and the component replaced. The data are from a single customer, with actual failure times for 41 failures. The two NHPP models discussed above will be applied for both failure time variables, 'Days' and 'Copies'.

In order to apply the NHPP model, we first test for possible trends in inter-failure times, using the Military handbook test (12.15) and the Laplace test (12.16). Table 12.2 shows the results of the statistical trend tests[9] for both 'Days' and 'Copies'. For both variables, both tests of the null hypothesis of no trend lead to rejection at the 1% level.

---

[8] According to [16], these tests appear to be the most frequently used trend tests.

[9] Minitab software also provides these tests for trend.

**Table 12.2** Trend tests for photocopier service data

| Test-value | Days | | Copies | |
|---|---|---|---|---|
| | MIL-Hdbk-189 | Laplace | MIL-Hdbk-189 | Laplace |
| Test statistic | 52.52 | 3.15 | 46.55 | 3.29 |
| *P*-Value | 0.009 | 0.002 | 0.001 | 0.001 |
| DF | 82 | | 82 | |

**Table 12.3** ML estimates of the parameters of two models

| Variable | Model | Parameter | Estimate | Standard error | 95% Confidence interval | | Log-likelihood |
|---|---|---|---|---|---|---|---|
| | | | | | Lower | Upper | |
| Days | Power law | $\beta$ | 1.561 | 0.241 | 1.089 | 2.033 | −189.0 |
| | | $\alpha$ | 152.919 | 58.358 | 38.54 | 267.298 | |
| | Log linear | $\gamma_0$ | −4.720825 | 0.406119 | −5.5168 | −3.92485 | −187.4 |
| | | $\gamma_1$ | 0.001086 | 0.000354 | 0.000392 | 0.001779 | |
| Copies | Power law | $\beta$ | 1.762 | 0.272 | 1.229 | 2.294 | −453.6 |
| | | $\alpha$ | 133528 | 45164 | 45007.8 | 222047 | |
| | Log linear | $\gamma_0$ | −11.28 | 0.412 | −12.09 | −10.47 | −453.5 |
| | | $\gamma_1$ | 0.00000171 | 0.00000054 | 0.00000066 | 0.00000276 | |

Rejection of the null hypothesis for the Military handbook test leads to the conclusion that there is trend in the data and an NHPP model with power law intensity is applicable. Rejecting the null hypothesis for the Laplace test indicates that an NHPP with log linear intensity is an appropriate model for the data. We proceed to estimate the two NHPP models for each of the variables and compare the results.

Table 12.3 shows the ML estimates and 95% confidence intervals for the parameters of the two models, obtained by the method discussed above. For the power law NHPP model, the ML estimate of the shape parameter is $\hat{\beta} = 1.561$ for Days and $\hat{\beta} = 1.762$ for Copies. Both are greater than 1, indicating that the failure rate is increasing for the two lifetime variables.

For the log linear NHPP model, the ML estimates of $\gamma_1$ are $\hat{\gamma}_1 = 0.001086$ for Days and $\hat{\gamma}_1 = 0.00000171$ for Copies, both of which are greater than zero, indicating that the intensity functions are increasing with Days and Copies.

The intensity and cumulative intensity functions for Days and Copies are estimated for the two models by substituting parameter estimates from Table 12.3 into (12.5) and (12.6) and the corresponding expressions for the cumulative functions. The ML estimates of the MCFs for the two variables for both models are shown in Fig. 12.3.

The estimated Log-likelihoods in Table 12.3 and Fig. 12.3 indicate that the model with Log linear intensity function fits better than the model with Power law intensity function for Days, and both of the models fit approximately equally for Copies.

Figure 12.4 shows plots of the ML estimates of the cumulative intensity functions for Days (left side) and Copies (right side) under the two NHPP models. For Copies, the figure indicates a smaller difference between the two fitted models

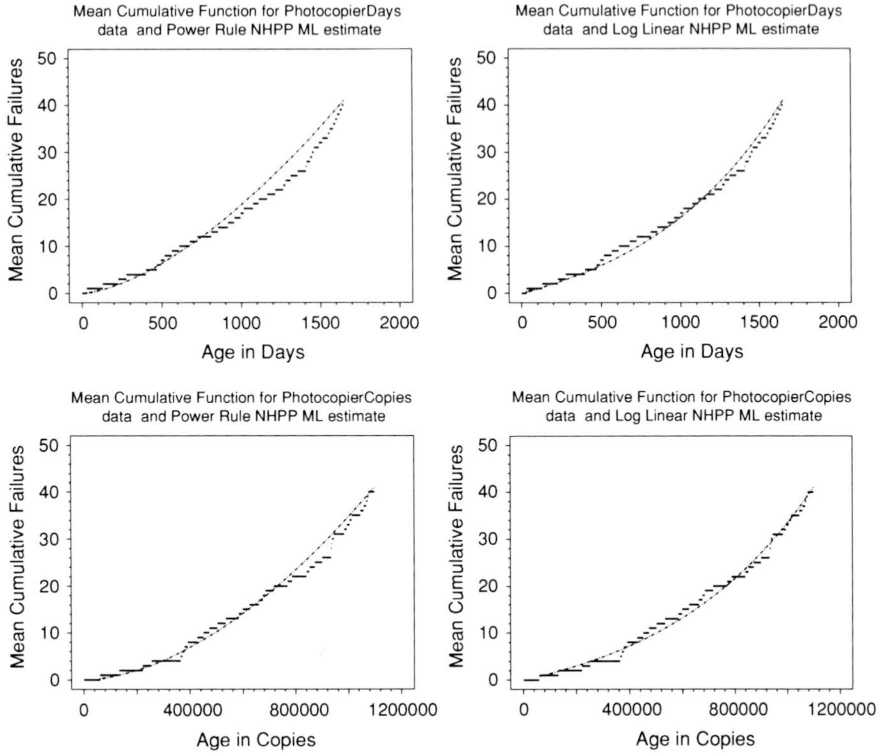


**Fig. 12.3** MCFs for days with power law (*upper-left*) and log linear (*upper-right*), and for copies with power law (*lower-left*) and log linear (*lower-right*)

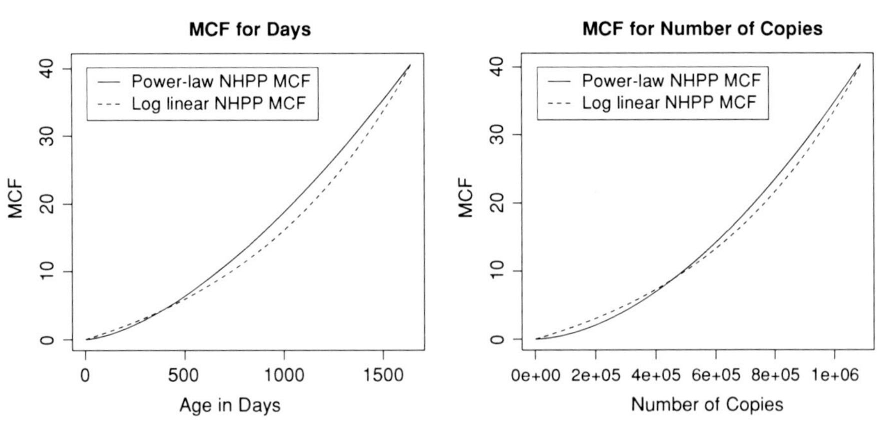


**Fig. 12.4** Fitted cumulative intensity functions for two variables under two models

than in the case of Days. The plots in this figure may be used to predict the MCF for any specified lifetime of the photocopier machine, either in days or number of copies, using the power law and log linear models.

### 12.3.3 Data Scenario 1.3

The data consist failure times for items that failed under warranty and censoring times for all other items. We assume a parametric model $f(t;\ \theta)$, with corresponding survival or reliability function $R(t;\ \theta)$, for the failure time variable $T$. Under Scenario 1.3, the likelihood function for data from all customers is given by

$$L(\theta) = \prod_{i=1}^{n_1} f(t_i; \theta) \prod_{j=1}^{n_2} R(\tilde{t}_j; \theta) \tag{12.17}$$

where $n_1$ denotes the total number of failed items (including the items sold as well as those used as replacements), with $t_i$ $(i = 1, 2, \ldots, n_1)$ denoting age at failure, and $n_2$ denoting the total number of censored items, with $\tilde{t}_j$ $(j = 1, 2, \ldots, n_2)$ denoting censoring ages. In the case of a non-renewing PRW policy, $n_1 \leq I$ and $n_2 = I - n_1$.

For a renewing PRW and both non-renewing and renewing FRW policies, $n_2 = I$, and the likelihood function (12.17) can be rewritten as

$$L(\theta) = \prod_{i=1}^{n} f(t_i; \theta)^{\delta_i} R(t_i; \theta)^{1-\delta_i}, \tag{12.18}$$

where $n = n_1 + n_2$ and $\delta_i$ is the failure-censoring indicator for $t_i$ (taking on the value 1 for failed items and 0 for censored). The maximum likelihood estimator of $\theta$ is obtained by maximizing the likelihood function (12.17) or (12.18), as appropriate.

*Example 12.3* In this example, we consider the battery failure data of Table F.3, for which $n = 54$, with $n_1 = 39$ failure times and $n_2 = 15$ censored times. The unit of measurement is days. Nonparametric and parametric analyses of this data set are given in Examples 8.8 and 8.12, respectively. Our objectives in this example are to analyze these data in more detail and to investigate the properties of the selected models. Example 8.12 indicates that the three-parameter loglogistic distribution with PDF

$$f(t; \mu, \sigma, \lambda) = \frac{\exp\left(\frac{\ln(t-\lambda)-\mu}{\sigma}\right)}{\sigma\left[1 + \exp\left(\frac{\ln(t-\lambda)-\mu}{\sigma}\right)\right]^2}, \quad t \geq \lambda$$

(where $\mu$ is the location parameter, $\sigma$ is a scale parameter, and $\lambda$ is the threshold parameter) is the best fit for the data among the eleven distributions listed in

**Table 12.4** Estimates of AD* and AIC for the eleven distributions for Data Set 03

| Distribution | AD* | AIC |
|---|---|---|
| Weibull (A.30) | 9.188 | 587.296 |
| Lognormal (A.44) | 9.988 | 597.072 |
| Exponential (A.21) | 11.809 | 606.398 |
| Loglogistic | 9.385 | 591.524 |
| 3-Parameter Weibull (A.46) | 9.143 | 589.180 |
| 3-Parameter Lognormal | 9.090 | 589.580 |
| 2-Parameter Exponential | 11.300 | 599.844 |
| 3-Parameter Loglogistic | 9.048 | 589.378 |
| Smallest Extreme Value (A.33) | 9.792 | 601.498 |
| Normal (A.26) | 9.127 | 590.152 |
| Logistic | 9.126 | 590.200 |

Table 12.4. Table 12.4 shows the calculated values of the adjusted AD and the AIC test statistics for the eleven distributions. The three-parameter loglogistic distribution and the Weibull distributions would be selected by the AD and AIC tests, respectively, as the best distributions for the data among the eleven distributions. Note, however, that many of the remaining distributions have AD* values and/or AIC values close to those of these two. The exponential and two-parameter exponential distributions should clearly be eliminated, but the remaining nine might all be candidate models. More likely, still other models would provide more adequate fits. Further investigation of this is needed.

We continue with analysis of the two models selected by the goodness-of-fit statistics. The maximum likelihood estimates of the parameters for 3-parameter loglogistic are $\hat{\mu} = 7.10774$, $\hat{\sigma} = 0.179736$ and $\hat{\lambda} = -524.225$. For 2-parameter Weibull, the MLEs are $\hat{\beta} = 1.9662$ and $\hat{\alpha} = 836.344$. For these two distributions, the estimates of MTTF are 764.567 and 741.441. Example 8.8 gives the nonparametric estimate of MTTF = 726.322. Both the fitted parametric models overestimate the mean life compared to the nonparametric estimate.

Figure 12.5 shows a comparison between the nonparametric (Product-limit) and parametric (with 3-parameter loglogistic and 2-parameter Weibull distributions) estimates of the reliability function $R(t)$ for the battery failure data. This figure shows that the ML estimates of the reliability functions for the two models are not very different from its nonparametric estimate. The loglogistic model is closer to the nonparametric estimate than is the Weibull model. The Weibull model estimate of $R(t)$ is lower than that of the loglogistic at early ages and higher at later ages. From Fig. 12.5, we can conclude that about 50% of the batteries survive past 700 days.

Since the three-parameter loglogistic and the two-parameter Weibull models do not show very different results for the reliability related quantities for this data set, the two-parameter Weibull model might be chosen because of its simplicity compared with the three-parameter loglogistic model, though, again, the search for a "best" model for this data set requires further investigation.

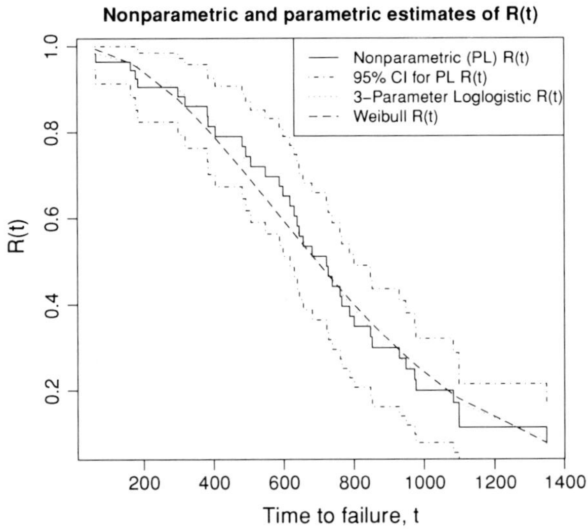


**Fig. 12.5** Nonparametric and parametric estimates of reliability function for battery failure data

### *12.3.4 Data Scenario 1.4*

Scenario 1.4 data consist of the ages at failure, as in Scenario 1.2, and supplementary data consisting of the censored ages $\tilde{t}_i$ for all customers. If the warranty of customer $i$ expired within the data collection period, then $\tilde{t}_i = W$. If it had not expired, $\tilde{t}_i$ is the age of item $i$ at the end of the data collection interval.

Based on the likelihood function (12.6) for Scenario 1.2 and using censored data, the likelihood function for Scenario 1.4 can be written as

$$L(\theta) = \prod_{i=1}^{I} \left[ \left\{ \prod_{j=1}^{n_i} \lambda(t_{ij}; \theta) \right\} \exp\{-\Lambda(\tilde{t}_i; \theta)\} \right] \tag{12.19}$$

where $t_{ij}$ is the age of the item at the time of the *jth* repair for the *ith* customer, and $\tilde{t}_i$ is the censoring age for the *ith* customer. The likelihood function (12.19) can be rewritten either as

$$L(\theta) = \prod_{i=1}^{n_1} \lambda(t_i; \theta) \prod_{j=1}^{n_2} \exp\{-\Lambda(\tilde{t}_j; \theta)\} \tag{12.20}$$

where $n_1$ and $n_2$ are, respectively, the total number of repairs and censored events for all customers, or as

$$L(\theta) = \prod_{i=1}^{n} \lambda(t_i; \theta)^{\delta_i} [\exp\{-\Lambda(t_i; \theta)\}]^{1-\delta_i} \tag{12.21}$$

where $n = n_1 + n_2$ and $\delta_i$ is the failure-censoring indicator.

**Table 12.5** ML estimates of the parameters of two NHPP models

| Model | Parameter | Estimate | Standard error | 95% Confidence interval | |
|---|---|---|---|---|---|
| | | | | Lower | Upper |
| Power law | $\beta$ | 1.39958 | 0.201 | 1.0066 | 1.79256 |
| | $\alpha$ | 553.643 | 57.864 | 440.232 | 667.054 |
| Log linear | $\gamma_0$ | −6.83239 | 0.32748 | −7.47423 | −6.19054 |
| | $\gamma_1$ | 0.001657 | 0.000801 | 0.000087 | 0.003228 |

*Example 12.4* In Example 11.2, we calculated the nonparametric estimate of the MCF for the valve seat replacement data of Table F.11. This data set can be considered Scenario 1.4 data, with multiple failures from multiple systems or from multiple customers ($I = 41$). In Table F.11, "Days observed" is the length of the observation period for the engine. Events that occur beyond this time are not observed, and the value recorded can be considered to be the censored time or retirement time of the system.[10]

In order to estimate the MCF of the NHPP model, we first apply the Military handbook test of (12.15) and the Laplace test of (12.16) to test for possible trend in the inter-failure time data. The results are $\chi^2_{\mathrm{MHB}} = 66.15$ with DF = 96 and $P$-value = 0.018; and $Z_{\mathrm{LP}} = 2.38$ with $P$-value = 0.017. At the 5% level of significance, both tests lead to rejection of the null hypothesis that there is no trend in the data. We therefore conclude that NHPP models for the failure process are required in analysis of the valve-seat data. As before, we use the power law and log linear intensity functions and compare the results with the nonparametric results given in Example 11.2.

Table 12.5 shows the ML estimates of the parameters with 95% confidence intervals for the two models. For the power law NHPP model, $\hat{\beta} = 1.39958$ is greater than 1, indicating that the failure rate is increasing. We are 95% confident that the interval (1.0066, 1.79256) contains the true value of the shape parameter.

For log linear NHPP model, $\hat{\gamma}_1 = 0.001657$, which is greater than zero, supporting the conclusion that the distribution is IFR. Both models, the power-law with $\hat{\beta} > 1$ and log-linear with $\hat{\gamma}_1 > 0$, indicate that the intensity is an increasing functions of age, i.e., the system degrades with age.

The ML estimates of cumulative intensity functions for the power law and log linear models are, respectively,

$$\hat{\Lambda}_1(t; \hat{\alpha}, \hat{\beta}) = (t/553.643)^{1.3996} = 0.000145t^{1.3996}$$

and

$$\hat{\Lambda}_2(t; \hat{\gamma}_0, \hat{\gamma}_1) = [\exp(-6.83239)][\exp(0.001657t) - 1]/0.001657$$

[10] In order to analyze the repairable system failure data by Minitab, each system must have a retirement time, which is the largest time for that system.

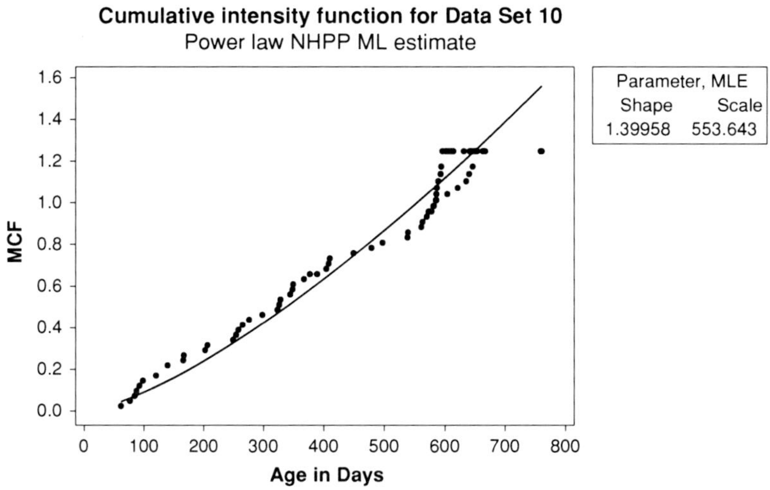


**Fig. 12.6** Mean cumulative intensity for power law NHPP model

**Fig. 12.7** Estimates of nonparametric MCF and fitted cumulative intensity functions for two models

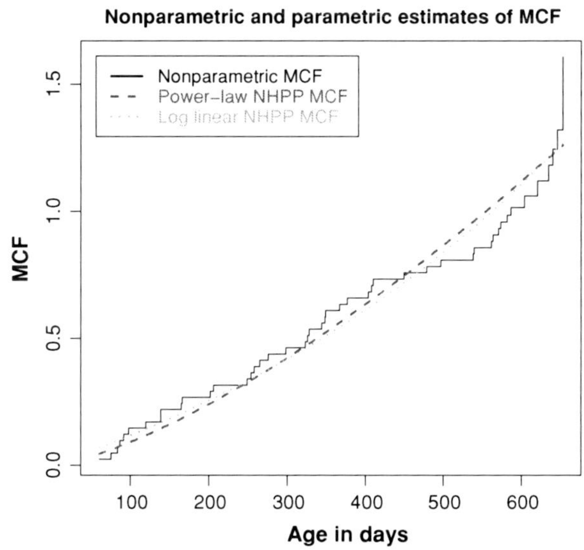


Figure 12.6 is a plot of the ML estimate of the cumulative intensity function versus age for the power law NHPP model, which shows a curve that is slightly concave upward. This plot is consistent with a shape parameter that is greater than one, or the system that is deteriorating with time.

Figure 12.7 shows the nonparametric estimate of the MCF and the fitted cumulative intensity functions for both the power law and log linear NHPP models. This figure indicates very little difference between the two fitted models.

Both NHPP models seem to follow roughly the pattern in the data (as indicated by the nonparametric estimate).

The slightly concave upward pattern of the plots of MCFs for both models indicates that the time between repairs is slightly decreasing over time, that is, the system reliability is deteriorating for increasing age. The values of the maximum log-likelihoods are $-346.5$ for the power law model and $-346.8$ for the log linear model, the close agreement of these values again indicating that the two models are equally appropriate for the data.

## 12.4 Analysis of Structure 2 Data

Structure 2 deals with counts of failures for each customer (or item) over different time intervals. Here we discuss analysis of Scenario 2.2 and Scenario 2.4 data, which are the data sets appropriate for estimating associated mean cumulative functions.

### 12.4.1 Data Scenario 2.2

We look at the likelihood functions for single and multiple customers. The ML estimators are obtained in the usual way.

*Single Customer* Suppose we observe a single customer (or system) for the time interval $(0, t_a]$ and $n_1, n_2, \ldots, n_m$ failures have been observed in non-overlapping time intervals $(t_0, t_1]$, $(t_1, t_2]$,..., $(t_{m-1}, t_m]$ (with $t_0 = 0$ and $t_m = t_a$). For Scenario 1.2, $t_a$ is the time of the last failure. Under the assumption that failed items are repaired minimally and the repair times are negligible, the likelihood for the NHPP model becomes [20]

$$L(\theta) = \prod_{i=1}^{m} \frac{[\Lambda(t_i;\theta) - \Lambda(t_{i-1};\theta)]^{n_i}}{n_i!} \exp[-\Lambda(t_a;\theta)] \tag{12.22}$$

*Multiple Customers* Let $n_{ij}$ denote the number of failures for customer $i$ observed in the interval $(t_{j-1}, t_j]$. The likelihood for multiple customers is

$$L(\theta) = \prod_{i=1}^{I} \left[ \prod_{j=1}^{m} \left\{ \frac{[\Lambda(t_j;\theta) - \Lambda(t_{j-1};\theta)]^{n_{ij}}}{n_{ij}!} \right\} \exp[-\Lambda(t_{a_i};\theta)] \right] \tag{12.23}$$

where $t_{a_i}$ is the upper limit of the last observed time interval for customer $i$.

### 12.4.2 Data Scenario 2.4

Under Scenario 2.4, for items with one or more warranty claims, $t_{a_i}$ is the upper limit of the last observed time interval for customer $i$ (which is either the end of the interval of observation if the item is still under warranty or the end of the warranty period). For items with no warranty claims, $t_{a_i}$ is the censoring time, as discussed earlier. The likelihood function (12.23), for multiple customers, is applicable for this data scenario by assuming that $t_{a_i}$ is the censoring time for customer $i$.

## 12.5 Analysis of Structure 3 Data

Structure 3 data consist of aggregated numbers of failures across customers and over discrete time intervals. In analyzing Structure 3 data, Scenario 3.3 provides information that is appropriate for estimating warranty claims rates. In this section, we discuss a method for investigating age-based claims patterns using a Poisson model (A.18), assuming that the expected number of claims per product (or customer) at age $t$ depends on the age of the product and is independent of other factors. The method is discussed in detail in [11–13, 18, 28]. We analyze Scenario 3.3 data under the assumption that failed items are replaced by new items.

### 12.5.1 Data Scenario 3.3

For data collected under Scenario 3.3, let $S_j$ be the number of items sold in interval $j$ and $n_{jt}$ the reported number of claims in age interval $t$ for those products sold in interval $j$ ($j = 1, 2, \ldots, J; t = 1, 2, \ldots, \min(K - j + 1, W)$, where $K$ denotes the number of intervals over the observation period. Here we assume that the length of the interval is one month[11] and that the $n_{jt}$'s and $S_j$'s are all known.[12] Let $\lambda_t$ be the expected number of claims for an item at age $t$. If claims occur according to a random process, the expected value of $n_{jt}$ is $S_j\lambda_t$. Then the moment estimator (See Sect. 9.4) of $\lambda_t$ is given by

$$\hat{\lambda}_t = \frac{\sum_{j=1}^{\min(J,K-t+1)} n_{jt}}{\sum_{j=1}^{\min(J,K-t+1)} S_j} = \frac{n_{\bullet t}}{R_t}, \quad t = 1, 2, \ldots, \min(W, K), \tag{12.24}$$

---

[11] If necessary, the length can be equal to a "week", "day" and so on.

[12] In some cases, manufacturers consider month of production instead of month of sale when investigating engineering changes, product design changes, manufacturing and assembly changes, etc. In these cases, the monthly sales amounts can be estimated using the sales-lag distribution of the claims data.

where $R_t = \sum_{j=1}^{\min(J,K-t+1)} S_j$ is the total number of units sold up to month $\min(J, K-t+1)$ and $n_{\bullet t} = \sum_{j=1}^{\min(J,K-t+1)} n_{jt}$ is the total number of age $t$ claims reported up to month $\min(J, K-t+1)$, for $t = 1, 2, \ldots, \min(W, K)$. The moment estimator of the cumulative claims rate is

$$\hat{\Lambda}_t = \sum_{u=1}^{t} \hat{\lambda}_u, \quad t = 1, 2, \ldots, \min(W, K) \tag{12.25}$$

The moment estimators (12.24) and (12.25) are the maximum likelihood estimators of $\lambda_t$ and $\Lambda_t$ respectively, when the $n_{jt}$'s are independent Poisson random variables [11, 18] with mean $S_j\lambda_t$. Under the Poisson model, since $V(n_{\bullet t}) = R_t\lambda_t$, an estimator of the variance of $\hat{\Lambda}_t$ is given by

$$\hat{V}\left\{\hat{\Lambda}_t\right\} = \sum_{i=1}^{t} \frac{\hat{\lambda}_i}{R_i}, \quad t = 1, 2, \ldots, \min(W, K) \tag{12.26}$$

This estimator is reasonable if units generate claims randomly and in an identical fashion. However, as pointed out by [11], there is often extra-Poisson variation in the claim frequency data and sometimes correlation as well. The authors discuss two approaches that allow for extra-Poisson variation. One gives the following estimator of the variance of $\hat{\Lambda}_t$:

$$\hat{V}\left\{\hat{\Lambda}_t\right\} = \hat{\sigma}^2 \sum_{i=1}^{t} \frac{\hat{\lambda}_i}{R_i}, \quad t = 1, 2, \ldots, \min(W, K) \tag{12.27}$$

where $\hat{\sigma}^2$ estimates *extra-Poisson variation*, which may arise because of variation in the robustness of units, variations in usage environments, non-Poisson claim patterns for individual units, and so on. The estimate of $\sigma^2$ is

$$\hat{\sigma}^2 = \frac{1}{m} \sum_{j=1}^{J} \sum_{t=1}^{\min(W,K-j+1)} \frac{(n_{jt} - \hat{\mu}_{jt})^2}{\hat{\mu}_{jt}}, \tag{12.28}$$

where $\hat{\mu}_{jt} = S_j\hat{\lambda}_t$ and $m$ is the number of terms in the summations in (12.28) minus the number of $\Lambda_t$'s that are estimated by (12.25). A $100(1-\alpha)\%$ confidence interval for $\Lambda_t$ based on the normal approximation is given by

$$\hat{\Lambda}_t \pm z_{(1-\alpha/2)} \sqrt{\hat{V}(\hat{\Lambda}_t)} \tag{12.29}$$

where $z_p$ is the $p$-quantile of the standard normal distribution (Table E.1). The above results can be used to provide information on the age-based claims pattern of the items.[13]

[13] Additional details on this method can be found in [11, 12, 18].

**Table 12.6** Estimated age-based claims rates and confidence intervals

| Age ($t$) | $\hat{\lambda}_t$ | $\hat{\Lambda}_t$ | 95% Confidence limits for $\hat{\Lambda}_t$ | |
|---|---|---|---|---|
| | | | Lower limit | Upper limit |
| 1 | 0.00009 | 0.00009 | 0.000021 | 0.000152 |
| 2 | 0.00021 | 0.00030 | 0.000177 | 0.000422 |
| 3 | 0.00063 | 0.00093 | 0.000710 | 0.001145 |
| 4 | 0.00081 | 0.00174 | 0.001438 | 0.002039 |
| 5 | 0.00064 | 0.00238 | 0.002027 | 0.002736 |
| 6 | 0.00052 | 0.00290 | 0.002503 | 0.003291 |
| 7 | 0.00032 | 0.00322 | 0.002800 | 0.003634 |
| 8 | 0.00043 | 0.00365 | 0.003199 | 0.004095 |
| 9 | 0.00027 | 0.00392 | 0.003454 | 0.004389 |
| 10 | 0.00040 | 0.00433 | 0.003829 | 0.004822 |
| 11 | 0.00044 | 0.00476 | 0.004234 | 0.005287 |
| 12 | 0.00036 | 0.00512 | 0.004569 | 0.005671 |
| 13 | 0.00020 | 0.00532 | 0.004754 | 0.005885 |
| 14 | 0.00011 | 0.00543 | 0.004860 | 0.006008 |
| 15 | 0.00007 | 0.00550 | 0.004924 | 0.006084 |
| 16 | 0.00007 | 0.00558 | 0.004992 | 0.006164 |
| 17 | 0.00016 | 0.00573 | 0.005135 | 0.006334 |
| 18 | 0.00010 | 0.00584 | 0.005230 | 0.006448 |

*Example 12.5* In this Example, we use the claims data on an automobile component given in Table F.12 and analyzed using nonparametric methods in Example 11.4. For this data set, $t = 1, 2, \ldots, 18$ and the values of $R_t$ are 104005, 103374, 101868, 98639, 94865, 91181, 87478, 83786, 80114, 76744, 73485, 69599, 65263, 60952, 57322, 54204, 51060, and 47829. Table 12.6 gives the estimates of age-based (in month) claims rates $\lambda_t$, cumulative claims rates $\Lambda_t$, and approximate 95% confidence intervals[14] for $\Lambda_t$. These estimates are also shown in Fig. 12.8.

The cumulative claims rates per item up to age 18 months is $\Lambda_{18} = 0.00584$, indicating a claims rate of 0.584%.

Figure 12.8 shows that the estimated claims rates at later ages (age $>$ 12 months) are low compared to those at earlier ages (age $\leq$ 12 months). One reason for this is that very few claims are reported after age 12 months (see Table F.12). This requires further investigation since it indicates a decreasing failure rate, which is very unlikely for most automotive parts. Tables 11.5 and 12.6 and Figs. 11.5 and 12.8 indicate that the estimated nonparametric CDF and cumulative claims rates are very similar. The smaller confidence intervals for the CDF based on the nonparametric analysis (Fig. 11.5) indicate that the CDF is estimated more precisely by the nonparametric approach. The age-based claims rate estimation method depends on the Poisson distributional assumption, whereas the nonparametric estimation of the CDF does not.

[14] Estimation of the variance using (12.27) and (12.28) requires the detailed claims data $\{n_{jt}\}$ for all $j$ and $t$, which are not available for publication but were used to obtain the result $\hat{\sigma} = 1.358$.

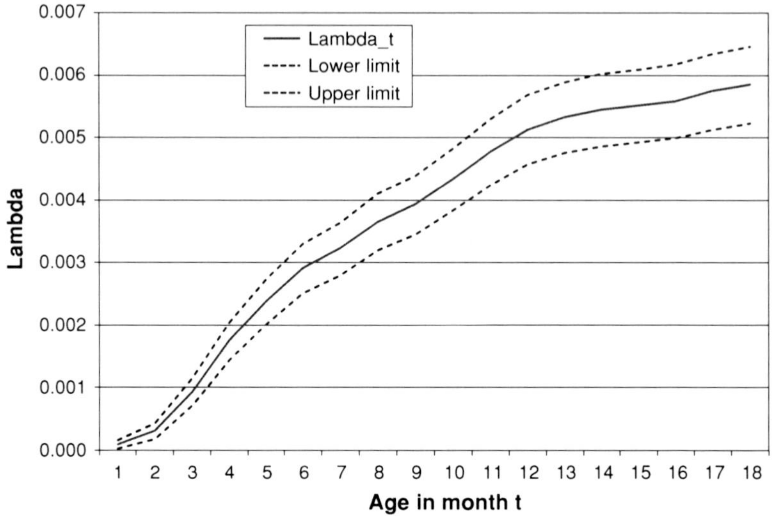


**Fig. 12.8** Estimated cumulative claims rates and approximate 95% confidence intervals

## 12.6 Predicting Future Warranty Claims and Costs

In this section, we discuss the prediction of warranty claims and costs, both of which depend crucially on the lifetime distribution or reliability function of the product. Predictions of future warranty claims and costs are ordinarily made for a specified calendar period. One may also predict the amount of usage based on observed claims and cost data and/or incorporate such data from previous model years into the prediction process. For example, an automobile manufacturer may wish to predict warranty claims and costs up to an age of 5 years or a mileage of 50,000 km, whichever comes first, based on the initial data observed during the first year or first 10,000 km. In general, to predict future claims and costs, we require a parametric statistical model to describe the claims distribution. This is typically done under the assumption that the failure modes, failure mechanisms, usage intensity, and other important characteristics of a product and its use do not vary substantially during the prediction period. The main purposes of estimating and predicting warranty claims and costs to the warranty practitioner are to assess product reliability and to better manage the warranty program associated with the given warranty policy.

Various models and method for estimating claims and costs for different warranty policies are discussed in Chap. 6. Some of the results will be used here for predicting claims and costs. Let $F(t;\ \theta)$, $M(t;\theta)$, and $\Lambda(t;\theta)$ denote, respectively, the CDF, renewal function and MCF for the lifetime variable $T$ of a product. Estimates of these functions will be applied here to prediction of warranty claims and costs for a variety of warranty policies.

### 12.6.1 Future Warranty Claims

The expected number of claims under warranty depends on several factors, such as the type of product (repairable or non-repairable), the type of warranty (FRW or PRW), the type of servicing (repair or replacement in the case of repairable products), and other factors, all of which are uncertain. We focus our attention on estimating the expected number of claims over time for a new item, based on parametric analysis of the warranty data collected.

*Single Item* We begin with claims associated with a single item warranted under FRW or PRW policies.

*Non-repairable product sold with 1-D non-renewing PRW policy* The expected number of claims per item over a period $W$ *is given by* $\hat{F}(W;\hat{\theta})$.

*Non-repairable product sold with non-renewing 1-D FRW policy* The expected number of claims per unit over a time period $W$ is $\hat{M}(W;\hat{\theta})$, the estimated renewal function associated with $\hat{F}(W;\hat{\theta})$.

*Repairable product sold with non-renewing FRW policy* We assume that all failures are repaired minimally. In this case the expected number of claims up to time $W$ is $\int_0^W \lambda(t;\hat{\theta})dt = \hat{\Lambda}(W;\hat{\theta})$.

*Group of Items* Suppose a single group of $N$ units is placed into service, and $r > 0$ units have failed in the interval (0, $V$). One is interested in point prediction of the number of additional failures, $K$, in a future interval ($V$, $W$), with $W > V$. Expressions for predicting $K$ along with the prediction bounds are given in ([20], Chap. 12) and are as follows:

The estimate of $K$ is given by $\hat{K} = (N - r)\hat{\rho}$, where

$$\hat{\rho} = [\hat{F}(W;\hat{\theta}) - \hat{F}(V;\hat{\theta})]/[1 - \hat{F}(V;\hat{\theta})]$$

is the conditional probability of failing in the interval, ($V, W$), given that a unit survived until $V$. The naive[15] $100(1-\alpha)\%$ upper prediction bound for $K$, $\tilde{K}(1-\alpha)$, is computed as the smallest integer $k$ such that BINCDF $(k; N-r, \hat{\rho}) \geq 1-\alpha$, where BINCDF is the binomial cumulative distribution function. This method can be extended for predicting future claims and costs for more general situations in which multiple groups of units enter into service over a longer period of time. For details, see [20].

[15] When the probability prediction intervals are computed on the basis of estimates from limited data, they are sometimes called "naive prediction intervals", and they can serve as a basis for developing more commonly needed statistical prediction intervals ([20], Chap. 12).

### 12.6.2 Future Warranty Costs

We look at predicting warranty cost per item for new items. Many approaches may be used to predict warranty costs, depending on the warranty policy and on the nature of the available data and information [4, 21, 22]. We omit the details and present the final expressions.

*Single Item* We again consider repairable and nonrepairable items under FRW and PRW.

*Non-repairable product sold with 1-D non-renewing PRW policy* We assume a linear rebate function $q(t) = (1 - t/W)c_b,\ 0 \le t \le W$. The predicted warranty cost per unit to the manufacturer for a warranty period of length $W$ is given by

$$\hat{C}_1(W) = c_b\left[\hat{F}(W;\hat{\theta}) - \frac{\hat{\mu}_W}{W}\right], \tag{12.30}$$

where $\hat{\mu}_W = \int_0^W tf(t;\hat{\theta})dt$ is the partial expectation [see (6.23)] and $c_b$ is the sale price.

*Repairable product sold with non-renewing 1-D FRW policy* The predicted warranty cost for a period of length $W$ is $\hat{C}_2 = c_m \times \hat{M}(W;\hat{\theta})$, where $c_m$ is the cost associated with supplying the new unit plus other servicing costs.

*Repairable product sold with non-renewing FRW policy* We assume that all failures are repaired minimally. The cost of minimal repair may be random, and is independent of failure time. Thus the expected cost is the product of the expected number of failures and the expected cost per failure. The latter can be estimated using warranty data. In this case the expected warranty cost for a period of length $W$ is $\hat{C}_3 = c_r \times \hat{\Lambda}(W;\hat{\theta})$, where $c_r$ is the expected cost of repairing items that fail under warranty.

*Group of Items* We consider warranty costs for a group of items as discussed in the previous sub-section. Following ([20], Chap. 12) we have the predicted additional warranty cost up to $W$ given by $c_s\hat{K}$, with the naive $100(1-\alpha)\%$ upper prediction bound given by $c_s\tilde{K}(1-\alpha)$.

### 12.6.3 Other Forecasting Methods

There is a limited literature on Bayesian approaches that may be used for forecasting warranty claims and costs. References [8, 9] discuss methods of forecasting warranty claims and costs by use of Markov mesh models. These can be represented as Bayesian dynamic linear models (DLM's). The authors proposed the DLM with leading indicators that are used to incorporate data from previous model years and presented examples for forecasting the number of repairs and the cost of

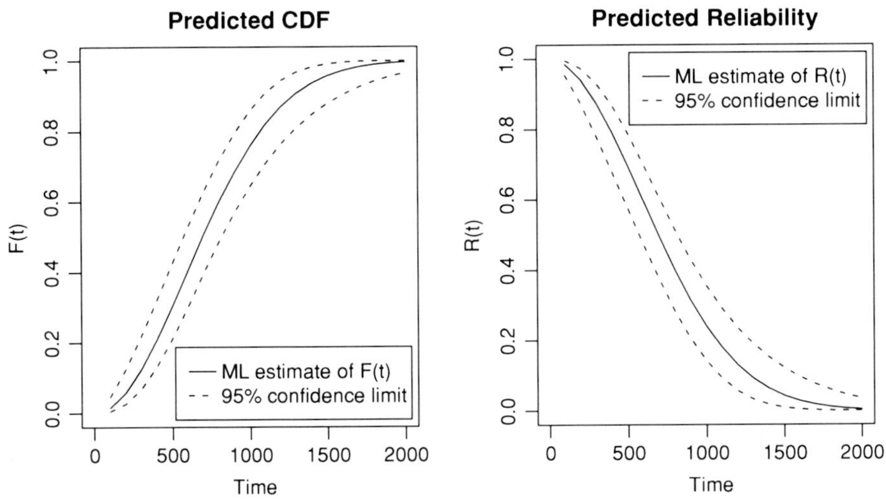


**Fig. 12.9** Plots of predicted $F(t)$ and $R(t)$ with 95% confidence intervals for battery failure data

repairing a unit. A Bayesian approach with Markov chain Monte Carlo (MCMC) sampling for predicting future warranty exposure is given in [27].

### *12.6.4 Examples*

In this section we present the prediction results for three examples discussed earlier in the book.

*Example 12.6* The data on battery life (given in Table F.3) are analyzed in Examples 8.8, 8.12 and 12.3. Examples 8.12 and 12.3 indicate that the Weibull distribution with the ML estimates of parameters $\hat{\beta} = 1.9662$ and $\hat{\alpha} = 836.344$ is a reasonable choice for the distribution of the lifetime of the battery. The predicted CDF and reliability function with 95% confidence intervals for lifetimes up to 2,000 days plotted in Fig. 12.9 are based on this distribution.

Since the battery is a non-repairable product and is sold with a non-renewing PRW policy, the predicted number of claims $\hat{N}$ for a batch of 50 items may be calculated by multiplying $\hat{F}(W)$ by 50. The resulting predictions along with 95% confidence intervals are given in Table 12.7 for $W = 90$, 180, 365 and 550.

Suppose that the unit sale price is $c_b = \$500$. The predicted expected warranty costs with linear rebate function are given in Table 12.8.

Note that warranty costs are over 10% of the sale price at $W = 550$ days. Thus a warranty of 18 months would lead to excessive costs. At one year, the cost of warranty would be slightly above 6%, which is still fairly high. Results such as these are important considerations in management decisions regarding warranty.

**Table 12.7** Predicted claims $\hat{N}$ for over different warranty periods

| Days ($W$) | $\hat{F}(W)$ | $\hat{N}$ | 95% Confidence interval for $\hat{N}$ | |
|---|---|---|---|---|
| | | | Lower | Upper |
| 90 | 0.0124 | 0.62 | 0.19 | 1.97 |
| 180 | 0.0476 | 2.38 | 1.04 | 5.38 |
| 365 | 0.1779 | 8.89 | 5.39 | 14.28 |
| 550 | 0.3551 | 17.75 | 12.80 | 23.92 |

**Table 12.8** Expected warranty costs of battery with linear rebate function

| Time $W$ (days) | $\widehat{F}(W)$ | $\widehat{\mu}_W$ | $\widehat{F}(W) - \widehat{\mu}_W/W$ | $\widehat{C}_1(W)$ |
|---|---|---|---|---|
| 90 | 0.0124 | 0.7393 | 0.0042 | 2.10 |
| 180 | 0.0476 | 5.6535 | 0.0162 | 8.10 |
| 365 | 0.1779 | 42.1815 | 0.0623 | 31.16 |
| 550 | 0.3551 | 123.6592 | 0.1303 | 65.13 |

**Table 12.9** Predicted renewal function $\hat{M}(W)$ for $W$ (km in thousand)

| $W$ | Parametric | Nonparametric | |
|---|---|---|---|
| | $\hat{M}_2(W)$ | $\hat{M}_1(W)$ | $\hat{M}_2(W)$ |
| 10 | 0.2257 | 0.1871 | 0.2083 |
| 20 | 0.7270 | 0.7329 | 0.7852 |
| 30 | 1.2849 | 1.2787 | 1.2530 |
| 40 | 1.8343 | 1.8245 | 1.8148 |
| 50 | 2.3812 | 2.3703 | 2.3480 |

*Example 12.7* In Example 11.1, the nonparametric renewal function was calculated for the automobile engine failure data (in thousand km) of Table F.2. Here we consider the same data and compare the parametric and nonparametric estimates of renewal function. If we assume a Weibull model for the lifetime variable $T$, we obtain the ML estimates of the parameters as $\hat{\alpha} = 20.5882$ and $\hat{\beta} = 1.9561$. Using these estimates of the parameters, $\hat{M}_2(.)$, defined in (11.4), can be obtained by direct calculation with an appropriate computer algorithm or by table look-up and interpolation.[16] The parametric (with Weibull) and the nonparametric obtained by two methods, defined in (11.1) and (11.4), estimates of the renewal function for $W = 10, 20, \ldots, 50$, are as shown in Table 12.9.

The nonparametric estimates obtained by Method 1, denoted $\hat{M}_1(W)$, show very similar results of the parametric estimates when $W$ is large. The performance of the another nonparametric estimator obtained by Method 2, denoted $\hat{M}_2(W)$, depends

[16] See [4] for more on calculation of the renewal function.

**Table 12.10** Predicted MCFs and warranty costs of photocopier machine for given days

| Time $W$ (in days) | Predicted claims | | Predicted costs $\hat{C}_3$ | |
|---|---|---|---|---|
| | $\hat{\Lambda}_1(W; \hat{\alpha}, \hat{\beta})$ | $\hat{\Lambda}_2(W; \hat{\gamma}_0, \hat{\gamma}_1)$ | $c_r \times \hat{\Lambda}_1(W; \hat{\alpha}, \hat{\beta})$ | $c_r \times \hat{\Lambda}_2(W; \hat{\gamma}_0, \hat{\gamma}_1)$ |
| 365 | 3.89 | 3.99 | 194.50 | 199.50 |
| 730 | 11.47 | 9.92 | 573.50 | 496.00 |
| 1095 | 21.61 | 18.74 | 1080.50 | 937.00 |

on the value of $m$ (see Sect. 11.3.1, here $m$ is considered as 5), but the amount of computation increases at a very rapid rate as $m$ and $W$ increase.

*Example 12.8* In Example 12.2, photocopier machine failure data were used to estimate NHPP models with power law and log linear intensity functions. The MLEs of MCFs for age in days with power law intensity function $\hat{\Lambda}_1(W; \hat{\alpha}, \hat{\beta})$and log linear intensity function $\hat{\Lambda}_2(W; \hat{\gamma}_0, \hat{\gamma}_1)$ are given in (12.16) and (12.18), respectively. If for this photocopier machine, $c_r = \$50$, then the predicted MCF and corresponding warranty costs for $W = 1, 2$ and 3 years (expressed in days) are as given in Table 12.10.

## 12.7 Conclusion

In this chapter, we discussed the parametric approach to analysis of 1-D warranty data under different data scenarios and data structures. The focus was on obtaining parametric estimates of the distribution function, reliability function, cumulative hazard function, renewal function, mean cumulative function, and age-based expected number of claims per item, based on claims and supplementary data.

The ML estimation method is used extensively to estimate the parameters of the models and the AD test statistic and AIC are employed for selecting the best approximating model among a set of alternatives. The fitted parametric models are used in forecasting warranty claims and costs for a given warranty period.

For warranty management purposes, the results of this chapter are particularly useful for gaining insight into the field reliability and quality of a product and into determination of possible changes in warranty and/or service plans.

## References


1. Akaike H (1973) Information theory and an extension of the maximum likelihood principle. In: Petrov BN, Csaki F (eds) Proceedings 2nd international symposium on information theory. Akademiai Kiado, Budapest, pp 267–281
2. Akaike H (1974) A new look at the statistical model identification. IEEE Trans Autom Control, AC 19:716–723

3. Basu AP, Rigdon SE (2001) The Weibull nonhomogeneous Poisson process. In: Balakrishnan N, Rao CR (eds) Handbook of statistic: advances in reliability. Elsevier, NY
4. Blischke WR, Murthy DNP (1994) Warranty cost analysis. Marcel Dekker, Inc, NY
5. Blischke WR, Murthy DNP (1996) Product warranty handbook. Marcel Dekker, NY
6. Blischke WR, Murthy DNP (2000) Reliability—modeling, prediction and optimization. John Wiley & Sons, Inc, NY
7. Burnham KP, Anderson DR (1998) Model selection and inference: a practical information theoretic approach. Springer, NY
8. Chen J, Lynn NJ, Singpurwalla ND (1995) Markov mesh models for filtering and forecasting with leading indicators. In: Koul HL, Deshpande JV (eds) Analysis of censored data, IMS lecture notes-monograph series, vol 27. Institute of Mathematical Statistics, pp 39–54
9. Chen J, Lynn NJ, Singpurwalla ND (1996) Chapter 31 Forecasting warranty claims. In: Blischke WR, Murthy DNP (eds) Product warranty handbook. Marcel Dekker, NY, pp 803–817
10. Hogg RV, Craig AT (1978) Introduction to mathematical statistics, 4th edn. Macmillan Publishing Co. Inc., NY
11. Kalbfleisch JD, Lawless JF (1996) Statistical analysis of warranty claims data. In: Blischke WR, Murthy DNP (eds) Product warranty handbook. Marcel Dekker, NY
12. Kalbfleisch JD, Lawless JF, Robinson JA (1991) Methods for the analysis and prediction of warranty claims. Technometrics 33:273–285
13. Karim MR, Yamamoto W, Suzuki K (2001) Statistical analysis of marginal count failure data. Lifetime Data Anal 7:173–186
14. Kendall MG (1951) The advanced theory of statistics, vol 2. 3rd edn. Charles Griffin & Co. Ltd., London
15. Kullback S, Leibler RA (1951) On information and sufficiency. Ann Math Statist 22:79–86
16. Kvaloy JT, Lindqvist BH (1998) TTT-based tests for trend in repairable systems data. Reliab Eng Sys Saf 60:13–28
17. Lawless JF (1982) Statistical models and methods for lifetime data. Wiley, NY
18. Lawless JF (1998) Statistical analysis of product warranty data. Int Statist Rev 66:41–60
19. Lindsey JK (1996) Parametric statistical inference. Clarendon Press, Oxford
20. Meeker WQ, Escobar LA (1998) Statistical methods for reliability data. John Wiley & Sons Inc., NY
21. Murthy DNP, Blischke WR (2001) Warranty and reliability. In: Balakrishnan N, Rao CR (eds) Handbook of statistics: advances in reliability, vol 20. Elsevier, Amsterdam, pp 541–583
22. Murthy DNP, Djamaludin I (2002) New product warranty: a literature review. Int J Prod Econ 79:231–260
23. Murthy DNP, Xie M, Jiang R (2004) Weibull models. Wiley & Sons, NY
24. Parzen E, Tanabe K, Kitagawa G (1998) Selected papers of Hirotugu Akaike, (eds), Springer series in statistics:perspectives in statistics. Springer, NY
25. Rigdon SE, Basu AP (2000) Statistical methods for the reliability of repairable systems. Wiley, NY
26. Sakamoto Y, Ishiguro M, Kitagawa G (1986) Akaike information criterion statistics. D. Reidel Publishing Co. KTK Scientific Publishers, Tokyo
27. Stephens D, Crowder M (2004) Bayesian analysis of discrete time warranty data. Appl Stat 53:195–217
28. Suzuki K, Karim MR, Wang L (2001) Statistical analysis of reliability warranty data. In: Rao CR, Balakrishnan N (eds) Handbook of statistic: advances in reliability. Elsevier, Amsterdam

# Chapter 13
# Complex Models for Parametric Analysis of 1-D Warranty Data

## 13.1 Introduction

In Chaps. 11 and 12 we looked at nonparametric and parametric approaches to analysis of component level warranty data based on the assumption that failure and censored data are all statistically similar and modeled by a single (standard) lifetime distribution. This is appropriate when the underlying population is homogeneous. As discussed in Chap. 3 this is often not the case. Because of quality variations the failure times do not follow standard distributions and must be modeled by more complex model formulations. In Sect. 3.6.3, we briefly discussed the use of competing risk and mixture models for modeling the effect of assembly errors and component non-conformance on component failures. Component failures may also be affected when usage varies across the consumer population. The effect of this on the component failure rate can be modeled by accelerated failure time (AFT) and proportional hazards (PH) models as discussed briefly in Sect. 3.6.5. For certain products (e.g., automobiles) the same component is used in different brands. In this case, component reliability may depend on brand (as the designs are different) and also on the operating environment of the automobile (e.g., temperature, humidity, roads, etc.), which can vary from region to region. Analysis of warranty data at the component level must take into consideration the effect of these variables (called *covariates*).

In this chapter we deal with analyses of 1-D warranty data for which the assumption of homogeneity is not valid. The data consist of lifetimes (failure/censored times) with additional information on one or more supplementary variables, such as failure modes, vendor, usage intensity, operating environment, and so forth. In some cases the required supplementary data might not be available. The focus is on component level analysis based on the parametric approach (discussed in Chap. 12). We consider the following six model formulations.

W. R. Blischke et al., *Warranty Data Collection and Analysis*,
Springer Series in Reliability Engineering, DOI: 10.1007/978-0-85729-647-4_13,

- Competing risk models (appropriate for modeling component failures with more than one mode of failure)
- Mixture models (appropriate for modeling quality variations across different component vendors)
- AFT models (appropriate for modeling the effect of variations in usage rate and operating environment across the population; effect of other covariates such as geographical location, etc.)
- PH models (Same as for AFT models)
- Regression models (Same as for AFT models)
- Imperfect repair models (appropriate when failed items can be repaired and the repair action affects item reliability)

The outline of the chapter is as follows. We begin with a discussion of the six models listed above in Sect. 13.2, where details of the model formulation are given. As discussed in Chaps. 4 and 5, warranty data consists of claims data and supplementary data. Proper analysis based on the models of this chapter requires additional supplementary data which may or may not be available. This issue is discussed in Sect. 13.3. Sections 13.4–13.9 deal with the parametric approach to the analysis of 1-D warranty data for each of the six model formulations. We conclude with some comments and recommendations in Sect. 13.10.

## 13.2 Model Formulations

### *13.2.1 Competing Risk Models*

A general $K$-fold competing risk model is given by

$$F(t) \equiv F(t;\theta) = 1 - \prod_{k=1}^{K} [1 - F_k(t;\theta_k)] \tag{13.1}$$

where $F_k(t) \equiv F_k(t;\theta_k)$ are the distribution functions of the $K$ sub-populations with parameters $\theta_k$, $1 \le k \le K$. Here $\theta \equiv \{\theta_k,\ 1 \le k \le K\}$ and we assume that $K \ge 2$.

This is called a "competing risk model" because it is applicable when an item (component or module) may fail by any one of $K$ failure modes, i.e., it can fail due to any one of the $K$ mutually exclusive causes in a set $\{C_1, C_2, \ldots, C_K\}$. Let $T_k$ be a positive-valued continuous random variable denoting the time to failure if the item is exposed only to cause $C_k$, $1 \le k \le K$. If the item is exposed to all $K$ causes at the same time and the failure causes do not affect the probability of failure by any other mode, then the time to failure is the minimum of these $K$ lifetimes, i.e., $T = \min\{T_1, T_2, \ldots, T_K\}$, which is also positive-valued, continuous random variable.[1]

[1] The competing risk model has also been called the *compound model*, *series system model*, and *multi-risk model* in the reliability literature.

Let $R(t)$, $h(t)$, and $H(t)$ denote the reliability, hazard, and cumulative hazard functions associated with $F(t)$, respectively, and let $R_k(t)$, $h_k(t)$, and $H_k(t)$ be the reliability function, hazard function and cumulative hazard function associated with the distribution function for $F_k(t)$, of the $k$th failure mode, respectively. It is easily shown that

$$R(t) = \prod_{k=1}^{K} R_k(t) \tag{13.2}$$

$$H(t) = \sum_{k=1}^{K} H_k(t) \tag{13.3}$$

and

$$h(t) = \sum_{k=1}^{K} h_k(t) \tag{13.4}$$

Note that for independent failure modes, the reliability function for the item is the product of the reliability functions for individual failure modes and the hazard function for the item is the sum of the hazard functions. The density function of $T$ is given by

$$f(t) = \sum_{k=1}^{K} \left\{ \prod_{\substack{j=1 \\ j \neq k}}^{K} [1 - F_j(t)] \right\} f_k(t) \tag{13.5}$$

which may be rewritten as

$$f(t) = R(t) \left\{ \sum_{k=1}^{K} \left[ \frac{f_k(t)}{R_k(t)} \right] \right\} \tag{13.6}$$

## 13.2.2 Mixture Models

A general $K$-fold finite mixture model is given by the distribution function

$$F(t;\theta) = \sum_{\mathrm{k}=1}^{\mathrm{K}} p_k F_k(t;\theta_k) \tag{13.7}$$

where $p_k \geq 0$, $k = 1, 2, \ldots, K$, is the mixture parameter, with $\sum_{K=1}^{K} p_k = 1$, and the CDF's $F_k(t;\theta_k) \geq 0$, $k = 1, 2, \ldots, K$, are the distribution functions associated with the $K$ subpopulations with parameters $\theta_k$. $\theta \equiv \{\theta_k,\ p_k;\ 1 \leq k \leq K\}$ is the set of parameters of the model.[2] The density, reliability and hazard functions are, respectively,

$$f(t;\theta) = \sum_{k=1}^{K} p_k f_k(t;\theta_k), \tag{13.8}$$

$$R(t) = 1 - F(t) = \sum_{k=1}^{K} p_k R_k(t), \tag{13.9}$$

where $R_k(t) = 1 - F_k(t)$ for $1 \leq k \leq K$, and

$$h(t) = \frac{f(t)}{1 - F(t)} = \sum_{k=1}^{K} w_k(t) h_k(t), \tag{13.10}$$

where $h_k(t)$ is the hazard function associated with subpopulation $k$, with

$$w_k(t) = \frac{p_k R_k(t)}{\sum_{k=1}^{K} p_k R_k(t)} \quad \text{and} \quad \sum_{k=1}^{K} w_k(t) = 1. \tag{13.11}$$

From (13.11), we see that the failure rate for the model is a weighted mean of the failure rates for the subpopulations.

There are situations where some components are produced over a period of time using different machines. Physical characteristics and the reliability of the components produced by different machines may be different, but it may be difficult to distinguish clearly between components made on different machines. In such situations, mixtures of distributions are often used in the analysis of warranty claims data for these components.

### 13.2.3 AFT Models

The reliability design of a product requires specifying the nominal operating conditions (environmental factors such as temperature, humidity, etc.) and usage factors (such as load, usage intensity, etc.) so that the stresses (electrical, mechanical, thermal, etc.) on the various components are within the ranges that will ensure that the product attains the desired reliability. Let the nominal stress on the $K$ components be $z^0 \equiv \{z_1^0, z_2^0, \ldots z_K^0\}$. Without loss of generality, the stresses can be normalized so that every element of the vector is one. Let the time to failure

---

[2] When all $F_k(t)$, $k = 1, 2, \ldots, K$ in (13.7) are either two- or three-parameter Weibull distributions, the model is called a *finite Weibull mixture model*. The Weibull mixture model has been referred to by many other names, including additive-mixed Weibull distribution, bimodal-mixed Weibull (for a two-fold mixture) mixed-mode Weibull distribution, Weibull distribution of the mixed type, multimodal Weibull distribution, and so forth [16].

for the item under nominal stress be $T_0$. When the stress level changes, the distribution of time to failure, MTTF, etc., also change, since the time to failure depends on the stress levels. In the AFT model, the time to failure $T_z$, with stress $z = \{z_1, z_2, \ldots, z_K\}$, is related to $T_0$ by the linear relationship

$$T_z = T_0 \phi(z), \tag{13.12}$$

with $\phi(z) \in [0, \infty)$, $\phi(1) = 1$, and the various partial derivatives $\partial\phi(z)/\partial z_k$ nonpositive. For obvious reasons, $\phi(z)$ is called the acceleration factor. If a component of the vector increases (decreases), then the time to failure decreases (increases). Many different forms have been proposed for the acceleration factor. In the following list, we give a few of these,[3] all of which have a linear form after the log transformation $\mu(z) = \log \phi(z)$:

- Linear: $\mu(z) = \beta_0 + \beta_1 z$, where $z$ is any accelerating variable other than temperature (e.g., usage-rate, voltage, cycle, loadings, etc.)
- Arrhenius: $\mu(z) = \beta_0 + \beta_1 z$, where the accelerating variable $z$ is temperature and $z =$ 11604.83/(temperature in °C + 273.16)
- Inverse power (or power): $\mu(z) = \beta_0 + \beta_1 z$, where $z$ is any accelerating variable other than temperature and $z = -\log$(accelerating variable)
- Eyring: $\mu(z) = \beta_0 + \beta_1 z_1 + \beta_2 z_2$, where $z_1$ is the log of any accelerating variable other than temperature and $z_2 = 1/$(temperature in °C $+ 273.16$).

The implication of (13.12) is that the failure distribution function $F(t; z)$ for $T_z$ has the same form as $F_0(t)$, the distribution function for $T_0$, except that the scale parameters are linked through a linear relationship given by (13.12). For more on the AFT models, see [14, 17, 19].

### 13.2.4 PH Models

In the AFT model, the effect of the covariate $z$ on the lifetime of the item is modeled by considering the scale parameter as a function of $z$. The proportional hazards model, on the other hand, models the effect of the covariate on the distribution function $F(t)$ through its hazard function $h(t)$ by relating the hazard functions at conditions $z$ and baseline conditions $z_0$. The result is

$$h(t|z) = h_0(t)\psi(z), \tag{13.13}$$

where $h_0(t) = h(t|z_0)$ is the baseline failure rate and $\psi(z)$ is a parametric function linking the two, with $\psi(z_0) = 1$ and $\psi(z) \geq 0$ for all $z$.

The Cox PH model proposed by Cox [5, 6] does not make any assumption about the form of the nonparametric part of the model, $h_0(t)$, but assumes a

---

[3] Many more can be found in [17].

parametric form for the effect of the predictors on the hazard function (the parametric part of the model). This model is therefore referred to as a semi-parametric model.

The PH model may also be expressed in terms of the reliability functions $R(t|z)$ and $R_0(t)$ corresponding to $h(t|z)$ and $h_0(t)$, respectively. The relationship is

$$R(t|z) = [R_0(t)]^{\psi(z)} \tag{13.14}$$

or, in terms of corresponding CDFs,

$$F(t|z) = 1 - [1 - F_0(t)]^{\psi(z)} \tag{13.15}$$

A review of the literature on PH models up to 1994 can be found in [11]. References [3, 7] discuss the use of proportional hazards models for the analysis of reliability data and [13] deals with the use of the Weibull proportional hazards model in the analysis of warranty claim data for automotive air conditioning compressors.[4]

Many different functional forms for $\psi(z) = \psi(z;\beta)$ have been proposed, where $\beta = (\beta_1, \beta_2, \ldots, \beta_K)'$ is the vectors of parameters. Some of these are:

- Linear: $\psi(z;\beta) = (1 + z\beta)$.
- Inverse linear: $\psi(z;\beta) = 1/(1 + z\beta)$
- Exponential: $\psi(z;\beta) = \exp(z\beta)$
- Logistic: $\psi(z;\beta) = \log(1 + \exp(z\beta))$

Among these, the exponential form is most widely used because of its simplicity. In this case we have

$$h(t|z) = h_0(t)\exp(\beta z) \tag{13.16}$$

The important characteristics of this model are

- The baseline hazard $h_0(t)$ depends on $t$, but not on the covariate $z$
- The hazard ratio $\exp(\beta z)$ depends on the covariate $z$ but not on $t$. (In this case, $z$ is known as a time-independent covariate).

*Special case*: Suppose that the covariate is scalar and the variable is scaled so that $z = 1$ corresponds to the unit being in its operating mode in which stress is being applied, and $z = 0$ corresponds to its being in its nominal or design usage condition (normal stress).[5] Then the field hazard rate $h(t|z)$ (which takes into account the influence of the operating environment) can be expressed in relation with the design hazard rate $h_0(t)$, as

$$h(t) = h(t; z = 1) = h_0(t)\exp(\beta) \tag{13.17}$$

---

[4] The Cox PH model is widely used in biomedical applications, especially in the analysis of clinical trial data and in its application in the reliability context is increasing [11].

[5] This scaling is different from that used in the AFT model.

**Table 13.1** Distributions of the lifetime and error variables for regression models

| Distribution of $w$ | Distribution of $T$ |
|---|---|
| Extreme value (one parameter) | Exponential (A.21) |
| Extreme value (two parameters) (A.33) | Weibull (A.30) |
| Normal (A.26) | Log-normal (A.44) |
| Logistic | Log-logistic |
| Log-Gamma | Gamma (A.24) |

This implies that the ratio of the hazards of two environments, given by $h(t)/h_0(t) = \exp(\beta)$, remains in the same proportion over time, and that the hazard is increased by a factor $\exp(\beta)$ for field operating conditions versus design conditions.

From (13.17), the reliability function for stress operating environment is

$$R(t) = [R_0(t)]^{\exp(\beta)} \tag{13.18}$$

It follows that testing whether or not $\beta = 0$ is equivalent to testing equality of the reliability functions under stress and design conditions.

### *13.2.5 Regression Models*

The general form of the parametric regression model is

$$y_{ij} = \mu(z_j) + \sigma w_{ij} \tag{13.19}$$

where $y_{ij}$ is the $i$th log failure time $[y_{ij} = \log(t_{ij})]$ for a given explanatory variable $z_j, \sigma > 0$, and $w_{ij}$ is a random variable that does not depend on $z_j (j = 1, 2, \ldots, K, i = 1, 2, \ldots, n_j)$. Expression (13.19) can be rewritten in terms of matrices so that the model is given by

$$Y = \mu(z) + \sigma w \tag{13.20}$$

where $\mu(z)$ is a location parameter and $\sigma$ is a scale parameter. This is known as the location-scale regression model with error $w$ [14]. In (13.20), the distribution of $T$ (where $Y = \log T$) depends on the assumed distribution of $w$ (see [10, 14, 17]). Table 13.1 is a list of the distribution of $w$ corresponding to various distributions of the lifetime variable $T$.

A variety of functional forms for $\mu(z)$ or $(\alpha(z) = \exp(\mu(z))$ in (13.20) have been proposed, but the most useful form is perhaps the log-linear model [14], for which

$$\mu(z) = z\beta \tag{13.21}$$

where $z = (z_1, z_2, \ldots, z_K)$ is the $1 \times K$ vector of independent (or *regressor*) variables and $\beta = (\beta_1, \beta_2, \ldots, \beta_K)'$ is a $K \times 1$ vector of regression coefficients.[6]

The basic linear regression method is given in Chap. 10, where we discussed the regression model and the least squares estimation technique. There are two major reasons why we consider general parametric regression models with ML estimation method rather than the regression analysis based on the normal distribution as discussed in Chap. 10 . First, the dependent variable of interest (survival/failure time) is most likely not normally distributed, which is a serious violation of an assumption for ordinary least squares regression analysis. Secondly, with lifetime data, it is necessary to deal with the problem of censoring. This requires a regression model and method of estimation that is applicable to censored data.

### 13.2.6 Imperfect Repair Models

In Sect. 3.7.2, we looked at imperfect repair, where all repaired items are statistically similar and have a failure distribution that is different from the failure distribution for new items. Imperfect repair can be modeled in many different ways, as discussed in Sect. 3.7.2. In the context of warranty data analysis, we look at a more general formulation, where the failure distribution of a repaired item depends on the number of times the item has been repaired. Let $F(t; \alpha_0)$ denote the, failure distribution of a new item, where $\alpha_0$ is a scale parameter, and assume that after the $j$th repair, the time to failure follows a distribution $F(t; \alpha_j)$, $j \geq 1$, with scale parameter $\alpha_j$. In general, every repair restores a failed item to its operational state, but with the reliability of the repaired item decreasing as the number of times it repaired increases. One way of ensuring this is that $F(t; \alpha_j)$ and $F(t; \alpha_0)$ have the same structure for $j \geq 1$, but the scale parameter $\alpha_j$ is a decreasing sequence in $j$. Two simple forms of $\alpha_j$ are the following:

$$\alpha_j = \gamma^j \alpha_0 \tag{13.22}$$

with $0 < \gamma < 1$, which implies that the reliability decreases after each repair in a geometric manner, and

$$\alpha_j = \alpha_0 / (1 + \gamma)^j \tag{13.23}$$

with $\gamma > 0$. Setting $\gamma = 1$ in (13.22) or $\gamma = 0$ in (13.23) implies that $\alpha_j = \alpha_0$, i.e., the repair restores the failed item to good-as-new.

Additional details on imperfect repair models can be found in [4, 18, 20, 21].

---

[6] This is the most widely used form in the literature on parametric regression, is comparatively simple to apply, and is available in many statistical software packages.

## 13.3 Data Collection and Analysis

The kind of data needed for proper analysis depends on the model formulation. Chapters 4 and 5 discussed warranty claims data and warranty supplementary data. The former refers to the data collected during the servicing of warranty claim and the latter to any other relevant data. In the context of some of the models discussed in this chapter, additional data are obtained through post-mortem analysis of failed components in special inspection and testing laboratories. Several new issues arise. For example, only a fraction of the failed items might be subjected to post-mortem analysis because of cost considerations. In addition, testing and inspection, in some cases may fail to reveal any additional useful information.

### *13.3.1 Data Collection*

The additional data that is required is grouped into three categories. These are listed below along with some potential problems.

*Relating to failure*

- Failure mode—usually identified through post mortem analysis. Possible outcomes are:
  - Failure mode for all items identified
  - Failure modes identified only for some items, because of the cost of testing or other factors
  - Additional potential problem: misclassification

*Production related*

- Component supplier (vendor), when there are two or more, may or may not be identifiable
- Production details for each vendor (date of manufacture, machine used, material source, operator, etc.) may be known, unknown, or known with some uncertainty

*Customer related*

- Usage intensity may be known with varying levels of uncertainty (e.g., three levels of usage—high, medium and low) and only known for some failed items
- Operating environment may be unknown or uncertain
- Follow-up actions revealing some additional information may or may not be performed

### 13.3.2 Data Analysis

In general, these complex models are used at the component level and, in some cases, at module level. At the component level, failed items may either not be repairable or too expensive to repair, so that failure results in replacement by a new item. In the analysis, we assume that this is the case for most models. For imperfect repair, it is important to note that the reliability characteristics of the item may be affected by such repairs.

In analysis of the data, we confine our attention to the parametric approach. Thus we assume a specific model formulation, e.g., a mixture model involving Weibull sub-populations. In the case of competing risk models, the post-mortem might suggest the number of failure modes. In some cases, however, this information is not available. The same is true for the mixture model. In these cases, a useful approach is to begin with $K = 2$ and then repeat the analysis by incrementing $K$. The final model used for decision making requires making some judgment as to the most appropriate value for $K$. We look at the following two scenarios:

*Scenario 1.3*: Failed units replaced (for competing, mixture, AFT and regression models)

*Scenario 1.4:* Failed unit repaired minimally (for PH models) and imperfect repair (for the imperfect repair model)

We will use the notation of Chaps. 4 and 5 to a large extent. Any additional notation needed will be defined as and when needed. Finally, we confine our attention to the method of maximum likelihood for estimation of the parameters of various complex models. For most all of the models, numerical maximization of the likelihood function is required because of the lack of closed form solutions for the ML equations.

## 13.4 Analysis of Data Using Competing Risk Models

Suppose that a component has $K$ failure modes and that the failure modes are statistically independent. In this section, we analyze warranty claims data of such a component using the competing risk model defined in (13.1) and (A.54). We look first at the general case in which the failure modes of some of the failed items are known and those of the remaining are unknown. In addition, we assume that it is not possible to determine the failure modes (or causes of failure) for the censored (non-failed) items.

Two special cases of interest are as follows:

*Case* (*i*): The failure modes are known for all failed item.

*Case* (*ii*): The failure modes are unknown for all failed items.

### 13.4.1 Warranty Data

We look at component level failure data with more than one mode of failure and failed components replaced by new components (data Scenario 1.3). The data are reordered into two sets, the first containing data for failed components and the second for censored components (similar to the set-up in Chaps. 11 and 12). Let $n_1$ be the number of failed units and $n_2$ the number of censored units. For the failed units, the post-mortem outcome is uncertain, that is, the failure modes for some units may not be known. Out of the $n_1$ failed items, let $n_{1k}$ denote the number of items with failure mode $k$, $1 \le k \le K$, and $n_{10} = n_1 - \sum_{k=1}^{K} n_{1k}$ the number of failures for which there is no information regarding the failure mode. Let $t_{kj}$ denote the lifetime of the $j$th item failing from failure mode $k$, and $\tilde{t}_i$ the $i$th censoring time.

*Note:* For Case (i), $n_{10} = 0$, and for Case (ii) $n_{10} = n_1$.

### 13.4.2 Statistical Analysis

For the general case, $n_{1k}$ units out of $n$ failed due to failure mode $k$, with failure times $\{t_{k1}, t_{k2}, \ldots, t_{kn_{1k}}\}$, and there are $n_{10}$ units with failure times $\{t'_1, t'_2, \ldots, t'_{n_{10}}\}$ for which there is no information regarding the failure mode. In addition, there are $n_2 = n - \sum_{k=1}^{K} n_{1k} - n_{10}$ censored units, with censoring times $\{\tilde{t}_1, \tilde{t}_2, \ldots, \tilde{t}_{n_2}\}$. The likelihood function in the general case is given by

$$L(\theta) = \prod_{k=1}^{K} \left[ \prod_{j=1}^{n_{1k}} f_k(t_{kj}) \prod_{\substack{l=1 \\ l \neq k}}^{K} R_l(t_{kj}) \right] \times \prod_{k=1}^{K} \left[ \prod_{j=1}^{n_{10}} f_k(t'_j) \prod_{\substack{l=1 \\ l \neq k}}^{K} R_l(t'_j) \right] \times \prod_{i=1}^{n_2} \prod_{k=1}^{K} R_k(\tilde{t}_i), \tag{13.24}$$

where $f_k(t)$ and $R_k(t)$ are as defined in Sect. 13.2.1.

The MLEs of the parameters are obtained by maximizing the likelihood function (13.24). For most distributions the ML estimation method requires numerical maximization because of the lack of closed form solutions for the estimators.

The results for the two special cases are as follows:

*Case (i)*: The expression for the likelihood function is given by (13.24) with the second term equal to unity, so that

$$L_1(\theta) = \prod_{k=1}^{K} \left[ \prod_{j=1}^{n_{1k}} f_k(t_{kj}) \prod_{\substack{l=1 \\ l \neq k}}^{K} R_l(t_{kj}) \right] \times \prod_{i=1}^{n_2} \prod_{k=1}^{K} R_k(\tilde{t}_i). \tag{13.25}$$

*Case* (*ii*): The expression for the likelihood function is given by (13.24) with the first term of equal to unity,

$$L_2(\theta) = \prod_{k=1}^{K}\left[\prod_{j=1}^{n_{10}} f_k(t'_j)\prod_{\substack{l=1\\ l\neq k}}^{K} R_l(t'_j)\right] \times \prod_{i=1}^{n_2}\prod_{k=1}^{K} R_k(\tilde{t}_i). \tag{13.26}$$

The cause-specific (or failure mode-specific) hazard function for cause $k$ can be written as

$$\tilde{h}_k(t) = \lim_{\Delta t\to 0}\frac{\Pr(t \le T < t+\Delta t,\ C = k|T \ge t)}{\Delta t} = \frac{f_k(t)}{R(t)}, \tag{13.27}$$

where $f_k(t)$ is the cause-specific PDF at time $t$ that represents the unconditional probability of failure of an unit at time $t$ from cause $k$, and $R(t)$ is the overall reliability function representing the probability of surviving from all causes up to time $t$. Relationship (13.27) implies that

$$f_k(t) = \tilde{h}_k(t)R(t). \tag{13.28}$$

Using (13.28) and (13.2), we can rewrite the likelihood functions (13.25) and (13.26), respectively as

$$L_1(\theta) = \prod_{k=1}^{K}\left[\prod_{j=1}^{n_{1k}} \tilde{h}_k(t_{kj})R(t_{kj})\right] \times \prod_{i=1}^{n_2} R(\tilde{t}_i) \tag{13.29}$$

and

$$L_2(\theta) = \prod_{k=1}^{K}\left[\prod_{j=1}^{n_{10}} \tilde{h}_k(t'_j)R(t'_j)\right] \times \prod_{i=1}^{n_2} R(\tilde{t}_i). \tag{13.30}$$

The MLEs of the parameters of the models are obtained by maximizing (13.25) or (13.29) for Case (i) and (13.26) or (13.30) for Case (ii).

*Exponential Lifetimes* Suppose that $K = 2$, and the lifetimes of failure modes 1 and 2 independently follow exponential distributions (A.21) with parameters $\lambda_1$ and $\lambda_2$, respectively. Time to failure is modeled by (13.1). We consider Case (i). The data consist of $n$ units, with $n_{11}$ units failing due to failure mode 1 with failure times $\{t_{11}, t_{12}, \ldots, t_{1n_{11}}\}$, $n_{12}$ units failing due to failure mode 2 with failure times $\{t_{21}, t_{22}, \ldots, t_{2n_{12}}\}$, and $n_2 = n - n_{11} - n_{12}$ units censored, with censoring times $\{\tilde{t}_1, \tilde{t}_2, \ldots, \tilde{t}_{n_2}\}$.

In this case, from (13.2), we have $R(\tilde{t}) = R_1(\tilde{t})R_2(\tilde{t}) = \exp -(\lambda_1 + \lambda_2)\tilde{t}$ and using this in (13.29), the log-likelihood function becomes

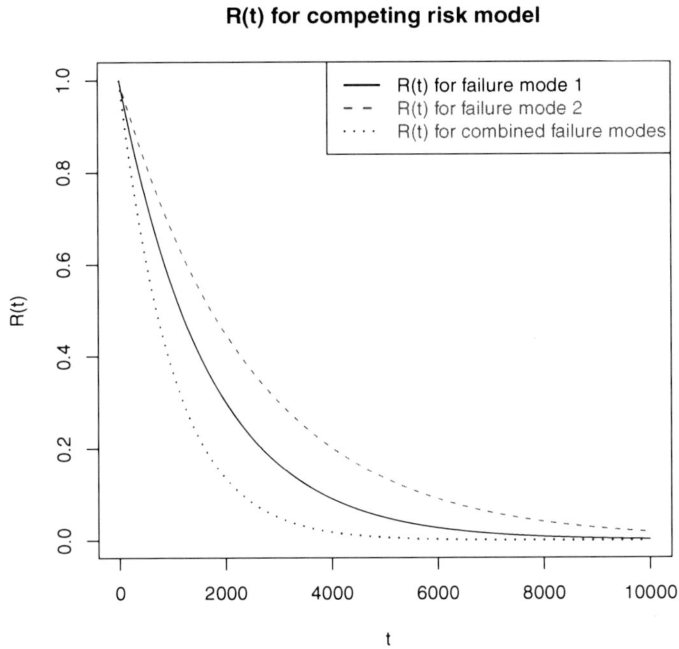


**Fig. 13.1** Comparison of ML estimates of reliability functions for competing risk model

$$\begin{aligned}\log L = n_{11}\log(\lambda_1) - (\lambda_1+\lambda_2)\sum_{j=1}^{n_{11}} t_{1j} + n_{12}\log(\lambda_2)\\ - (\lambda_1+\lambda_2)\sum_{j=1}^{n_{12}} t_{2j} - (\lambda_1+\lambda_2)\sum_{i=1}^{n_2}\tilde{t}_i\end{aligned} \tag{13.31}$$

From this, the ML estimators of $\lambda_1$ and $\lambda_2$ are found to be

$$\hat{\lambda}_i = \frac{n_{1i}}{\sum_{j=1}^{n_{1j}} t_{1j} + \sum_{j=1}^{n_{2j}} t_{2j} + \sum_{i=1}^{n_2}\tilde{t}_i}, \quad i = 1, 2 \tag{13.32}$$

It follows from (13.2) that the maximum likelihood estimate of the reliability function of the component is

$$\hat{R}(t) = \exp\Big(-(\hat{\lambda}_1 + \hat{\lambda}_2)t\Big), \quad t \geq 0 \tag{13.33}$$

*Example 13.1* We consider an electronic component for which lifetimes follow an exponential distribution (A.21). The component exhibits a new mode of failure due to mounting problems. If incorrectly mounted, it can fail earlier, and this is also modeled by an exponential distribution. The parameters of the exponential distributions for failure modes 1 and 2 are $\lambda_1 = 0.0006$ and $\lambda_2 = 0.0004$ per day.

From (13.33), the maximum likelihood estimate of the reliability function of the component is $\hat{R}(t) = \exp(-(0.0006 + 0.0004)t) = \exp(-0.001t), t \geq 0$. Figure 13.1 displays a comparison of the estimated reliability functions for failure mode 1, failure mode 2 and combined failure modes 1 and 2 for $0 \leq t \leq 10{,}000$ days.

This figure can be used to assess reliability of the component for given days. For example, the figure indicates the reliabilities of the component at age 2,000 days are 0.30 for failure mode 1, 0.45 for failure mode 2 and 0.14 for the combined failure modes. Based on (13.33), the estimated MTTF of the component is found to be $\hat{\mu} = \int_0^{\infty} \hat{R}(t)dt = 1/(\hat{\lambda}_1 + \hat{\lambda}_2) = 1{,}000$ days.

This example dealt a simple problem with just two failure modes. Applications based on field data with more than two failure modes are discussed in Chaps. 17 and 18.

## 13.5 Analysis of Data Using Mixture Models

In this section, we look at the analysis of warranty claims data based on the mixture model defined in (13.7). The failure data can, in some instances, be classified into the different groups or sub-populations that are the components of the mixture model based on a proper postmortem analysis [9]. In other instances, no such postmortem is possible and the failed item cannot be assigned to one specific sub-population.

We look at the general case where the subpopulations for some of the failed items are known, and those for the remaining failed items unknown. Note that there is no information on the subpopulations for the censored items, as they have not yet failed.

The two special cases are as follows:

*Case* (*i*): It is possible to assign each failed unit to the appropriate subpopulation, i.e., all failed items are subjected to post-mortem analysis and there is no uncertainty in the classification.

*Case* (*ii*): It is not possible to assign any failed unit to its appropriate subpopulation, i.e., no items are subjected to a post-mortem analysis.

### 13.5.1 Warranty Data

Here the data are the component level failure data from more than one subpopulation and the failed components are replaced with new components (Scenario 1.3). The data are reordered in two sets, the first containing the $n_1$ failed items and the second containing the $n_2$ censored items. A further classification of the items from the first set based on the results of the postmortem analysis results in $n_{1k}$ items failed being assigned to subpopulation $k$, $1 \le k \le K$, and $n_{10}$ failed items that cannot be assigned to a particular subpopulation. The warranty claims data consist of $n_{1k}$ failure times from subpopulation,$n_{10}$ failure times from unknown subpopulations, and $n_2$ censoring times.

*Note:* For Case (i) $n_{10} = 0$, and for Case (ii) $n_{10} = n_1$.

### 13.5.2 Statistical Analysis

The likelihood function for the general case is given by

$$L(p,\theta) \propto \prod_{k=1}^{K}\left[p_k^{n_{1k}}\prod_{j=1}^{n_{1k}} f_k(t_{kj}|\theta_k)\right] \times \prod_{i=1}^{n_{10}}\left[\sum_{k=1}^{K} p_k f_k(t_i'|\theta_k)\right] \times \prod_{i=1}^{n_2}\left[\sum_{k=1}^{K} p_k R_k(\tilde{t}_i|\theta_k)\right] \quad (13.34)$$

where $f_k(t|\theta_k)$ is the PDF of the failure time random variable $T$ with parameters $\theta_k$ for subpopulation $k$, $R_k(t|\theta_k)$ is the corresponding reliability function, and $p_k \ge 0$ is the mixing parameter, with $\sum_{k=1}^{K} p_k = 1$.

The results for the two special cases are as follows:

*Case (i)*: The likelihood function for this case is given by (13.34) with the second term of (13.34) equal to unity. The result is

$$L_1(p,\theta) \propto \prod_{k=1}^{K}\left[p_k^{n_{1k}}\prod_{j=1}^{n_{1k}} f_k(t_{kj}|\theta_k)\right] \times \prod_{i=1}^{n_2}\left[\sum_{k=1}^{K} p_k R_k(\tilde{t}_i|\theta_k)\right], \quad (13.35)$$

where $\sum_{k=1}^{K} n_{1k} = n_1$.

*Case (ii)*: The likelihood function for this case is given (13.34) with the first term equal to unity. The result is

$$L_2(p,\theta) \propto \prod_{i=1}^{n_{10}}\left[\sum_{k=1}^{K} p_k f_k(t_i'|\theta_k)\right] \times \prod_{i=1}^{n_2}\left[\sum_{k=1}^{K} p_k R_k(\tilde{t}_i|\theta_k)\right], \quad (13.36)$$

where $n_{10} = n_1$.

The MLEs of the parameters $\{\theta_k, p_k\}$ for the two cases are obtained by maximizing (13.35) and (13.36), respectively.[7]

---

[7] Reference [9] presents an algorithm for finding MLEs of the parameters of a Weibull mixture model with right censored data. Reference [1] presents a procedure for finding the MLEs of the parameters of two-fold Weibull mixture models.

**Table 13.2** Estimated AIC for seven mixture models for Data Set F.16

| Mixture Models | Models forms and parameters | AIC |
|---|---|---|
| 1. Weibull-Weibull | $p \times \text{Weib}(\beta_1, \alpha_1) + (1-p) \times \text{Weib}(\beta_2, \alpha_2)$ | 350.159 |
| 2. Weibull-Exponential | $p \times \text{Weib}(\beta, \alpha) + (1-p) \times \text{Exp}(\lambda)$ | 348.260 |
| 3. Weibull-Normal | $p \times \text{Weib}(\beta, \alpha) + (1-p) \times \text{Nor}(\mu, \sigma)$ | 349.710 |
| 4. Weibull-Lognormal | $p \times \text{Weib}(\beta, \alpha) + (1-p) \times \text{Lnor}(\mu, \sigma)$ | 351.235 |
| 5. Normal-Exponential | $p \times \text{Nor}(\mu, \sigma) + (1-p) \times \text{Exp}(\lambda)$ | 351.513 |
| 6. Normal-Lognormal | $p \times \text{Nor}(\mu_1, \sigma_1) + (1-p) \times \text{Lnor}(\mu_2, \sigma_2)$ | 351.579 |
| 7. Lognormal-Exponential | $p \times \text{Lnor}(\mu_2, \sigma_2) + (1-p) \times \text{Exp}(\lambda)$ | 349.301 |

**Table 13.3** MLEs of parameters for seven mixture models

| Mixture Models | MLEs of parameters |
|---|---|
| 1. Weibull-Weibull | $\{\hat{\beta}_1 = 1.249, \hat{\alpha}_1 = 0.245, \hat{\beta}_2 = 2.777, \hat{\alpha}_2 = 3.485, \hat{p} = 0.017\}$ |
| 2. Weibull-Exponential | $\{\hat{\beta} = 2.768, \hat{\alpha} = 3.484, \hat{\lambda} = 4.052, \hat{p} = 0.983\}$ |
| 3. Weibull-Normal | $\{\hat{\beta} = 7.359, \hat{\alpha} = 4.481, \hat{\mu} = 2.303, \hat{\sigma} = 0.868, \hat{p} = 0.387\}$ |
| 4. Weibull-Lognormal | $\{\hat{\beta} = 1.246, \hat{\alpha} = 0.395, \hat{\mu} = 1.075, \hat{\sigma} = 0.440, \hat{p} = 0.029\}$ |
| 5. Normal-Exponential | $\{\hat{\mu} = 3.053, \hat{\sigma} = 1.220, \hat{\lambda} = 24.003, \hat{p} = 0.995\}$ |
| 6. Normal-Lognormal | $\{\hat{\mu}_1 = 0.302, \hat{\sigma}_1 = 0.182, \hat{\mu}_2 = 1.073, \hat{\sigma}_2 = 0.443, \hat{p} = 0.027\}$ |
| 7. Lognormal-Exponential | $\{\hat{\mu}_2 = 1.080, \hat{\sigma}_2 = 0.435, \hat{\lambda} = 1.729, \hat{p} = 0.965\}$ |

*Example 13.2* We consider the aircraft windshield data of Table F.16. For this data set, [16] fitted the two-fold Weibull mixture model, applying the WPP plotting procedure. Here we consider seven two-fold mixture models,[8] $f(t;\theta) = pf_1(t;\theta_1) + (1-p)f_2(t;\theta_2)$, (listed in Table 13.2) to model the data set. The ML estimation method is used to estimate the model parameters and the Akaike Information Criterion (AIC) defined in Sect. 12.2.2 is used to select one of these as the best-fitting model of the seven.

The calculated AIC and the parameter estimates for the seven models are shown in Tables 13.2 and 13.3, respectively.[9] Based on the lowest value of the AIC, the Weibull-exponential mixture is the best model among the seven models. The lognormal-exponential, Weibull-normal and Weibull-Weibull mixture models also may be viewed as reasonable models since the AICs for these models are fairly close to the minimum AIC value.

The plots of the CDFs of the Weibull-Exponential, lognormal-exponential, Weibull-normal and Weibull-Weibull mixture models and the nonparametric (KM) estimates of CDF are displayed in Fig. 13.2. From Fig. 13.2, we see that the CDFs of the four mixture models are very close to the nonparametric CDF.

[8] Two-fold mixture refers to a mixture model with two components, that is, $K = 2$ in (13.8) and (A.53).

[9] The function "mle" given in the "stats4" library of R-language is used to find the MLEs of the parameters. It is very sensitive to initial values of parameters of these models.

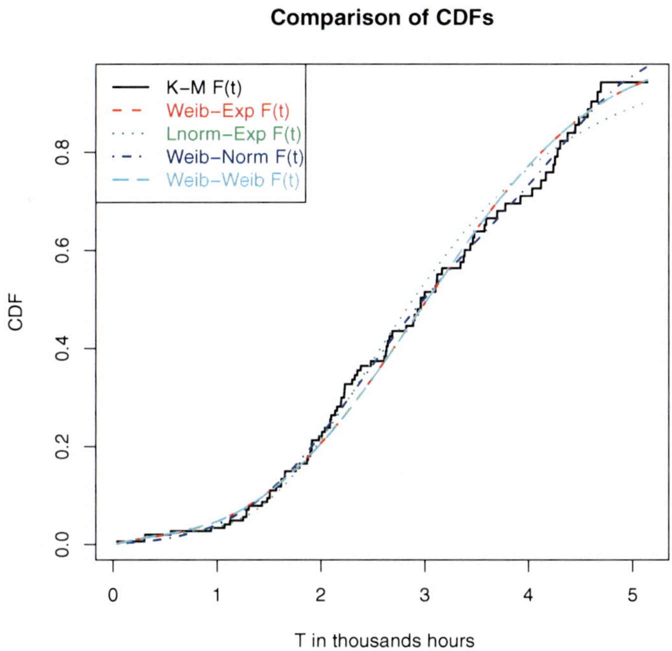


**Fig. 13.2** Comparison of ML estimates of CDFs of mixture models and nonparametric estimate of CDF for Windshield failure data

However, the CDF of the Weibull-normal mixture model is a bit closer to the nonparametric CDF than are those of the remaining models.

The justification for considering a mixture model for the data is that failures may occur by more than one failure mode, and these may have different failure distributions. The two-fold model is a first attempt at taking this into account in modeling the overall pattern of failures. A more thorough analysis would involve additional mixture components or separate analyses for each mode.

## 13.6 Analysis of Data Using Accelerated Failure Time Models

As discussed earlier, the AFT model can be used for many purposes, and testing is only one of these. If the data are limited to field data (claims and censored), then test data are not included in the analysis. On the other hand, if data from the development program are included, then test data are ordinarily available as well. In the case of warranty claims data, we may have information on usage. This may be considered to be equivalent to a stress variable. In this case, in applying the AFT model, usage is grouped into several categories. This situation is similar to testing at different stress levels.

The first step in performing an accelerated-test data analysis is to select suitable lifetime distributions that fit the failure time data of each group. This results in overstress life distributions for each stress or usage group. The second step is to choose a life-stress or life-group relationship[10] (Sect. 13.2.3) in order to estimate the lifetime distribution, and a variety of functions that relate component reliability at normal usage to the characteristics of the overstress distributions.

### 13.6.1 Warranty Data

We look at component level data where the data have been categorized into $K$ groups based on a set of supplementary variables[11] $z$ (e.g., use-rate, ambient temperature) that may be expected to cause early failures. The variable $z$ may be either a scalar or a $K$-dimensional vector. In the case of test data, the accelerating variables are known for both failure and censored components. In the case of warranty claims data, the accelerating variables (or covariates) are known for failed components but some accelerating variables (such as use rate, environment, etc.) are unknown for censored components.

We assume that the usage pattern for a user does not change with time and that failed components are replaced by new. As a result, the failure and censored times for each group are statistically similar in the sense they have the same failure distribution. Let the data are reordered in two sets, with $n_1$ failure units in the first and $n_2$ censored units in the second, and let $n_{1k}$ and $n_{2k}$ denote the number of failed and censored units, respectively, under stress level $k$.

### 13.6.2 Statistical Analysis

The data set is $\{t_{jk}, z_j, \delta_{jk}\}$, where the $t_{jk}$ are mutually independent, with each either a failure time or censoring time, $z_j = (z_{j1}, z_{j2}, \ldots, z_{jK})$ is a stress or group vector and $\delta_{jk}$ is the censoring indicator (equal to 1 for failure and 0 for censored) for $t_{jk}$. The likelihood function can be written as

[10] The life-stress relationship describes a characteristic point or a life characteristic of the distribution from one stress level to another. For example, for the Weibull distribution, the scale parameter, $\alpha$, is considered to be life characteristic that is stress dependent and thus the life-stress relationship is assigned to $\alpha$. For the exponential and lognormal distributions, the mean life and median life, respectively, are considered to be life characteristics that are stress dependent.

[11] These may be accelerating variables or stresses such as use-rate, temperature, voltage, humidity, pressure, etc.

$$L(\theta|T,z) \propto \prod_{k=1}^{K}\prod_{j=1}^{n_{1k}} [f_z(t_{jk}|z_j)]^{\delta_{jk}} [R_z(t_{jk}|z_j)]^{1-\delta_{jk}}, \tag{13.37}$$

where $R_z(u|z) = \int_u^\infty f_z(x|z)dx$.

The parameter vector $\theta$ can be estimated by maximizing the likelihood (13.37). A numerical maximization method such as the Newton–Raphson method is required to obtain the MLE of $\theta$ and its asymptotic variance–covariance matrix. Once the parameters are estimated, a variety of functions related to component reliability at normal operating conditions can be estimated.

In Minitab, the accelerating life testing program allows for seven distributions (Weibull, exponential, normal, lognormal, logistic, loglogistic and smallest extreme value) for the lifetime variable $T$ in (13.37) and for one or two predictors. The first predictor is an accelerating variable and the second predictor can be either a second accelerating variable or another factor. Minitab considers four types of life-stress relationships—linear, Arrhenius, inverse temperature and log-power—with MLEs of the model parameters given as part of the output.

### 13.6.3 Weibull Distribution

For the $k$th group, the failure distribution is given by a Weibull distribution with shape parameter $\gamma$ and scale parameter $\alpha_k$, with

$$\alpha_k = \phi(z) = \exp(\mu(z)) = \exp(\beta_0 + \beta_1 z_k). \tag{13.38}$$

As a result, the failure density function for items from Group $k$ is given by

$$f(t_{kj};\alpha_k,\gamma) = \frac{\gamma}{\alpha_k}\left(\frac{t_{kj}}{\alpha_k}\right)^{\gamma-1} \exp\left[-\left(\frac{t_{kj}}{\alpha_k}\right)^{\gamma}\right], \quad t_{kj} \geq 0,\ \alpha_k,\gamma > 0. \tag{13.39}$$

*Note*: To avoid confusion with the parameters of (13.38), $\gamma$ is used in (13.39) to denote the shape parameter of the Weibull distribution instead of $\beta$.

The log-likelihood function based on (13.37) and (13.39) is

$$\log L(\theta) \propto \sum_{k=1}^{K}\sum_{j=1}^{n_{1k}} \left[\delta_{kj}\log\{f(t_{kj};\alpha_k,\gamma)\} + (1-\delta_{kj})\log\{R(t_{kj};\alpha_k,\gamma)\}\right]. \tag{13.40}$$

Putting the values of $\alpha_k$ from (13.38) into (13.40), we obtain the log-likelihood as

$$\log L(\beta_0, \beta_1, \gamma) \propto \sum_{k=1}^{K} \sum_{j=1}^{n_{1k}} \left[ \delta_{kj} \log \left\{ \frac{\gamma (t_{kj})^{\gamma - 1}}{\exp(\beta_0 + \beta_1 z_k)^{\gamma}} \exp\left\{ - \left( \frac{t_{kj}}{\exp(\beta_0 + \beta_1 z_k)} \right) \right\}^{\gamma} \right\} \right]$$
$$+ \sum_{k=1}^{K} \sum_{j=1}^{n_{1k}} \left[ (1 - \delta_{kj}) \log \left\{ \exp\left\{ - \left( \frac{t_{kj}}{\exp(\beta_0 + \beta_1 z_k)} \right) \right\}^{\gamma} \right\} \right] \tag{13.41}$$

This form of the model is considered by some software package for the AFT model and the parameters $\{\beta_0, \beta_1, \gamma\}$ are estimated numerically by maximizing (13.41).

#### 13.6.3.1 Graphical Approach for Model Checking

Under the Weibull transformation, the Weibull CDF for the $k$th group (or items subjected to stress level $k$) becomes

$$\log\left[ - \log\left\{ R(t_{kj}) \right\} \right] = \gamma \log t_{kj} - \gamma \log[\alpha_k]. \tag{13.42}$$

As a result, the WPP plot is a straight line for a given $\alpha_k$, and different values for $\alpha_k$ result in parallel straight lines. For a given data set, if the generated plots for all of the groups are roughly parallel straight lines, then the Weibull AFT model is appropriate for modeling the data. The common slope is a graphical estimate of the shape parameter, and the intercepts for lines are estimates of the scale parameters for the different groups.

*Example 13.3* We illustrate the approach using simulated data. Increasing use-rate during the warranty period is an effective method of acceleration for products such as electrical motors, relays and switches, photocopiers, printers, toasters, dish-washers. Here we simulate warranty claims data for such a product under 1-D warranty with limit $W = 365$ days, whose lifetime depends, in part, on use rate. Suppose that the age-based life distribution of the product is Weibull and the relationship with the accelerating variable (use-rate) is linear, that is $\mu(z) = \beta_0 + \beta_1 z$. Assume that the normal usage rate of the product is 2 times per day. To accelerate failures, we consider three stress levels $z = 3.0$, 3.5 and 4.0 times per day and take $\beta_0 = 10.0$, $\beta_1 = -1.0$, and the Weibull shape parameter $\gamma$ to be 2. For each stress level, the number of customers is 100. Under the above setting, we generate warranty claims data for a period of 365 days. In the simulated data, the number of failures and the number censored, respectively, are (7, 14, 56) and (93, 86, 44), corresponding to three stress levels (3.0, 3.5, 4.0).

Figure 13.3 shows the scatter plot of the lifetime in days against use-rate stress for the simulated data. The triangles in the plot indicate right-censored observations. This figure indicates clearly that failures occur early at higher use-rates.

Figure 13.4, created by Minitab, is a probability plot based on the fitted model. The plot may be used to determine whether or not the distributions,

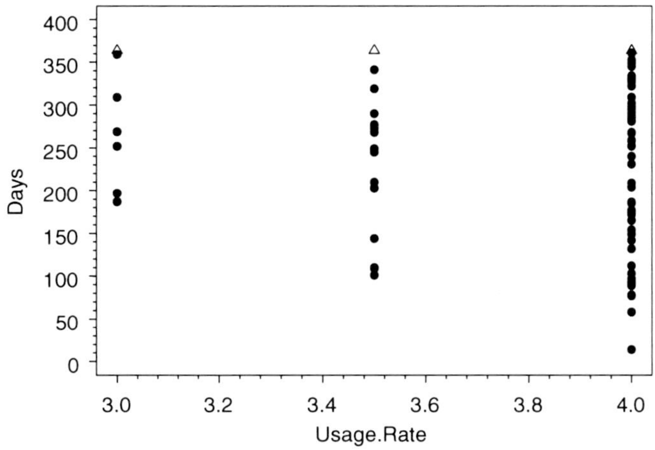


**Fig. 13.3** Scatter plot of days against use-rate for the simulated data

transformations, and assumption of equal shape parameter at each level of the accelerating variable are appropriate. In this figure, as expected, the points lie approximately on parallel lines, thereby verifying that the assumptions of the model are appropriate for the accelerating variable levels.

The shape and scale parameters of the fitted model at assumed normal usage stress are 1.93 and 7123.45. These estimates can be used to investigate the properties of the product at normal usage stress. For example, the median life (50th percentiles) of the product is 5892.70 days (about 16 years) with 95% confidence intervals (2396.45, 14489.70) days.[12]

The estimated coefficients for the regression model are given in Table 13.4. We can compare these estimates with the true values of the parameters used for generating data. This table describes the relationship between usage rate and failure time of the product for the Weibull distribution.

The accelerated life testing command in Minitab provides several probability plots that may be used to assess whether or not a particular distribution fits the data. In general, the closer the points fall to the fitted line, the better the fit. Figures 13.5 and 13.6 show the exponential probability plots of Cox–Snell residuals and the smallest extreme value probability plots of standardized residuals respectively, created by Minitab. These figures do not indicate any serious departure from the fitted model or the Weibull assumption. The only anomaly is that the data point at the lower end of the plot appears to be an outlier.

[12] These and any other percentiles can be estimated by Minitab.

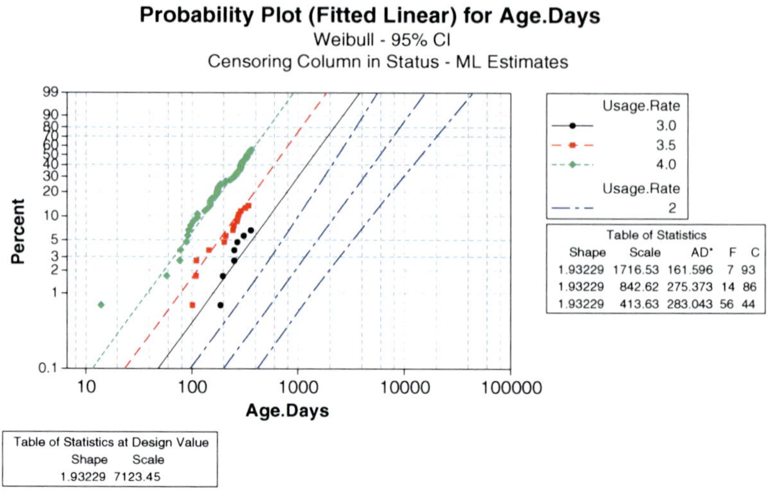


**Fig. 13.4** Weibull probability plots of days at each use-rate

**Table 13.4** Regression table for AFT model

| Parameters | ML estimate | Standard Error | *Z* | *P* | 95% Normal CI | |
|---|---|---|---|---|---|---|
| | | | | | Lower | Upper |
| $\beta_0$ | 11.7173 | 0.950608 | 12.33 | 0 | 9.85417 | 13.5805 |
| $\beta_1$ | −1.42309 | 0.240854 | −5.91 | 0 | −1.89516 | −0.95102 |
| $\gamma$ | 1.93229 | 0.205651 | | | 1.56848 | 2.38047 |

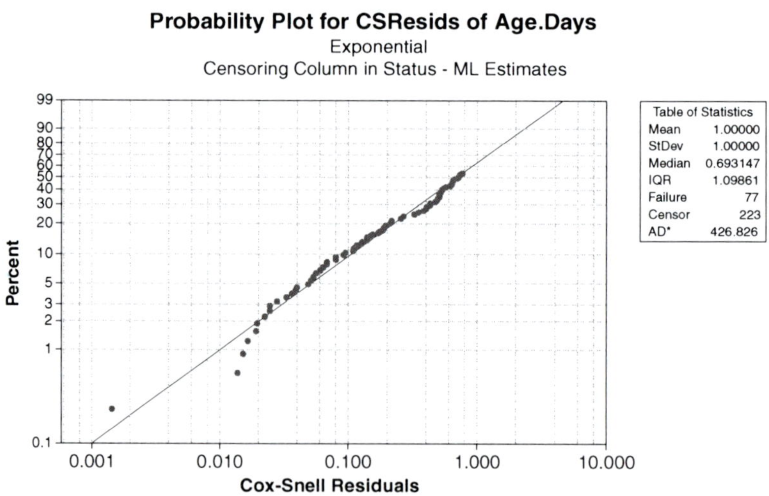


**Fig. 13.5** Exponential probability plots of Cox–Snell Residuals

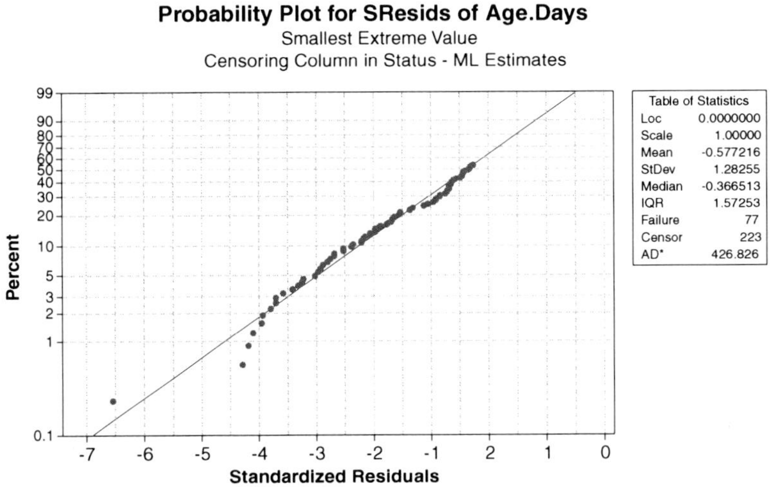


**Fig. 13.6** Smallest extreme value probability plots of standardized residuals

## 13.7 Analysis of Data Using Proportional Hazards Models

In this section, we analyze warranty claims data of a component or product using the proportional hazard (PH) model defined in (13.13). In employing the PH model, the effects of covariates (e.g., usage intensity, operating environment, etc.) on the distribution of lifetimes $T$ are modeled through its associated hazard function $h(t)$.

### *13.7.1 Warranty Data*

The proportional hazards model is appropriate for data that are very similar to those used for AFT modeling discussed in Sect. 13.6. The data may be component level with failed components replaced (Scenario 1.3) or product level with failed products repaired minimally (Scenario 1.4).

### *13.7.2 Statistical Analysis*

Suppose we observe independent observations $(t_i, \delta_i, z_i)$ for individual $i$, where $t_i$ is the lifetime, $\delta_i$ the failure/censoring indicator ($1 = \text{fail}$, $0 = \text{censor}$), and $z_i$ represents a set of covariates for individual $i$, $i = 1, 2, \ldots, n$. The likelihood function for the PH model of (13.13), including both $\beta$ and $h_0(t)$, is

$$\begin{aligned} L(\beta, h_0(t)) &= \prod_{i=1}^{n} [f(t_i)]^{\delta_i} [R(t_i)]^{1-\delta_i} = \prod_{i=1}^{n} [h(t_i)]^{\delta_i} R(t_i) \\ &= \prod_{i=1}^{n} [h_0(t_i)\psi(z_i; \beta)]^{\delta_i} \exp(-\Lambda_0(t_i)\psi(z_i; \beta)) \end{aligned} \tag{13.43}$$

where $\Lambda_0(t_i)$ is the baseline cumulative hazard function. If we assume the exponential form $\psi(z; \beta) = \exp(\beta z)$, the PH model has the form given in (13.16). For this model, the log-likelihood function, based on (13.43) is

$$\log L(\beta, h_0(\cdot)) = \sum_{i=1}^{n} [\delta_i\{\log(h_0(t_i)) + \beta z_i\} - \Lambda_0(t_i) \exp(\beta z_i)] \tag{13.44}$$

In fitting the Cox proportional hazards model, we estimate $h_0(t)$ and $\beta$. This may be done either by means of a distribution-free approach or a parametric model (e.g., a Weibull hazard function) may be assumed for $h_0(t)$. One approach is to maximize the log-likelihood function (13.44) simultaneously with respect to $h_0(t)$ and $\beta$. A simpler approach is to use Cox's partial likelihood function (proposed by Cox [6] that depends on $\beta$, but not on $h_0(t)$.

Cox [6] suggested using

$$L(\beta) \propto \prod_{i=1}^{n} \left[ \frac{\exp(\beta z_i)}{\sum_{j \in \Re(t_i)} \exp(\beta z_j)} \right]^{\delta_i}$$

as a partial or quasi likelihood function, where $\Re(t_i)$ denotes the set of subjects at risk just before $t_i$. The log-likelihood then becomes

$$\log L(\beta) \propto \sum_{i=1}^{n} \delta_i \left[ \beta z_i - \log \left\{ \sum_{j \in \Re(t_i)} \exp(\beta z_j) \right\} \right], \tag{13.45}$$

and this can be used in the usual way to estimate and test hypotheses about $\beta$.

*Comment:* For data at the system or product level, with failed systems or products repaired minimally, failures over time are modeled by a point process with the ROCOF under baseline conditions given by an intensity function $\lambda_0(t)$.

## 13.8 Analysis of Data Using Regression Models

The parametric regression model discussed in Sect. 13.2.5 relates the failure time distribution to explanatory variables. This involves specifications for the distribution $f(T|z)$ of a lifetime variable, $T$, given a vector of explanatory variables or covariates, $z$, upon which the lifetime may depend.[13]

[13] [2] deals with a mixed-Weibull regression model to analyze automotive component warranty data. The distribution function for each subpopulation depends on a vector of covariates that characterize specific operating conditions.

### 13.8.1 Warranty Data

Data for which the regression model is appropriate are the similar to the data used for AFT modeling in Sect. 13.6.

### 13.8.2 Statistical Analysis

Suppose that independent observations $(y_i, z_i, \delta_i)$, are available, where $y_i$ is either a log lifetime or log censoring time, $z_i = (z_{i1}, z_{i2}, \ldots, z_{iK})$ is a regression vector and $\delta_i$ is the censoring indicator. For the location-scale or log-location-scale family of models,[14] the likelihood function is

$$L(\beta,\sigma) \propto \prod_{i=1}^{n} \left[\sigma^{-1} f(u_i)\right]^{\delta_i} \left[R(u_i)\right]^{1-\delta_i}, \tag{13.46}$$

where $u_i = (y_i - z_i\beta)/\sigma$ and $R(u) = \int_u^\infty f(x)dx$.

For example, for the Weibull regression model (with $\alpha$ but not $\gamma$ depending on $z$), the PDF of $Y = \log T$ given $z$ follows the extreme value distribution (A.33) with density function

$$f(y|z) = \frac{1}{\sigma}\exp\left[\frac{y-\mu(z)}{\sigma} - \exp\left(\frac{y-\mu(z)}{\sigma}\right)\right], \quad -\infty < y < \infty, \tag{13.47}$$

where $\sigma = 1/\gamma$ and $\mu(z) = \log\alpha(z)$.

Under model (13.47), (13.46) becomes

$$L(\beta,\sigma) \propto \prod_{i=1}^{n} \left[\frac{1}{\sigma}\exp\left\{\frac{y_i - z_i\beta}{\sigma} - \exp\left(\frac{y_i - z_i\beta}{\sigma}\right)\right\}\right]^{\delta_i} \left[\exp\left\{-\exp\left(\frac{y_i - z_i\beta}{\sigma}\right)\right\}\right]^{1-\delta_i}. \tag{13.48}$$

The maximum likelihood equations for estimating the parameters of this model and for some other commonly used models can be found in [10, 14, 17].

*Example 13.4* Data Set 14 in Appendix F consists of warranty claims data for an automobile component. The data-set includes age (in days), mileage (kilometers), failure mode, region, types of automobile that used the unit, and other factors. We analyze the failure data (498 claims) using the regression model.

The aim of the analysis is to investigate how the usage-based lifetime of the unit differs with respect to age ($z_1$) and three categorical covariates: region

[14] Many of the widely used statistical models are either location-scale or log-location-scale families of distributions. The analytical methods developed for these families can be applied easily to any of its members. (For more on these families of distributions, see [14, 15].)

[($z_2$: Region1 (R1), Region2 (R2), Region3 (R3), Region4 (R4)], type of automobile in which the unit is used [$z_3$: Auto1 (A1), Auto2 (A2)], and failure modes [$z_4$: Mode1 (M1), Mode2 (M2), Mode3 (M3)]. The number of observed failures in R1, R2, R3, R4, A1, A2, M1, M2, and M3 are, respectively, 29, 105, 172, 192; 143, 355; 364, 106, and 28.

Without loss of generality, {R1, A1, M1} is taken as the reference or baseline level, the level against which other levels are compared. The covariate vector $z = (1, z_1, z_2, z_3, z_4)$ can then be rewritten as $z = (1, z_D, z_{R2}, z_{R3}, z_{R4}, z_{A2}, z_{M2}, z_{M3})$ under the assumption that, except for $z_D$, the other six dichotomous covariates take on the values 1 or 0 to indicate the presence or absence of a characteristic. A Weibull model $f(x|z, \beta, \sigma)$ is assumed for mileage $X$, with scale parameter $\sigma$ and location parameter dependent on covariates $z$, namely $\mu(z) = z\beta = (\beta_0 + z_D\beta_D + z_{R2}\beta_{R2} + z_{R3}\beta_{R3} + z_{R4}\beta_{R4} + z_{A2}\beta_{A2} + z_{M2}\beta_{M2} + z_{M3}\beta_{M3})$. Table 13.5 summarizes the numerical results for the Weibull regression model obtained by using Minitab.[15] In this table, very small $p$ values for all of the regression coefficients except $\beta_{M2}$ ($p = 0.867$), provide strong evidence of the dependency of average lifetime on those covariates.

The log-likelihood of the final model is −5479.3, while the log-likelihood of the null model (with intercept only) is −5628.4. The likelihood ratio chi-square statistic is $-2[-5628.4 - (-5479.3)] = 298.2$ with 7 degrees of freedom and the associated $p$-value is 0. Thus we reject the null hypothesis that all regression parameters are zero.

*Comment:* A set of models (Smallest extreme value, exponential, Weibull, normal, lognormal, logistic and log logistic) were fitted to the data. It was found, based on the AIC values and the plots of residuals, that the Weibull is the best model for the data among these alternatives.

The estimates of Table 13.5 can be used to estimate and compare other reliability-related quantities (e.g., B10 life, MTTF, etc.) at specified levels of the covariates. For example, when the covariates of age, region, auto type, and failure mode are fixed, respectively, at 365 days, Region1, Auto1, and Mode1, the ML estimate and 95% confidence intervals of B10 life are 12,072.7 km and [9,104.48, 16,008.7]. These estimates become 23,433.6 km and [16,890.3, 32,511.8] for covariate values of age 365 days, Region3, Auto2, and Mode3. Under the first combination of levels of covariates, the estimates imply that there is 95% confidence that 10% of the units are expected to fail between usages 9,104 and 16009 km. Estimates of Bp life for other values of the covariates may be estimated and interpreted similarly.

The probability plot for standardized residuals (Fig. 13.7) is used to check the assumptions of a Weibull model with assumed parameters for the data. The plotted points do not fall on the fitted line perfectly, but the fit appears to be adequate, with the possibility of one or a few outliers. This suggests that the residual plot does not

[15] This may also be done with S-plus and R-language.

**Table 13.5** Estimates of parameters $\beta$ and $\gamma$ for the weibull regression model for usage

| Parameters | ML estimates | Standard error | Z | P | 95% Normal CI | |
|---|---|---|---|---|---|---|
| | | | | | Lower | Upper |
| $\beta_0$ | 8.9713 | 0.1516 | 59.18 | 0.000 | 8.6741 | 9.2684 |
| $\beta_D$ | 0.0052 | 0.0003 | 16.03 | 0.000 | 0.0045 | 0.0058 |
| $\beta_{R2}$ | 0.3860 | 0.1432 | 2.70 | 0.007 | 0.1055 | 0.6666 |
| $\beta_{R3}$ | 0.5678 | 0.1377 | 4.12 | 0.000 | 0.2980 | 0.8376 |
| $\beta_{R4}$ | 0.5027 | 0.1319 | 3.81 | 0.000 | 0.2442 | 0.7613 |
| $\beta_{A2}$ | −0.1638 | 0.0690 | −2.37 | 0.018 | −0.2991 | −0.0286 |
| $\beta_{M2}$ | 0.0127 | 0.0758 | 0.17 | 0.867 | −0.1359 | 0.1614 |
| $\beta_{M3}$ | 0.2593 | 0.1304 | 1.99 | 0.047 | 0.0037 | 0.5149 |
| Shape $\gamma = 1/\sigma$ | 1.5376 | 0.0495 | | | 1.4435 | 1.6377 |

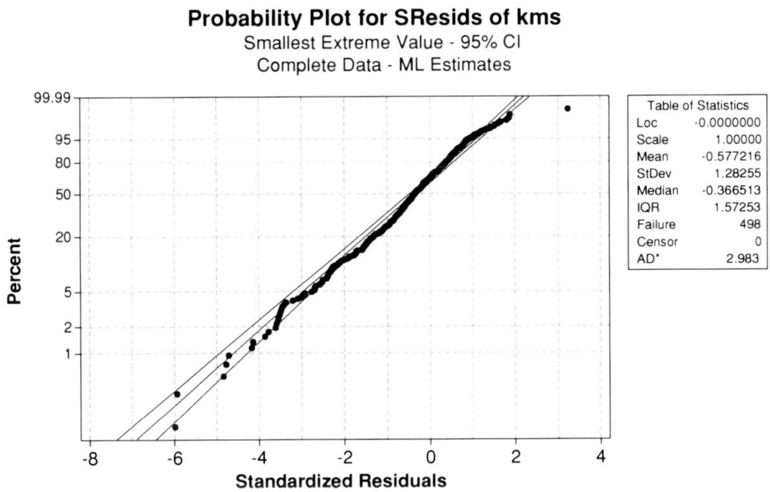


**Fig. 13.7** Smallest extreme value probability plots for standardized residuals

represent any serious departure from the Weibull distributional assumption in the model for the observed data.

## 13.9 Analysis of Data Using Imperfect Repair Models

The imperfect repair models defined in (13.22) and (13.23) are also applicable to analysis of warranty claims data for a repairable product or system. We discuss the MLEs of the parameters of the models. The parameters to be estimated are those of the distributions $F(t;\alpha_j)$, $j \geq 0$.

### 13.9.1 Warranty Data

We consider Scenario 1.4, with failed products repaired imperfectly. Let $n$ denote the number of items sold (or customers) and $n_i$ the number of failures (and repairs) under warranty for item $i$. Further, let $t_{ij}$ and $\tilde{t}_i$ be the age at $j$th failure for item $i$ based on the age clock (with the clock reset after each repair) and censor time for item $i$, $i = 1, 2, \ldots, n$, respectively.

### 13.9.2 Statistical Analysis

The likelihood function based on the data of Sect. 13.9.1 with either model (13.22) or (13.23) is given by

$$L(\theta) = \prod_{i=1}^{n} \left[ \left\{ \prod_{j=0}^{n_i} f(t_{ij}; \theta_j) \right\} R(\tilde{t}_i; \theta_j) \right] \tag{13.49}$$

where $\theta_j = \{\alpha_j, \beta\}$ is the parameter set for the models $F(t; \alpha_j)$, $j \geq 0$, defined in (13.22) and (13.23). Here the shape parameters for all models $F(t; \alpha_j)$ are equal to $\beta$. The ML estimates of the parameters are obtained by maximizing (13.49).[16]

## 13.10 Concluding Comments

Recently many researchers and practitioners are pursuing the development of reliability models for products whose reliability depends on explanatory variables such as usage intensity, operating environment, manufacturing periods, vendors, and so forth. Global competition and increasing customer expectations for reliable and safe products in a variety of usage environments are driving this interest. As a result, warranty data collection, modeling, and estimation in situations involving several subpopulations are of necessary. The results of this chapter are useful in this regard as follows:

- For modeling component failures with more than one mode of failure, competing risk models are appropriate.
- In modeling quality variations across different component vendors, produced by using different machines, etc., the mixture models are applicable.

[16] ML estimation for a different type of imperfect repair model is discussed in [8, 12].

- For modeling the effects of variations in usage rate, operating environment, manufacturing periods, and other covariates such as vendors, geographical locations, etc., the AFT, PH, and/or regression models can be applied.
- When failed items are repaired and the repair action affects the reliability of the item, the imperfect repair models are appropriate.

## References


1. Ahmad KE, Abdelrahman AM (1994) Updating a nonlinear discriminant function estimated from a mixture of 2-Weibull distributions. Math Comput Model 18:41–51
2. Attardia L, Guidab M, Pulcinic G (2005) A mixed-Weibull regression model for the analysis of automotive warranty data. Reliab Eng Sys Saf 87:265–273
3. Bendell A (1985) Proportional hazards modelling in reliability assessment. Reliab Eng 11(3):83–175
4. Brown M, Proschan F (1983) Imperfect repair. J Appl Probab 20(4):851–859
5. Cox DR (1972) Regression models and life tables (with discussion). J Royal Statist Soc B 34:187–220
6. Cox DR (1975) Partial likelihood. Biometrika 62:269–276
7. Dale CJ (1985) Application of the proportional hazards model in the reliability field. Reliab Eng 10:1–14
8. Doyen L, Gudaion D (2004) Classes of imperfect repair models based on reduction of failure intensity and virtual age. Reliab Eng Sys Saf 84:45–56
9. Jiang S, Kececioglu D (1992) Maximum likelihood estimates, from censored-data, mixed-Weibull distributions. IEEE Trans Reliab 41:248–255
10. Kalbfleisch JD, Prentice RL (1980) The Statistical analysis of failure time data. John Wiley & Sons Inc., NY
11. Kumar D, Klefsjo B (1994) Proportional hazards model: A review. Reliab Eng Sys Saf 29:177–188
12. Kvam PH, Singh H, Whitaker LR (2002) Estimating distributions with increasing failure rate in an imperfect repair model. Lifetime Data Anal 8:53–67
13. Landers TL, Kolarik WJ (1987) Proportional hazards analysis of field warranty data. Reliab Eng 18:131–139
14. Lawless JF (1982) Statistical models and methods for lifetime data. John Wiley & Sons Inc., NY
15. Meeker WQ, Escobar LA (1998) Statistical methods for reliability data. John Wiley & Sons Inc., NY
16. Murthy DNP, Xie M, Jiang R (2004) Weibull models. Wiley, NY
17. Nelson W (1990) Accelerated Testing. Wiley, NY
18. Pham H, Wang H (1996) Imperfect maintenance. European J Oper Res 94(3):425–438
19. Watkins AJ (1994) Review: likelihood method for fitting Weibull log-linear models to accelerated life-test data. IEEE Trans Reliab 43:361–365
20. Yun WY, Kang KM (2007) Imperfect repair policies under two-dimensional warranty. J Risk Reliab 221:239–247
21. Yun WY, Murthy DNP, Jack N (2008) Warranty servicing with imperfect repair. Int J Prod Econ 111(1):159–169

# Chapter 14
# Analysis of 2-D Warranty Data

## 14.1 Introduction

In Chaps. 4 and 5, we discussed warranty claims and warranty supplementary data, focusing mainly on data relating to products sold with 1-D warranties. In Chaps. 11–13, we looked at the analysis of the 1-D warranty data. When products are sold with 2-D warranties, the warranty claims data are two-dimensional. Usually the first dimension refers to the age of the item at the time of failure and the second to usage. Supplementary data in the case of 2-D warranties pose additional challenges, as discussed in Sect. 5.3.2. The analysis of 2-D warranty data raises several new issues that are different from those encountered when dealing with 1-D warranty data. The first involves the different approaches to modeling 2-D failures. In Chap. 7, we looked at three different approaches to modeling. In this chapter, we look at the analysis of 2-D warranty data.

We begin with a discussion of warranty data collection and alternative scenarios under which the data may have been collected, in Sect. 14.2. We discuss two different data structures (Structures 4 and 5) and the four alternative scenarios (Scenarios 5–8) that result from this and whether the data used in the analysis are only claims data or claims as well as supplementary data.[1] A general analysis of the data is discussed in Sect. 14.3. The analysis depends on the approach to modeling failures and, as indicated earlier, there are three different approaches. The analysis of the data for the three approaches is discussed in Sects. 14.4–14.6, respectively. We conclude the chapter with a discussion of forecasting of warranty claims and warranty costs in Sect. 14.7.

[1] In the analysis if 1-D warranty data, we looked at three data structures (Structures 1–3) and four different scenarios (Scenario 1–4). These are discussed in Chap. 5.

W. R. Blischke et al., *Warranty Data Collection and Analysis*,
Springer Series in Reliability Engineering, DOI: 10.1007/978-0-85729-647-4_14,

## 14.2 Data Collection and Alternative Scenarios

As discussed in Chap. 4, warranty data consist of claims data and supplementary data. In this chapter, we confine our attention to warranty data for products sold with 2-D non-renewing FRW or PRW warranties. We restrict our attention to 2-D warranties that are characterized by a rectangular region with age limit $W$ and usage limit $U$.

### 14.2.1 Claims Data

We look at two structures, detailed data and aggregated data, and list typical data elements for each.

#### 14.2.1.1 Structure 4 [Detailed Data]

Detailed data are the data collected as a result of claims over time for each customer. The typical history of warranty claims for a customer (sale) is as shown in Fig. 14.1, with time measured using the calendar clock.

The notation we use to characterize the claims data is as follows:

$I$ Number of customers (or sales)
$I_1$ Number of customers filing at least one claim
$I_2$ Number of customers with no warranty claims ($=I–I_1$)
$n_i$ Number of warranty claims exercised by customer $i$ ($1 \leq i \leq I$)

*Comments*:

1. $n_i = 0$ or 1 for products sold with non-renewing PRW policies and $n_i \geq 0$ for products sold with FRW policies.
2. We renumber the customers so that the first $I_1$ customers have one or more claims and the remaining $I_2$ have no claims.

$t_{i0}$ Time of purchase for customer $i$
$(t_{ij}, x_{ij})$ Age and usage at $j$th ($j \geq 1$) warranty claim for customer $i$ ($1 \leq i \leq I_1$)
$(\tilde{t}_{ij}, \tilde{x}_{ij})$ Time and usage between failures for customer $i$
$x_i(t)$ Usage up to age $t$ for customer $i$

As a result, we have.

$$\tilde{t}_{ij} = \{t_{ij} - t_{i(j-1)}\} \quad \text{for } j \geq 1, \quad \tilde{x}_{ij} = \left\{ \begin{array}{ll} x_{ij} & \text{for } j = 1 \\ x_{ij} - x_{i(j-1)} & \text{for } j > ; 1 \end{array} \right\}; \quad 1 \leq i \leq I_1$$

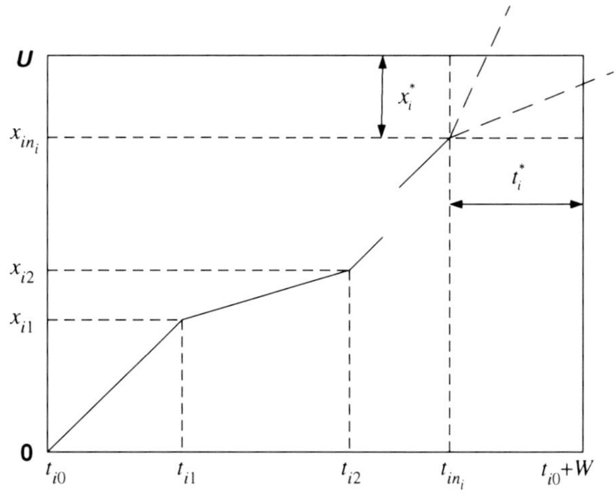


**Fig. 14.1** Typical pattern for 2-D warranty claim data for a customer

As discussed in Chap. 5, other data that may be collected include the following:

- Identification of failed component
- Failure mode
- Operating environment
- Symptoms leading to failure, as reported by customers
- Possible cause of failure, as reported by the service agent

#### 14.2.1.2 Structure 5 [Aggregate Data]

Here the claims data are aggregated over discrete time intervals. We use the following notation.

| | |
|---|---|
| $\delta_1$ | Interval for discretization of time (usually one month) |
| $\delta_2$ | Interval for discretization of usage |
| $i\ (i \geq 1)$ | Indexing of discrete time periods |
| $j\ (j \geq 1)$ | Indexing of discrete time periods for age of item, with the same interval as for time (If the time interval is in months, this corresponds to MIS[2]) |
| $k\ (k \geq 1)$ | Indexing for discrete usage intervals |
| $n_{ijk}$ | Number of failures in time interval $i$ with MIS $j$ and usage interval $k$ |
| $n_{i\bullet k} = \sum_j n_{ijk}$ | Number of items failing in time interval $i$ with usage in interval $k$ |
| $n_{\bullet jk} = \sum_i n_{ijk}$ | Number of items failing in age interval $j$ with usage in interval $k$ |
| $n_{ij\bullet} = \sum_k n_{ijk}$ | Number of items failing in time interval $i$ with MIS in interval $j$ |
| $n_{i\bullet\bullet} = \sum_{j,k} n_{ijk}$ | Number of items failing in time interval $i$ |

[2] MIS is months in service. See Sect. 11.5 for more details.

**Table 14.1** Two-way table of aggregated warranty counts

| Age (in days) | Usage [Mileage (km)] | | | |
|---|---|---|---|---|
| | 0–10,000 | 10,000–20,000 | | 80,000–100,000 |
| 0–30 | $n_{\bullet 11}$ | $n_{\bullet 12}$ | ... | ... |
| 30–60 | $n_{\bullet 21}$ | $n_{\bullet 22}$ | ... | ... |
| ... | ... | ... | ... | ... |
| 1,770–1,800 | ... | ... | ... | ... |

Table 14.1 is a typical table of aggregated claim counts for an automobile, where $\delta_1$ is 30 days and $\delta_2$ is 10,000 km.

Claims data are often aggregated into several groups based on factors such as age, usage, and usage rate, as indicated below:

- Based on age: A1 (low, less than 12 months), A2 (normal, 12–24 months), A3 (moderately high, 24–36 months), and A4 (high, 36 months or older).
- Based on Usage (mileage): U1 (low, less than 20,000 km), U2 (normal, 20,000–40,000 km), U3 (moderately high, 40,000–60,000 km), and U4 (high, 60,000 km or higher).
- Based on usage rate: LU (light users, usage rate Z in [Z1, Z2)), NU (normal users, Z in [Z2, Z3)), HU (heavy users, Z in [Z3, Z4)), and VU (very heavy users, Z in [Z4, Z5)).

## 14.2.2 Supplementary Data

### 14.2.2.1 Structure 4 [Detailed Data]

Supplementary data typically consist of the following:

| | |
|---|---|
| $t_{i0}$ | Date of sale for the item bought by customer $i$ $(1 \leq i \leq I)$ |
| $(\tilde{t}_i^*, \tilde{x}_i^*)$ | Age and usage when the warranty expires for customer $i$ $(1 \leq i \leq I)$ $i$ |
| *Comment* | $\tilde{t}_i$ and $\tilde{x}_i$ are uncertain for reasons discussed in Sect. 5.3.2 |
| $(t_i^*, x_i^*)$ | Age and usage remaining after the last failure for customer $i$ |

Note that $t_i^* = t_{i0} + W - \sum_{j=1}^{n_i} \tilde{t}_{ij}$ and $x_i^* = U - \sum_{j=1}^{n_i} \tilde{x}_{ij}$ (See Fig. 14.1.)

### 14.2.2.2 Structure 5 [Aggregate Data]

The data are

$s_i$ Sales in time period $i$

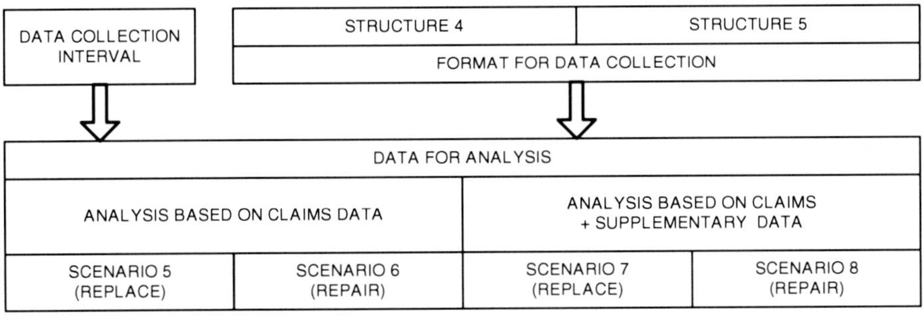


**Fig. 14.2** Alternate data structures and scenarios

### *14.2.3 Alternative Scenarios*

The numbers of failures and claims depend on the type of action taken by the manufacturer to rectify the failure. As with the 1-D warranty, we consider the following two types of actions:

- Replace by new (appropriate for component level data in the case of repairable products or at the product level for non-repairable products). In this case, the $(\tilde{t}_{ij}, \tilde{x}_{ij})$ constitute a set of iid random variables.
- Minimal repair (appropriate for complex repairable products where the failure is due to one or few components failing and replacing them has no effect on the reliability of the product after repair).

As a result we have four different scenarios (Scenarios 5–8) for the analysis of 2-D warranty data, as indicated in Fig. 14.2.

## 14.3 Data Analysis

The appropriate method of data analysis depends on the approach to modeling failures over time. In Chap. 7, we looked at three different approaches. The kind of data available determines which of the four scenarios discussed in the previous section is appropriate. Either nonparametric or parametric approaches may be used for the analysis. As a result, there are a number of different combinations that must be considered.

Another complicating factor is the uncertainty in the censored data, as discussed in Sect. 5.3.2. Because of this uncertainty, ad hoc methods may be required in order to extract useful information from the data. In this section, we summarize model formulations for three approaches to modeling and discuss various analytical problems that will be addressed in the remainder of the chapter.

Some other issues relating to data, such as data collected through periodic inspections (e.g., in the case of cars brought in for regular servicing), follow-up surveys to collect post-warranty information, and so forth, further complicate the analysis. These are beyond the scope of this chapter.

### 14.3.1 Approaches to Modeling Data

The underlying model formulations for the three approaches discussed in detail in Chap. 7 are given below. We only look at first failures. Subsequent failures depend on the rectification action—repair or replace, type of repair, and so forth.

Approach 1. This approach assumes a constant usage rate for each customer, with the rate varying across the population of customers. Usage rate is modeled as a random variable $Z$ with distribution function $G(z) = P\{Z \leq z\}$ and density function $g(z)$.[3] The time to first failure, conditional on the usage rate, is given by the conditional failure distribution $F(t|z)$.

Approach 2. In this approach the two scales, usage $x$ and time $t$, are combined to define a composite scale $v$ and the time to first failure is modeled by a distribution function $F_V(v)$.

Approach 3. The time to first failure is modeled by a bivariate distribution function $F(t,x)$. The density and hazard functions associated with this are given by $f(t,x)$ and $h(t,x)$, respectively.

### 14.3.2 Nonparametric Approach to Data Analysis

In the nonparametric approach, the data alone are used to estimate the CDF. In Approach 1, this is done by use of the usage rate data which are used to obtain a nonparametric estimate of $g(z)$ by dividing data into different intervals of usage. The conditional failure distribution $F(t|z)$ is then obtained for each interval. In Approach 2, the composite variable $V$ can be used to estimate the EDF. In this case, special attention should be given to interpreting the EDF. In Approach 3, the analysis is based on nonparametric estimates of the bivariate distribution, density and hazard functions and the bivariate renewal function.

### 14.3.3 Parametric Approach to Data Analysis

In the parametric approach, we begin with a specific model formulation and determine the MLEs of the model parameters. Further statistical analysis and testing is then needed to check the adequacy of the selected model.

---

[3] For notational ease, we omit the parameters of the functions.

## 14.4 Data Analysis: 1-D Approach [Based on Usage Rate]

In this approach, the two-dimensional warranty data are effectively reduced to one-dimensional data by treating usage as a function of age or age as a function of usage.

In this section, we assume that the usage rate for a customer is constant over the warranty period but varies across the customer population. Under this assumption, the usage rate $Z$ is a random variable, which can be modeled either as a discrete variable (for example, low, medium and high users) or as a continuous variable with distribution function $G(z) = \Pr(Z \leq z)$ and density function $g(z)$. Modeling of failures under warranty is then done using 1-D models by conditioning on the usage rate.

Under Scenarios 5 and 7, with failed units replaced by new ones, the model is obtained as follows:

- The time to first failure conditional on the usage rate is modeled by the conditional failure distribution $F(t|z)$. The corresponding conditional hazard function is denoted $h(t|z)$.
- The expected number of subsequent failures conditional on the usage rate can be modeled by the conditional renewal function $M(t|z)$ associated with the conditional CDF $F(t|z)$.

Under Scenarios 6 and 8, if failed units are repaired minimally and repair times are negligible, then failures over time, conditional on usage rate, can be modeled by a counting process with conditional intensity function $\lambda(t|z)$.

Both the nonparametric and parametric approaches for analysis of data from Scenarios 5–8 will be discussed below.

### *14.4.1 Usage Rate*

We assume that usage rate is different for each different customers. Let $(t_{ij}, x_{ij})$ denote the age and usage at the time of the $j$th warranty claim for customer $i$. The usage rate for customer $i$ with a single failure is

$$\hat{z}_i = \frac{x_{ij}}{t_{ij}}, \quad j = 1,\ i = 1, 2, \ldots, I_1 \tag{14.1}$$

The usage rate for customer $i$ with two or more failures can be

$$\hat{z}_i = \frac{x_i(t_i)}{t_i}, \tag{14.2}$$

where $x_i(t_i)$ is the cumulative usage to age $t_i$ for customer $i$, or

$$\hat{z}_i = \frac{\sum_{j=1}^{n_i} [\tilde{x}_{ij}/\tilde{t}_{ij}]}{n_i}, \quad i = 1, 2, \ldots, I_1, \tag{14.3}$$

which is the average usage rate over the warranty period for customer $i$ (where $n_i$ is the number of claims experienced by customer $i$), or it may be the median usage rate for customer $i$, i.e.,

$$\hat{z}_i = \text{Median of } \left(\frac{\tilde{x}_{i1}}{\tilde{t}_{i1}}, \frac{\tilde{x}_{i2}}{\tilde{t}_{i2}}, \ldots, \frac{\tilde{x}_{in_i}}{\tilde{t}_{in_i}}\right), \quad i = 1, 2, \ldots, I_1 \tag{14.4}$$

Estimation of usage rates depends on the nature of the available data on age and usage (e.g., single or multiple failures, detailed or cumulative usage, etc.).

## 14.4.2 Nonparametric Approaches

We look at nonparametric approaches for analysis of data collected under Scenarios 5–8. Inferences may be based on estimates of the conditional distribution function $F(t|z)$, the conditional renewal function $M(t|z)$, and the conditional mean cumulative function $\mu(t|z)$ at the product, component or some intermediate level.

### 14.4.2.1 Scenario 5

We use the usage rate data to obtain a nonparametric estimate of $g(z)$. This is done by dividing the data into different intervals of usage rate and then obtaining the conditional failure distribution $F(t|z)$ for each group. We divide usage rates into $K$ disjoint intervals by selecting a set of ($K$-1) delimiters $v_k$, with $0 < v_1 < v_2 < \cdots < v_{k-2} < v_{k-1} < \infty$, $v_0 = 0$ and $v_K = \infty$. Thus the kth interval is given by $[v_{k-1}, v_k)$, $k = 1, 2, \ldots, K$. Define $\aleph_k$ as follows: $\aleph_k \equiv N\{i : v_{k-1} \le z_i \le v_k\}$, with $z_i$ computed as indicated in (14.1)–(14.4), where $N\{A\}$ is the number of elements in the set $A$. The data in $\aleph_k$ can be used to estimate the EDF (see Sect. 8.5.1) for $\tilde{z}_k = (v_{k-1} + v_k)/2$. In this case, the usage rate can assume one of the $K$ values $\tilde{z}_k$, with proportions $p_k$ equal to the ratio of the frequency of $\tilde{z}_k$ and $I_1$.[4] In addition, the $t_i$ and corresponding and values of $\tilde{z}$ can be used to obtain the nonparametric estimate of the conditional hazard function $h(t|\tilde{z})$.

The times of subsequent failures conditional on the usage rate can be modeled by the conditional renewal function $M(t|\tilde{z})$ associated with the conditional CDF $F(t|\tilde{z})$ by applying the nonparametric method for renewal function estimation as discussed in Sect. 11.3.2.

---

[4] Additional details can be found in [1].

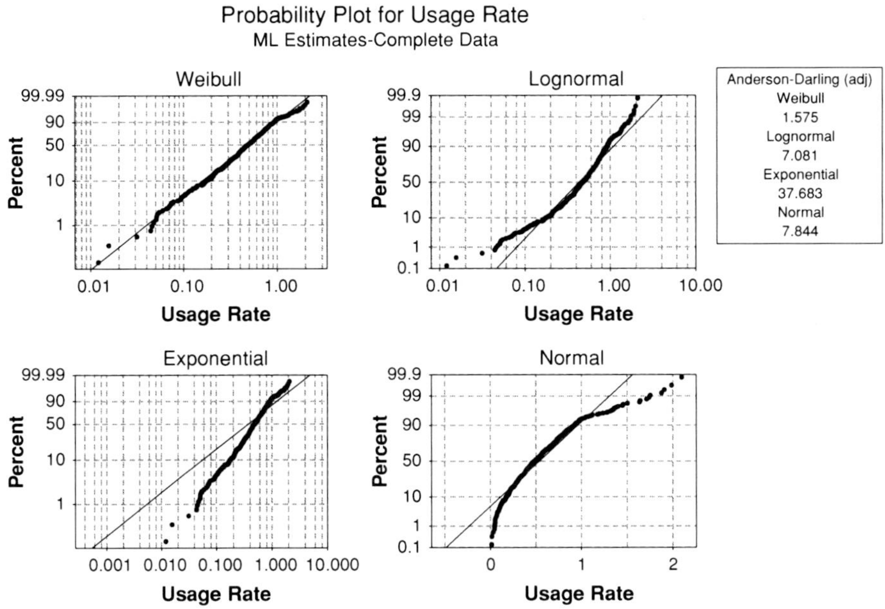


**Fig. 14.3** Probability plots for usage rate

### 14.4.2.2 Scenario 6

For this scenario, the nonparametric estimation of MCF given in Sect. 11.3.3 can be used to model the failure time conditional on usage rate by the conditional MCF $\mu(t|\tilde{z})$.

### 14.4.2.3 Scenario 7

The unavailability of the usage times of non-failed items leads to difficulties in estimation of usage-based lifetime distributions from warranty claims data only. In this case, supplemental data consisting of both failed and non-failed items is required to estimate the usage rate distribution. This information on age and usage may be obtained from follow-up survey data or recall data.[5] We discuss the parametric approach for analysis of Scenario 7 data in detail in Sect. 14.4.3.

### 14.4.2.4 Scenario 8

The data-related problem is the same as in Scenario 7 and the analysis requires information on age and usage for censored items.

[5] Detailed discussions of this are given in [16, 24–30].

### 14.4.3 Parametric Approaches

In this section, we look at the parametric approach, as discussed in Chap. 12. Here we assume specific forms for $F(t|z)$, $M(t|z)$ and $\mu(t|z)$, and obtain MLEs of the model parameters estimated under the various data scenarios. Reliability is then assessed using the assumed model and the estimated parameters.

#### 14.4.3.1 Scenario 5

Time to first claim is modelled by an appropriate parametric conditional failure distribution $F(t|z)$. The corresponding conditional hazard function $h(t|z)$ may have different forms that are non-decreasing functions in both $t$ and $z$, as discussed in Sect. 7.5.1.

As discussed in Sect. 11.3.1, subsequent failure times are modelled by the parametric form of the conditional renewal function $M(t|z)$ associated with $F(t|z)$. The parameters of the model are estimated using any standard estimation method (e.g., method of moments, maximum likelihood).

*Example 14.1* In this Example, we use the warranty claims data of Table F.15. The data are age and usage at time of claim for an automobile component. We discuss one-dimensional modelling, treating age as a random function of usage rate and assuming that all customers experience only a single failure. For the 498 customers, usage rates are $UR_i = z_i$, with units of 10,000 km per month. The usage rate for one customer is excluded in the analysis as it is exceptionally high as compared to other customers and is assumed to be an outlier.

Figure. 14.3 shows probability plots of the usage rates for four distributions (normal, lognormal, exponential and Weibull). From the plots, we see that the Weibull distribution appears to provide a very good fit to the sample of $z$-values.

The Weibull distribution overview plot shown in Fig. 14.4 gives the following ML estimates: shape parameter $\hat{\beta} = 1.695$, scale parameter $\hat{\alpha} = 0.603$, mean mileage accumulation rate 0.538 (equivalent to 5,380 km/month or 179 km/day).

Next the usage rates (minimum = 0.012, maximum = 2.10) are grouped into three groups as follows:

- Group 1 (Low usage rate): $0 \leq z_i < 0.75$;
- Group 2 (Medium usage rate): $0.75 \leq z_i < 1.50$; and
- Group 3 (High usage rate): $1.50 \leq z_i < 2.25$

The numbers of customers in groups 1, 2, and 3 are 396, 91 and 10, respectively.

Figure 14.5 shows four probability plots of age in month (Weibull, lognormal, exponential, and normal) for the three groups of customers. Among these four choices, the Weibull distribution is the best fitting for each of the three groups.

The Weibull distribution overview plots for age with the maximum likelihood estimates of the parameters for each of the three groups are shown in Fig. 14.6

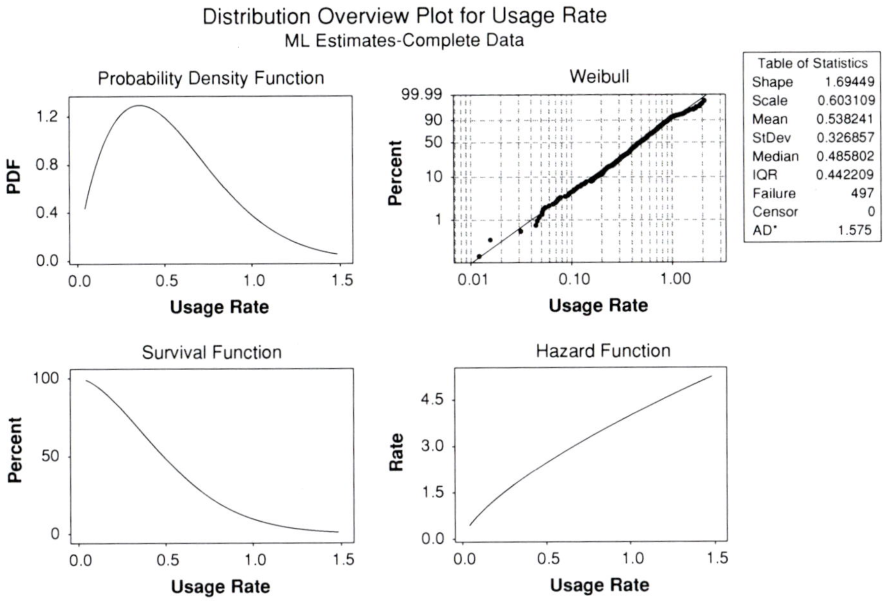


**Fig. 14.4** Weibull distribution overview plot for usage rate

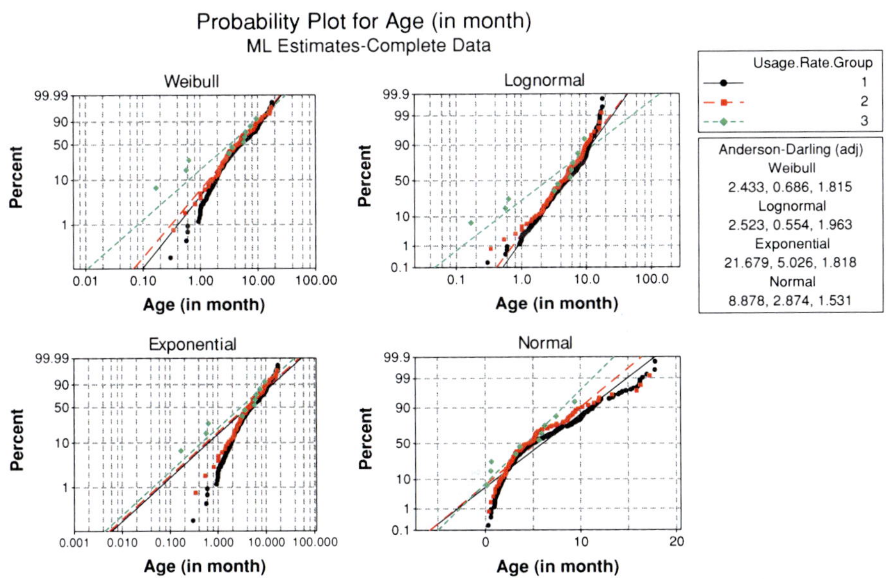


**Fig. 14.5** Four distributions probability plots for age (*in month*) for 3 groups of usage rate

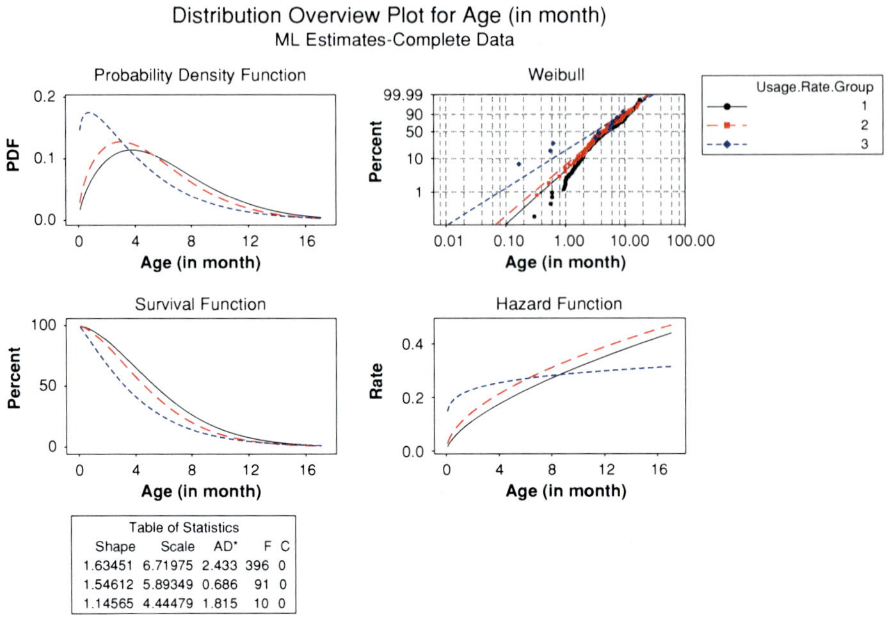


**Fig. 14.6** Weibull distribution overview plots for age for 3 groups

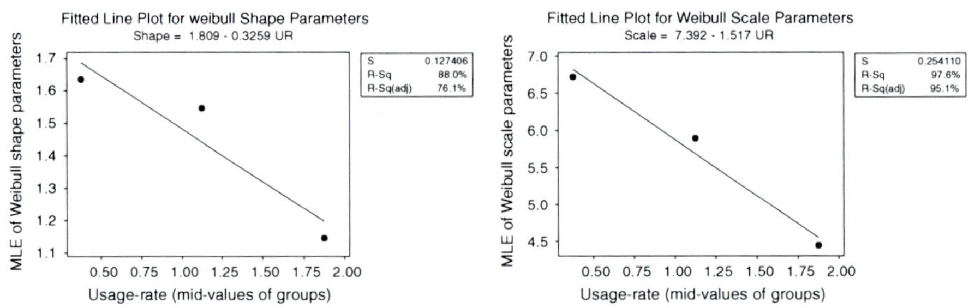


**Fig. 14.7** Plots of fitted shape and scale parameters of Weibull distribution versus the usage-rate

We next examine the relationships between the ML estimates of the parameters and the usage rates. Figure 14.7 shows plots of the ML estimates of the shape and scale parameters of Weibull distributions (obtained from Fig. 14.6) against the mid-values of each usage rate group. From the figure, the shape and scale parameters of Weibull distribution can be expressed roughly as a function of usage rate $z$ using a linear regression model with parameters

$$\alpha_z = 7.392 - 1.517z \quad \text{and} \quad \beta_z = 1.809 - 0.3259z.$$

Based on these relationships and Fig. 14.6, a preliminary model relating the age-based lifetime $T$ to usage rate $z$ is a Weibull distribution (A.30) with shape parameter $\beta_z$ and scale parameter $\alpha_z$, where $\beta_z$ and $\alpha_z$ are given above.

This is a one-dimensional model for $T$ conditional on $z$. The model for $X$ conditional on $z$, $f(x|z)$, can be developed similarly.

*Comment:* In this example, the number of customers and the number of groups are small and the results given are preliminary and intended only to demonstrate the technique. More customers with a larger number of groups are necessary for an adequate study of the relationships.

#### 14.4.3.2 Scenario 6

Here the counting process (Appendix B.1.1) with a parametric form of conditional intensity function (Sect. 12.3.2) for age given usage rate $\lambda(t|z)$ is used to model the counts of warranty claims of repairable items. Since failures are repaired minimally, and repair times are negligible, $\lambda(t|z) = h(t|z)$.[6]

#### 14.4.3.3 Scenario 7

To analyze Scenario 7 data, an appropriate parametric distribution conditional on usage rate $F(t|z)$ must be selected. The data contain both failure and censored data and special methods for dealing with the censored items for which the usage rates are unknown must be devised. Here we look at the following two methods:

*Method 1* This method [1] is based on a model formulation takes into account the uncertainty in usage rate across the customer population. The method involves the following three steps.

*Step I* [Selecting the distribution function $G(z;\phi)$ and estimating $\phi$]: A histogram of the number of customers in $\Im_k$ is used to determine the appropriate form for the density function $g(z;\phi)$. The parameter set $\phi$ is then estimated based on the $z_i$. Let $\hat{\phi}$ denote the estimate.[7]

*Step II* [Selecting $F(t:\alpha_0,\beta)$ and estimating $\{\alpha_0,\beta,\gamma\}$ using only the failure data pertaining to customers belonging to $I_1$]: The average usage rate for customers in interval $k$is given by[8]

$$\hat{z}_k = \int_{v_{k-1}}^{v_k} \frac{zg(z)}{[G(v_k) - G(v_{k-1})]} dz \tag{14.5}$$

---

[6] See Sects. 3.7.2 and 12.3.2.

[7] If we model the usage rate is expressed as a discrete random variable, then $v_k = z_k$ and the histogram yields estimates of the proportions $p_k$.

[8] One can use $\hat{z}_k = (v_{k-1} + v_k)/2$ instead of (14.5)

We assume that the usage rate of all customers in $\Im_k$ can be approximated by $\hat{z}_k$. This implies that the component failure distribution for all customers in $\Im_k$ is given by the CDF $F(t; \alpha(\hat{z}_k), \beta)$.

An estimate of $z$ for customers for whom only censored data are available may be obtained as follows[9]: If Customer $i$ data are censored and $n_i > 0$, the estimate is $\{W, \bar{z}_i W\}$ if $\bar{z}_i \le U/W$, and $\{U/\bar{z}_i, U\}$ if $\bar{z}_i > U/W$. On the other hand, if $n_i = 0$, then one cannot estimate the censored age and usage at the instant when the warranty expires. This gives an estimate of $\tilde{t}_{i,n_i}$ for all $i$ $(i = 1, 2, \ldots, I_1)$. Define $\tau_{ij} = t_{ij} - t_{i,j-1}$, $1 \le j \le n_i$, and $\tilde{\tau}_{in_i} = \tilde{t}_{in_i} - t_{in_i}$ for $i = 1, 2, \ldots, I_1$.

One can use the failure data and estimates of censored data to estimate the parameters $\{\alpha_0, \beta, \gamma\}$. Denote the MLEs of these parameters $\{\hat{\alpha}_0, \hat{\beta}, \hat{\gamma}\}$. The likelihood function is given by

$$L_1\{\alpha_0, \beta, \gamma\} = \prod_{k=1}^{K}\left[\prod_{i \in \aleph_k}\left\{\prod_{j=1}^{n_i} f(\tau_{ij}; \alpha_0(\hat{y}_k), \beta)\bar{F}(\tilde{\tau}_{in_i}; \alpha_0(\hat{y}_k), \beta)\right\}\right] \tag{14.6}$$

*Step III* [Revise the estimate of the scale parameter using the number of customers $I_2$ who did not experience any failures]: In estimating the parameters of the distribution in Step II, we ignored the information about no failures associated with customers $I_2$. Ignoring this information underestimates the component reliability. Often $I_2 \gg I_1$, which implies that this information is necessary for assessing component reliability. The following two approaches may be used to obtain the information.

*Approach* (a): Note that the probability of no component failure under warranty is

$$\psi = \int_0^\infty [1 - F(\min\{U/z, W\}; \alpha(z), \beta]g(z)dz \tag{14.7}$$

An estimate of $\psi$ is given by

$$\hat{\psi} = \frac{I_2}{I_1 + I_2} \tag{14.8}$$

Using $\hat{\beta}$ and $\hat{\gamma}$ obtained from Step II and $\hat{\psi}$ from (14.8) in (14.7) results in an equation with the scale parameter $\alpha_0$ as the only unknown. Solving for this yields a new estimate $\hat{\hat{\alpha}}_0$, which takes into account all the data available.

*Approach* (b): For the $I_2$ customers who did not experience failures, the likelihood function is given by

$$L_2(\phi, \alpha_0, \beta, \gamma) = \left[\int_0^\infty \bar{F}(\min(U/z, W); \alpha(z), \beta)dG(z)\right]^{I_2} \tag{14.9}$$

[9] Details are given in [1]

The overall likelihood function is $L_1(\Theta)L_2(\Theta)$. Estimates of the parameters can be updated by maximizing this likelihood function to obtain new estimates $\hat{\hat{\alpha}}_0$, $\hat{\hat{\beta}}$ and $\hat{\hat{\gamma}}$.

*Comments:*

(i) In Approach (a), only the scale parameter is updated. In contrast, in Approach (b) all of the parameters are updated.
(ii) If Approach (b) is used, it is not necessary to obtain the estimates in Step II.

*Method 2* This method [10] assumes that the usage accumulation is represented by a linear usage function [16], i.e.,

$$x_i(t) = z_i t, \quad t \geq 0, \tag{14.10}$$

where usage rate $z_i$ is a positive random variable with distribution function $G(z)$. Using (14.10), we obtain the probability that item $i$ does not fail in the warranty region $(0, W] \times (0, U]$ as

$$\begin{aligned} 1 - \mathrm{P}(T < W, X < U) &= \mathrm{P}\{T_i > \min(W, U/z_i)\} \\ &= R_T(W) \int_0^{U/W} dG(z) + \int_{U/W}^{\infty} R_T(U/z) dG(z) \end{aligned} \tag{14.11}$$

The log-likelihood function for age is given by

$$\begin{aligned} \log L(\theta|T) &= \sum_{i \in I_1} \log f_T(t_i; \theta) \\ &+ \sum_{i \in I_2} \log \left[ R_T(W; \theta) \int_0^{U/W} dG(z) + \int_{U/W}^{\infty} R_T(U/z; \theta) dG(z) \right] \end{aligned} \tag{14.12}$$

If the lifetime of the item is measured in usage, the probability that item $i$ does not fail in the warranty region can be rewritten using the reliability function of $X$ and the relationship between age and usage in (14.11). The result is:

$$\begin{aligned} 1 - \Pr(T < W, X < U) &= \Pr\{X_i > \min(z_i W, U)\} \\ &= \int_0^{U/W} R_X(zW) dG(z) + R_X(U) \int_{U/W}^{\infty} dG(z) \end{aligned} \tag{14.13}$$

From (14.13), the log-likelihood function for usage can be written as

$$\begin{aligned} \log L(\theta|X) &= \sum_{i \in I_1} \log f_X(x_i; \theta) \\ &+ \sum_{i \in I_2} \log \left[ \int_0^{U/W} R_X(zW; \theta) dG(z) + R_X(U; \theta) \int_{U/W}^{\infty} dG(z) \right] \end{aligned} \tag{14.14}$$

Likelihood functions (14.12) and (14.14) can be used to estimate the univariate lifetime models from the 2-D warranty claims data. Unless a known distribution for the usage rate is available, the following two-stage method may be used to obtain MLEs: (i) a distribution of usage rate is selected by fitting the failure data, and (ii) a numerical method is used to maximize the log-likelihood function.

*Special Case* If $T$ follows a Weibull distribution with reliability function $R_T(t;\alpha_1,\beta_1)=\exp\left\{-(t/\alpha_1)^{\beta_1}\right\}$,

then the age-based log-likelihood function (14.15) becomes

$$\log L(\theta|T)=\sum_{i\in I_1}\left[\log(\beta_1)-\beta_1\log(\alpha_1)+(\beta_1-1)\log(t_i)-(t_i/\alpha_1)^{\beta_1}\right]$$
$$+\sum_{i\in I_2}\log\left[\exp\left\{-(W/\alpha_1)^{\beta_1}\right\}\int_0^{U/W}dG(z)+\int_{U/W}^{\infty}\exp\left\{-(U/(z\alpha_1))^{\beta_1}\right\}dG(z)\right]. \tag{14.15}$$

Similarly, if $X$ follows a Weibull distribution with parameters $\alpha_2$ and $\beta_2$, then the usage-based log-likelihood function (14.14) becomes

$$\log L(\theta|X)=\sum_{i\in I_1}\left[\log(\beta_2)-\beta_2\log(\alpha_2)+(\beta_2-1)\log(x_i)-(x_i/\alpha_2)^{\beta_2}\right]$$
$$+\sum_{i\in I_2}\log\left[\int_0^{U/W}\exp\left\{-(zW/\alpha_2)^{\beta_2}\right\}dG(z)+\exp\left\{-(U/\alpha_2)^{\beta_2}\right\}\int_{U/W}^{\infty}dG(z)\right]. \tag{14.16}$$

The MLEs of the parameters of age-based and usage-based models are obtained independently by maximizing log-likelihood functions (14.15) and (14.16), respectively.

#### 14.4.3.4 Scenario 8

Under the assumption that the failures are repaired minimally and repair times are negligible, the conditional intensity function becomes $\lambda(t|z)=h(t|z)$, which can be estimated by the method used in the previous data scenario.

## 14.5 Data Analysis: 1-D Approach [Composite Scale]

As indicated in Chap. 7, Approach 2 involves forming a new variable $V$ that is a combination of usage $X$ and age $T$. Here we follow the approach suggested by Gertsbakh and Kordonsky (called the *GK approach*), in which the composite variable is a linear combination of the form

$$V = (1 - \varepsilon)T + \varepsilon X \tag{14.17}$$

where $\varepsilon$ $(0 < \varepsilon < 1)$ is chosen such that the coefficient of variation of $V$ is minimized, with the minimizing value obtained using a search routine.[10]

In practice, there are several problems in applying this approach and interpreting the results. These include

- Use of the method when supplementary data are not available
- Dealing with incomplete data for which usage information is not known
- Dealing with data aggregated at different levels
- Determination of an appropriate distribution for $V$
- The scale of measurement of the combined measure $V$

In the following sections, we look at each of these.

### *14.5.1 Data Needs*

In principle, the approach is applicable to data from Scenarios 5 and 7. The analysis may be done at the component or product level. For simplicity, we let $N$ denote the total number of items in the data set, including original purchases and replacements, and let $n$ be the total number of failures. Observed values are denoted $t_i$, $x_i$, and $v_i$. Renumber the data so that $(t_i, x_{i,})$ correspond to failures for $i = 1, 2, \ldots, n$, and to age and usage of unfailed items for $i = n + 1, \ldots, N$.

In practice, a thorough analysis using the GK approach can only be done if supplementary data are available, i.e., if we are dealing with Scenario 7. Scenario 5 data, however, may be used in a preliminary analysis of claims data alone. The results may be interpreted as conditional on item failure or as preliminary information concerning $\varepsilon$ and the distribution of $V$. In the remainder of this section, we deal primarily with Scenario 7.

Supplementary data needed for application of the GK approach are sales data, which are often aggregated on a monthly basis,[11] and age and usage at failure for each warranty claim. In addition, age and usage data are needed for unfailed items, both under warranty and out of warranty.

Age data for all items are easily obtained from the sales data (at an aggregated level, if necessary). Usage data, however, are rarely available for unfailed items. There are a number of ways of dealing with this, all of which involve approximating usage-based on age information. These include the following methods:

---

[10] The approach is discussed in detail in [6] and [12–15]. This and other approaches to scaling in 2-D or higher dimensional problems are discussed in [5]

[11] Claims data may be aggregated as well. In this case, all of the data must be aggregated to a common level.

1. Use claims data to estimate the average usage per day. Estimate the unknown usages by multiplying daily usage by age in days (or by length of period for aggregated data).
2. Calculate the median daily usage from the claims data and use this as in Method 1.
3. Use the usage limit for all items for which usage is unknown.
4. Conduct a survey of owners who have not filed a claim to obtain usage information. Use the mean or median of these results as an estimate of usage as above for all items for which no claim has been made.

Method 4 would be expected to yield the most accurate result, but this is not often done because of the cost of the survey. Of the remaining, Method 2 will typically be the most conservative, but it may be instructive to also try Methods 1 and 3 as well as any other reasonable approach.

### 14.5.2 Parameter Estimation

We assume that data have been aggregated at some level, with $k$ = number of periods for which data are available, and that the following data are available for $j = 1, 2, \ldots, k$:

$N_j$ = sales in period $j$
$A_j$ = number of valid claims on the $N_j$ units sold in period $j$
$M_j$ = number of items sold in period $j$ for which no claims were made = $N_j - A_j$
$W_j$ = age at censoring for items sold in period $j$
$K$ = total number of valid claims = $\Sigma A_j$
$U_j$ = usage at censoring for items sold in period $j$

If usage is not known, it may be estimated as indicated in the previous section. We now identify the data for analysis. Let

$$\Omega_j = (t, x) : t \in [0, U_j), \quad x \in [0, W_j) \tag{14.18}$$

Valid warranty claims resulting from item failures must be such that $(t_i, x_i) \in \Omega_j$. $\Omega_j$ may be viewed as service ages and usages for unfailed items sold in period $j$. Data for analysis are failure data, namely those for which $(t_i, x_i) \in \Omega_j$, and censored data, i.e., unobserved random variables $(T_j, X_j)$ which are not in the region $\Omega_j$ for any $j$.

Let $W_j$ and $U_j$ denote the values of censored age and usage (where usage is approximated by any of the four methods indicated above) for items sold in period $j$. The corresponding censored values of $V_j$, denoted $\tilde{V}_j$, are calculated as

$$\tilde{V}_j = (1 - \varepsilon)W_j + \varepsilon U_j \tag{14.19}$$

Failure data are given on the $v$-scale by $v_i = (1 - \varepsilon)t_i + \varepsilon x_i$.

The value of $\varepsilon$ is determined by a search, as indicated above. A method for estimation of $\varepsilon$ and the model parameters begins with a set of observed values of $T$ and $X$ and involves the following steps[12]:

- Step 1: Select a set of $\varepsilon$-values spanning the interval (0,1). For each $\varepsilon$, calculate $V$ and $\tilde{V}$ for all data, using (14.17) and (14.19).
- Step 2: Calculate the CV for each $\varepsilon$-value.
- Step 3: Select the $\varepsilon$ that minimizes the CV.[13]
- Step 4: Select a distribution $F$ and use the unidimensional data $V$ and $\tilde{V}$ associated with the minimum $\varepsilon$ to calculate MLEs of the parameters of $F$ based on the likelihood function

$$L(\theta) = \prod_{i=1}^{K} f(v_i;\theta) \prod_{j=1}^{K} [1 - F(\tilde{V}_j;\theta)]^{M_j} \tag{14.20}$$

The Step 2 above uses the nonparametric estimate of the CV. Step 4 requires selection of a distribution for $V$. In the GK approach, the Weibull distribution is usually used for this purpose. Alternatively, goodness-of-fit tests may be used to select from among a number of candidate distributions (e.g., those available in Minitab or other packages, or those listed in [22]). Denote the CDF and density function of the chosen distribution $F(v;\theta)$ and $f(v;\theta)$, respectively.

MLEs of the elements of the parameter $\theta$ are obtained by maximizing logL($\theta$) given in (14.20) for the selected $\varepsilon$, using differentiation and/or numerical methods for solution. The asymptotic normality of the MLEs may be used to obtain confidence intervals for the estimated parameters. This is done by forming the matrix with elements given by (D.4), inverting the matrix, and substituting the estimated values of unknown parameters into the result. This provides estimated variances and covariances of the estimators and the confidence intervals are obtained by use of the standard normal distribution, as shown in Sect. 9.5.

### *14.5.3 Interpretation and Use of Results*

The $v$-scale is a weighted average of time units (for age) and usage units. As such, measures on this scale are not directly interpretable. They have no physical meaning, but capture the joint effects of age and usage. Since the coefficient of variation is unit free, however, the value of $\varepsilon$ is independent of the scales of measurement of age and usage, so that the transformation to $v$ will lead to the same essential result regardless of the units of the original variables.

---

12 Another procedure, similar to this method, is given in [9].

13 If the CV plotted as a function of $\varepsilon$ is relatively flat in the region of the minimum, it may be desirable to select a finer grid of values in that region and begin again at Step 1.

In spite of the difficulty in interpreting the results directly, the transformed data can be used in warranty analysis in much the usual way, as long as a reasonable fit to the data on the $v$-scale has been found. This requires transforming elements of a given analysis to the $v$-scale as well (e.g., age and usage limits). If this is done properly, it enables the analyst to evaluate various alternative warranty policies and the warranty parameters (e.g., length of the warranty, refund policy, etc.).

It is also meaningful to analyze warranty costs on this scale. Again, if the distribution of $V$ is satisfactorily determined, elements of the cost models, such as those involving partial expectations or renewal functions, can be estimated and the results used in the cost analysis.

A detailed study of cost models and warranty policies based on 2-D warranty data on a motorcycle is given in a case study by [7]. The analysis uses the GK approach to investigate alternative warranty policies and estimated cost models for the PRW and FRW.

*Example 14.2* In Example 14.1 we used the warranty claims data for an automobile component (Table F.15) and discussed one-dimensional modelling of age as a function of usage rate, assuming the 2-D warranty data. Using the same data, we derive a composite scale model based on both age $T$ and usage $X$ at failure. Assuming Scenario 5 data, only the failure data for 498 customers are considered here.[14] For convenience, in this example measurement units are changed from days to months and km to 10,000 km. For a set of values of $\varepsilon$, we first combine $T$ and $X$ into samples of composite measures $V$ using (14.20). The CV's are then calculated for each $\varepsilon$ ($0 < \varepsilon < 1$), and the value of $\varepsilon$ that minimizes the coefficient of variation of $V$ is determined. For the auto component data, the estimate of $\varepsilon$ is 0.138 and the corresponding estimate of the CV is 0.650.

Figure. 14.8 shows the four probability plots used previously for the composite scale variable $V$. This figure indicates that the lognormal, loglogistic or Weibull distributions are the best choices of $F(v;\theta)$ among these distributions, but none of the fits is particularly good.

Application of the method for the case of censored data can be found in Chap. 18.

## 14.6 Analysis Based on Approach 3 [Bivariate Model Formulations]

We confine our attention to Scenarios 5 and 7 and look at both parametric and nonparametric analysis of two-dimensional warranty data.

[14] An application of the method for Scenario 7 data (with both failure and censored data) is given in Chap. 18.

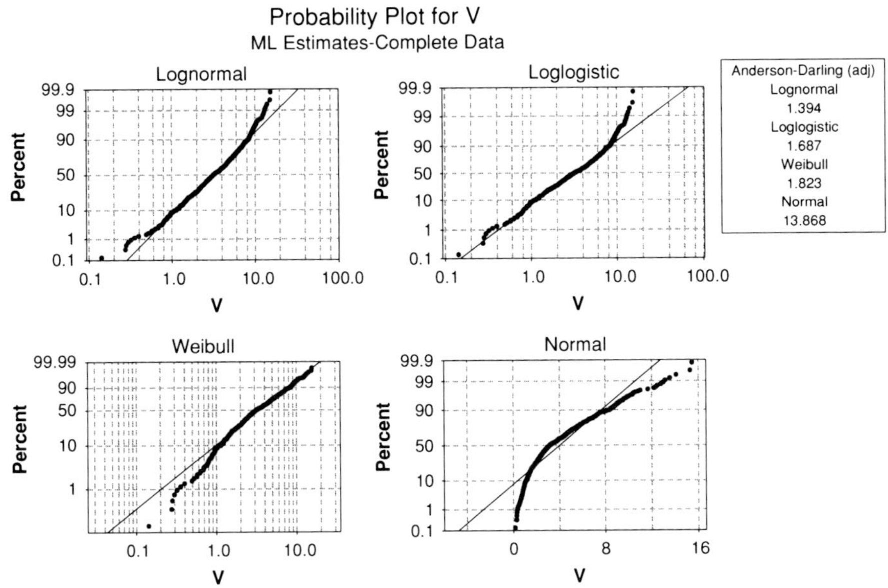


**Fig. 14.8** Four distributions probability plots for the composite scale data ($V$)

**Fig. 14.9** Division of warranty region into $J \times K$ cells

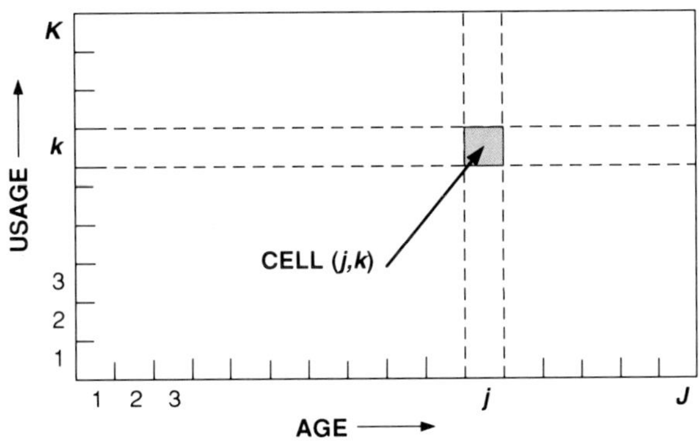


## 14.6.1 Nonparametric Approach

Divide the warranty region into $J \times K$ cells as shown in Fig. 14.9 by choosing $\delta_1$ and $\delta_2$ such that $J = W/\delta_1$ and $K = U/\delta_2$ are both integers.

### 14.6.1.1 Scenario 5

Here the failed components are replaced by new ones and the claims data available for analysis are given by $(t_{ij}, x_{ij})$, the age and usage at the time of the $j$th warranty claim $(1 \leq j \leq n_i)$ for customer $i$ $(1 \leq i \leq I_1)$. We divide the data among the cells based on age and usage at failure and let $n_{jk}$ denote the number of items in cell $(j,k)$. From this we compute the following:

$n_{jk}^{sw}$: Number of failure times in cells to the lower-left of cell $(j,k)$

$$\equiv \sum_{a=1}^{j-1} \sum_{b=1}^{k-1} n_{ab}$$

$n_{jk}^{ne}$: Number of failure times in cells to the upper-right of cell $(j,k)$

$$\equiv \sum_{a=j+1}^{J} \sum_{b=k+1}^{K} n_{ab}$$

The non-parametric estimates of the density, hazard and renewal functions are as follows:

*Estimate of density function*

$$\hat{f}_{j\bar{k}} = \hat{f}([j-0.5]\delta_1, [k-0.5]\delta_2) = \frac{n_{jk}}{\sum_{i=1}^{I_1} n_i}, \quad 1 \leq j \leq J;\ 1 \leq k \leq K \tag{14.21}$$

*Estimate of hazard function*

$$\hat{h}_{ij} = \hat{h}([j-0.5]\delta_1, [k-0.5]\delta_2) = \frac{n_{jk}}{n_{jk}^{ne}}, \quad 1 \leq j \leq J-1;\ 1 \leq k \leq K-1 \tag{14.22}$$

*Estimate of renewal function*

$$\hat{M}_{jk} = \hat{M}([j-0.5]\delta_1, [k-0.5]\delta_2) = \frac{n_{jk}^{sw}}{\sum_{i=1}^{I_1} n_i}, \quad 1 \leq j \leq J-1;\ 1 \leq k \leq K-1 \tag{14.23}$$

*Comment*: Since the estimates are based solely on claims data and ignore information on non-failed items, they are conditional estimates of the functions.

#### 14.6.1.2 Scenario 7

We first look at the special case where the censored data $(\tilde{t}_i, \tilde{x}_i)$ (the age and usage at the instant the warranty ceases due to exceeding either the age or usage limit) are known. We divide the data among the cells based on the age and usage at failure. $\tilde{n}_{jk}$ denotes the number of items in cell $(j,k)$.[15] From this we compute the following:

[15] These are the items which have survived for a time and usage corresponding to the north–east point of the cell.

$\tilde{n}_{jk}^{sw}$: Number of censored items in cells to the lower-left of cell $(j,k) \equiv \sum_{a=1}^{j-1} \sum_{b=1}^{k-1} \tilde{n}_{ab}$.

$\tilde{n}_{jk}^{ne}$: Number of censored items in cells to the upper-right of cell $(j,k) \equiv \sum_{a=j+1}^{J} \sum_{b=k+1}^{K} \tilde{n}_{ab}$.

The non-parametric estimates of the density, hazard and renewal functions are as follows:

*Estimate of density function*

$$\hat{f}_{j\bar{k}} = \hat{f}([j-0.5]\delta_1, [k-0.5]\delta_2) = \frac{n_{jk}}{I + \sum_{i=1}^{I_1} n_i}, \quad 1 \le j \le J;\ 1 \le k \le K \tag{14.24}$$

*Estimate of hazard function*

$$\hat{h}_{ij} = \hat{h}([j-0.5]\delta_1, [k-0.5]\delta_2) = \frac{n_{jk}}{n_{jk}^{ne} + \tilde{n}_{jk}^{ne}}, \quad 1 \le j \le J-1;\ 1 \le k \le K-1 \tag{14.25}$$

*Estimate of renewal function*

$$\hat{M}_{ij} = \hat{M}([j-0.5]\delta_1, [k-0.5]\delta_2) = \frac{n_{jk}^{sw}}{I + \sum_{i=1}^{I_1} n_i}, \quad 1 \le j \le J-1;\ 1 \le k \le K-1 \tag{14.26}$$

*Supplementary Data*

The following additional notation is needed:

$L_t$: The length of the time period over which data are collected
$L = L_t/\delta_1$ = Number of intervals over which data are collected
$S = \sum_{i=1}^{L} s_i$ = Total sales over the period of data collection

The supplementary data that are available are not the censored data discussed above, but the following:

- $(t_i^*, x_i^*)$: Age and usage limit remaining after the last failure for customer $i$
- $s_i$: Sales in interval $i$

From the supplementary data, we calculate estimated values of the censored data $(\tilde{t}_i, \tilde{x}_i)$ for the $I$ customers. It is important to note that the censoring may be due to the time of observation, the age limit being exceeded or the usage limit being exceeded. We consider the following two cases:

- When $L$ exceeds the number of age groups $J$, then items sold in interval $i$, $1 \le i \le L-J$, are no longer under warranty (as they have all exceeded the time limit). Of the items sold in interval $i$, $L-J+1 \le i \le L$, some may be still under warranty (provided the usage limit is not exceeded).
- When $L \le J$, items on which the usage limit has not been exceeded will still under warranty, while the remainder will not.

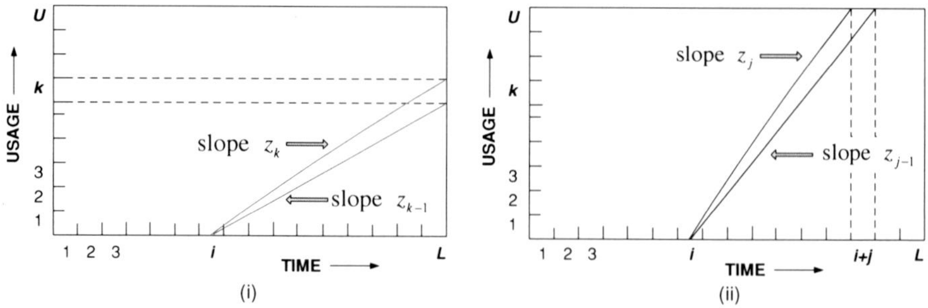


**Fig. 14.10** Two sub-cases for Case (b). Sub-case (i), Sub-case (ii)

When lifetimes are censored because the warranty limit has been exceeded, then either $\tilde{t}_i = t_i^*$ and $\tilde{x}_i < x_i^*$ or $\tilde{t}_i < t_i^*$ and $\tilde{x}_i = x_i^*$. Three cases must be considered:

a) Censored age and usage for items with one or more warranty claims
b) Censored age and usage for items sold in period $i$, $(L - J) \leq i \leq L$. Note that in this case no item has exceeded the age limit. Some that have exceeded the usage limit are no longer under warranty but the time at which the warranty ceased is uncertain. The remaining are still under warranty, with age known but usage unknown.
c) Censored age and usage limits for items sold in period $i$, $1 \leq i \leq (L - J - 1)$. Note that none of the items is under warranty. The warranty can have ceased either due to exceeding the time limit or the usage limit.

Let $s_i^c$ $[s_i^f]$, $1 \leq i \leq L$ denote the number of items sold in period $i$ that had no warranty [one or more] warranty claims. (Note: $s_i^c + s_i^f = s_i$)

*Case (a)* As can be seen from Fig.14.1, the average usage rate for customer $i$, $1 \leq i \leq I_1$, is $z_i = (U - x_i^*)/(W - t_i^*)$. Assuming that this usage rate remains the same then the values for censored age and usage are estimated by

- $\tilde{t}_i = t_i^*$ and $\tilde{x}_i = z_i \tilde{t}_i$ if $z_i < x_i^*/t_i^*$
- $\tilde{t}_i = x_i^*/z_i$ and $\tilde{x}_i = x_i^*$ if $z_i \geq x_i^*/t_i^*$

*Case (b)* For items sold in period $i$ we take the sale date for each to be the midpoint $(i + 0.5)\delta_1$. The censored time and usage depends on the usage rate. We look at two sub-cases:

*Sub-case (i):* [Censoring due to exceeding the time limit] For items with $z < J/\{(L - i - 0.5)\}$, the censored time is $\tilde{t}_i = \{L - i - 0.5\}\delta_1$. The censored usage is in the interval $[(k - 1)\delta_2, k\delta_2)$ if the usage rate in the interval $[z_{k-1}, z_k)$, where $z_{k-1} = [(k - 1)\delta_2/\{L - i - 0.5\}\delta_1]$ and $z_k = [k\delta_2/\{L - i - 0.5\}\delta_1]$. This is illustrated in Fig. 14.10.

The number of items with usage rate in this interval is the largest integer less than $[G(z_k) - G(z_{k-1})]s_i^c$. As a result, the estimated values of censored age and

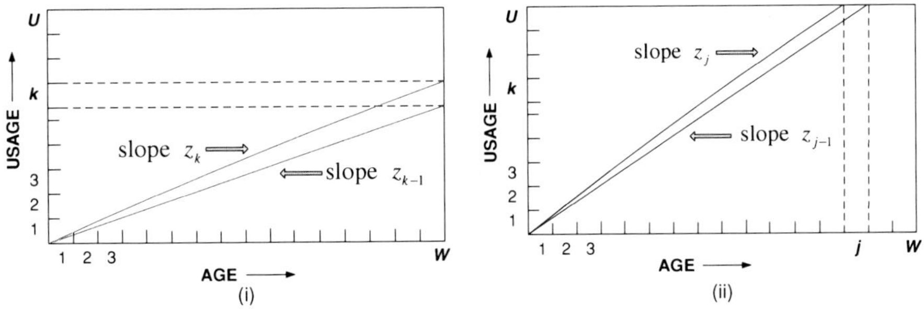


**Fig. 14.11** Two sub-cases for Case (c). scenario (i), scenario (ii)

usage for these $s_{ik}$ items are given by $\tilde{t}_i = \{L - i - 0.5\}\delta_1$ and $\tilde{x}_{ik} = (k - 0.5)\delta_2$, $1 \leq k \leq K, (L - J) \leq i \leq L$.

*Sub-case (ii):* [Censoring due to exceeding the usage limit] For items with $z > J/\{(L - i - 0.5)$, the censored usage is $\tilde{x}_i = U$ and the censored time is in the interval $[(j - 1)\delta_1, j\delta_1)$ if the usage rate is in the interval $[z_{j-1}, z_j)$, $i \leq j \leq L$, where $z_{j-1} = [U/\{j + 0.5\}\delta_1]$ and $z_j = [U/\{j - 0.5\}\delta_1]$ (see Fig. 14.10). The number of items with usage rate in this interval is the largest integer less than $[G(z_j) - G(z_{j-1})]s_i^c$. As a result the estimated values of the censored age and usage for these items are given by $\tilde{t}_{ij} = \{j + 0.5\}\delta_1$ and $\tilde{x} = U$.

*Case (c)* The approach is the same as in Case (b). Also, we can combine all items with no warranty claims over the periods $i$, $1 \leq i \leq (L - J - 1)$ so that the total number of such items is $s^C = \sum_{i=1}^{L-J-1} s_i^c$. As indicated in Fig. 14.11, two sub-cases must again be considered. The analysis follows the method outlined in Case (b).

From Cases (a)–(c), we obtain estimated values for the censoring times for all $I$ customers. These are then used in place of the true values and, following the procedure outlined in the Special case discussed previously, we obtain the non-parametric estimates of the density, hazard and renewal functions given by (14.24)–(14.26).

## 14.6.2 Parametric Approach

In parameter analyses, we assume a specific form for the distribution function and focus on estimating the model parameter $\theta$. We again look at Scenarios 5 and 7.

### 14.6.2.1 Scenario 5

The warranty claims data $(\tilde{t}_{ij}, \tilde{x}_{ij})$ are assumed to be iid random variables. As a result, the likelihood function is

$$L(\theta) = \prod_{i=1}^{I_1} \prod_{j=1}^{n_i} f(\tilde{t}_{ij}, \tilde{x}_{ij}) \tag{14.27}$$

The MLE $\hat{\theta}$ is obtained by maximization of (14.27).

#### 14.6.2.2 Scenario 7

It is assumed that supplementary data consist of the following:

- Sale dates $t_{i0}$, $I_1 < i \leq I$, of items for which no warranty claim has been received.
- Dates $t_{in_i}$, $1 \leq i \leq I_1$, of the last replacement under warranty for customers who have had one or more warranty claim.

Note that some of the items that have not failed under warranty may still be covered by warranty, whereas the remaining are no longer under warranty because either the age limit or the usage limit has been exceeded.

The following additional notation is required:

$$\tau_i = \begin{cases} \min\{t_i^*, t_{i0} + W - t_{in_i}\} & \text{for } 1 \leq i \leq I_1 \\ \min\{W, t_{i0} + W - t_{i0}\} & \text{for } I_1 < i < I \end{cases} \tag{14.28}$$

The contribution to the likelihood function of items that resulted in valid warranty claims is given by (14.27). In order to obtain the contribution of items that did not result in warranty claims, we consider the following two cases:

*Case (a)* For items that have one or more warranty claim, the information that we have is that the last replacement under warranty that did not fail in the region $[0, \tau_i) \times [0, x_i^*)$ for $1 \leq i \leq I_1$.[16]

The contribution to the likelihood function is

$$L_{c1}(\theta) = \prod_{i=1}^{I_1} [1 - F(\tau_i, x_i^*)], \tag{14.29}$$

with $\tau_i$ given by (14.28).

*Case (b)* For items with no warranty claim, some (those sold during the time interval $[0, L - W)$) have exceeded the time limit $W$, so that there was no failure in the region $[0, W) \times [0, U)$ for these items. For other items (sold over the time interval $[L - W, L)$), there is no failure in the region$[0, \tau) \times [0, U)$, where $\tau$ is the age of the item at the end of the observation interval. The contribution of these to the likelihood function is given by

[16] $x_i^*$ has defined in Sect. 14.2.2 and it denotes the usage remaining after the last failure for customer $i$.

$$L_{c2}(\theta) = \prod_{i=I_1+1}^{I} [1 - F(\tau_i, U)] \tag{14.30}$$

with $\tau_i$ given by (14.28).

As a result, the likelihood function is

$$L(\theta) = L_f L_{c1} L_{c2} = \left\{ \prod_{i=1}^{I_1} \prod_{j=1}^{n_i} f(t_{ij}, x_{ij}) \right\} \left\{ \prod_{i=1}^{I_1} [1 - F(t_i^*, x_i^*)] \right\} \left\{ \prod_{i=I_1+1}^{I} [1 - F(\tau_i, U) \right\} \tag{14.31}$$

The estimate $\hat{\theta}$ is the value of $\theta$ that maximizes the likelihood function.

## 14.7 Forecasting Expected Warranty Claims

Let $N(W, U)$ denote the number of failures of a item within the warranty coverage region $(W, U)$, and $C(W, U)$ the corresponding cost. Here we are interested in estimating the expected values of $N(W, U)$ and $C(W, U)$ for given values of $W$ and $U$. These depend on the action taken to rectify the failures—repair or replace. In the following, we omit the details of the derivation and present the final expressions for forecasting warranty claims and costs by the three approaches discussed in Sects. 14.4–14.6.[17]

### 14.7.1 Forecasting by Approach 1

Forecasting by Approach 1 is done using the usage rate distribution for the various data Scenarios.

*Scenarios 5 and 7* The expected number of failure within the warranty period $(W, U)$ is given by

$$\begin{aligned} E[N(W, U)] &= \int_0^{U/W} M_z(W) g(z) dz + \int_{U/W}^{\infty} M_z(U/z) g(z) dz \\ &= M_z(W) G(U/W) + \int_{U/W}^{\infty} M_z(U/z) g(z) dz, \end{aligned} \tag{14.32}$$

[17] Details of the derivations, properties of the estimators, and applications can be found in Chap. 7 and in [3].

where $M_z(.)$ is the renewal function defined in (7.20). The corresponding expected warranty cost is $E[C(W,U)] = C_s \times E[N(W,U)]$, where $C_s$ is the cost of providing a new replacement under warranty. (See Sect. 7.6.1.1 for details.)

**Scenarios 6 and 8** For Scenarios 6 and 8, the expected number of failures over the warranty region ($W$, $U$) is

$$E[N(W,U)] = \int_{0}^{U/W} \Lambda_z(W)g(z)dz + \int_{U/W}^{\infty} \Lambda_z(U/z)g(z)dz, \tag{14.33}$$

where $\Lambda_z(.)$ is the cumulative intensity function defined in (7.22). The corresponding expected warranty cost is $E[C(W,U)] = c_r \times E[N(W,U)]$, where $c_r$ is the average cost of a minimal repair. (See Sect. 7.6.1.2.)

### 14.7.2 Forecasting by Approach 2

Approach 2 is applicable for forecasting warranty claims and costs for data from Scenarios 5 and 7. For this approach, the variable $V$ is a combination of age and usage defined in (14.20), and $F_V(v)$ is the CDF of $V$. Under a 2-D warranty with only first failure coverage, the expected cost within the warranty coverage region ($W$, $U$) is given by $E[C(W,U) = c_s F_V(v)$, where $v = (1-\varepsilon)U + \varepsilon W$.

For the nonrenewing free replacement warranty on nonrepairable items, the expected warranty cost within the warranty coverage region ($W$, $U$) is given by $E[C(W,U) = c_s M_V(v)$, where $M_V(v)$ is a the renewal function associated with $F_V(v)$. Additional details on the application of Approach 2 to forecasting warranty costs can be found in [7].

### 14.7.3 Forecasting by Approach 3

The bivariate distribution, renewal and intensity functions are utilized for forecasting claims and costs by Approach 3.[18]

**Scenarios 5 and 7** The expected number of failure over the warranty region ($W$, $U$) is $E[N(W,U)] = M(W,U)$, where $M(W,U)$ is a bivariate renewal function associated with $F(W,U)$, defined in (7.29). The corresponding expected warranty cost is $E[C(W,U)] = C_s \times M(W,U)$.

**Scenarios 6 and 8** For Scenarios 6 and 8 data, the expected number of failures over the warranty region ($W$, $U$) is given by $E[N(W,U)] = \Lambda(W,U)$, where $\Lambda(W,U)$ is the bivariate cumulative intensity function [2].

---

[18] Several bivariate models for 2-D warranty policies are discussed in the literature. See, for example, [2–4, 8, 11, 17–21, 23].

## References

1 Baik J, Murthy DNP (2008) Reliability assessment based on two-dimensional warranty data. Int J Reliab Saf 2:190–208
2 Baik J, Murthy DNP, Jack N (2004) Two-dimensional failure modelling and minimal repair. Nav Res Logist 51:345–362
3 Blischke WR, Murthy DNP (1994) Warranty cost analysis. Marcel Dekker, Inc., New York
4 Blischke WR, Murthy DNP (eds) (1996) Product warranty handbook. Marcel Dekker, Inc., New York
5 Duchesne T, Lawless JF (2000) Alternative time scales and failure time models. Lifetime Data Anal 6:157–179
6 Gertsbakh IB, Kordonsky KB (1998) Parallel time scales and two-dimensional manufacturer and individual customer warranties. IIE Trans 30:1181–1189
7 Iskandar BP, Blischke WR (2003) Reliability and warranty analysis of a motorcycle based on claims data. In: Blischke WR, Murthy DNP (eds) Case studies in reliability and maintenance. Wiley, New York, pp 623–656
8 Iskandar BP, Murthy DNP (2003) Repair-replace strategies for two-dimensional warranty policies. Math Comput Model 38:1233–1241
9 Jiang R, Jardine AKS (2006) Composite scale modeling in the presence of censored data. Reliab Eng Sys Saf 91:756–764
10 Jung M, Bai DS (2007) Analysis of field data under two-dimensional warranty. Reliab Eng Sys Saf 92:135–143
11 Kim HG, Rao BM (2000) Expected warranty cost of a two-attribute free-replacement warranties based on a bi-variate exponential distribution. Comput Ind Eng 38:425–434
12 Kordonsky KB, Gertsbakh IB (1993) Choice of best time scale for reliability analysis. Euro J Ops Res 65:235–246
13 Kordonsky KB, Gertsbakh IB (1995a) System state monitoring and lifetime scales I. Reliab Eng Sys Saf 47:1–14
14 Kordonsky KB, Gertsbakh IB (1995b) System state monitoring and lifetime scales II. Reliab Eng Sys Saf 49:145–154
15 Kordonsky KB, Gertsbakh IB (1997) Multiple time scales and lifetime coefficient of variation: engineering applications. Lifetime Data Anal 2:139–156
16 Lawless JF, Hu XJ, Cao J (1995) Methods for the estimation of failure distributions and rates from automobile warranty data. Lifetime Data Anal 1:227–240
17 Manna DK, Pal S, Sinha S (2006) Optimal determination of warranty region for 2D policy: A customers' perspective. Comput Ind Eng 50:161–174
18 Manna DK, Pal S, Sinha S (2007) A use-rate based failure model for two-dimensional warranty. Comput Ind Eng 52:229–240
19 Manna DK, Pal S, Sinha S (2008) A note on calculating cost of two-dimensional warranty policy. Comput Ind Eng 54:1071–1077
20 Moskowitz H, Chun YH (1994) A Poisson regression model for two-attribute warranty policies. Nav Res Logist 41:355–376
21 Murthy DNP, Iskandar BP, Wilson RJ (1995) Two-dimensional failure free warranties: Two-dimensional point process models. Oper Res 43:356–366
22 Murthy DNP, Xie M, Jiang R (2004) Weibull models. Wiley, New York
23 Pal S, Murthy GSR (2003) An application of Gumbel's bivariate exponential distribution in estimation of warranty cost of motorcycles. Int J Qual Reliab Manag 20:488–502
24 Rai B, Singh N (2003) Hazard rate estimation from incomplete and unclean warranty data. Reliab Eng Sys Saf 81:79–92
25 Rai B, Singh N (2005) A modeling framework for assessing the impact of new time/mileage warranty limits on the number and cost of automotive warranty claims. Reliab Eng Sys Saf 88:157–169

26 Rai B, Singh N (2006) Customer-rush near warranty expiration limit and nonparametric hazard rate estimation from the known mileage accumulation rates. IEEE Trans Reliab 55:480–489
27 Suzuki K, Karim MR, Wang L (2001) Statistical analysis of reliability warranty data. In: Balakrishnan N, Rao CR (eds) Handbook of statistics: Advances in reliability, vol 20. Elsevier Science, New York, pp 585–609
28 Suzuki K (1985a) Estimation of lifetime para-meters from incomplete field data. Technometrics 27:263–271
29 Suzuki K (1985b) Nonparametric estimation of lifetime distribution from a record of failures and follow-ups. J Am Statist Assoc 80:68–72
30 Yang G, Zaghati Z (2002) Two-dimensional reliability modeling from warranty data. Proc Annu Reliab Maintainab Symp 272-278

# Part V
# Warranty Management

# Chapter 15
# Use of Warranty Data for Improving Current Products and Operations

## 15.1 Introduction

Product reliability performance depends on several factors, some under the control of the manufacturer, e.g., the decisions made during the design, development and production stages of the product life cycle, and others under the control of customers, e.g., usage intensity, operating mode and environment, maintenance, etc. These, in turn, impact a number of factors, including customer satisfaction/dis-satisfaction levels, warranty claims, and warranty costs. During the new product development process, the manufacturer defines target limits for these and other performance measures.

Warranty data, including both claims and supplementary data, allow the manufacturer to evaluate performance measures and assess them relative to the defined targets. Problems arise when the measured performance levels do not meet the target levels. Improvements are needed to address these problems and avoid their potentially adverse affect on profits. Even if the targets are met, the data may suggest opportunities for changes to the product and/or operations that would improve the overall business performance.

Total Quality Management (TQM) provides an approach for continuous improvement. In this chapter, we look at the use of warranty data for improving a company's current products and operations based on the TQM paradigm. This corresponds to Stage-2 of Warranty Management discussed in Sect. 2.12. The outline of the chapter is as follows. We begin with a brief discussion of the TQM approach in Sect. 15.2. In Sect. 15.3, we look at problem detection for improvement based on warranty data, categorize the problems into four different categories, and illustrate the approach by means of two case studies. In the next four Sects. 15.4–15.7, we briefly look at solutions to these four categories of problems. Section 15.8 deals with effective warranty management for continuous improvement. In this context, we highlight the need for an effective organizational structure and a well-designed warranty management system.

W. R. Blischke et al., *Warranty Data Collection and Analysis*, Springer Series in Reliability Engineering, DOI: 10.1007/978-0-85729-647-4_15,

## 15.2 The TQM Approach

There are many different definitions of Total Quality Management (TQM). According to the International Standards Organization [22]:

> TQM is a management approach for an organization, centered on quality, based on the participation of all its members and aiming at long-term success through customer satisfaction, and benefits to all members of the organization and to society.

TQM evolved to facilitate a total orientation of all members of the staff of an organization toward achieving high quality, and an emphasis on zero defects and no reworking ("right the first time"). TQM also emphasizes continuous improvement ("kaizen" in Japanese).

TQM is a set of management practices to be applied throughout the organization and geared to ensure that customer requirements are consistently met or exceeded. TQM began in the 1950s in Japanese industrial organizations and became popular in the West, beginning in the early 1980s and growing steadily since. Over the years, TQM has become a very important tool for improving a firm's process capabilities sop that they are consistent with business objectives and sustain competitive advantages. TQM focuses on encouraging a continuous flow of incremental improvements throughout the organization. The process involves the following:

- Pursuing strategic thinking
- Knowing the customers
- Defining customer requirements
- Concentrating on prevention as opposed to correction
- Reducing chronic waste
- Pursuing a continuous improvement
- Using a structured methodology

The three key elements of TQM are (1) The PDCA cycle, (2) Problem solving and (3) Root cause analysis.

### *15.2.1 PDCA Cycle*[1]

This cycle symbolizes the process of quality improvement and involves executing the four steps indicated in Fig. 15.1 in an iterative manner.

The four steps are as follows:

---

[1] The concept of the PDCA Cycle was originally developed by Walter Shewhart and is often referred to as the "Shewhart Cycle". It was promoted very effectively by W Edwards Deming and is also referred to as the "Deming Wheel." Discussion on PDCA can be found in many books on quality. See, for example, [8, 19].

**Fig. 15.1** PDCA cycle

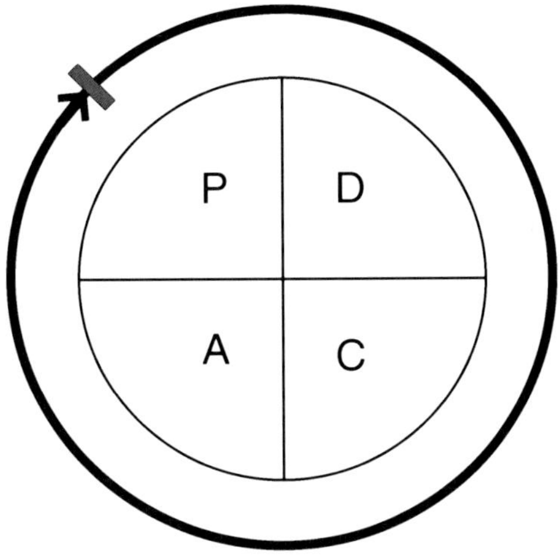


P: Plan the change by finding out what things are going wrong (that is, identify the problems faced), and develop ideas for solving these problems

D: Do changes designed to solve the problems, on a small-scale to begin with, to test whether or not the changes will work without causing undue interruptions to operations

C: Check whether or not the small-scale changes are achieving the desired results and identify any new problems that may crop up

A: Act to implement changes on a larger scale if the small-scale changes are successful

### *15.2.2 Problem Solving Methodology*

There are many approaches to problem solving, depending on the nature of the problem. The rational approach involves several steps, as indicated in Fig. 15.2. We briefly discuss each below.[2]

*Step 1: Define the Problem.* A common definition of a problem is that it is a situation that needs fixing—e.g., the engine of a car failing to start. A different definition is that it is an opportunity to improve. In the context of TQM, one tackles problems of either kind–the former being reactive and the latter being proactive.

*Step 2: Identify and Document the Process.* This requires a deep understanding of the underlying processes that are relevant to solving the problem. In the case of component failures, this step involves understanding the different modes of failure and the mechanisms that may be responsible for each failure mode.

[2] There are a many books that discuss problem solving; see for example [1, 33].

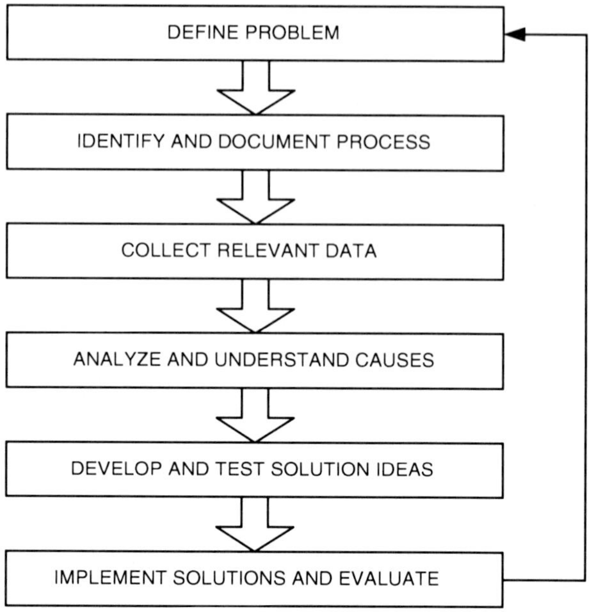


**Fig. 15.2** Problem solving methodology

*Step 3: Collect Relevant Data*. The data that need to be collected must be credible and relevant. As an example, in the case of pump failures, one needs to know whether the pump was used continuously or intermittently, the flow rate and the pressure at which it was operated, the chemical properties of the fluid being pumped, and so forth.

*Step 4: Analyze and Understand Causes*. The analysis can be based solely on data (empirical analysis) or in combination with relevant theories (combining empirical and theoretical) and is problem specific. In the case of reliability-related problems, this involves reliability science combined with testing. The objective is to achieve a better understanding of the causes of failure. This stage involves a mix of scientific approaches and includes inductive and deductive reasoning.

*Step 5: Develop and Test Solutions*. In some cases the problem is solved using known solutions from the past, whereas in other cases highly innovative and non-standard solutions are required. This stage involves various kinds of thinking, such as innovative, creative and lateral thinking.[3]

*Step 6: Implement Solutions and Evaluate*. One usually implements the solution to a problem in a controlled manner, as the analysis and solution are based on limited data and knowledge and not all of the relevant factors are fully understood. The outcome can be that the solution is either effective or it is not. In the latter case, it is necessary to revert back to the first step and repeat the process.

[3] There are many books on thinking. See, for example [7, 12].

### 15.2.3 Root Cause Analysis

Root Cause Analysis (RCA) is a technique that seeks to answer the question "why did the problem occur in the first place?" RCA seeks to identify the origin of the problem and uses a specific set of steps, with associated tools, to find the primary cause of the problem, so that one can address the following:

- Determine what happened.
- Determine why it happened.
- Determine what must be done to reduce the likelihood that it will happen again.

RCA is not the same as problem solving. It is merely the diagnostic part of the process ("Analyze and Understand Causes" in Fig. 15.2). If this is done incorrectly, it will result in solutions that have little value. RCA means finding the specific source(s) that created the problem so that effective action can be taken to prevent its recurrence. This must be done in an iterative manner that combines divergent and convergent thinking, using a variety of tools and techniques. The following are some well known tools for implementation of RCA[4]:

- Fault tree analysis: In this where one starts at the top (system) level and then proceed downward to part (component) level and link failure at system level to failures at the part level.
- Logic tree diagram: Breaking down a system/process into functional subsystems/components and identifying the logical cause and-effect relationships.
- Bayesian inference: Use of prior knowledge in statistical analysis of data.[5]
- Flowcharts: Allow understanding the operational steps involved in the process in order to identify their possible contributions to the cause of the problem.
- Cause and effect analysis: This is based on the fact that for every effect there is a cause and that there can a fairly long chain of relationships between a cause and its effects.
- Ishikawa diagram (also known as the fishbone diagram or cause and effect diagram): This is very similar to cause-and-effect analysis with the relationships indicated in the form of the skeleton of a fish.
- Run charts: Analysis of data over time to look for trends/patterns that may indicate the root cause.
- 5 Whys: This involves thinking "WHY" consecutively five times. Through these five "WHYS" one can break down causes to a very specific level of detail.
- Brainstorming: A mechanism for identifying all possible causes and then eliminating the least likely, so that attention is focused on the most likely.

[4] There are many books that cover most of the techniques listed. See, for example [3, 29, 34, 41]. Books that deal primarily with one method are [35] dealing with the 5 WHY'S, [20] with Fishbone diagrams, [39] with failure mode and effects analysis, and [5] with fault tree analysis.

[5] See Sect. 9.4.4.

## 15.3 Problem Detection for Improvement

As mentioned in Sect. 15.1, a problem arises when warranty data indicate that one or more of the performance measures do not meet their desired target levels. In these instances, the data may, in fact, suggest avenues to improvement. The process of detecting and solving problems through the use of warranty data is indicated in Fig. 15.3. The process begins with warranty data collection. The next step is data analysis for problem detection using the concepts and techniques discussed in Chaps. 8–14. Once the problem is identified, one proceeds to finding solutions following the PDCA cycle for improvement discussed in the previous section.

### *15.3.1 Data Collection*

As discussed in Chaps. 4 and 5, the two types of data needed are 1. Warranty claims data and 2. Appropriate supplementary data. The main sources for warranty claims data are the following:

- Call centers
- Service agents, when warranty servicing is out-sourced
- Technician and engineer reports, when warranty servicing is done in-house
- Customers

The main sources for supplementary data are the following:

- Design, development and production departments of the manufacturer
- Component vendors
- Retailers
- Customers

Each warranty claim generates a report which may contain detailed information on a number of important variables, including some or all of the following:

- Customer identification number
- Retailer identification number
- Sale date
- Customer description of the problem
- Identification of failed component
- Failure mode
- Actions to fix the problem (spares used, repairs, etc.)[6]
- Cost of repair

---

[6] The different actions are also referred to as "labor codes"—see, for example [45], or as "defect codes".

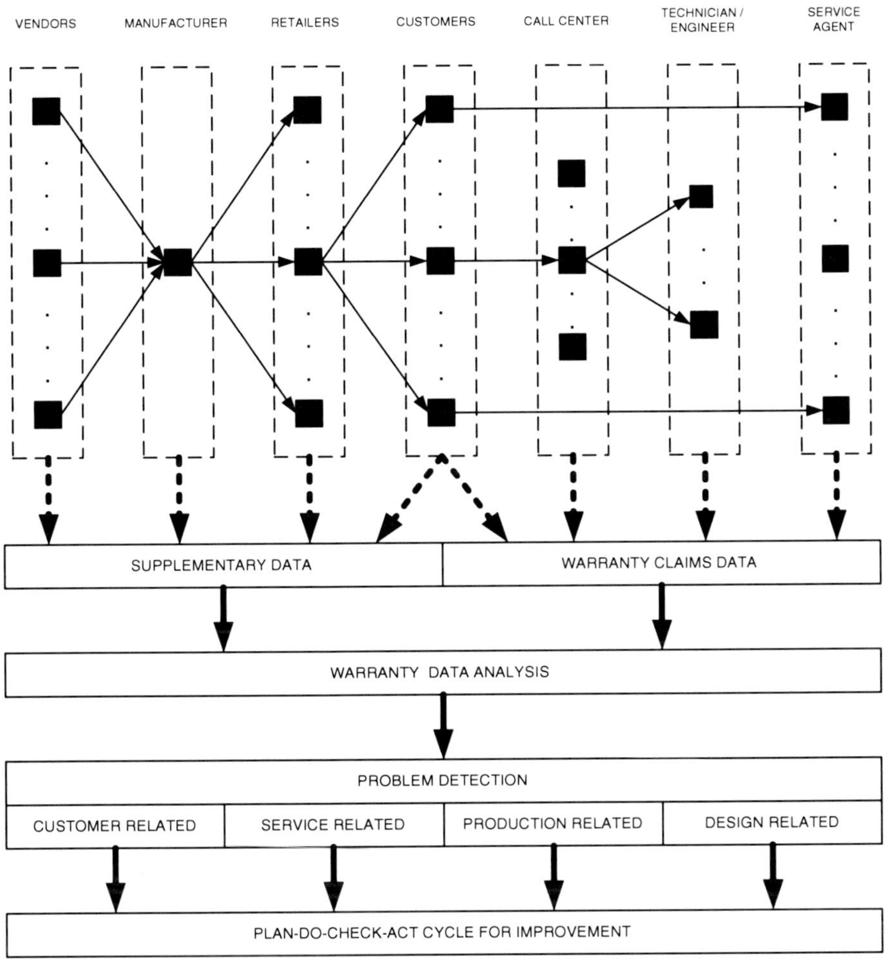


**Fig. 15.3** Problem detection for improvement

- Usage[7]
- Operating environment
- Usage mode
- Other customer, dealer or service information

Customer surveys are another important source of data, especially that relating to customer satisfaction/dissatisfaction with product performance and/or product support service. This is best done by means of a properly designed questionnaire, usually based on scales with discrete levels ranging from strongly-satisfied to

[7] See Sect. 3.6 for various notions of usage.

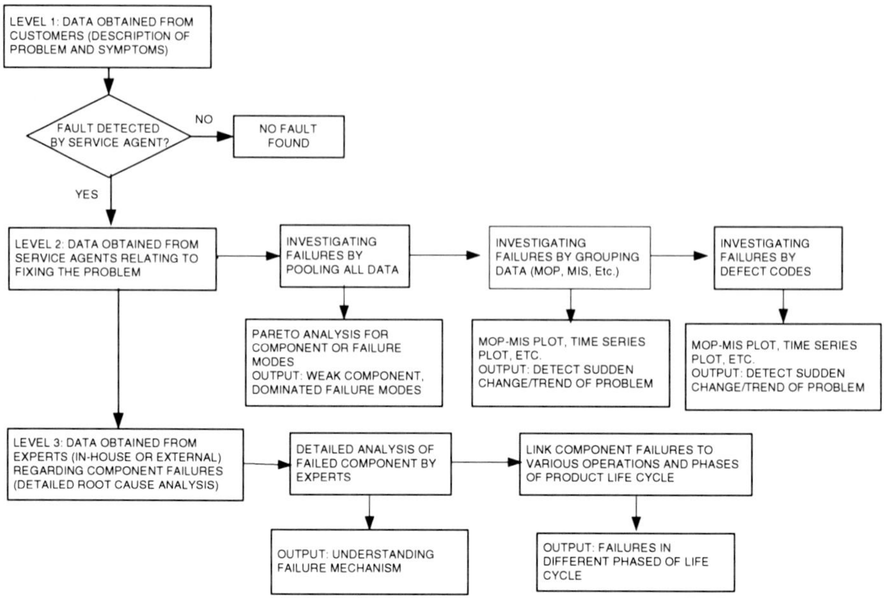


**Fig. 15.4** Three levels of data analysis

strongly-dissatisfied, and a statistically valid survey design.[8] Another approach is through critical incidents, that is, events that are out of the ordinary in the mainstream of events that may occur [11]. Such an incident may cause a positive or negative adjustment to a customer's opinion of the quality of a product or its support services. The critical incident approach has been used to study consumer dissatisfaction and product quality in the case of automobiles [4].

## *15.3.2 Data Analysis*[9]

In order to obtain meaningful and effective results, data analysis must be done at three different levels and must be carried out by an analyst with a strong background in statistics and a good understanding of the warranty process. The three different levels of analysis are indicated in Fig. 15.4.

[8] Measuring satisfaction for consumer durables is different from that for industrial products. For more details regarding satisfaction with consumer durable products, see [36]. For industrial products, see [16].

[9] In the IT-warranty literature, the term "Early Warning System" is used to denote software packages for carrying out this kind of analysis.

#### 15.3.2.1 Level 1 Analysis

Customer complaints are usually in text format. Text mining approaches provide a methodology for identification and classification of the problems expressed in the complaints. It is not uncommon to find that a customer reports a problem, but the service agent is unable to find any fault.[10]

#### 15.3.2.2 Level 2 Analysis

This can involve several stages. The first stage involves looking at failures at the product level by pooling the data without differentiating MOP, MIS,[11] and carrying out a Pareto analysis based on failed components and/or failure mode. These analyses give an indication of which components are weak (with low reliability) and identify the dominant failure modes. The second stage involves plotting failure data at the product level by grouping the data based on MOP, MIS, or other factors, without differentiating by defect code. These results are used to prepare plots such as time series plot, with time being linked to MOP, MIS-MOP plots (discussed in Sect. 11.5), etc. The aim of these analyses is to detect sudden changes or noticeable trends, which may be indicators of a problem. The third stage is to carry out an analysis similar to that in Stage 2, but in a more detailed manner, for example, for each different defect code or any other meaningful classification.

Problem Detection

The detection of a problem is based on specified rules for detection, which are often similar to the rules used for detecting quality variation based on control charts.[12] Any detection rule suffers from the following two types of errors:

- Type 1 error: The rule indicates no problem when there is an underlying problem
- Type 2 error: The rule indicates that there is a problem when there is none

These errors occur mainly because of limited data and the resulting uncertainties associated with small samples.[13]

#### 15.3.2.3 Level 3 Analysis

There are two different types analysis at Level 3. The first is a detailed analysis of failed components carried out by in-house or external experts to better

---

[10] This is referred to as "No Fault Found". For more on this, see [43].

[11] MOP (month of production); MIS (month in service before failure).

[12] There are many papers that deal with this—for example [45] deals with MIS-MOP plots and [25] looks at time-series plots.

[13] See [6] for further discussion.

understand the causes of the various types of failures. This requires understanding the mechanism of failure and the factors that influence the failure (for example, material used, vendor, design). The second is to link component failure to various operations in different phases of the product life cycle (such as design approach used, quality control in production, etc.). In either case, information at the detailed level is required, and the tools used for detecting the problem more complex.

## 15.3.3 Classification of Problems

Problems may be classified into the following four groups:

- Customer-related: Usage, misuse, fraud, etc.
- Service related: Incompetence of service agent, improper servicing, poor servicing logistics, over-servicing and other types of fraud, etc.
- Production related: Assembly errors, component nonconformance (in-house or vendor-supplied), etc.
- Design related: Incorrect targets in the design phase because of lack of understanding of customer needs, usage profiles, component specifications, material selection, etc.

A simple problem is a problem belonging to one group, whereas a complex problem involves several simple problems belonging to more than one group.

### 15.3.3.1 Customer Related Problems

The primary data for detecting customer related problems are (1) The data collected by the service agents from customers when a warranty claim is initiated, and (2) Through customer surveys. Upward trends in the numbers of customers who are not satisfied with product performance and/or warranty service provided indicate needs and opportunities for improvement.

### 15.3.3.2 Service Related Problems

The primary data for detecting service related problems are customer complaints relating to warranty servicing and claims for reimbursement. For optimal impact, the analysis of data of this type must be done for each service agent in detail, Information needed includes labor cost, service time for different labor or defect codes, and other important measures, many of which are product specific. Service agents are then compared to determine if some have significantly higher or lower values. Significant differences indicate that there is a service-agent-related problem.

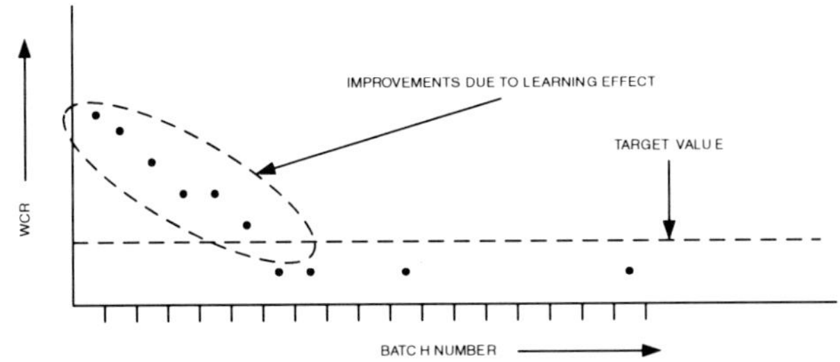


**Fig. 15.5** Improvement in quality due to learning effect

#### 15.3.3.3 Production Related Problems

The primary data in this area are those relating to the servicing of failed units provided by the service agents. The data is obtained from the production department and from component vendors. The data must allow a failed component to be traced to the month of production or batch number, and the component manufacturer, which may be either the product manufacturer or an external vendor.

In the first stage of Level 2 analysis, one typically views production in terms of batches, with a batch defined as all of the items produced in a given month. The Warranty Claims Rate (WCR) is then plotted as a function of MOP.[14] Typically, the plot would be expected to be similar to that in Fig. 15.5, where the initial downward trend is due to the fact that production processes are often initially relatively uncertain and out of control, but after some time settle down, due to what is called the "learning effect," to the point that most items conform to design specifications. If there were no production problems, WCR would stay below some defined target value that is acceptable.

If the WCR value for a batch exceeds the target value it is an indication of some production-related problem associated with that batch. This is shown in Fig. 15.6, where there are two batches for which the WCR is well above the target value.[15]

In later stages of analysis, a more detailed analysis would be done. This may include, for example, plots based on MIS similar to those in Figs. 15.5 and 15.6 for each MIS.

#### 15.3.3.4 Design Related Problems

The primary source of data regarding design aspects is the detailed analysis carried out at Level 3. The purpose of the analysis is to understand the failure mechanism and the design flaws that failed to take this mechanism into account.

---

[14] WCR can be defined in many different ways. One definition is the following: WCR = [Cumulative number of claims/Number of items sold] × 1000.

[15] Sometimes the target value is determined by the average value of the WCRs, based on batches produced after to the process has reached its steady state.

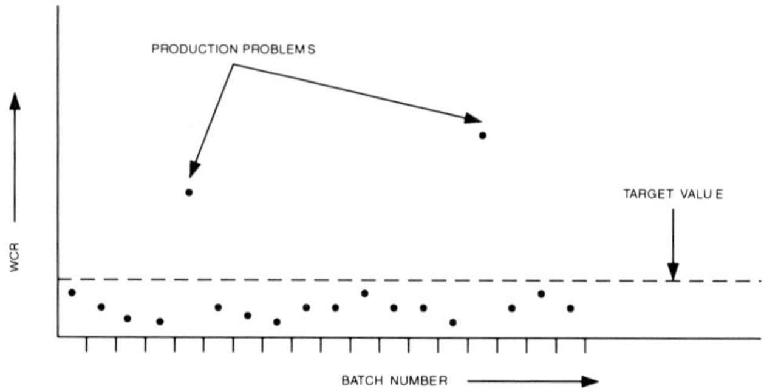


**Fig. 15.6** Production problems

**Table 15.1** Level of analysis appropriate for detecting problems

| Problem | Level of analysis |
|---|---|
| Customer | Level 1 |
| Service agent and production | Levels 1 and 2 |
| Production and design | Levels 1, 2 and 3 |

#### 15.3.3.5 Level of Analysis

The level of analysis required to detect the various problems discussed above is indicated in Table 15.1.

### *15.3.4 Some Complicating Factors*

There are several issues that must be addressed when dealing with warranty data. These include the following:

#### 15.3.4.1 Aggregation

In some cases, the data collected are aggregated over time or other factors. Examples are sales, claims, and other variables aggregated within discrete time intervals. In other cases, data are collected at the individual unit level. The former is acceptable for preliminary analysis for problem detection, but is not adequate when a deeper analysis is needed since aggregation results in loss of information.

#### 15.3.4.2 Uncertainty

Various kinds of uncertainty affect warranty data. This is discussed in some detail in Chaps. 4 and 5. Two critical problems of this type are (1) The lag between failure and reporting of the failure, and (2) The number of items under warranty at any given time.

#### 15.3.4.3 Traceability

For production related problems, one needs to trace the failed component to its production batch. For expensive products (such as an automobile), the manufacturer can track this information if he has access to vendor production systems. For many products, however, tracing a component to its original source is not possible. This can become an issue in analyzing and interpreting the data, especially if there is more than one vendor supplying the same component.

#### 15.3.4.4 Cause of Component Failure

A component can fail for one of the following three reasons:

1. The component is defective
2. A non-defective component is damaged during assembly
3. Customer misuse

In some cases, it is possible to differentiate between causes 1 and 2, and in other cases it is not. This becomes an issue when components are supplied by a vendor, since it has implications with regard to cost sharing. When the service operation decides whether the failure cause is 1 or 2, incorrect classification can become an issue.[16]

### *15.3.5 Illustrative Cases*

In this section, we discuss two case studies. The first deals with motor cycles produced and sold in Indonesia, and illustrates Level 1 analysis. The second deals with a component of an automobile sold in Asia and illustrates Level 2 analysis. Cases of Level 3 analysis are discussed later in the chapter.[17]

#### 15.3.5.1 Case 15.1: Motor Cycles[18]

A motorcycle is a multi-component system, with many items supplied by vendors. This case deals with motor cycles produced and sold in Indonesia. Not all of the

---

[16] This is discussed in detail in [42].

[17] A case study involving Level 2 analysis for several electronic products produced by Philips can be found in [37].

[18] For details of this case, see [21].

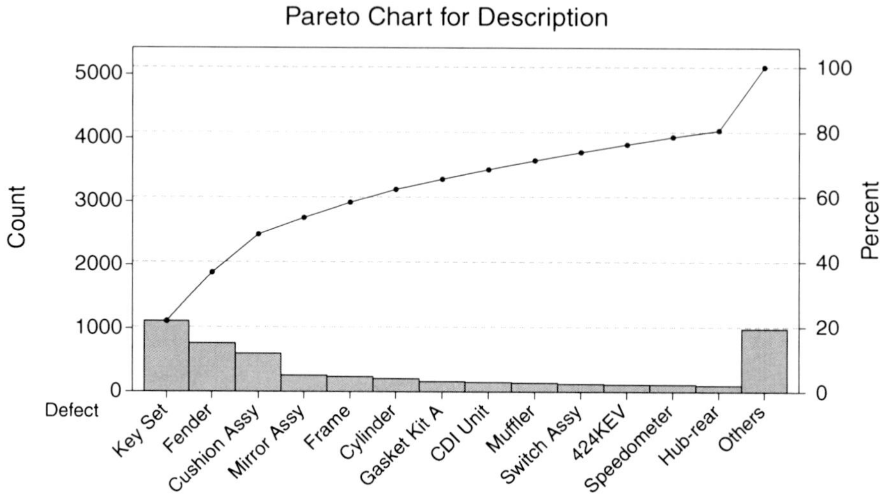


| | | | | | | | | | | | | | | |
|---|---|---|---|---|---|---|---|---|---|---|---|---|---|---|
| Count | 1118 | 763 | 601 | 261 | 241 | 211 | 161 | 152 | 144 | 129 | 120 | 119 | 103 | 996 |
| Percent | 22 | 15 | 12 | 5 | 5 | 4 | 3 | 3 | 3 | 3 | 2 | 2 | 2 | 19 |
| Cum % | 22 | 37 | 48 | 54 | 58 | 62 | 66 | 69 | 71 | 74 | 76 | 79 | 81 | 100 |

**Fig. 15.7** Pareto chart based on component responsible for warranty claim

components are covered by the manufacturer's warranty. Components and materials not covered are those that deteriorate or wear out with usage, such as spark plugs, fuel filter, oil filter, brake pads, light bulbs, tires, and so forth. (Many of these, e.g., tires, are covered by warranties provided by the suppliers). In this context, failure of the motorcycle is defined to be failure of one or more of the components that are covered under the Company's warranty. Figure 15.7 is a Pareto chart indicating the count of failures of different components during the warranty period.

As can be seen, the first three components are responsible for more than half of the total warranty claims.

#### 15.3.5.2 Case 15.2: Automobile Component

This is a follow up of the Example 11.5 discussed in Sect. 11.5.3.

As can be seen from Fig. 11.6, the warranty claims rate is initially decreasing and there is a significant increase at $t = 5$ and $t = 6$ (the months of May and June), and another increase at the end of the year. This suggests that there are some problems with the June month of production and possibly December as well. Further study is needed to resolve this issue.

The warranty data provided in this case included the cause of failure. These were coded as defect codes A–H, indicating eight possible failure causes. The data were divided into eight categories based on the defect code and the analysis was repeated for each defect category. The results are shown in Fig. 15.8. As can be

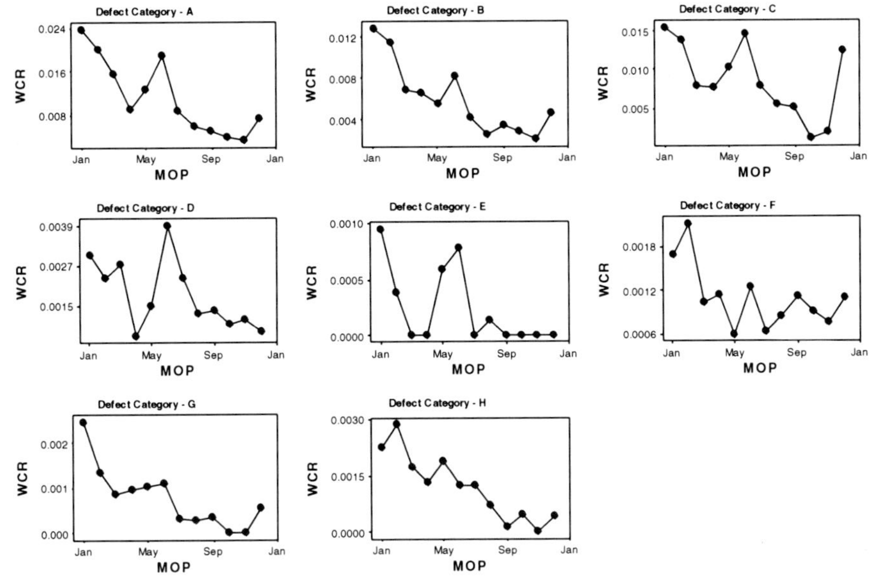


**Fig. 15.8** WCR($t$),$t$ = 1, 2, ... , 12, as a function of MOP for the eight defect codes

seen, the increase in failures for items produced in June is mostly due to defect codes C, D and E.

## 15.4 Customer-Related Problems

A proper study of customer-related problems requires an understanding of consumer behavior. Key elements of this are: (1) Attribution theory and (2) Consumer satisfaction and dissatisfaction.

### 15.4.1 Consumer Behavior

#### 15.4.1.1 Attribution Theory

Attribution theory deals with a consumer's perception in ascribing an effect (in our case, failure) to a particular source or cause.[19] Most consumers perceive a product failure as being caused by (1) The product itself (2) Inappropriate usage by the consumer, or (3) An external influence such as the situation or environment.

[19] Reference [24] defines attribution theory as "a theory about how people make causal explanations for events of which they have knowledge, about how they answer questions beginning with 'why'.".

Attribution theory provides a framework that permits investigation of observed behavioral phenomena. Reference [23] use this theory to investigate how consumers impute product failure to the causes of failure.

#### 15.4.1.2 Customer Satisfaction and Dissatisfaction

A customer is satisfied when his/her perception of the ratio of benefits from the product and associated services, to the costs paid to obtain the benefits, are met or exceeded. This involves the following three notions of quality:

- *Perceived Quality*: Perceived quality is a result of the consumer's (the buyer's) subjective assessment of the quality of a given product. Perceived quality thus differs from objective quality, which depends on the technical and functional specifications of a product. Perceived or subjective quality, on the other hand, is linked to the consumer and her/his perception of the quality of a product.
- *Value-based Quality*: The concept of value has its roots in the microeconomic tradition. Here the quality of a product is determined in relation to its price and not solely on its own merits.
- *Quality of Service*: Service, in the context of product warranty, refers to all activities associated with the servicing of warranty claims. Customers also have expectations regarding the quality of service. According to *SERVQUAL* the five dimensions of service quality are[20]:
  - Tangibles: physical aspects—look and feel of premises, staff, etc.
  - Responsiveness: provision of speedy service, willingness to serve, etc.
  - Reliability: delivering promised service dependably and accurately
  - Assurance: professional and courteous service which conveys trust
  - Empathy: individualized, personal service.

The benefits customers expect depend primarily on how they perceive product and service quality, and whether or not their perceptions are valid.

For consumer durables, customer satisfaction and customers' perception of product quality are complicated issues. Different individuals may have different views on a product's quality. A customer's perception of a product's external properties may be the result several factors and include the following:

- Cultural background (e.g., different cultures have different views on aesthetics)
- Physical and cognitive capabilities (cognitive capabilities determine, for example, how easy it is for the customer to operate a product)
- Individual experiences and preferences (a customer having had a bad experience with a manufacturer's product will consider another manufacturer)

[20] For more details regarding *SERVQUAL*, see [46].

- Basic functional needs, that can vary depending on the customer and/or product.

Satisfaction (or dissatisfaction) is linked to an evaluation or discrepancy between prior expectations and the actual (or perceived) product performance and quality of service. A customer is satisfied when performance exceeds expectations. The reverse situation leads to a dissatisfied customer.[21]

## *15.4.2 Problem Classification*

Customer-related problems can be broadly grouped into the following three categories:

- Dissatisfaction with product and/or warranty service
- Misuse of product
- Fraud

### 15.4.2.1 Dissatisfaction

A customer may be dissatisfied with the product itself or with the warranty service provided in the event of product failure. Failure of a product to perform as expected can be due to:

1. Incorrect specification of the consumer usage profile during design,
2. Failure to do a proper reliability analysis during design,
3. Poor quality control in manufacturing, or
4. Customer usage.

The first three of these depend on the actions of the manufacturer. They are design and production problems and are discussed later in the chapter. Usage depends on the consumers and when the usage mode or intensity differs significantly from nominal values, product performance will also differ. Dissatisfaction with servicing depends on the actions of the service agent. This is discussed in the next section.

### 15.4.2.2 Misuse of Product

Misuse can be either deliberate or unintentional. In either case, the manufacturer has no obligations to service the failed item under warranty. However, the

[21] Several books on consumer behavior (e.g., Reference [32]) discuss consumer satisfaction in detail.

manufacturer may choose to service the item in the latter case in order to retain customer goodwill. Failures that are unintentional can occur for a variety of reasons, including inadequate information concerning proper usage, a poorly written operating manual, and so forth.

#### 15.4.2.3 Fraud

Fraud arises when a customer makes a claim that is not valid. Some of the ways in which this can occur are the following:

- Tampering with the product (e.g., altering the reading of the odometer)
- False claim of sale date
- False reporting of the cause of failure
- False reporting of the date of failure

### *15.4.3 Problem Solutions*

Solutions to the three problems discussed above require different approaches. We briefly discuss each problem area.

#### 15.4.3.1 Dissatisfaction and Complaint Handling

Complaint handling is an important element of service. It is often the manufacturer's or service agent's response to a failure, rather than the failure itself, that triggers discontent which, in turn, leads to dissatisfaction. It is important that the response be perceived as just. This has a significant impact on satisfaction with the product, the service provided, and with the firm itself. The following three types of justice are involved [28]:

*Distributive justice*: This focuses on the role of "equity," where individuals assess the fairness of exchange by comparing their inputs to outcomes. It is defined as the extent to which customers feel that they have been treated fairly with respect to the final outcome. The outcomes of distributive justice can be refunds, discounts, and other remedies to compensate for product or service failure.

*Procedural justice:* This refers to perceived fairness of policies and procedures involved obtaining compensation. An example of lack of procedural justice is a situation in which the manufacturer or service agent provides a refund but the customer encounters a great deal of difficulty in obtaining it.

*Interactional justice:* This indicates the extent to which customers feel that they have been treated fairly regarding their personal interaction with service agents throughout the recovery process. Factors involved include features such as honesty, courtesy, interest in fairness, as perceived by the customer. Interactional justice is more influential in forming satisfaction with service than with the durable good itself.

#### 15.4.3.2 Misuse

One way of avoiding misuse is to make the product "idiot-proof" ("Poko Yoko" in Japanese). Revising the manual or brochure to provide better information and to educate the customers on proper usage is an effective approach to avoiding non-deliberate failures.

#### 15.4.3.3 Fraud

Some forms of fraud are easy to establish; others are more difficult. Having an effective warranty management system is critical for determining whether or not an item is still under warranty and for discouraging customers from filing claim outside warranty. Detecting tampering requires proper examination and documentation, so that the manufacturer has the necessary evidence should the customer litigate.

### *15.4.4 Illustrative Cases*

#### 15.4.4.1 Customer Dissatisfaction and Complaint Handling

Distributive justice has been used in the automobile industry to recover from problems associated with a product failing to perform as claimed. The following are a few examples:

- A Mazda RX-8 rotary-engine sports car failed to reach the advertised engine power (155 hp) causing customer dissatisfaction. Mazda attributed the drop in power (to 142 hp) to a last-minute change in engine tuning needed to meet emission rules, but needed to act quickly to prevent damage to its reputation. It offered to buy back the car at full sticker price plus taxes and other fees, irrespective of the mileage, or to provide free scheduled maintenance for the fourth-year, 50,000-mile warranty period, plus $500 [15].[22]
- Nissan claimed that its 2002 Infiniti would accelerate to 60 mph in 5.9 s. It achieved this result only in ideal weather conditions, using a light base model and lightweight driver. It did not offer any compensation to customers to counteract the resulting customer dissatisfaction [14].

---

[22] See http://www.usatoday.com/money/autos/2003-09-03-carbuyback_x.htm.

- Ford Motors recalled 8,100 Ford Mustang Cobras after owners found the engines did not produce the advertised 320 hp. Ford blamed changes in mufflers and intake manifolds for the problem and installed new ones free on nearly all Cobras [14].
- Some models produced by Honda (2000–01 Honda Odyssey minivans, 2000–2001 Honda Accord cars with automatic transmission, 2000–2001 Prelude and 2000–2003 Acura CL and TL cars) had problems with transmissions slipping out of gear or not going into gear, abruptly downshifting, or refusing to shift. Honda offered to extend the warranty from the usual three-year/36,000 miles to seven-years/100,000 miles to placate dissatisfied customers [13].[23]

## 15.5 Service Related Problems

### *15.5.1 Service Providers*

There are three possible scenarios for warranty servicing, namely:

- Service carried out by the manufacturer
- Service carried out by independent retailers (e.g., car dealers)
- Service carried out by an independent service agents

In the last two cases, there is a contractual agreement between the manufacturer and the service provider regarding payment for the warranty services carried out. It is important to note that disputes can arise, as the interests of the two parties involved are different.

### *15.5.2 Classification of Service Agent Related Problems*

Service agent related problems can be broadly grouped into the following three categories:

- Dealing with customers
- Servicing warranty claims
- Fraud

#### 15.5.2.1 Dealing with Customers

As discussed earlier, customer dissatisfaction with a product depends on the performance of the item purchased. Product performance depends on the actions of the manufacturer (decisions made during the design and manufacture of the

[23] See http://www.usatoday.com/money/autos/2002-09-20-honda-warranty_x.htm.

product) and of the customers (factors such as operating environment, usage intensity, etc.). Problems arise when the manufacturer fails to deal adequately with customer dissatisfaction. This may include lack of empathy, lack of professionalism, quality of service provided as perceived by customers, and so forth.

From a customer perspective, the main concern in warranty servicing is the time needed to carry out this activity. This, in turn, depends on the logistics of the process, which includes availability of spare parts, availability and competence of the repair crew, and so forth. Poor performance on this front leads to customer dissatisfaction with the warranty servicing provided. From the manufacturer's point of view, the two important issues in this area are 1. The effect of warranty servicing on the manufacturer's reputation, and 2. The cost of warranty servicing. If the contract specifies a fixed payment for fixing each different type of failure (as identified by labor code or defect code), then the service agent may rush the servicing, which, in turn, may not fully fix the problem and lead to additional claims and greater customer dissatisfaction. On the other hand, if the manufacturer reimburses for each failure, the service agent may be tempted to over-service, and this results in higher warranty costs to the manufacturer. Another concern of importance to the customer is the ease of obtaining service.

Other factors of importance in dealing satisfactorily with customers include:

- Stock levels for spares,
- Training of repair personnel,
- The sharing of costs between the manufacturer, vendor and the service agent,
- Ineffective servicing leading to multiple visits to have a failure repaired, and
- Poor collection and recording of warranty data.

#### 15.5.2.2 Fraud

Some of the problems in this category are the following:

- Bogus claims
- Servicing items no longer under warranty and billing the manufacturer
- Unnecessary repairs:
- Replacing components that have not failed
- Duplicate claims
- Excessive labor costs
- Use of inferior, less costly components rather than OEM-specified components.

The last of these may not only enable the repair facility to reduce warranty servicing cost, it may provide an opportunity to sell the costlier OEM parts to service failures beyond the warranty period.

### 15.5.3 Problem Solutions

As mentioned earlier, the main reason that service-agent-related problems occur is that the interests of the manufacturer and the service agent are different, and the actions of the service agent are influenced by the contract between the two parties. Agency theory provides a framework for dealing with the issues involved in devising effective contracts that reduce the likelihood of problems. A second source of difficulties in this regard is associated with issues relating to warranty logistics.[24] The best deterrent is effective data collection and monitoring to identify any anomalies in the claims rate.

#### 15.5.3.1 Agency Theory

Agency theory deals with the relationship that exists between two parties (a principal and an agent), where the principal delegates work to the agent who performs that work, and a contract defines the relationship between the two. Agency theory is concerned with resolving two problems that can occur in agency relationships. The first problem arises when the two parties have conflicting goals and it is difficult or expensive for the principal to verify the actual actions of the agent and whether or not the agent has behaved properly. The second problem involves the risk-sharing that takes place when the principal and agent have different attitudes to risk. Both of these are relevant in the context of warranty servicing by an external agent. A properly designed incentive contract overcomes some of the potential problems that can arise.[25]

### 15.5.4 Illustrative Cases

The following cases involving fraud were reported in *Warranty Week*[26]:

- A study in 1997 reported that the warranty fraud cost to the high-tech industry was $1.9 billion in 1996 and predicted that this cost would reach $3.4 billion by 2000.
- HP has more recently confirmed that 6 to 8% of its warranty claims are fraudulent and that this equates to an estimated $142 to $189 million in warranty fraud costs per year. One authorized service provider of HP submitted warranty claims using serial numbers that were secretly gathered during legitimate customer site visits. Over a 24-month period, they frequently re-used

---

[24] See [31] for a review of the issues involved in warranty logistics.

[25] For more on agency theory and incentive contracts, see [9, 44].

[26] Issue of April 19, 2005.

those serial numbers to submit additional bogus claims, resulting in a loss of $1 million to HP.

- Warranty fraud is a significant problem for vehicle service contract administrators. Surprisingly, however, the biggest culprits are usually the mechanics and repair shops, not the consumers.[27]

## 15.6 Production Related Problems

### *15.6.1 Production Process*

Production involves several stages. The process begins with raw materials that are used to produce components. The components can either be produced in-house or outsourced to external vendors who are under contract to produce the items to the manufacturer's requirements and/or specifications. The components are then assembled to produce sub-modules, modules, sub-systems, and the final product.

Depending on the volume produced, the production process may be a batch or a continuous process. Continuous production may be treated as a batch process by considering the production in a specified period of time, for example a day or shift as a batch.

### *15.6.2 Classification of Problems*

Figure 15.9 shows the different types of production related problems a company may encounter. The components supplied by the vendor may or may not conform to the required specifications. This depends on the materials used and the process, either of which may lead to a problem. Process problems arise when the process is out-of-control. The same is true for components produced in-house. The second class of production related problems is assembly errors. The following are a few examples:

- Dry soldering joints in the case of electronic products
- Misalignment in the case of transmission assembly for automobiles
- Inadequate tolerances
- Use of wrong components

[27] *Warranty Week*, April 12, 2005.

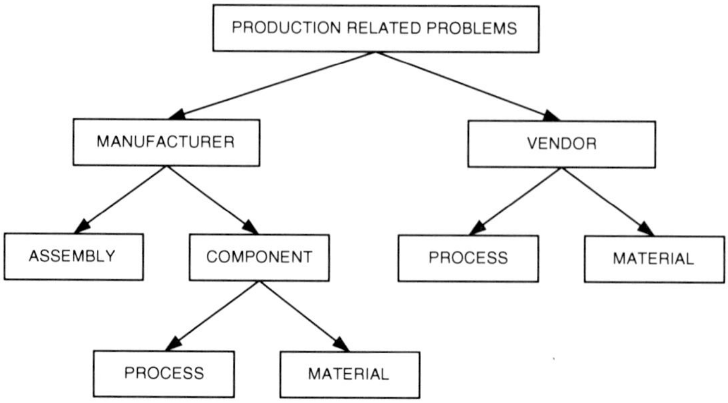


**Fig. 15.9** Production related problems

## *15.6.3 Problem Solutions*

There are several possible causes of assembly errors and component nonconformance. These include raw materials and components from vendors not meeting specifications; quality control problems with the production process (out-of-control, operators not trained properly, inadequate inspection and testing, etc.). Finding a fix to a problem of this type requires a detailed analysis to identify the problem(s) responsible for poor product performance. This process is often referred to as a "drill-down" process, in which the root causes in production responsible for problems with a particular batch are traced through a logical process, as shown in Fig. 15.10.[28]

## *15.6.4 Illustrative Cases*

In the USA, the Federal Transport Authority (FTA) monitors safety issues relating to automobiles. The TREAD legislation[29] requires manufacturers and suppliers to collect warranty data and report failures to the FTA on a regular basis. The FTA has the authority to issue recall notices when it deems that a component of an automobile is not sufficiently reliable and that the car is unsafe as a result of this.

This case study deals with seat belts for which the manufacturer changed the material used in manufacture of the belts in order to reduce costs. This decision

[28] Figure 15.10 is based on an approach proposed by professor suzuki. Reference [40] uses the same approach to solving design and service related problems with "process standard" replaced by "design standard" or "service standard," respectively.

[29] The TREAD Act was introduced into Congress as H.R. 5164, and signed into law by President Bill Clinton on Nov. 1, 2000 as Public Law 106–414.

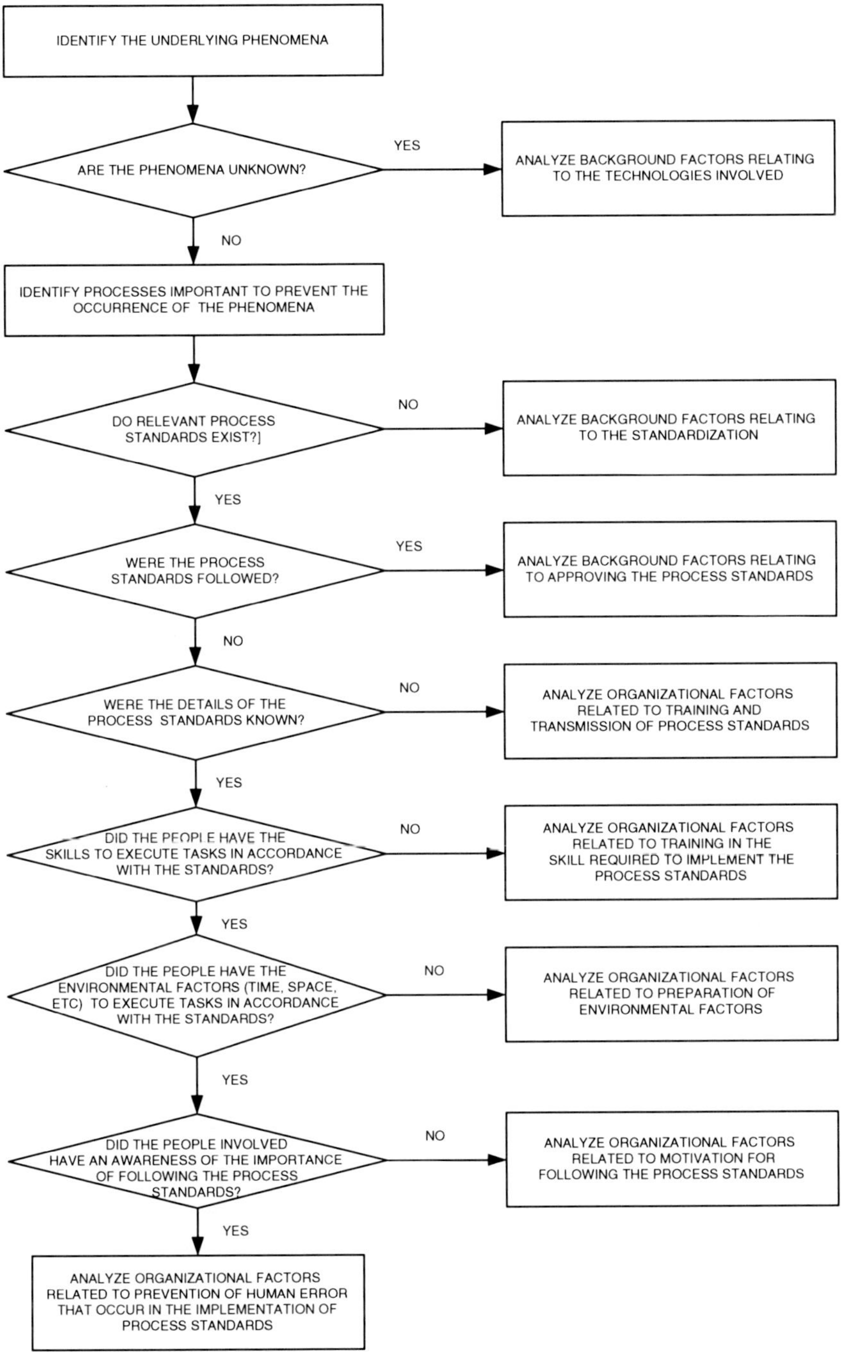


**Fig. 15.10** Methodology for fixing production process problem

was a costly mistake, as it led to large number of seat belt failures under warranty and the possibility of a recall action being issued by the FTA.

#### 15.6.4.1 Case 15.3: Seat Belt Failures[30]

Seat belts are safety devices used in cars to protect passengers in case of accidents. Takata Corporation is one of the largest seat belt manufacturers, supplying seat belts to several Japanese and US car manufacturers. In 1995, a particular brand of seat belts manufactured by Takata became a safety concern, and a mandatory recall was being considered. Since its seat belts were installed in 8.8 million cars, a recall represented a potential cost of more than a billion dollars to the company. This situation was avoided by a voluntary recall by Takata and the automobile manufacturers involved. The following excerpt is from a letter sent by Honda to the owners of the affected vehicles:

> The Reason for This Notice: Honda has determined that front seat belt buckle release buttons have broken, and others may break in the future, in some (1986–1991) Honda cars equipped with seatbelts made by the Takata Corporation. These seat belt buckle release buttons are made of red plastic, and are marked PRESS. If a button breaks, pieces may fall into the buckle assembly. If this occurs, the buckle may not operate properly, thereby creating a safety risk. To prevent this problem from occurring, Honda will replace all broken front seat belt buckles, free of charge. In addition Honda will modify all unbroken buckles manufactured by Takata to prevent future button breakage.

Under the terms of the voluntary recall, owners of affected vehicles were asked to take their cars to their dealer, who would perform an inspection and then either replace or modify the seat belts.

Failure of the seat belt is due to the failure of the "polymeric release button" that is adjacent to the slot into which the seat belt clasp fits when the belt is engaged. The purpose of the button is to release the clasped belt when pressed. The failures were the result of degradation and fracture of the polymeric release button. When fragments break away from the buttons, they can become lodged within the seat belt mechanism in a variety of locations and lead to failure of the seat belt by one or more of the following three failure modes:

- Failure mode 1: The belt fails to latch
- Failure mode 2: The latched belt will not unlatch
- Failure mode 3: The belt appears to be latched but is not.

Takata Corporation produced these buttons using injection molding of ABS (acrylonitrile–butadiene–styrene) copolymer. The failure of the release buttons involved a combination of (1) repeated, low-level impact damage and (2) degradation of the material due to the combined effects of radiation and oxidation (photo-oxidative degradation).

---

[30] This section is based on material from [15].

The industry devised short- and long-term solutions to fix the problem. The short-term solution was part of the recall process and involved a visual inspection of the front seat belt receptacles when a vehicle was returned to the dealer. If there was no breakage of the release buttons, a small plastic impact guard was inserted on each seat belt receptacle. The plastic guard installed on the seat belt receptacle is intended to reduce impact damage to the release button. If evidence of severe degradation of the release buttons was noted, the seat belt receptacles were replaced with new assemblies from the floorboards up. The long-term solution involved replacing the ABS with a more environmentally resistant polymer.

## 15.7 Design-Related Problems

### *15.7.1 The Design Process*

In its simplest characterization, design can be viewed as a two-step process. The first step is conceptual design. This is concerned with determining the product architecture—the physical arrangement of and interaction between each sub-system, assembly, sub-assembly and components, in turn. The second step deals with the detail design. This involves defining the component specifications such as forms, dimensions, tolerances and materials that will ensure the desired product performance.[31]

In some instances, designs of some of the components are outsourced. In this case, a proper flow of information between the manufacturer and the external designers is critical.

### *15.7.2 Classification of Problems*

Design problems can be grouped into the following three categories:

- Customer needs and usage profile not correctly identified
- Design process not correct
- Component specifications not correct

#### 15.7.2.1 Customer Needs and Usage Profile Not Correctly Identified

If a product does not meet customer needs, there is very little demand for it. This results in sales falling well below the target values used in making decisions during

[31] The design process also deals with the bill of materials (BOM) for production planning, all of the drawings (assembly and detail), and other production documents.

the new product development stage.[32] Product performance depends in part, on customer usage. If the usage profile assumed in the design process is not correct, higher failure rates and greater customer dissatisfaction can result, and these can have a negative impact on sales.

#### 15.7.2.2 Design Process Not Correct

Problems in this area arise if the manufacturer does not have a proper system for manage the design and development processes. This type of problem manifests itself in terms of higher WCR across many different cells for a given MIS. The reasons for this can be many and include the following:

- Not carrying out a proper design review before the design is released for production
- Details needed for the production phase are not properly documented
- Incorrect standards are used

Another factor that may lead to problems of this type is the lack of effective communication between the manufacturer and external design houses.[33]

#### 15.7.2.3 Component Specifications Not Correct

Problem of this type are mainly due to material selection and typically arises when insufficient prior testing is done. Detecting this problem requires a detailed analysis of failed components.[34] The problem can also arise because of new failure modes of which the designers were unaware during the design process.

### *15.7.3 Problem Solutions*

Solution of the problem involves further drill-down analysis using a format similar to that in Fig. 15.10. The solution can include the following:

- Upgrading the skills of design engineers (Case 16.1 is an example of this)
- Changes to the design and development process
- Changes to the current design or withdrawing the product from some or all markets

---

[32] The new product development process is discussed in Chap. 16.

[33] The most famous case of this is the well documented Ford-Firestone debacle. For an overview, see Wikipedia. Scores of articles can be found on Google.

[34] This is called "metallurgical analysis" in the case of metallic components.

## 15.7.4 *Illustrative Cases*

We discuss two cases. The first deals with leaks from oil seals in transaxles in automobiles. The data for detecting this problem came from customer complaints. Although the fraction of seals that leaked was low, the manufacturer felt that the consequence of failure would be a very significant increase in customer dissatisfaction and that the necessary fix was a change in the design of the seals. This requires first understanding the mechanism leading to seal degradation and failure and then devising a new design.

The second case describes the process used by Ford to improve product reliability. In our discussion, we focus on the process. Application to improving the reliability of automotive wheel bearings can be found in [27].

### 15.7.4.1 Case 15.4: Oil Seals for Transaxles Used in Toyota Cars[35]

The drive train in an automobile is used to transmit power from the engine to the axle. Its reliability depends on the reliability of its components. One such component is the oil seal for the transaxle. The oil seal prevents leaking of the oil and is comprised of a rubber lip molded onto a round metal casing. The rubber lip grips the surface of the shaft around its entire circumference, thus creating a physical barrier to prevent leakage. Failure of this results in oil-leaking from the transaxle, which results in high repair costs as well as high customer dissatisfaction.

Toyota uses seals produced by a component manufacturer, NOK Corporation. Prior to 1995, oil seal units having leakage were recovered and analyzed. Corrective measures were incorporated into the design, but with no permanent solution, as the failure mechanism was not fully understood. Toyota, in partnership with NOK, implemented an approach based on the TQM paradigm to fix this problem.

An important issue was the need for a proper understanding of the variation in macroscopic pressure distribution resulting from shaft eccentricity, and the variation of microscopic pressure distribution due to foreign matter in the oil. The effort to find a solution began with an analysis of failed seals returned from service agents. The analysis revealed that one of the causes of failure was the accumulation of foreign matter between the oil seal lip and the contact point of the transaxle shaft, resulting in insufficient sealing. Oil leaks were found not only during running, but also at rest. There was insufficient understanding of the root cause and no proper quantitative analysis, critical for high reliability design, was done.

The improvement process involved the following:

[35] The material for this case is based on [2] and is used with their permission.

- Clarifying the failure analysis processes through investigation of the parts.
- Carrying out a factor analysis of the oil leakage process.[36]
- Changing the design and manufacturing processes based on the new understanding.

The study of the oil leak mechanism involved looking at the different factors affecting the leak to explore the causal relationship of these with oil leakage. This revealed that the wear of the transaxle engagement part (differential case) during running increased the fine metal particles in the lubrication oil, which accelerated the wear of the oil seal lip unexpectedly, and this, in turn, caused oil leakage due to a decrease in the sealing margin of the oil seal lip. A device was developed to visualize the dynamic behavior of the oil seal lip to better understand the process. This indicated that very fine foreign matter (which was previously thought not to be impacting the oil leakage) grew at the contact section. From the component analysis, it was confirmed that the fine foreign matter was the powder produced during gear engagement inside the transaxle gear box and this, combined with the microscopic irregularities on the lip sliding surface, resulted in a microscopic pressure distribution that eventually led to degrading the sealing performance. As the amount of foreign matter increased, the oil sealing balance position of the oil seal lip moved more toward the atmospheric side and caused oil leaks at low speeds or even when at rest. This was the previously unknown information that had not been not been taken into account in the design of the oil seals.

The influence of the five dominant wear-causing factors (period of use, mileage, margin of tightening, hardness of rubber and lip average wear width) was studied by means of a two-group linear discriminant analysis using oil-leaking and non-oil-leaking parts collected in the past. The results showed that the most significant factor was the hardness of the rubber at the oil seal lip.

Based on the new knowledge, the following design modifications to improve the transaxle reliability were implemented:

- Improvement in wear resistance was achieved by increasing the gear surface hardness through changes to the gear material and heat treatment.
- The mean value of the oil seal lip rubber hardness was increased and the specification allowance range narrowed. This, in combination with improvements in oil seal lip production technology (including the rubber compound mixing process to suppress deviation between production lots), led to improving the process capability.

[36] Factor analysis is a technique in multivariate statistical analysis that is an attempt to model the variability in a large number of observed variables in terms of a smaller number of unobservable variables called factors. The objection is to reduce the data set to a manageable number of variables. See [30] or books on multivariate analysis for details.

#### 15.7.4.2 Case 15.5: Improvement Process at Ford[37]

The Ford reliability improvement process consists of the following nine phases, which may be used generally, with slight modifications, depending on the problem:

Phase 1: Initial analysis of warranty claim data to define the problem; obtain approval from top management to initiate the program; select team champion

Phase 2: Team leader sets up a cross-functional team with the expertise needed and obtains management commitment for each team member

Phase 3: Detail failure modes and identify uncontrollable or noise variables. This involves the following three steps:

- Step 1: Perform a comprehensive analysis of warranty and/or survey data
- Step 2: Review actual field failures or customer identified annoyance parts
- Step 3: Determine the best in class product

Phase 4: Analyze failure mechanisms to determine root causes of field failures by drawing on past experience and judgment and uncovering potential causes using the 8D problem solving process which involves the following 8 steps:

- D1: Establish team
- D2: Describe the problem
- D3: Develop interim containment action
- D4: Define and verify root cause and escape point
- D5: Choose and verify permanent corrective actions for root cause and escape point
- D6: Implement and validate permanent corrective actions
- D7: Prevent recurrence
- D8: Recognize team and individual contributions

Phase 5: Establish system design specification and worldwide corporate requirements compliance so that they conform to design standards, etc.

Phase 6: Develop test and establish performance of current design to reproduce failure modes; detailed documentation of bench test; bench test of the performance of the current design

Phase 7: Develop noise factor management strategy for isolating the product from the adverse affects of uncontrollable (noise) variables to obtain a product design robust to environmental variables

Phase 8: Develop new design and verify using reliability disciplines

Phase 9: Sign-off design and update lessons learned

---

[37] This case study is based on material from [27], where the approach is used to address the issues raised in complaints from customers regarding excessive noise from the bearings. The warranty data indicated that roughly 3% of the bearings suffered from this problem, and it was deemed to be a design problem. Ford reliability engineers initiated a reliability improvement process to fix this problem and reduce warranty cost and customer dissatisfaction.

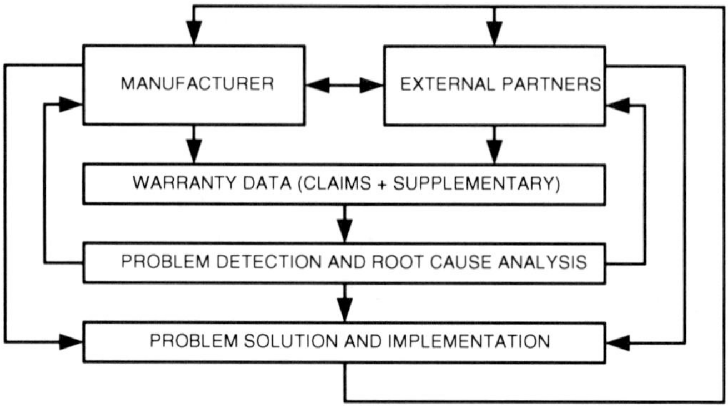


**Fig. 15.11** Feedback for improvement

Ford's reliability improvement process uses many different tools and techniques. For more details, see [27].

## 15.8 Effective Management of Continuous Improvement

Effective management of the continuous improvement process involves the following four elements:

- Proper warranty data collection
- A methodology or approach for detecting problems for improvement.
- A warranty management system
- An organizational structure to carry out the improvement process.

The first element is discussed in Chaps. 4 and 5 and the second in the earlier sections of this chapter. In this section we focus on the warranty management system. The last element is discussed in Chap. 16.

### *15.8.1 Warranty Management System*

A proper warranty management system is needed for using warranty data to organize and implement improvements to current products and operations. The system should have a closed loop structure as shown in Fig. 15.11, and must have features that will assist in decision-making.

The purpose of the warranty management system is to assist in the improvement process. The key modules of an effective warranty management system are the following:

- Automatic Claims Processing Module: Automate all aspects of the claims business process, including submission, authorization and, payment and supplier cost recovery.

- Operational Data Storage (ODS): A unified location for storage of relevant data from diverse and disparate sources.
- Parts Traceability: To provide comprehensive parts tracking and tracing throughout the product lifecycle.
- Service Parts Management: To optimize the inventory, order and distribution of service parts, taking into account costs and desired service levels.
- Early Warning System: Analytic tools to detect problems for improvement. This consists of advanced statistical and mathematical algorithms to detect anomalies and trends from multiple layers of data and to provide clear visualization of specific problem areas across the warranty chain.
- Warranty Dashboard and Key Performance Indicators: For detailed reporting in an easy-to-read format for decision-making at all levels.

There are many commercial software packages that will perform some of these functions available on the market. Most focus only on automated claim processing to minimize warranty cost through detection of fraud. These packages are appropriate for businesses in Stage 1 of warranty management, as discussed in Sect. 2.12. There are a few packages that deal with the use of warranty data for product improvements as discussed in this chapter, and contain most of the modules discussed above. See [10, 17, 18] and [38] for more details. See *Warranty Week* for more recent information.

## References


1. Ackoff RL (1978) The art of problem solving. Wiley, NY
2. Amasaka K, Osaki S (2003) Reliability of oil seal for transaxles–a science SQC approach at toyota. In: Blischke WR, Murthy DNP (eds) Case studies in reliability and maintenance. John Wiley and Sons, NY
3. Andersen B, Fagerhaug T (2006) Root cause analysis. Quality Press, Milwaukee, Wisconsin
4. Archer NP, Wesolowsky GO (1996) Consumer response to service and product quality: a study of motor vehicle owners. J Oper Manag 14:103–118
5. Barlow RE, Fussell JE, Singpurwalla ND (1975) Reliability and fault tree analysis. SIAM, Philadelphia
6. Batson RG, Jeong Y, Fonseca DJ, Ray PS (2006) Control charts for monitoring field failure data. Qual Reliab Eng Int 22:733–755
7. de Bono E (1970) Lateral thinking. Harper and Row Pub, NY
8. Deming WE (1989) Out of the crisis. MIT Pres, Cambridge, Mass
9. Eisenhardt KM (1989) Agency theory: assessment and review. Acad Manag Rev 14:57–74
10. Entigo (2003) Identifying and reducing warranty costs in your organization. Entigo white paper series
11. Flanagan JC (1954) The critical incident technique. Psychol Bul 54:327–358
12. Harrison AF, Bramson RM (1982) The art of thinking. The Berkeley Pub Group, NY
13. Healey JR (2002) USA Today, Sept. 20
14. Healey JR (2003) USA Today, Sept. 3
15. Henshaw JM, Wood V, Hall AC (1999) Failure of automobile seat belts caused by polymer degradation. Eng Fail Anal 5:13–25
16. Homburg C, Rudolph B (2001) Customer satisfaction in industrial markets: dimensional and multiple role issues. J Bus Res 52:15–33

17. HP (2006) HP warranty solutions. A white paper from HP (www.hp.com/go/manufacturing/warranty)
18. IBM Automotive (2004) Component business modeling. IBM Global Business Services
19. Imai M (1986) Kaizen. McGraw-Hill, NY
20. Ishikawa K (1990) Introduction to quality control. 3A Corporation, Tokyo
21. Iskandar BP, Blischke WR (2002) Reliability and warranty analysis of a motorcycle based on claims data. In: Blischke WR, Murthy DNP (eds) Case studies in reliability and maintenance. John Wiley and Sons, NY
22. ISO-8402 (1986) Quality vocabulary. International Standards Organization, Geneva
23. Jolibert AJP, Peterson RA (1976) Causal attributions of product failure: an exploratory investigation. J Acad Mark Sci 4:446–455
24. Kelly HH (1973) The process of casual attribution. Am Psychol 28:107–128
25. Lee SH, Kim CM, Yeom SJ, Kim G, Moon K, Kim BG (2007) A study on development of time series warning module in warranty claims database. Proceeding of 6th international conference Adv language proc web informaton technology, IEEE computer society pp 553–558
26. Majeske KD, Lynch-Caris T, Herrin G (1997) Evaluating product and process design changes with warranty data. Int J Prod Econ 50:79–89
27. Majeske KD, Riches MD, Annadi HP (2003) Ford's reliability improvement process–A case study on automotive wheel bearing. In: Blischke WR, Murthy DNP (eds) Case studies in reliability and maintenance. John Wiley and Sons, New York
28. Maxham JG III, Netemeyer RG (2002) Modeling customer perceptions of complaint handling over time: the effect of perceived justice on satisfaction and intent. J Retail 78:239–252
29. Mears P (1995) Quality improvement tools and techniques. McGraw Hill, NY
30. Mulaik SA (2009) Foundations of factor analysis. Chapman and Hall, London
31. Murthy DNP, Solem O, Roren T (2004) Product warranty logistics: issues and challenges. Euro J Oper Res 156:110–126
32. Neal C, Quester P, Hawkins D (1999) Consumer behavior. McGraw Hill, NY
33. Newman V (1995) Problem solving for results. Gower Pub. Ltd., Hampshire, UK
34. Oakes D (2009) Root cause analysis: the core of problem solving and corrective actions. Quality Press, Milwaukee, WI
35. Ohno T (1988) Toyota production system: beyond large scale. Productivity Inc., Portland, OR
36. Oliver RL (1996) Satisfaction: a behavioral perspective on the consumer. McGraw Hill, NY
37. Sander PC, Toscano LM, Luitjens S, Petkova VT, Huijben A, Brombacher AC (2003) Warranty data analysis for assessing product reliability. In: Blischke WR, Murthy DNP (eds) Case studies in reliability and maintenance. John Wiley and Sons, NY
38. Sprague BM (2005) An end-to-end strategy for improved warranty management capabilities. White Paper, Accenture
39. Stamatis DH (2003) Failure mode and effect analysis. ASQ Quality Press, Milwaukee, WI
40. Suzuki K (2008) A map for assurance of reliability and safety for industrial products. J Jap Soc Qual Cont 38:405–412 (in Japanese)
41. Tague NR (2004) The quality toolbox. Quality Press, ASQ, Milwaukee, Wisconsin
42. Teng SG, Ho SM, Shumar D (2005) Enhancing supply chain through effective classification of warranty returns. Int J Qual Reliab Manag 22:137–148
43. Thomas DA, Ayers K, Pecht M (2002) The trouble not identified phenomena in automotive electronics. Microelectro Reliab 42:641–651
44. van Ackere A (1993) The principal/agent paradigm: its relevance to various fields. Euro J Oper Res 70:88–103
45. Wu H, Meeker WQ (2002) Early detection of reliability problems using information from warranty databases. Technometrics 44:120–133
46. Zeithaml VA, Parasuraman A, Berry LB (1990) Delivering quality service. The Free Press, NY

# Chapter 16
# Role of Warranty Data in New Product Development

## 16.1 Introduction

Modern industrial societies are characterized by the introduction of new products into the marketplace at an ever increasing pace. Some of the reasons for this are 1. Rapid advances in technology, 2. Increasing consumer expectations, and 3. Global competition. With each new generation, the complexity of the product ordinarily increases and this, in turn, has implications for product reliability. Customers need assurance that the product purchased will perform satisfactorily and, as discussed in Chap. 2, warranty provides this assurance. Offering warranty results in additional costs, as discussed in Chaps. 6 and 7. This, combined with customer dissatisfaction if the reliability of the item is not adequate, has an impact on overall profitability and, finally, on the share value of the company on the stock market.

Product reliability depends on decisions made during the design, development and production phases of the new product development process. In most businesses, reliability decisions are made independently of warranty decisions during the early stages of new product development, and warranty decisions are made prior to launch of the product. Since reliability and warranty are closely linked, they must be considered jointly for effective management of both and to assure optimal profitability.

In this chapter, we begin with a historical perspective with regard to warranty decisions on new products. This is discussed in Sect. 16.2. We define three epochs and discuss the approaches used in each of them. The last epoch looks at new product warranty decision making from a strategic management perspective. This involves formulating strategies for each of the different phases of the new product development process, as part of the overall business strategy. The starting point is an understanding of the different phases of business strategy, and the decisions that must be made in each phase. This is discussed in Sect. 16.3. Section 16.4 looks at the formulation of strategies and highlights the role of warranty claims data and

W. R. Blischke et al., *Warranty Data Collection and Analysis*,
Springer Series in Reliability Engineering, DOI: 10.1007/978-0-85729-647-4_16,

supplementary data in this process. We conclude with a brief discussion of warranty management in Sect. 16.5.

## 16.2 Deciding on New Product Warranty

One can define three distinct epochs with regard to warranty decision making for new products. These are:

First Epoch: Pre 1960
Second Epoch: 1960–2000
Third Epoch: Post 2000

### 16.2.1 The First Epoch

Prior to about 1960, manufacturers used warranty primarily to protect their interests.[1] Warranty periods were short (for example, an automobile warranty was typically for only 90 days) and the aim was to discourage customers from exercising warranty claims. This period saw the introduction of warranty legislation—the Uniform Commercial Code (around 1920) and, in turn, subsequent legislation in the second epoch, such as the Magnuson Moss Act (1975), Lemon Laws passed by many states in the U.S, and the Tread Act (2002).

Warranty was simply viewed as a cost to be avoided as much as possible, and the multiple roles of warranty and the link between warranty and reliability was not fully appreciated. Warranty was an afterthought, decided prior to the launch of a product and done on an ad-hoc basis.

### 16.2.2 The Second Epoch

During this epoch, warranty was viewed as part of a marketing strategy and warranty decisions were made by the marketing staff. Warranty, from a marketing perspective, was viewed as one or more of the following:

1. A promotional tool to inform customers with regard to product reliability,[2]
2. A marketing tool to differentiate the company's product from those of competitor's, and
3. A way to increase sales by offering better warranty terms.

[1] This is the basis for the "exploitative theory" of warranty proposed by [27].

[2] This is the basis for the "signaling theory" theory of warranty proposed by [29] and leading to a considerable amount of subsequent research papers in economic journals.

**Table 16.1** Warranty program mix for offensive and defensive warranty strategies

| Elements | Strategy | |
|---|---|---|
| | Offensive | Defensive |
| Warranty type | FRW | PRW |
| Warranty length | Medium/long | Short/medium |
| Warranty breadth | Broad | Narrow |
| Product scope | All components covered over the warranty period | Some components not covered for the whole period |
| Market scope | Global | Limited (country, region) |
| Coverage | Parts and labor | Parts or labor |
| Conditions | Loose | Strict |

Warranty strategy was formulated prior to launch of the product.[3] Reference [16] proposed offensive and defensive approaches to warranty strategy formulation. The elements involved in each of these, also called the "warranty program mix," are as indicated in Table 16.1.

Elements of the warranty program mix are as follows

- Warranty type: FRW, PRW or Combination; renewing or non-renewing.
- Warranty length: Duration of the warranty period.
- Warranty breadth: This defines the extent to which a product is warranted. Coverage may vary from full to limited. There many examples of the latter in which some components of the product are covered for a longer period and the rest for a shorter period (e.g., the compressor in a refrigerator, the picture tube in a television, the drive train in an automobile).
- Product scope: This deals with base warranty offered with all products and extended warranties (or service contracts) for some, sold at a higher price.
- Market scope: This is relevant when the product is sold in different markets (for example, different countries). In this case the warranty must take into account local warranty legislation, consumer heterogeneity, etc.
- Coverage: This deals with issues such as whether material, parts, or both are covered under warranty and whether there are any deductibles.
- Conditions: These state the conditions to be met in order to invoke a warranty claim. An example is that consumers must register their product within a specified period subsequent to the purchase.

Reference [25, 26] looks specifically at extended warranties and their role in the segmentation of the market based on consumer heterogeneity. The managerial implications are similar to those discussed above but the focus is on providing a

[3] Reference [32] was one of early papers dealing with warranty as an element of competitive strategy. Since then, several papers have appeared on this topic in the marketing literature. See the review by [18] for examples.

clear understanding of the role of the base warranty and the segmentation role of the extended warranty.

In the Second Epoch, warranty was still viewed as an afterthought, decided prior to the launch of a product, and it played no role in the design, development and production of the product. Warranty decisions were made based on competitors' actions without fully understanding the cost implications, sometimes leading to serious fiscal consequences.

### *16.2.3 The Third Epoch*

Reference [17] marks the beginning of the third epoch. In this, warranty on a new product is based on strategic management and the product life cycle perspective. Strategic management involves the following five steps[4]:

- Step 1: Formulating a Mission statement—Long term vision for the company
- Step 2: Translating the Mission into long and short term business objectives (targets for performance)
- Step 3: Crafting a strategy to achieve the targeted performance
- Step 4: Implementing and executing the chosen strategy
- Step 5: Constant reviewing and initiating corrective actions as needed

The basic underlying principle is that decisions with regard to warranty for a new product must begin at the Front-end phase of the new product development process (and not as an afterthought just prior to the launch stage) and linked effectively with the decisions in subsequent phases.

This begins with the business strategy adopted at the Front-end phase. Goals for the new product are then formulated based on targets such as (1) Total revenue, (2) Return on investment, (3) Market share, and so forth. These, in turn, define both technical and commercial targets for subsequent phases, including design & development; production; marketing and post-sale support, of the new product development process, and the formulation of strategies to achieve them. The outcome of the strategy formulation process is the overall business strategy, in which all of the lower level strategies, as shown in Fig. 16.1, are cohesive and well integrated.

Typically a company in its long range planning will be considering a number of possible new products or product lines as a part of its overall business strategy. From this will emerge a set of goals with regard to new product development. These goals may be quite broad, but will ordinarily include market share, profit objectives, per unit production cost, and so forth.

In planning to achieve these goals, strategies for addressing the technological issues, as shown of the left side of Fig. 16.1, and the commercial issues, shown on the right side, must be developed. From the point of view of warranty strategy, technology issues include the engineering aspects of product design and

---

[4] Details of these can be found in most books on strategic management. See for example, [31].

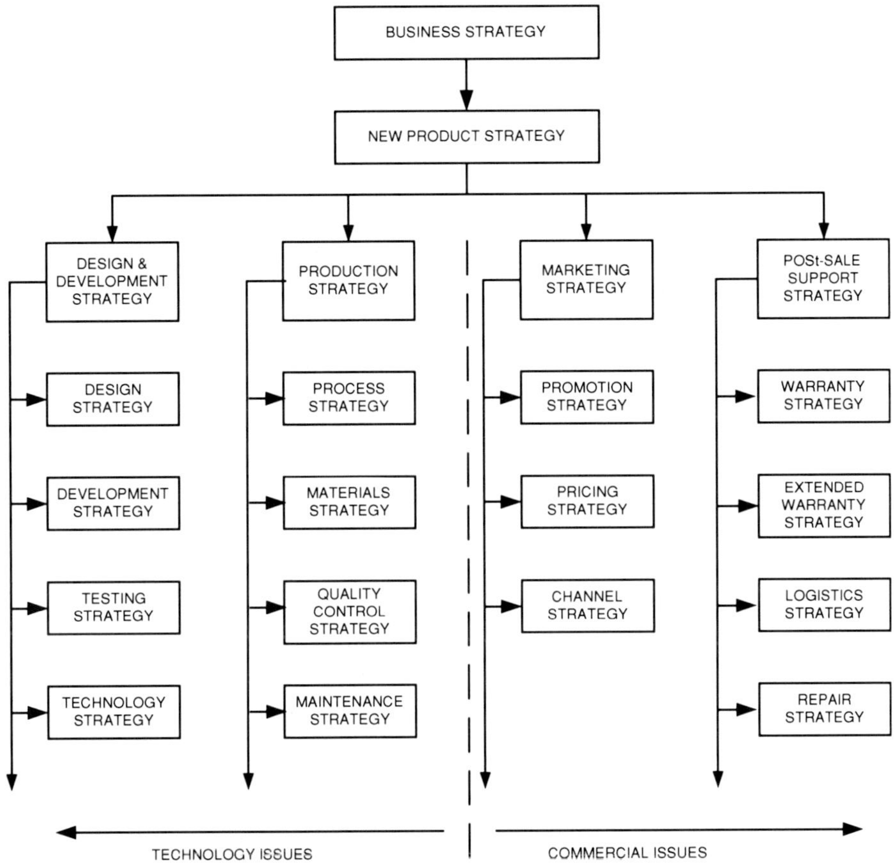


**Fig. 16.1** Hierarchy of strategies

manufacturing.[5] Commercial issues in this context involve marketing and servicing aspects. To be effective, these strategies must be coherent and integrated.

### *16.2.4 Data and Information*

In the first epoch, warranty data collection was limited, data analysis was virtually non-existent and the data played no role in warranty decisions for new products. In the early period of the second epoch, data collected were mainly warranty claims data (see Chap. 4). This information was stored on punch cards, and used primarily

[5] For more on technology management, see [4] and [6].

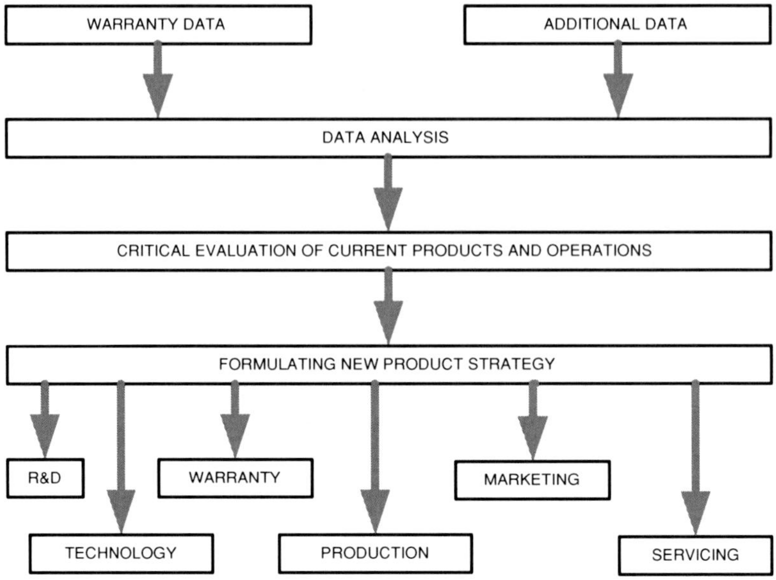


**Fig. 16.2** Data collection and analysis for strategy formulation

for monitoring and controlling fraud. Toward the end the epoch, data collection involved both warranty claims data and warranty supplementary data (see Chap. 5), which were used in the continuous improvement process, as discussed in Chap. 15.

The strategic management approach for deciding on the warranty to be offered on a new product requires not only the warranty data (claims + supplementary data), but also some further data and information, which we shall refer to as "Additional Data". Figure 16.2 shows the data collection and analysis process needed for strategy formulation.

Additional Data are grouped into the following categories:

*Technology Related*. These relate to advances in technology, including new materials, new design procedures, new processes, etc. The main sources for such data are:

- Scientific and technical journals
- Technical conference proceedings
- Reports from research laboratories
- Patent office

*Market Related*. This includes data and information relating to competitors and consumers. Competitor-related items include product attributes and characteristics,[6]

[6] A product is designated by its characteristics and attributes. The distinction between these two is best explained by the statement "product characteristics physically define the product and influence the formation of product attributes; product attributes define consumer perceptions and are more abstract than characteristics" [30]. Consumers view products in terms of attributes whereas design engineers are more concerned with characteristics.

production processes, marketing factors such as price, warranty, promotions, etc. Consumer-related items include expectations and satisfaction with the products on the market, consumer needs, etc. Some competitor-related data are available freely in the public domain (annual reports, brochures for promotion, etc.) and others may be gathered by indirect means. Sources for consumer-related data are consumer reports and well planned consumer surveys (especially for identifying needs).

*Industry Related.* These data are intended to provide snapshots of the industry as a whole. The main sources are the annual reports from various government agencies (indicating total sales, trends, etc.) and industry magazines.

*Historical.* The data and information can be either internal (relating to earlier products, processes, etc.) or external (obtained from archives).

*Other Data.* These include economic and financial data that may be obtained from periodic reports from banks and treasury departments in different countries, new legislation relating to the product, environment, etc.

Further discussion on data and information sources can be found in [9, 19].

### 16.2.5 Current Status

There are a few manufacturing businesses that use the first epoch approach for determining the warranty for new products. Nearly all manufacturers decide on the warranty for their new products based on the second epoch approach. Both of these approaches are reactive, as warranty is viewed as an afterthought and the decision is made just prior to launch of the product. A few manufacturers are recognizing the need for the third epoch approach, based on strategic management, for deciding on the warranty for their new products. This approach, in contrast to the earlier two approaches, is a proactive approach, with warranty issues being on addressed in all phases of the product life cycle. The focus is on formulating a warranty strategy that is part of the overall business strategy and is compatible with other functional strategies.[7]

### 16.2.6 An Illustrative Case [Automobile Warranty]

Ordinarily, the warranty for an automobile covers all parts of the vehicle (called bumper-to-bumper coverage). Warranty coverage varies from manufacturer to manufacturer and by brand. The power train, which includes the engine, transmission and other parts of the drive train only, is usually covered by a longer warranty. Warranty terms have changed significantly over time. The following is a brief history:

[7] For a case study on warranty planning and development framework in a high-tech firm, see [7].

- Circa 1930: Coverage was for 90 days
- Until 1960: Basic warranty was 3 months or 4000 miles, whichever came first
- 1960: Increased to 1 year or 12,000 miles.
- 1962: Increased to 2 years or 24,000 miles on most brands.
- 1962: Chrysler increased its warranty to 5 years or 50,000 miles in an attempt to reverse its sagging sales. Warranty costs increased dramatically and Chrysler was forced to reduce its warranty terms to those of its competitors. This led to a considerable amount of consumer dissatisfaction.[8]
- 1969: The auto industry reduced the warranty on automobiles to 1 year or 12,000 miles, and an additional warranty on the power train (which includes engine, transmission and other parts of the drive train) was made optional.
- 1981: The power train warranty increased to 2 years or 24,000 miles in most of the industry. Chrysler increased it to 5 years or 50,000 miles.
- 1987: GM increased power the train warranty from 3 years and unlimited miles to 6 years or 60,000 miles and warranty protection against corrosion was increased from 3 years and unlimited miles to 6 years or 100,000 miles.

In 2002, Suzuki offered a 7 years, 10,0000 mile warranty on the power train and a 3 years, 36,000 mile bumper-to-bumper warranty. A report in the *New York Times*[9] commented that it was the best auto warranty then available. The *Times* also reported that "Hyundai sales have quadrupled over the last four years because of affordable prices and a 10 years or 100,000 miles warranty".

## 16.3 New Product Development Process

The US based Product Development & Management Association (PDMA) defines New Product Development (NPD) as

> A disciplined and defined set of tasks and steps that describe the normal means by which a company repetitively converts embryonic ideas into saleable products or services [5]

The NPD process begins either with an idea to build a product that meets specific needs defined by customers and/or the manufacturer, or to create new needs for a radically innovative product, and ends when the product is launched on the market. The process involves several phases. Many different models have been proposed for the NPD process. The number of phases in the process and the descriptions of the phases vary from model to model. Reference [20], gives a comparison of some of these models and proposes the following eight-phase model involving three stages and three levels which includes the NPD process. This is shown in Fig. 16.3. A brief description of the stages, levels and phases is as follows:

---

[8] See [10] for further discussion as well as additional references.

[9] *New York Times*, August 4, 2002.

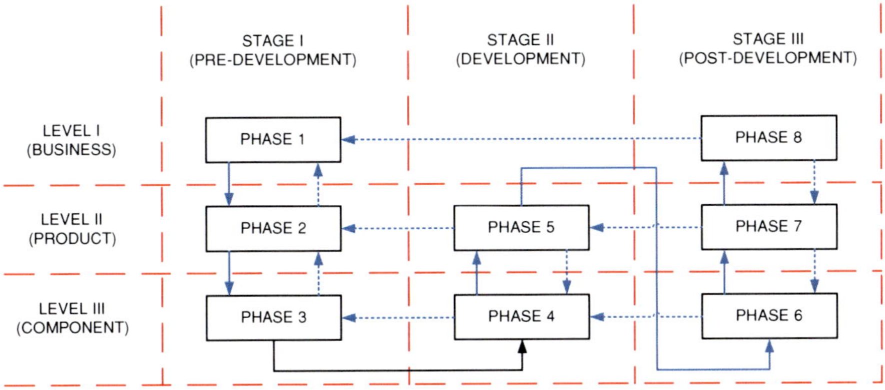


**Fig. 16.3** Eight-phase new product development process

### 16.3.1 Stages

- Stage-I [Pre-development]: This stage is concerned with a *non-physical,* abstract conceptualization of the product, with increasing levels of detail.
- Stage-II [Development]: This stage deals with the physical embodiment of the product through R&D and prototyping.
- Stage-III [Post-development]: This stage is concerned with the remainder of the product life cycle (e.g., production, sale, use) subsequent to the NPD.

### 16.3.2 Levels

- Level-I [Business level]: This level is concerned linking the business objectives for a new product to desirable product attributes.
- Level-II [System (Product) level]: This level links product attributes to product characteristics. The product is treated as a black-box.
- Level-III [Component level]: This level is concerned with linking product characteristics to lower level component characteristics, at an increasing level of detail.

### 16.3.3 Phases

*Phase 1* [*Stage-I Level-I*]: In this phase, the need for a new product is identified and decisions regarding product attributes (customer's view of product) are made in the context of the overall strategic management of the business.

*Phase 2* [*Stage-I Level-II*]: In this phase, the product attributes are translated into product characteristics (engineer's view of product).

*Phase 3* [*Stage-I Level-III*]: In this phase the detail design, proceeding from product to component, is carried out. The objective is to arrive at a set of specifications that will ensure that the product has the required characteristics.

*Phase 4* [*Stage-II Level-I*]: This phase deals with product development, proceeding from component to product, and ending up with a product prototype.

*Phase 5* [*Stage-II Level-II*]: In this phase, the prototype is released to a limited number of consumers to evaluate a customer's assessment of product features.

*Phase 6* [*Stage-III Level-III*]: This phase deals with production of products, starting from component and ending with the product for release to customers.

*Phase 7* [*Stage-III Level-II*]: This phase looks at field performance of the product, taking into account the variability in usage intensity, operating environment, etc., from the perspective of the customer.

*Phase 8* [*Stage-III Level-I*]: Here the performance of the product released for sale is evaluated from an overall business perspective.

The NPD process is an iterative process in which it is often necessary to return to an earlier phase (indicated by dotted lines in Fig. 16.3). For more details, see [20].

## 16.4 Formulating a New Product Development Strategy

### *16.4.1 Framework*

A proper framework is needed for formulating new product strategies. Figure 16.4 shows the key elements of the process and the interactions between them.

The elements can be grouped into the following categories:

*Business Related Elements*

- Objectives
- Reputation of products and of business
- Stock value
- Market share
- Investments

*Technology Related Elements*

- Product development
- Process development
- Technology acquisition

*Market Related Elements*

- Marketing (promotion, advertisement, etc.)
- Sale price

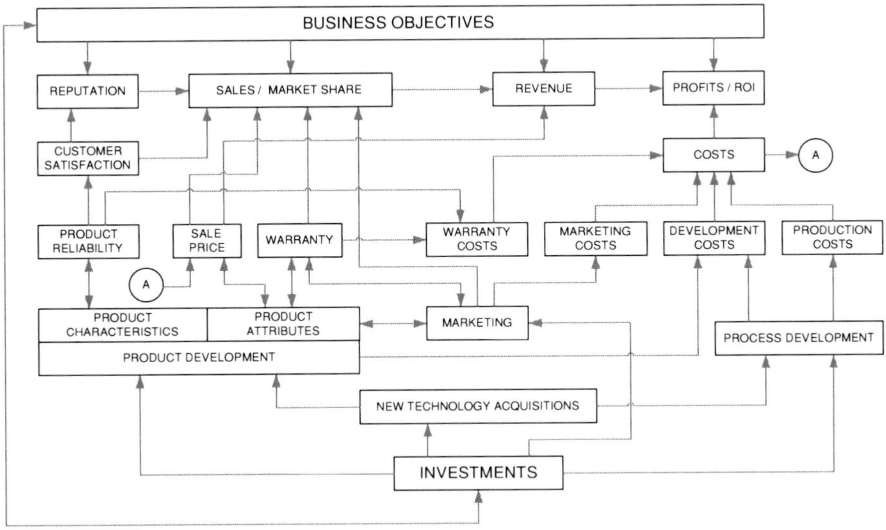


**Fig. 16.4** Framework for formulating new product warranty strategy

- Warranty/extended warranty
- Sales

*Cost Related Elements*

- Product development
- Process development
- New technology acquisition
- Production
- Warranty costs

For further discussion of this framework, see [21].

## 16.4.2 Some Complicating Factors

The two main complicating factors are 1. Outsourcing of design, production and/or servicing, and 2. Flexible warranties. Both of these factors bring external parties into the strategy formulation process, as indicated in Fig. 16.5 and discussed below.

*Outsourcing.* The actions of external parties impact business performance. If the externally designed components are poor or have design faults, the reliability of the product is lowered, resulting in higher warranty costs, greater customer dissatisfaction and a negative impact on product and business reputations. If components bought from external component manufacturers have quality related

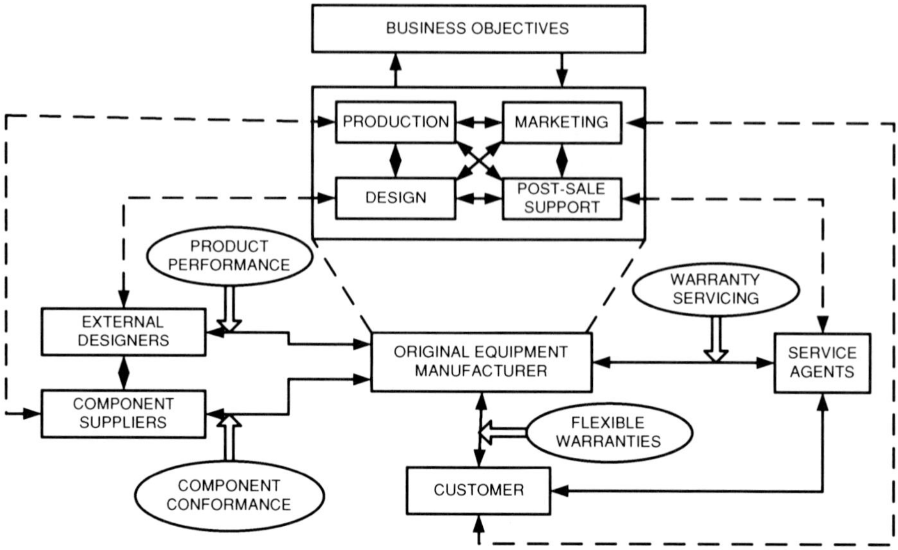


**Fig. 16.5** Outsourcing of operations

problems, the effect is the same. This can be controlled through better quality control strategies, but this again results in additional costs.

These additional costs need to be shared between the OEM—the brand name on the product — and either the component manufacturer, the external designers, the component suppliers, or some combination of these parties. The question of responsibility for product failure is further complicated by the existence of a chain of suppliers and sub-suppliers. Failures should be governed by a workmanship warranty.

Warranty servicing is often carried out by a service agent, e.g., the retailer or some other external agent. Warranty costs depend on how efficiently the service agent carries out this task. Poor servicing can lead to higher warranty costs and greater customer dissatisfaction.

All of this implies that a manufacturer must have proper contracts with all of the various external parties. These include reliability improvement warranty contracts with external designers, group warranty contracts for items bought from component suppliers, and service contracts with external service agents.[10]

*Flexible Warranties*. Customers vary in terms of the type of assurance that they require for the product they buy. Some prefer not to purchase extended warranties, whereas others prefer to choose from a range of warranty possibilities. The manufacturer must have the capability of analyzing different warranty structures

[10] Contracts involving multiple parties with different interests and objectives are the concern of *agency theory*. For more on this, see [1, 12].

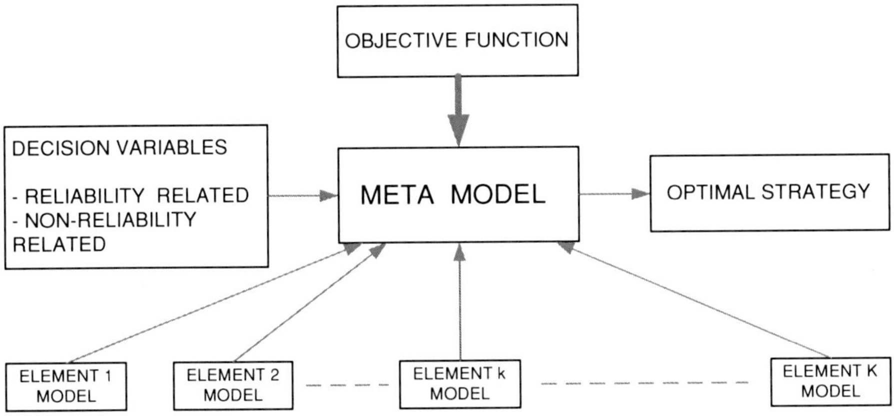


**Fig. 16.6** Meta-model for strategy formulation

for different customers. Significantly, there exists a high revenue potential in extended, flexible and transferable warranties.

### *16.4.3 Role of Models*

The strategic management approach requires deciding on product characteristics and attributes that will achieve the stated business objectives. This translates into selecting the best option from a range of alternative options in each of the eight phases of the NPD process. Mathematical models play a vital role in this and both warranty and additional data are very critical in building the models. As shown in Figs. 16.4 and 16.5, the number of elements that must be incorporated into the model depends on the objective function, which, in turn, defines the decision variables that need to be selected optimally as part of the strategy formulation. This implies that it is necessary to build a meta-model that incorporates the models for the various elements (numbered 1 through $K$) that are relevant, as indicated in Fig. 16.6.

A large number of models have been developed for the various elements of Figs. 16.4 and 16.5. The structure and complexity of the models vary from simple to complex. The selection of an appropriate model is a challenging task, requiring a trade-off between complexity and reality, and taking into account the data requirements for estimating the model parameters. We give a short list of references where additional details of the models are given.

- Reliability: [9]
- Warranty cost analysis: [8]
- Customer Satisfaction: [19]
- Reputation: [28]

- Marketing: [13]
- New product sales: [15]
- Product development: [33, 34]
- Life cycle costs: [2]
- Cost estimation: [3, 14]

## 16.5 Use of Warranty Data in Strategy Formulation

In Chaps. 4 and 5, we looked at warranty data collection. Warranty data play the following two important roles in the context of strategy formulation:

1. Assessing the performance of current products at three different levels: 1. Business level, 2. Product level and 3. Component level.
2. Updating the models and knowledge base.

Chapters 8 through 14 dealt with assessing product performance at each of the three levels mentioned above. In this section, we focus our attention on the use of warranty data in making decisions, as part of strategy formulations, at the different phases of the new product development process. We indicate the nature of the decision problems and the role of warranty data in selecting the best options.[11]

### *16.5.1 Phase 1 [Feasibility]*

The business strategy adopted at the Front End phase determines the goals for the new product through targets such as 1. Total revenue, 2. Return on investment, 3. Market share, and so on, as indicated in the top second line of Fig. 16.4. This in turn leads to defining targets (technical and commercial) for the subsequent phases discussed in Sect. 16.3, and formulating strategies to achieve them in a manner so that all the lower level strategies are cohesive and well integrated. Figure 16.7 indicates this in a schematic manner and also shows some of the other factors that need to be taken into account.

The process begins with the analysis of warranty data at the business level. This yields an assessment of the current product performance in terms of the following:

- Customer dissatisfaction with the product (% of dissatisfied customers at different dissatisfaction levels)
- Warranty cost per unit

[11] The data and information aspects for the different phases of the NPD process are discussed in [20].

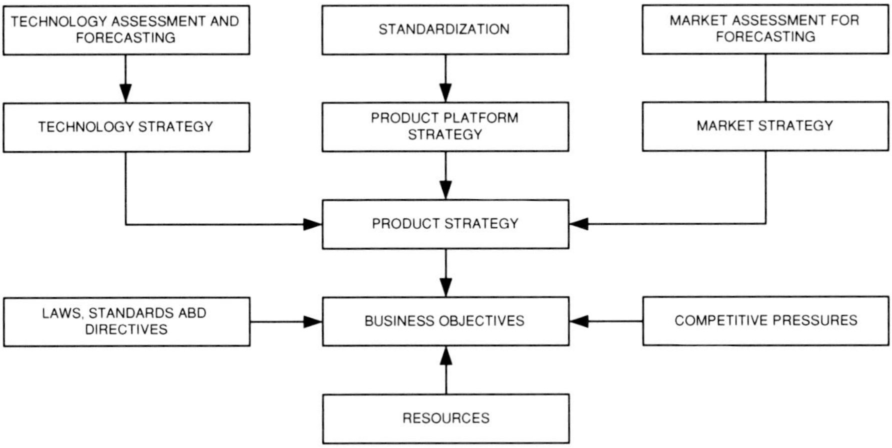


**Fig. 16.7** Factors influencing business objective

- Number of changes made to the production process (as part of the improvement process discussed in Chap. 15)

These measures are then compared with the initial target values to identify the magnitude of any gaps and their technical implications (e.g., research and development needed) and commercial implications (e.g., warranty costs) for the new product. In this connection, an important part of decision making is an evaluation of the investment in product reliability [21].

## *16.5.2 Phases 2 and 3 [Design]*

Phase 2 deals with conceptual design. In this, decisions are made regarding product attributes (e.g., sale price, warranty, etc.) and product characteristics (e.g., product reliability) that will ensure that the product will achieve the desired business targets defined in Phase 1. Decisions must be made regarding product variety, level of reliability for the different products, etc. Several scenarios are considered in [22]. The meta-model for one of these is shown in Fig. 16.8. For this model, the associated decision variables are:

y: The scale parameter of the failure distribution for the product
W: Warranty period
P: Unit sale price

Phase 3 deals with detail design. The decisions to be made are the component specifications that will ensure that the product meets the targets for its attributes and characteristics defined in Phase 2. The process leading to component

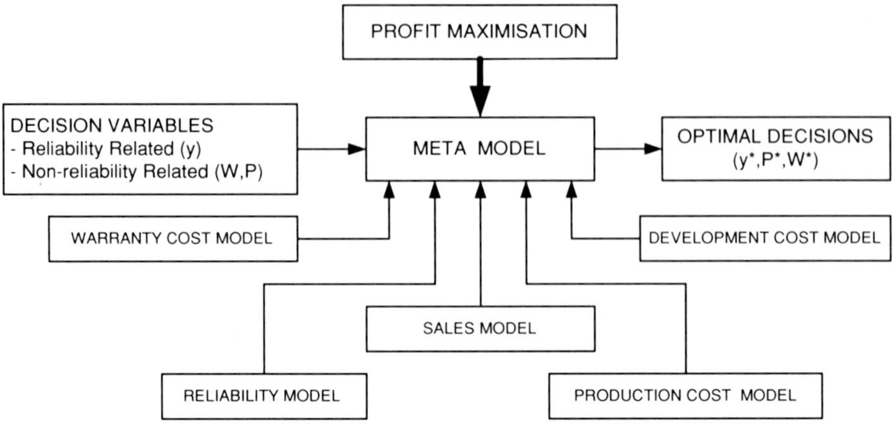


**Fig. 16.8** An illustrative scenario for decision making in Phase 2

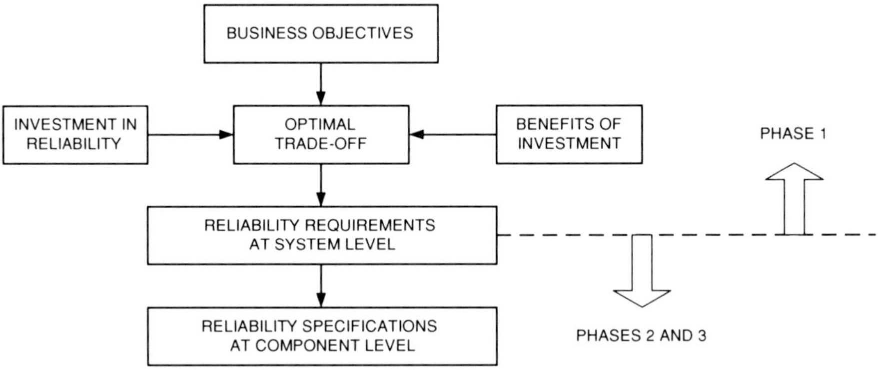


**Fig. 16.9** Investment in product reliability

reliability specification involves comparison of alternative ways of ensuring that the desired component level reliability and optimal decision making occur [23].

In the context of product reliability, the links between the decisions made in Phases 1–3 are shown in Fig. 16.9.

Warranty data and its analysis at both product and component level are important for the design process. Component level analysis provides information regarding the reliability of different components in the current product and on whether or not they should be used in the new product as is or with some specified improvements. The customer usage profile and operating environment provide information for designing more robust products. The gap between the desired reliability (at product, component and intermediate levels) and the reliability of existing products, which may be estimated using warranty data, allows one to evaluate the resources of time, cost, personnel, etc., needed for research and development.

### 16.5.3 Phases 4 and 5 [Development]

The development phase involves testing an item at the component, product or some intermediate level, to failure; identifying the root cause and making design changes to eliminate the cause for failure or the failure mode.

Warranty data provide information about failure modes that must be eliminated. In addition, the data help in evaluating the testing program used for current products and, in conjunction with design changes, provide information for the testing and development program for the new product.

### 16.5.4 Phase 6 [Production]

Phase 6 deals with implementation of the production process for manufacturing the new product. Warranty data relating to production problems on earlier products (discussed in Chap. 15) are useful for initiating changes to the production process and for the selection of component vendors. This information is also useful for making changes to the quality control process to be employed internally to ensure the desired quality, and for making necessary changes to contracts with external suppliers.

### 16.5.5 Phase 7 and 8 [Post-sale]

Phase 7 deals with product performance after launch and the logistics required to service warranty claims. Warranty service logistics involves several elements—call centers, warehouses for storing spares, location of repair facilities, out-sourcing of service, transportation of material, etc. [24]. Warranty data provide useful information regarding the existing warranty servicing processes, the performance of service agents, etc., so that changes can be made to these for the new product, if necessary.[12]

## 16.6 Warranty Management

Integral to strategic management of the NPD process are an effective organizational structure and the development and use of a sophisticated Management System.[13]

[12] For Phases 6–9, warranty data are useful for initiating changes to the process and operations. Warranty data collected after the product is launched are used for continuous improvements, as discussed in Chap. 15.

[13] Reference [11] deals with some issues relating warranty management.

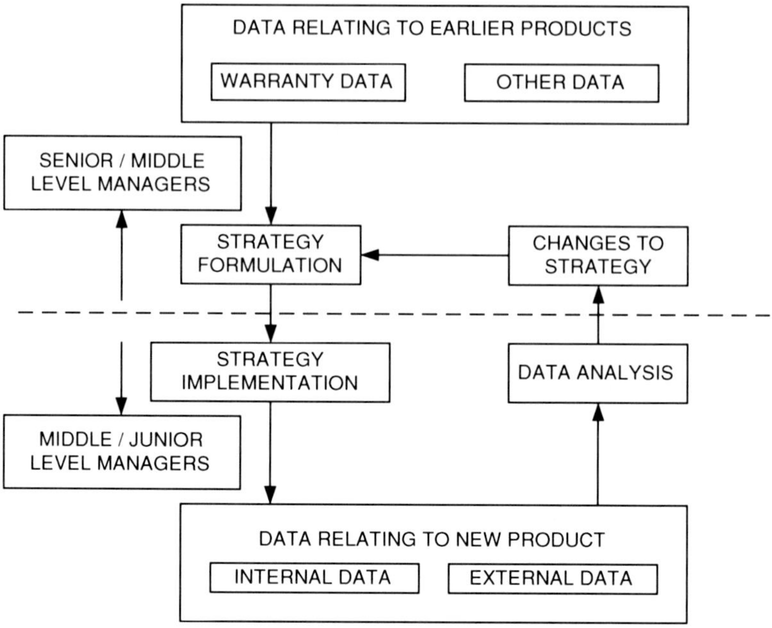


**Fig. 16.10** Warranty management tasks

### *16.6.1 Organizational Structure and Management Tasks*

Strategic management of the NPD process involves two issues—strategy formulation and strategy implementation. Strategy formulation must be done jointly by senior level managers (CEO, functional managers such as Marketing Manager, Production Manager, etc.) and middle level managers (responsible for sections such as material acquisition, product development, etc.). Strategy implementation should be done jointly by middle level managers and junior level managers (responsible for management of operational activities). Different types of data are needed for strategy formulation and strategy modification and refinement over time. This is shown in Fig. 16.10.

### *16.6.2 Warranty Management System*

An effective management system has four interacting components, as shown in Fig. 16.11 and briefly described below.

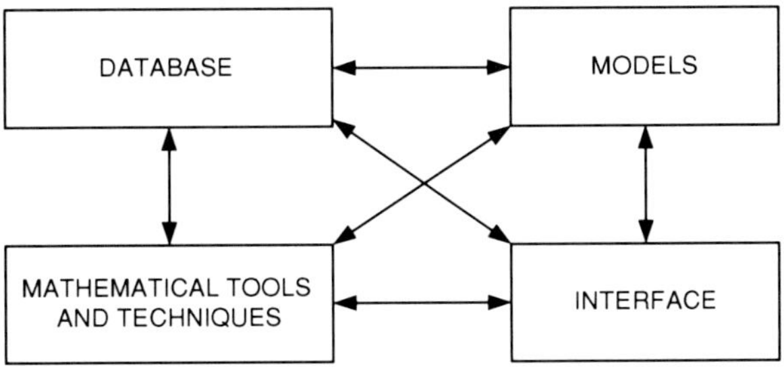


**Fig. 16.11** Key elements of management systems

#### 16.6.2.1 Database

Data from many different sources are needed for managing the NPD process. The warranty management system (WMS) must be able to incorporate data and information from these sources. In the earlier stages of product development, much of the data and information will be subjective (e.g., based on the judgment of experts) or will come from historical records of similar systems or components. As product development progresses, better data become available as a result of design definition, prototype testing, and trial manufacturing runs. Once the product enters the market, new data relating to product performance, sales, etc., may be analyzed.

For each generation of products, many types of in-house data are generated over the product life cycle and should be included in the database for a new product. This information can then be combined with data from external sources such as vendors and industry groups, to make proper decisions.

External data are data relating to partners in the supply chain such as component suppliers, service agents, dealers, or to competitors, customers, etc. Internal data are the data generated by groups responsible for the different phases of the NPD process.

#### 16.6.2.2 Models

Many different types of models are needed to assist the decision making for each phase of the product life cycle. This component of the management system takes the form of a library of available models.

#### 16.6.2.3 Mathematical Tools and Techniques

This component contains packages needed for data analysis, model building, general analysis and optimization. Advances in computer technology have resulted in packages that can handle large and complex data sets and carry out the analysis using modern techniques such as data mining, expert systems, artificial intelligence, etc. These are important in looking for underlying patterns in the data set.

#### 16.6.2.4 Interfaces

The management system should have a user interface and an applications interface. The user interface facilitates the flow of information from the user to the system and back, while the applications interface provides a link between a variety of external databases and programs which may be used for analysis, and for transferring data to and from the management system.

## References

1. Ackere A van (1993) The principal/agent paradigm: its relevance to various fields. Eur J Oper Res 70:88–103
2. Asiedu Y, Gu P (1998) Product life cycle cost analysis: state of the art review. Int J Prod Res 36:883–908
3. Bashir HA, Thomson V (2001) An analogy-based model for estimating design effort. Des Stud 22:157–167
4. Bayraktar BA (1990) On the concepts of technology and management of technology. In: Khalil TM, Bayraktar BA (eds) Management of technology II. Industrial Engineering & Management Press, Norcross
5. Belliveau P, Griffin A, Somermeyer S (eds) (2002) The PDMA Tool Book 2 for New Product Development. Wiley, NY
6. Betz F (1993) Strategic technology management. McGraw–Hill, NY
7. Ben-Oumlil A (2008) Warranty planning and development framework: a case study of a high-tech multinational firm. J Bus Ind Mark 23:507–517
8. Blischke WR, Murthy DNP (1994) Warranty cost analysis. Marcel Dekker, NY
9. Blischke WR, Murthy DNP (2000) Reliability. Wiley, NY
10. Burton JR (1996) Warranty protection: A consumerist perspective. In: Blischke WR, Murthy DNP (eds) Product warranty handbook. Marcel Dekker, NY, pp 719–757
11. Byrne PM (2004) Making warranty management manageable. Logist Manag 43:31–32
12. Eisenhardt KM (1989) Agency theory: An assessment and review. Acad Manag Rev 14:57–74
13. Kotler P, Armstrong G (1996) Principles of marketing. Prentice Hall Englewood Cliffs, New Jersey
14. Layer A, Brinke ET, Van-Houden F, Haasis S (2002) Recent and future trends in cost estimation. Int J Comput Integr Manuf 56:499–510
15. Mahajan V, Muller E, Wind Y (eds) (2000) New product diffusion models. Springer, NY
16. Menezes MAJ, Quelch JA (1990) Leverage your warranty program. Sloan Manag Rev 31:69–80
17. Murthy DNP, Blischke WR (2000) Strategic warranty management - a life cycle approach. IEEE Trans Eng Manag 47:40–54
18. Murthy DNP, Djamaludin I (2002) Product warranty—A review. Int J Prod Econ 79:231–260
19. Murthy DNP, Blischke WR (2005) Warranty management and product manufacturing. Springer, London
20. Murthy DNP, Rausand M, Osteras T (2008) Product reliability—specification and performance. Springer, London
21. Murthy DNP, Hagmark PE, Virtanen S (2009) Product variety and reliability. Reliab Eng Sys Saf 94:1593–1600
22. Murthy DNP, Osteras T, Rausand M (2009) Component reliability specification. Reliab Eng Sys Saf 94:1601–1608

23. Murthy DNP, Rausand M, Virtanen S (2009) Investment in new product reliability. Reliab Eng Sys Saf 94:1609–1617
24. Murthy DNP, Solem O, Roren T (2003) Product warranty logistics: issues and challenges. Eur J Oper Res 156:110–126
25. Padmanabhan V (1995) Usage heterogeneity and extended warranties. J Econ Manag Strat 4:33–53
26. Padmanabhan V (1996) Extended warranties. In: Blischke WR, Murthy DNP (eds) Product warranty handbook. Marcel Dekker, NY
27. Priest GL (1981) A theory of the consumer product warranty. Yale Law J 90:1297–1352
28. Purohit D, Srivastava J (2001) Effect of manufacturer reputation, retailer reputation, and product warranty on consumer judgments of product quality: a cue diagnosticity framework. J Consumer Psychol 10:123–134
29. Spence M (1977) Consumer misperceptions, product failure and producer liability. Rev Econ Stud 44:561–572
30. Tarasewich P, Nair SK (2000) Design for quality. Eng Manag Rev 18:401–425
31. Thompson AA, Strickland AJ (1990) Strategic management. BPI Irwin, Homewood, IL
32. Udell JG, Anderson EE (1968) The product warranty as an element of competitive strategy. J Mark 32(4):1–8
33. Ulrich KT, Eppinger SD (1995) Product design and development. McGraw–Hill, NY
34. Wind YJ (1982) Product policy: concepts methods and strategy. Addison–Wesley, Reading MA

# Part VI
# Case Studies

# Chapter 17
# Case Study 1: Analysis of Air Conditioner Claims Data

## 17.1 Introduction

In this study, we analyze claims data for an air conditioner (AC) used to cool a room or small office. The unit is manufactured and sold in a country in which it is used essentially all year round. The units are covered under an FRW with a warranty period of $W = 18$ months. The warranty lists a number of conditions and exclusions.

The AC consists of many parts and components. Of these, 20 components, each of which may fail by one of a number of failure modes, are included in the study. Analyses are done at the unit and component levels. Warranty claims data and production data are available for analysis. The period covered is of approximately 78 weeks duration, from January, 1998 to June, 1999.

The objectives of the study are to use the warranty claims data to identify key components and failure modes, estimate field reliability, estimate and predict warranty costs, and investigate alternative warranty policies.

The outline of the chapter is as follows: In Sect. 17.2, we discuss the company, the product and its components, the warranty terms, and the objectives of the study in detail. Section 17.3 is devoted to a detailed description of the data that are available for analysis as well as some comments regarding additional data needed. A preliminary analysis of the data, including editing, graphical descriptions of the final data set used in the analysis, and descriptive statistics for the variables included are given in Sect. 17.4. A detailed analysis of the data, including modeling of failures at the product and component levels and parametric and non-parametric comparisons of key characteristics is given in Sect. 17.5. Reliability estimation is discussed in Sect. 17.6. Section 17.7 deals with policy issues and cost analyses. Finally, conclusions of the study and recommendations to practitioners and theorists are given in Sect. 17.8.

W. R. Blischke et al., *Warranty Data Collection and Analysis*,
Springer Series in Reliability Engineering, DOI: 10.1007/978-0-85729-647-4_17,

## 17.2 Context and Objectives of the Study

In this section, we discuss the company, product description, including a list of components, the warranty, and the objectives of the study.

### 17.2.1 Company and Product Description

The company is the producer of one of the most popular AC units in the country. All units are produced in a single plant located near a large city. The item in question is one of many AC models produced by the company. The model chosen for the study can be used for cooling and ventilation and can be operated at two speeds, with a maximum cooling capacity of 7,500 BTU/h. The AC systems are used throughout the year and are operated almost continuously.

The AC system is recommended for comfort regulation in rooms between 12 and 18 $m^2$, for a maximum of two persons. The system is also used in small offices. Installation of the equipment is the responsibility of the customer. This may be done by the manufacturer's authorized representative, at additional cost. If the customer chooses to install the unit himself, he must follow manufacturer-supplied guidelines to ensure that the installation is done in accordance with technical specifications.

Production of the model used in this study began in 1997. The period of observation included only units produced between January 1, 1998 and June 30, 1999.

### 17.2.2 List of Components

The main AC system consists of 17 major components and a few additional components that are not critical to the performance of the product, but may contribute to warranty costs. The 17 major components (major in the sense that the failure of any one causes the system to fail) and other sources of warranty claims are listed in Table 17.1. $C_1$–$C_{17}$ are the major components.

Components $C_{18}$ and $C_{19}$ are not critical to operation of the AC, but are contributors to the cost of warranty. $C_{20}$ is not a single component, but a set of items for which information is unknown or incomplete. $C_{21}$ represents systems for which the entire unit was replaced on the second failure of the unit.

### 17.2.3 Warranty

The warranty offered on this model was a nonrenewing free replacement warranty with coverage of $W = 18$ months from date of purchase. On failure of a unit under

**Table 17.1** List of components of AC units

| Symbol | Description |
|---|---|
| $C_1$ | Control buttons |
| $C_2$ | Selection key |
| $C_3$ | Thermostat |
| $C_4$ | Electric network |
| $C_5$ | Helix |
| $C_6$ | Turbine |
| $C_7$ | Motor |
| $C_8$ | Capacitor |
| $C_9$ | Compressor |
| $C_{10}$ | Pipeline |
| $C_{11}$ | Condenser |
| $C_{12}$ | Capillary |
| $C_{13}$ | Shell |
| $C_{14}$ | Curl |
| $C_{15}$ | Evaporator |
| $C_{16}$ | Wire |
| $C_{17}$ | Thermister |
| $C_{18}$ | Plastic Front |
| $C_{19}$ | Vanes |
| $C_{20}$ | Other |
| $C_{21}$ | Replacement |

warranty, the AC is repaired by replacement of the failed component. If a repaired AC unit fails a second time, the entire unit is replaced free of charge. This occurred less than 0.03% of the time.

Protection is provide to the manufacturer by the following provisions: The warranty is cancelled if (1) the unit is modified; (2) the data that identify the unit are altered; (3) the unit is used under conditions that are different from technical specifications; (4) the equipment is installed with ducts; or (5) the equipment is damaged because of fire, flooding, and other fortuitous causes. In addition, the warranty is restricted to repairs or replacement of defective components.

The manufacturer also requires that the unit be installed according to stated specifications, as discussed in Sect. 17.2.2, and be properly maintained.

### *17.2.4 Objectives of the Study*

The objectives of this study are to describe to data graphically and numerically, to determine which components fail most frequently and the most frequent modes of failure, and to study warranty costs and field reliability of the product. Specifically, the objectives are to:

- Edit, summarize and describe the claims and supplementary data
- Identify the components that are the major sources of claims

- Determine the most frequently occurring failure modes for the components
- Select failure distributions that best describe the failure patterns for each component and use claims and supplementary to estimate the parameters of these distributions
- Select an overall failure distribution to model failures for the product as a whole and use claims and supplementary data to estimate its parameters
- Use the estimated overall failure distribution to estimate product reliability and future warranty claims rates
- Use this information to estimate expected warranty costs
- Select other warranty policies that may be considered for future products of this type and compare predicted warranty costs of the selected policies

Each of these items will be discussed below, and analyses will be pursued at various levels of detail.

## 17.3 Data

Data available for analysis include claims data over an eighteen month period and a limited amount of supplementary data. These, along with some data problems and additional data needed for a thorough analysis with regard to the above objectives, are discussed below.

### *17.3.1 Claims Data*

Production of the AC system began in 1997. The data used in this study were collected in the period January, 1998, through June, 1999, a period of approximately 78 weeks. The longest observed time in operation was 78 weeks. The product has since been replaced by a more advanced model. Since some information concerning the system remains proprietary, however, we do not identify the company, the specific product, or the country of origin.

The total production during the eighteen month period in question was 92,530 AC units. Data are given on a total of 4,081 claims. Thus the percentage of claims received during this portion of the warranty period was 4.41% of production.

The components for which data are available are listed in Table 17.1. The category "Other" in the table includes some miscellaneous parts and a few for which only one or two failures were observed (e.g., solenoid), but is mostly composed of claims for which information is unknown or incomplete. These do, however, have an entry for time to failure (TTF), and hence will be included in the overall product analysis. "Replacement" indicates that the entire unit was replaced free of charge to the customer. This occurred when the original unit experienced a second component failure. There were 24 instances of this. Since the data set does not include information about the components that failed or the times of individual

**Table 17.2** Failure codes/modes for AC components

| Component | No. of claims | Failure codes (Modes) |
|---|---|---|
| $C_1$ | 62 | 571/625(damaged); 626(missing); 500; 501; 572; 626 |
| $C_2$ | 15 | 230(fails to turn); 231(broken); 234(short circuit) |
| $C_3$ | 314 | 205(fails to start); 204(fails to stop); 206(broken); 186/217(moved bulb); 182/202; 183/203; 703; 263 |
| $C_4$ | 363 | 397,409(loose); 398/410; 399/411; 404/416 |
| $C_5$ | 1,284 | 429/443(broken); 436; 430/437; 431/438; 432/439; 440/441; 435/442 |
| $C_6$ | 144 | 441/446(loose); 451/451(broken); 448(unbalanced) |
| $C_7$ | 337 | 215/248(does not start); 223/256(short); 233/266 (broken clamp); 249; 225/258; 226/259; 274 |
| $C_8$ | 67 | 236/281(interrupted); 239/284(damaged); 280 |
| $C_9$ | 282 | 161/167(noisy); 177(burned); 154/157/192; 196; 160/166; 164; 169/175; 171; 173/179; 402 |
| $C_{10}$ | 40 | 577/617(friction); 129/130(leak); 110; 112; 113; 115; 119; 120; 121; 122; 124; 126; 128/131; 127; 518 |
| $C_{11}$ | 86 | 105(curves); 104; 106; 109; 133; 141/144 |
| $C_{12}$ | 100 | 198/219(broken); 218/194(clogged); |
| $C_{13}$ | 218 | 527/528/707(placement); 536/709(broken); 537; 538; 610; 694; 706 |
| $C_{14}$ | 29 | 491/510(broken) |
| $C_{15}$ | 47 | 102(solder); 103(connector); 108; 155 |
| $C_{16}$ | 13 | 279/348(interrupted); 352(short) |
| $C_{17}$ | 49 | 193/704 (adaptation) |
| $C_{18}$ | 68 | 463/465(broken); 461; 582 |
| $C_{19}$ | 33 | 514(loose); 489; 482; 483 |
| $C_{20}$ | 144 | 999(others); 13; 37; 51; 55; 144; 184; 185; 187; 254; 263; 505; 509; 530; 701 |
| $C_{21}$ | 24 | 705 |

failures, these are not included in the analysis. They must be accounted for, however, in the Company's final evaluation of warranty costs. For purposes of analysis, the total number of failures is taken to be $4{,}081 - 24 = 4{,}057$.

The data made available for the study were in separate spreadsheets for each of the 21 components listed in Table 17.1 as well as for the data set as a whole. Each spreadsheet included the following information:

- Claim number
- Item number
- Week of production
- Failure mode code
- Date of failure
- Time to failure (TTF)

Date of failure was transformed to week of failure and TTF was calculated as the difference between week of failure and week of production.

Separate and distinct failure codes are used for each component. Table17.2 lists several failure modes for several of the 21 components to illustrate the types of

**Table 17.3** Monthly production and claims counts

| Month | Production | No. Failed |
|---|---|---|
| Jan 1998 | 9,101 | 1,159 |
| Feb 1998 | 7,411 | 619 |
| Mar 1998 | 5,546 | 460 |
| Apr 1998 | 2,600 | 256 |
| May 1998 | 4,150 | 347 |
| Jun 1998 | 3,000 | 214 |
| Jul 1998 | 4,800 | 223 |
| Aug 1998 | 3,994 | 186 |
| Sep 1998 | 5,998 | 222 |
| Oct 1998 | 13,381 | 204 |
| Nov 1998 | 8,439 | 64 |
| Dec 1998 | 3,140 | 97 |
| Jan 1999 | 7,200 | 21 |
| Feb 1999 | 4,750 | 9 |
| Mar 1999 | 2,300 | 0 |
| Apr 1999 | 2,050 | 0 |
| May 1999 | 3,160 | 0 |
| Jun 1999 | 1,510 | 0 |
| Total | 92,530 | 4,081 |

failures observed. Also listed are failure codes for all components and modes and the number of AC claims received as a result of failure of each type of component. (Note: more than on code is used for some modes, indicating different forms of failure by that mode, e.g., two types of damage. These are treated as separate modes in the analysis).

Information concerning failure times was obtained by the Post Sales Department of the Company. In the original data set, failure times are expressed as month of failure. To translate this to week of failure, failure time is approximated by assuming that all failures occurred in the second week of the month.[1]

## *17.3.2 Supplementary Data*

For analysis of the claims data in the context of the objectives of the project, additional data are necessary. The only available supplementary data available are censored observations and weekly production data. These data are used in the calculation of TTF as indicated above.

To illustrate the production trend, the production data are aggregated to monthly totals. These are given in Table17.3. For purposes of comparison the numbers of failures linked to each month's production are also given in the table.

[1] An alternative approach is to divide the failures equally over all weeks of the month.

### 17.3.3 Data Problems

In this section, we discuss some of the problems in using the information available for estimation of reliability and warranty analysis, and suggest changes in data collection that may obviate these difficulties in future analyses.

*Existing Data* Visual examination of the spreadsheets and some tallies of results showed that the data are apparently unusually clean. No missing entries were found and only a few data errors were noted. These included 3 items that were miscoded with regard to week of failure (e.g., 10 instead of 11). This would have negligible impact on the analysis. There were also several items for which the failure code was apparently incorrect. These were included in the "Other" category.

There appear to be some outliers in the data. This will be investigated more fully and discussed in the next section.

Some more serious problems affecting the analysis and interpretation of the results are the following:

- The data do not include the actual date of sale, and production date will be used as a proxy for this in the analysis. This clearly affects the results, since there is obviously a lag between production and sale. In addition, there may be a lag between sale and beginning of use of the product. There is no information that would enable us to estimates these lags. The effect of this is that the calculated TTF will be larger than the true TTF, and, as a result, the reliability of the product will be overestimated.
- We do not have data on the number of items actually sold. Using production data for this purpose effectively assumes that all units were sold, and, in fact, sold during the 18-month period of observation. This again biases the estimated reliability upward.
- Very few claims are made in the last six months of production. In fact, none at all are recorded for the production of the last four months. This may reflect the lag between production and sale mentioned above. An alternative to the results given below would be to analyze just the first 12 months of data. This has not been pursued.

*Additional Data Needed* In order to address the objectives of the project effectively, at least the following additional data are needed in future studies of this type:

- Claims data should contain the actual date of failure, not just the month
- The data bank should include data on repair activities, including times and cost of each repair/replacement
- Service facility should be indicated (if there is more than one)
- Service personnel should be identified for each claim processed
- Service costs are needed to calculate the cost of warranty
- Sales data should include the actual date of sale

- Other relevant supplementary data, such as quality assurance data, R&D test data, and any other information pertaining to reliability, quality, and costs would enhance and help to affirm the results of the study.

## 17.4 Preliminary Data Analysis

In this section, we look at basic summaries of the data at the component and item levels. Preliminary analyses include a search for outliers, histograms of the data, Pareto charts for determining key failure modes, descriptive statistics, including means, standard deviations and other descriptive measures, and a Month-of-Production/Time-to-Failure (MOP-TTF) diagram.

### *17.4.1 Component-Level Analysis*

Preliminary analysis at the component level will include, in order, histograms of TTF, descriptive statistics, checks for outliers, Pareto charts for determining most frequently occurring failure modes, and comparisons of MTTF's for the different failure modes of the various components. The analyses in this section will be of claims data only (Chap. 5, Scenario 1.1). In many cases, component $C_{21}$ will be omitted. Limited results will be given for components $C_2$, $C_{14}$, $C_{16}$, $C_{17}$, and $C_{19}$ for which relatively few failures occurred.

*Histograms* Figure 17.1 shows histograms of TTF, arranged in numerical order, for all components except $C_{21}$. Some observations regarding the charts:

- These histograms estimate the conditional distributions of TTF, given that failures occurred during the observation period. They do not necessarily reflect the unconditional distributions of TTF.
- Note the considerable diversity of shapes—some skewed right as expected for failure time distributions, some bimodal or multi-modal, indicating possible mixtures, etc.
- It is doubtful that any single distribution or family of distributions will be appropriate for modeling TTF for all components
- Some of the diversity is likely due to the fact that each component has a unique (or nearly unique) set of failure modes. (The most important modes for selected components will be listed below). This is an important aspect requiring further study in any detailed engineering analysis of failures.

*Descriptive statistics* Table 17.4 gives descriptive statistics for all components except for Replacement. These were calculated by use of Minitab and are discussed in detail in Chap. 8. Some observations regarding the results:

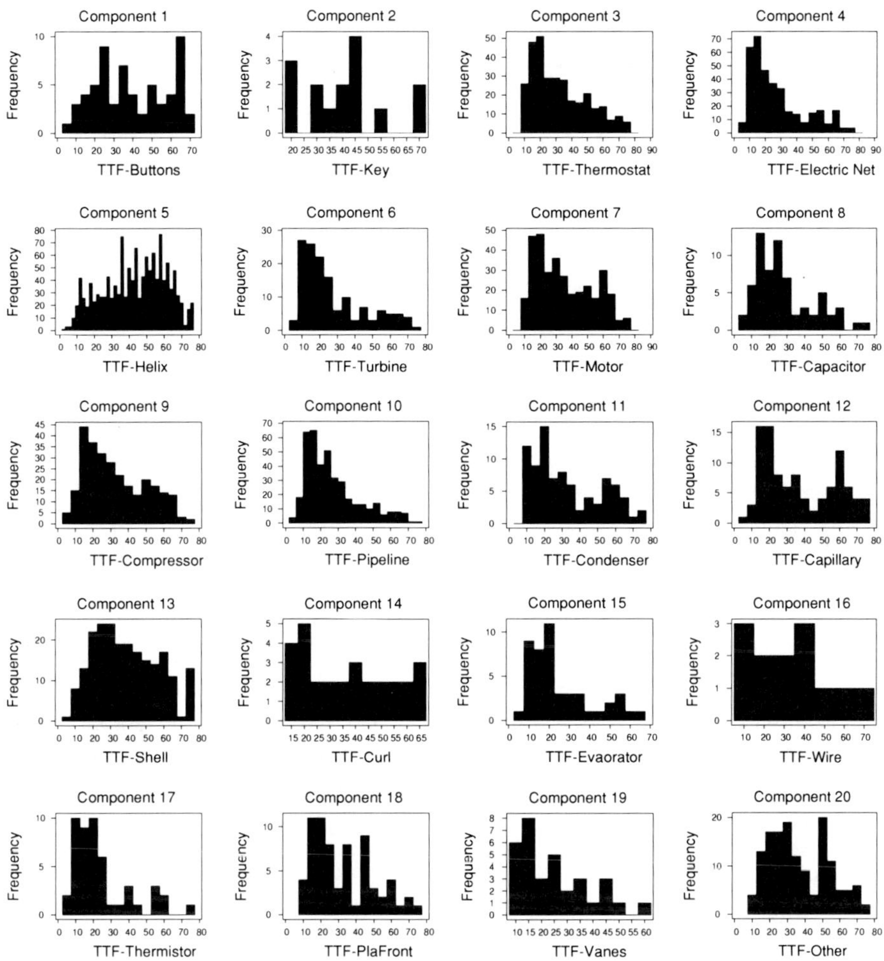


**Fig. 17.1** Histograms of TTF for components 1–20

- Again, these are conditional results, given that the failure occurred during the eighteen-month observation period, since censoring is not taken into account in the calculations.
- In most cases, the mean exceeds the median, indicating skewness to the right, as would be expected and as noted in most of the histograms above.
- It is apparent from the results that MTTF's differ for the different components. This is not surprising since these are a mix of simple and relatively complex mechanical, electrical and other parts, of various materials, stresses, and so forth.
- The standard deviations are remarkably consistent. This may be a result of the fact that all distributions are restricted to the observation.

**Table 17.4** Descriptive statistics for components 1–20

| Component | $n$ | Mean | S.D. | Min | $Q_1$ | Med | $Q_3$ | Max |
|---|---|---|---|---|---|---|---|---|
| Buttons | 62 | 39.65 | 18.63 | 6 | 24 | 36.5 | 62 | 70 |
| Selection key | 15 | 40.67 | 16.13 | 18 | 28 | 40 | 47 | 70 |
| Thermostat | 314 | 32.18 | 17.27 | 8 | 18 | 28 | 44 | 76 |
| Electric net | 363 | 27.19 | 17.43 | 3 | 13 | 21 | 36 | 75 |
| Helix | 1,284 | 42.78 | 17.66 | 1 | 29 | 44 | 57 | 76 |
| Turbine | 144 | 27.38 | 17.91 | 4 | 13 | 22 | 36.75 | 74 |
| Motor | 337 | 35.35 | 18.04 | 9 | 19 | 32 | 51 | 76 |
| Capacitor | 67 | 28.30 | 16.49 | 4 | 16 | 23 | 39 | 73 |
| Compressor | 282 | 32.77 | 17.09 | 4 | 18 | 29 | 46.25 | 75 |
| Pipeline | 402 | 26.10 | 15.45 | 3 | 14 | 22 | 33 | 75 |
| Condenser | 86 | 32.65 | 18.89 | 8 | 17.75 | 27.5 | 51 | 75 |
| Capillary | 100 | 37.47 | 20.05 | 7 | 21 | 33 | 58 | 75 |
| Shell | 218 | 39.04 | 17.82 | 4 | 23 | 37 | 53 | 76 |
| Curl | 29 | 37.07 | 17.36 | 13 | 22 | 35 | 52 | 66 |
| Evaporator | 47 | 24.87 | 15.15 | 7 | 13 | 19 | 34 | 63 |
| Wire | 13 | 34.00 | 19.75 | 7 | 17 | 32 | 47.5 | 74 |
| Thermister | 49 | 23.76 | 16.84 | 5 | 12.5 | 18 | 29 | 74 |
| Plastic front | 68 | 32.43 | 17.01 | 5 | 21 | 27.5 | 43.75 | 74 |
| Vanes | 33 | 24.79 | 13.36 | 9 | 13.5 | 22 | 34 | 60 |
| Other | 144 | 37.06 | 16.87 | 11 | 23 | 33 | 51.75 | 76 |

*Outlier tests* We look at two approaches to checking for outliers, boxplots on which outliers are indicated and Grubbs' test, discussed in Sect. 10.2.2. Since it is not possible to verify the correctness of the data by examining the original source, we adopt a conservative approach[2] in checking outliers by either approach. This provides some protection against discarding some valid data.

Figure 17.2 shows boxplots for all components except $C_{21}$. In each boxplot, outliers are indicated by stars lying beyond the whiskers. As discussed in Sect. 10.2.1, outliers in the Minitab version of boxplots are observations that lie more than 1.5(IQR) from the box. By this standard, three outliers are indicated for component $C_4$, one for $C_6$, nine for $C_{10}$, and five for $C_{17}$. In general, outliers in this region are considered mild outliers, unless they are, in fact beyond 3(IQR) from the box, in which case they are considered to be extreme outliers. The conservative approach is to reject only extreme outliers. Note that in all cases, outliers lie above the box.

To pursue this further, we look at the normal test for outliers given in Sect. 10.2.2. Since the distributions appear not to be normal, this will be considered to be an approximate test. The conservative approach taken is to test at $\alpha = 0.01$.

[2] Note, however, that these ignore the fact that we are dealing with censored data. The properties of Grubbs' test and the outlier plots in this context are not known.

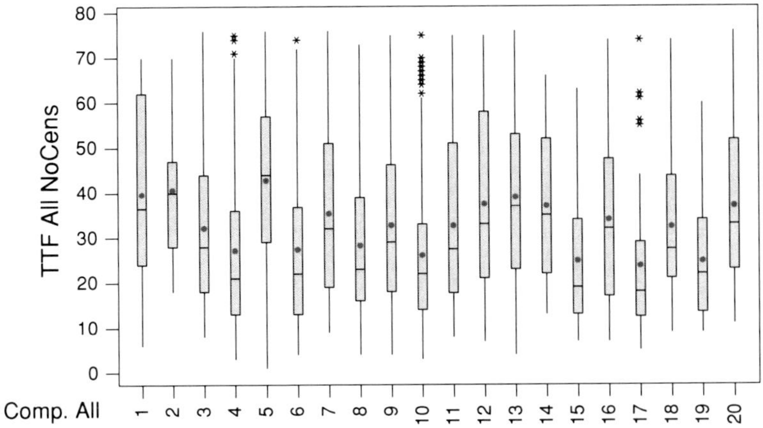


**Fig. 17.2** Boxplots of data, components 1–20 (means are indicated by *solid circles*)

We consider the four components for which the boxplots identified possible outliers:

*Component* $C_4$ From Table 17.4, we find the IQR to be $36 - 13 = 23$. The box covers the interval [13, 26] and the whiskers extend 34.5 units on either side. Mild outliers are observations between 39 and 78 units above the box, i.e., between 70.5 and 105 units above the box. Since the max in 75, all three outliers are mild outliers and none will be deleted from the data.

To perform Grubbs' test, we use the test statistic of (10.1), obtaining (from Table 17.4):

$$G = \frac{\max |y_i - \bar{y}|}{s} = \frac{75 - 27.19}{17.43} = 2.74.$$

A formula for determining the critical region for the test is given in (10.2). Here the level of significance is $\alpha/2n = .01/726 = 0.00001377$. The corresponding tabulated $t$ is 4.2475.[3] From (10.2), the critical region is found to be $G > 4.145$. The max observation is not found to be significant, and no observations are deleted by this test as well.

*Component* $C_6$ The one observation marked as an outlier in the boxplot is clearly not an extreme outlier since it is just above the whisker. The Grubbs' statistic is $G = 2.60$. The critical value is found to be 3.876, and the observation is not deleted.

*Component* $C_{10}$ From Fig. 17.2, it can be seen that all nine outliers are just above the whisker, so none are identified as extreme. The Grubbs' test applied to the

[3] Calculated by Minitab. Note that the absolute value of $t$ is used in (10.2).

maximum observation does not lead to rejection. We conclude that none of the observations should be deleted.

*Component* $C_{17}$ Inexplicably, the whisker is printed incorrectly in Fig. 17.2 for this sample. The IQR is 29 − 12.5 = 16.5, so 1.5IQR is 24.825 and the whisker should extend to 54 instead of 45. In any case, we find that 3IQR = 49.5, so extreme observations are those that exceed 78.5, which is not possible here since the maximum observation must be less than 78. Thus we again find no extreme outliers in the sample. The Grubbs' test gives a non-significant result for the max.

Since none of the possible outliers is rejected, all of the data will be used, as appropriate, in the analyses to follow.

*Pareto charts* For purposes of illustration, Pareto charts are given in Fig. 17.3 for components $C_1$, $C_3$–$C_9$, and $C_{11}$. The key use of the charts is to identify and visually display the most important failure causes for each component. As is apparent from the charts, in some cases there are one or a few modes that account for most of the failures—e.g., buttons, helix, condenser; in other cases, there is no predominant mode, but several that jointly account for most failures–motor and compressor are examples.

These results are useful in efforts to improve the reliability of the product and thereby decrease warranty costs. Engineering changes in design and/or production to reduce or eliminate key modes for some or all components, easily identified in the Pareto charts, are usually required. Key failure modes identified in the Pareto charts for all components are highlighted in Table 17.2. We briefly consider this again in the next section.

## 17.4.2 Product-Level Analysis

In this section, we address the issues discussed in 17.4.1 at the product level. In addition to histograms, descriptive statistics and Pareto charts, we give a Month-of-Production/Time-to-Failure (MOP-TTF) diagram.

*Histogram* A histogram of TTF for all 4081 claims is given in Fig. 17.4. The overall pattern is skewed to the right, indicating that one or more of the standard life distributions may fit the data. Again, this is a chart of failure data conditional on failure within the observation period. The ragged appearance evident in the plot may be a result of coding of the failure time. In that process, some months were represented as four-week periods and some as five.

*Pareto charts* Figure 17.5 is a Pareto chart of frequency of claims by component. It is easily seen from the chart that the greatest gains in product reliability could be achieved by reduction of failures for the helix. Other significant contributors to claims are pipeline, electrical network, motor, thermostat, and compressor.

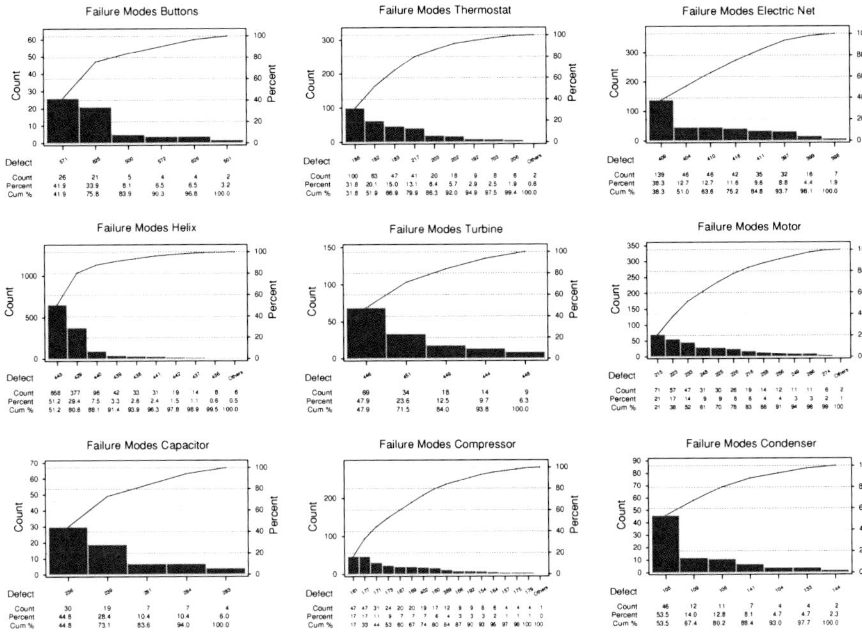


**Fig. 17.3** Pareto charts of failure modes for selected components

**Fig. 17. 4** Histogram of TTF, all claims

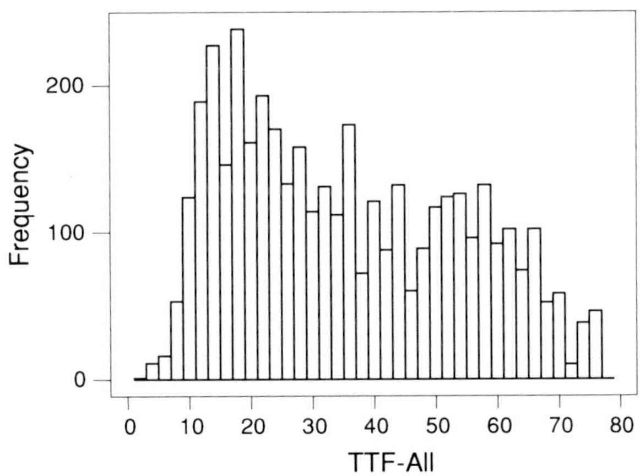


Addressing the causes of key failure modes for these components, as indentified in the previous section, could significantly lower the claims rate and reduce warranty costs.

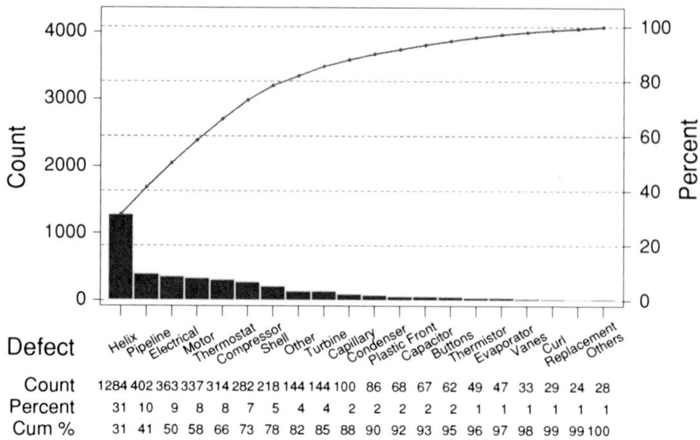


**Fig. 17.5** Pareto chart of failure counts by component

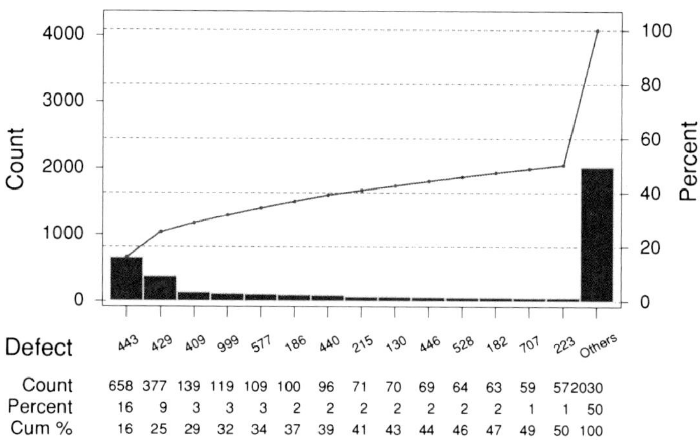


**Fig. 17.6** Pareto chart of failure modes

A Pareto chart of all modes for all components, given in Fig. 17.6, verifies this conclusion. Engineering effort should be especially directed to improvements in design and production of the helix.

*Descriptive statistics* Table 17.5 contains descriptive statistics of TTF for all claims except those that resulted in replacement of the entire unit. Note that the mean is larger than the median, reflecting the right-skewness of the conditional distribution of TTF given failure in the observation period.

*MOP-TTF diagram* The purpose of the MOP-TTF chart is to relate failure times to month of production. In particular, the marginal totals relate the monthly

**Table 17.5** Summary statistics for claims

| $n$ | Mean | $s$ | Min | $Q_1$ | Median | $Q_3$ | Max |
|---|---|---|---|---|---|---|---|
| 4057 | 35.182 | 18.517 | 1 | 19 | 32 | 51 | 76 |

**Table 17.6** MOP-TTF diagram for all claims

| MOP | Prod Quan | TTF and claims counts | | | | | | | | | | | |
|---|---|---|---|---|---|---|---|---|---|---|---|---|---|
| | | 1 | 2 | 3 | 4 | 5 | 6 | ... | 15 | 16 | 17 | 18 | Total |
| 1 | 9,101 | 6 | 10 | 99 | 70 | 130 | 57 | ... | 56 | 42 | 94 | 125 | 1,101 |
| 2 | 7,411 | 0 | 10 | 47 | 51 | 21 | 44 | ... | 64 | 64 | 38 | 27 | 670 |
| 3 | 5,546 | 0 | 9 | 20 | 17 | 18 | 27 | ... | 25 | 14 | 22 | | 276 |
| 4 | 2,600 | 1 | 9 | 36 | 39 | 15 | 22 | ... | 25 | 56 | | | 433 |
| 5 | 4,150 | 0 | 1 | 14 | 16 | 25 | 23 | ... | 54 | | | | 348 |
| 6 | 3,000 | 0 | 0 | 2 | 5 | 1 | 2 | ... | | | | | 31 |
| 7 | 4,800 | 0 | 5 | 16 | 32 | 23 | 13 | ... | | | | | 254 |
| 8 | 3,994 | 0 | 4 | 2 | 16 | 7 | 12 | ... | | | | | 93 |
| 9 | 5,998 | 3 | 7 | 23 | 10 | 29 | 28 | ... | | | | | 175 |
| 10 | 13,381 | 0 | 4 | 16 | 21 | 45 | 21 | ... | | | | | 205 |
| 11 | 8,439 | 0 | 4 | 8 | 27 | 27 | 33 | ... | | | | | 209 |
| 12 | 3,140 | 0 | 2 | 10 | 24 | 18 | 33 | ... | | | | | 117 |
| 13 | 7,200 | 0 | 4 | 4 | 7 | 13 | 9 | ... | | | | | 47 |
| 14 | 4,750 | 2 | 0 | 12 | 26 | 12 | 39 | | | | | | 91 |
| 15 | 2,300 | 0 | 0 | 2 | 4 | 15 | | | | | | | 21 |
| 16 | 2,050 | 0 | 0 | 0 | 8 | | | | | | | | 8 |
| 17 | 3,160 | 0 | 0 | 2 | | | | | | | | | 2 |
| Total | | 12 | 69 | 313 | 373 | 399 | 363 | ... | 224 | 176 | 154 | 152 | 4,081 |

production to the time of failure within the observation period. The MOP-TTF table for the claims data is given in Table 17.6.

From the table, we can calculate the proportion of each month's production that ultimately failed within the observation period. These are given in Table 17.7. As expected, the early months of production show the largest percentage of warranty claims, and the latest months the least, since these had relatively little exposure.

It is instructive to look at this on a monthly basis. The results (percent claims by month of failure and MOP) are given in Fig. 17.7. It is apparent from the Fig. 17.7 and Table 17.7 that production in month 4 resulted in proportionally more failures and claims than in any other month. This may occur for any of a number of reasons, including a batch of inferior raw materials, new production crew, and so forth. For understanding of the causes of the increase in failure percent and prevention, additional investigation is needed.

Additional results with regard to individual components and the product as a whole will be discussed in the next section.

**Table 17.7** Percent of claims by production month

| MOP | Percent |
|---|---|
| 1 | 12.01 |
| 2 | 9.04 |
| 3 | 4.98 |
| 4 | 16.65 |
| 5 | 8.39 |
| 6 | 1.03 |
| 7 | 5.292 |
| 8 | 2.328 |
| 9 | 2.918 |
| 10 | 1.532 |
| 11 | 2.477 |
| 12 | 3.726 |
| 13 | 0.65 |
| 14 | 1.92 |
| 15 | 0.91 |
| 16 | 0.39 |
| 17 | 0.06 |
| Total | 4.41 |

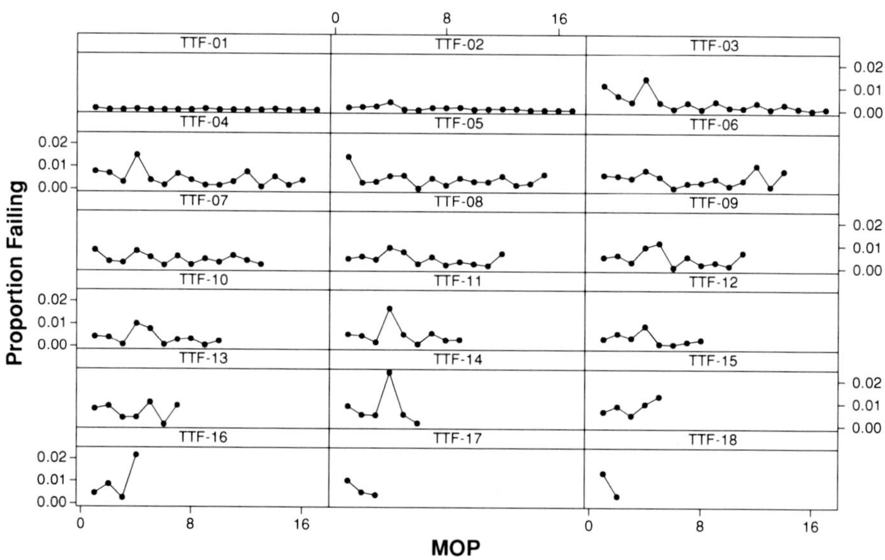


**Fig. 17.7** Monthly failure proportions

**Table 17.8** Product level ANOVA

| Source | *df* | SS | MS | *F* | *p* |
|---|---|---|---|---|---|
| Components | 19 | 169,128 | 8901 | 29.42 | 0.000 |
| Error | 4,037 | 1,221,556 | 363 | | |
| Total | 4,056 | 1,390,684 | | | |

## 17.5 Detailed Data Analysis

In this section, we look at comparisons of conditional MTTF's of modes of failure for each component. The techniques used are Analysis of Variance (ANOVA) and the nonparametric Kruskal–Wallis test. This continues the analysis of Scenario 1.1 data at a more detailed level and is of value in identifying components and failure modes that are important contributors to warranty costs. The same two procedures will then be applied at the product level to compare MTTF's of components.

The remainder of the section will deal with claims plus supplementary data (Scenario 1.3 of Chap. 5). Parametric and nonparametric approaches to determining distributions of TTF are considered. Models will be selected and MTTF's estimated based on the selected models.

### *17.5.1 Comparisons of Means and Medians*

Conditional means of failure modes of selected components will be compared by ANOVA (Sect. 10.3), which is an overall test, and pairwise by Tukey's multiple comparison procedure (Sect. 10.3.1). ANOVA and the Tukey test are parametric procedures (assuming normality). We also apply the nonparametric Kruskal–Wallis procedure (Sect. 9.8.3), which is a test that compares medians. We begin with the analysis at the product level, since these results will be used along with those of the previous section to select the components for detailed analysis.

#### 17.5.1.1 Product level comparisons

*ANOVA* Since the structure of the component failure data is basically that of a Completely Randomized Design, the conditional MTTF's are compared by the ANOVA discussed in Sect. 10.5.1. We analyze the 20 components and 4,057 failures selected previously. The result is given in Table 17.8. The calculated F is significant at $\alpha = 0.001$. Before continuing to the Tukey procedure, we look at the assumptions made in performing the ANOVA.

*Tests of assumptions* The key assumptions of ANOVA are that the error term in the linear ANOVA model are independent and normally distributed with identical variances in all of the groups. It is reasonable to assume independence since the measurements are taken on separate items, and proceed to test the latter two

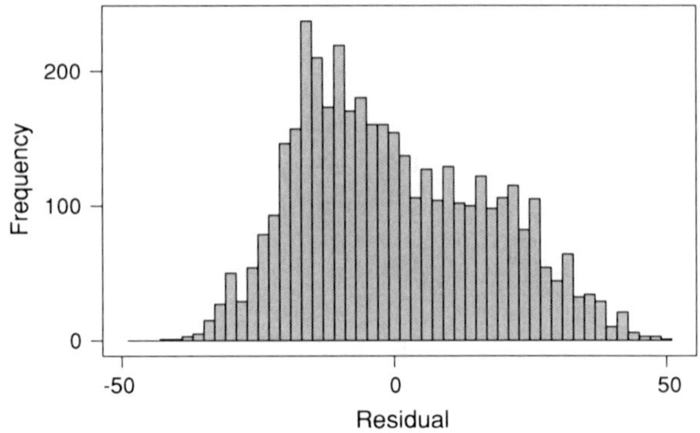


**Fig. 17.8** Histogram of ANOVA residuals, all claims except $C_{21}$

assumptions. As discussed in Sect. 10.8.3, these may be tested by analysis of residuals. Figure 17.8 is a histogram of the residuals obtained in the above ANOVA. Note that these do not appear to follow a normal distribution. Instead the distribution is skewed to the right, much as one would expect for life distributions, and a test of goodness-of-fit to the normal leads to rejection.[4]

Minitab output provides two tests for equality of variance, Bartlett's Test, which is sensitive to the assumption of normality, and Levene's Test, which is based on sample medians instead of means and is more appropriate for skewed distribution such as that of Fig. 17.8. The Minitab output for these tests is shown in Fig. 17.9. Note from the output that the test statistic for Bartlett's test is not significant at even the 10% level ($p = 0.145$). Levene's test, however, leads to significance at 0.001 or lower, leading to rejection of equality of variance.

We conclude that there is evidence of both nonnormality and unequal variances. We will look briefly at the Tukey procedure and then at the Kruskal–Wallis nonparametric procedure.

The failure of assumptions to hold is due in part to the fact that the data for each component is, itself, a mixture of TTF's for different failure modes. ANOVA at the failure mode level (not shown here) shows less departure from assumptions. In fact, the histogram of residuals is only slight skewed right, and the normal plot of residuals is nearly a straight line. With 154 separate modes, however, the results are at too detailed a level to be easily interpreted.

*Tukey's multiple comparison procedure* The Tukey procedure is based on the assumption of normality, equal variances, and, in its original form, equal sample sizes. In that situation, if mean $i$ differs from mean $j$, it will differ from all means that are more different from mean $i$ than is mean $j$. Equal $n$'s is not the case in this analysis. In fact, the sample sizes differ by two orders of magnitude, the smallest being 13, and the largest, 1,284. As a result, the procedure leads to somewhat

[4] Tests of goodness-of-fit will be pursued in more detail in Sect. 17.5.2.

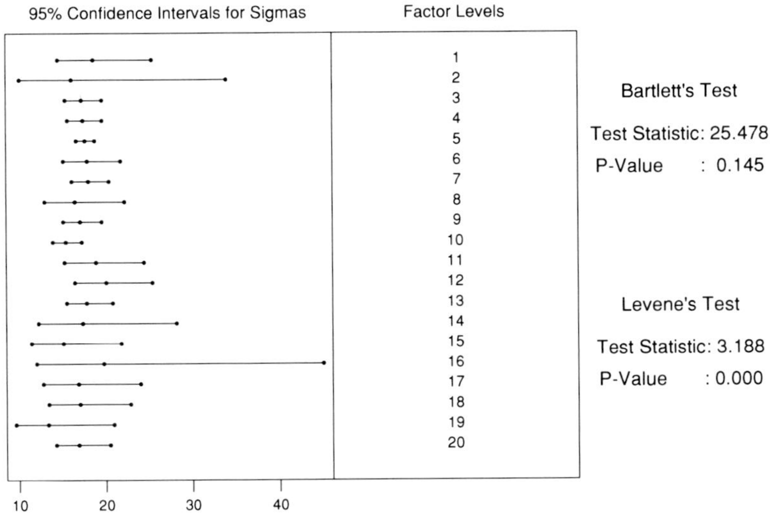


**Fig. 17.9** Tests of equality of variance of TTF for components (*factor levels*)

**Table 17.9** Summary of differences identified by Tukey's test of MTTF's for components

| Comp | Mean | Differs from Comp | Comp | Mean | Differs from Comp |
|---|---|---|---|---|---|
| 1 | 39.65 | 4,6,8,10,15,17,19 | 11 | 32.65 | 5 |
| 2 | 40.67 | None | 12 | 37.47 | 4,6,10,15,17,19 |
| 3 | 32.18 | 4,5,10,13 | 13 | 39.04 | 3,4,6,8–10,15,17,19 |
| 4 | 27.18 | 1,3,5,7,9,12,13,20 | 14 | 37.07 | None |
| 5 | 42.78 | 3,6-11,15,17–20 | 15 | 24.87 | 1,5,7,12,13,20 |
| 6 | 27.37 | 1,5,7,12,13,20 | 16 | 34.00 | None |
| 7 | 35.35 | 4–6,10,15,17 | 17 | 23.76 | 1,5,7,12,13,20 |
| 8 | 28.30 | 1,5,13 | 18 | 32.43 | None |
| 9 | 32.77 | 4,5,10,13 | 19 | 24.79 | 1,5,12,13,20 |
| 10 | 26.10 | 1,3,5,7,9,12,13,20 | 20 | 37.06 | 4–6,10,15,17,19 |

confusing results and we omit details and briefly summarize the differences identified by the Tukey procedure in Table 17.9.

The results given are for a *family error rate* of 0.05, meaning that the total error rate for each group of comparisons (largest versus smallest, largest versus second smallest and second largest versus smallest, etc.) is 5%. In this case, this results in an individual error rate of 0.00040, which is relatively very conservative.

It is apparent from Table 17.9 that, in spite of the conservative nature of the test, many differences have been identified by the Tukey procedure. To further pursue this, we look next at the nonparametric approach.

**Table 17.10** Kruskal–Wallis test results, comparison of components 1–20

| Comp. | $n$ | Median | Ave Rank | $z$ |
|---|---|---|---|---|
| 1 | 62 | 36.5 | 2329.8 | 2.04* |
| 2 | 15 | 40.0 | 2426.2 | 1.32 |
| 3 | 314 | 28.0 | 1848.5 | −2.84** |
| 4 | 363 | 21.0 | 1486.9 | −9.24** |
| 5 | 1,284 | 44.0 | 2512.9 | 17.91** |
| 6 | 144 | 22.0 | 1495.4 | −5.57** |
| 7 | 337 | 32.0 | 2051.6 | 0.37 |
| 8 | 67 | 23.0 | 1595.6 | −3.05** |
| 9 | 282 | 29.0 | 1896.6 | −1.97* |
| 10 | 402 | 22.0 | 1445.0 | −10.53** |
| 11 | 86 | 27.5 | 1856.0 | −1.38 |
| 12 | 100 | 33.0 | 2157.5 | 1.11 |
| 13 | 218 | 37.0 | 2293.6 | 3.43** |
| 14 | 29 | 35.0 | 2177.2 | 0.68 |
| 15 | 47 | 19.0 | 1348.7 | −4.01** |
| 16 | 13 | 32.0 | 1944.2 | −0.26 |
| 17 | 49 | 18.0 | 1246.1 | −4.71** |
| 18 | 68 | 27.5 | 1871.5 | −1.12 |
| 19 | 33 | 22.0 | 1361.5 | -3.39** |
| 20 | 144 | 33.0 | 2182.1 | 1.60 |
| Total | 4,057 | | 2029.0 | |

*Kruskal–Wallis test* As noted, this test is based on rank sums and is a comparison of medians (See Sect. 9.8.3). The results of application of the test to the 4,057 observations of TTF are given in Table 17.10. The table includes the sample size $n$, the median, and the average rank, as well as a $z$-test for each component.

Note that the overall average rank is 4,058/2 = 2,029, as shown. The $z$-test is a test of whether or not the average rank for a given component is different from the overall average rank. The test is based on the asymptotic normality of the rank sum. For a two-tailed test, the critical value for a test at the 5% level is 1.96; at the 1% level, the value is 2.576, and the null hypotheses is rejected if $|z|$ exceeds the critical value at the selected level of significance. The null hypothesis in each case is that the median for the component is not different from the overall median. In the table, the calculated $z$'s are marked, in the usual fashion, with a "*" to indicate significance at the 5% level, and a "**" to indicate significance at the 1% level.

Positive values of $z$ indicate that the median for that component is greater than the overall median, indicating that the component has a longer MTTF (meaning in this case "median time to failure"). It is useful to know if this is the case, since it indicated that the component is not a major cause of warranty claims. If the $z$-value is negative, the component has a lower MTTF, i.e., would be expected to fail more frequently.

From the table, we see that the medians of components 1 and 9 differ from the average, testing at the 5% level, with that of component 1 being higher and that of component 9 being lower. The differences between the medians of components 3–6, 8, 10, 13, 15, 17, and 19 differ from the overall median testing at the 1% level, with the medians of components 3, 4, 6, 10, 15, 17, and 19 being lower than the average.

This information will be used in the next section in the selection of components for additional analysis.

*Selection of components for detailed analysis* We look at the following two criteria for selection of components: (1) significantly low MTTF, and (2) large number of failures leading to warranty claims.

The results of Table 17.10 are relevant for the first criterion. We will analyze all components for which the median is found to be significantly lower than the overall median, testing at the 5 or 1% level. Those with negative $z$-values that satisfy this criterion are components 3, 4, 6, 8–10, 15, 17, and 19.

We next look at number of claims. Curiously, the component with the most claims, helix ($C_5$) is the one with the largest observed conditional MTTF. (One would expect that it would have the smallest.) Because of the large number of claims, however, we will include this component in the analysis. Another component included as a result of a large numbers of claims is component 7.

*Component level comparisons* Here we analyze the 11 components identified above as the most important contributors to product failure and warranty cost. The analysis will include the same techniques that were used at the product level. At the component level, the objective is to identify the most frequently occurring failure modes. We omit details and give summaries of the results of the analyses for the selected components.

Summaries of the results of the ANOVA and Kruskal–Wallis procedures are given in Table 17.11. The results include the values of the test statistics as well as identification of the key failure modes for each component. These are selected on the basis of smallest means that are indicated by Tukey's test to be significantly different from those of other modes, and by the $z$-tests in the Kruskal–Wallis analysis for the corresponding result regarding medians.

From Table 17.11, we see that the ANOVA and nonparametric tests basically agree with regard to the overall tests of equality of means/medians. There is also considerable agreement with regard to key failure modes, i.e., those that need to be addressed in order to reduce warranty costs.

Related analyses and results (details not given here) are as follows:

- Histogram of TTF: the histogram is skewed right for Modes 3, 4, 6, 10, 15, 18, and 19. It is bimodal for Mode 2, and multimodal for Modes 4, 6, 7, 8, and 9.
- Residuals in ANOVA: skewed right for all modes except 5 and 7; skewed left for Mode 5.

**Table 17.11** Comparison of MTTF for failure modes of selected components

| Comp. | Procedure | | | | | |
|---|---|---|---|---|---|---|
| | ANOVA | | | | Kruskal–Wallis | |
| | *F* | *df* | *p* | Key modes | *p* | Key modes |
| 3 | 9.19 | 10,303 | 0.000 | 182,186 | 0.000 | 182,186 |
| 4 | 42.20 | 7,355 | 0.000 | 409,410 | 0.000 | 409,410 |
| 5 | 96.45 | 11,1272 | 0.000 | 488,449,442 | 0.000 | 436,438–440,442,443 |
| 6 | 35.84 | 4,139 | 0.000 | 446 | 0.000 | 446 |
| 7 | 23.14 | 12,324 | 0.000 | 6 modes | 0.000 | 215,225,226,223 |
| 8 | 7.88 | 4,62 | 0.000 | 236,239 | 0.000 | 236 |
| 9 | 5.21 | 16,215 | 0.000 | None | 0.000 | 164,171 |
| 10 | 5.75 | 14,307 | 0.000 | 577,617 | 0.000 | 577 |
| 15 | 2.29 | 3,43 | 0.092 | None | 0.138 | None |
| 17 | 16.82 | 1,47 | 0.000 | 43 | 0.002 | 704 |
| 19 | 2.21 | 4,28 | 0.043 | None | 0.223 | 25 |

- Pareto charts: agree with selections of most important failure modes in terms of most frequently occurring.
- Boxplots: indicate that there are possible outliers for all modes except 7, 9, and 19. Almost all appear to be mild outliers. The analysis has not been pursued in detail.

Conclusions with regard to the analyses of this section are as follows:

- Distributions of TTF are generally skewed to the right, indicating forms typical of failure distributions. The exception is component 5, for which the distribution is skewed left. The same is true of residuals.
- Many distributions of TTF are multimodal, indicating that modes have different distributions and that, as a result, the distribution of TTF is a mixture.
- For components 15, 17, and 19, the median TTF's are significantly smaller than the overall median. In spite of this, since the n's are relatively small (each about 1% of claims), these will be not be analyzed further.
- Most importantly, reduction or elimination of the frequency of occurrence of the key failure modes identified above could lead to a significant reduction in warranty claims.

### *17.5.2 Selection of Failure Distributions*

Here we look at nonparametric and parametric models for the distribution of TTF, and at the characteristics of the best fitting distributions. The nonparametric approach is based on the EDF, discussed in Sects. 8.5 and 11.5. The parametric approach is based on goodness-of-fit tests, discussed in Sect. 10.3. In both cases,

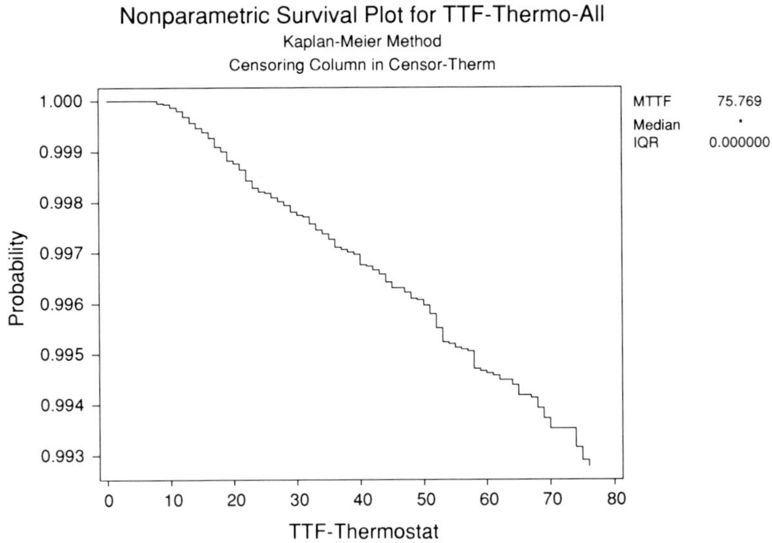


**Fig. 17.10** Nonparametric survival plot for component 3 (*Thermostat*)

we analyze all information, including censored data (Scenario 1.3) in the selection of a best fitting distribution. We look briefly at the component level, and then address the problem at the product level, which is the more important for our further analyses.

**Component Level** We look at component 3, the thermostat, in detail, and then summarize the results for the remaining components analyzed. Models used in attempting to represent the failure distribution of component 3 include the non-parametric EDF, standard parametric distributions, and mixture and competing risk models.

*EDF* Figure 17.10 is the Minitab survival plot for the component. The EDF is one minus the values shown on the plot. The plot appears to be reasonable, but the estimated MTTF shown is 75.769 weeks, or 1.5 years, which is not a reasonable result, given the failure patterns shown in the data and discussed above.

*Parametric models* In analyzing the claims data on component 3, fits to all of the distributions given as options in Minitab as well as mixtures of two Weibull distributions and the Weibull competing risk model were investigated.

The Minitab output for basic distributions (Weibull, exponential, normal, extreme value, lognormal, loglogistic, logistic) is given in Fig. 17.11. None of these distributions provides an acceptable fit to the data. Except for the exponential, for which the plot is not interpretable, all of the plots show a concave shape, with a modest to steep decline on the left.

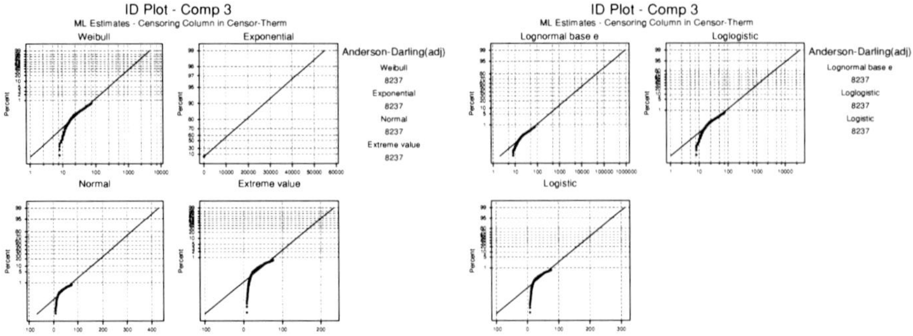


**Fig. 17.11** Distribution ID plots, basic distributions, component 3 (*Thermostat*)

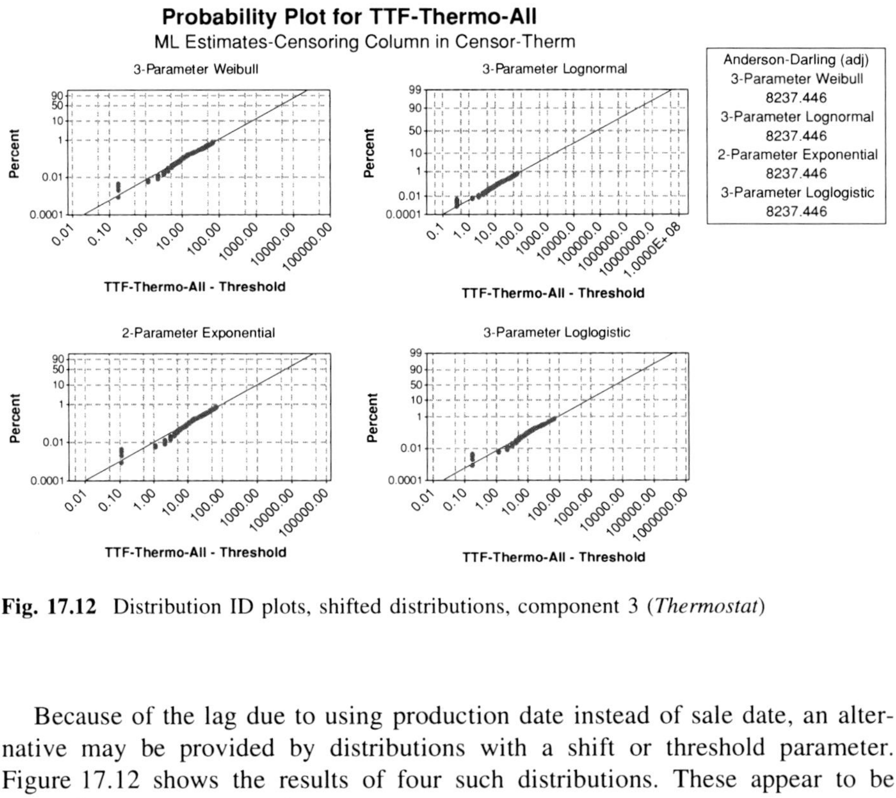


**Fig. 17.12** Distribution ID plots, shifted distributions, component 3 (*Thermostat*)

Because of the lag due to using production date instead of sale date, an alternative may be provided by distributions with a shift or threshold parameter. Figure 17.12 shows the results of four such distributions. These appear to be better fits.

The concave nature of the plots pointed out above suggests other possibilities, as discussed in Appendix D. In particular, the approximate linearity in the two segments of the plots in Fig. 17.11 suggests that a two-component mixture with well separated means or a competing risk model might provide a better fit.

**Table 17.12** MLE's of MTTF's for failure models, component 3 (Thermostat)

| Distribution | Estimated MTTF (years) |
|---|---|
| EDF | 1.5 |
| Two-parameter Weibull | 29.9 |
| Exponential | 229 |
| Normal | 4.9 |
| Extreme Value | 3.5 |
| Lognormal | 1,252 |
| Logistic | 3.8 |
| Loglogistic | 74.7 |
| Three-parameter Weibull | 12.5 |
| Three-parameter lognormal | 85,370 |
| Two-parameter exponential | 187.3 |
| Three-parameter loglogistic | 1,665 |
| Mixture of two Weibulls | 23.8 |
| Weibull–Weibull competing risk | 19.3 |

The fits to a two-component Weibull mixture model and a Weibull–Weibull competing risk model (not shown) also appear to provide better fits.

As a further check, we look at the estimated MTTF's for all of the above distributions. The results (MLE's) are given in Table 17.12.

The results are extremely variable, some seeming logically reasonable (three-parameter Weibull, normal), some remotely possible (logistic, Weibull, mixture of two Weibulls), some extremely unlikely (EDF), and some plainly impossible (exponential, lognormal, three-parameter lognormal). A look at individual failure modes is no more enlightening.

Similar patterns of poor fits and extreme variability in MTTF's are found for components 4, 6–10. In these, the lognormal appears graphically to provide a good fit, but the estimated MTTF is very large; the Weibull appears to fit in some cases, with a somewhat large MTTF. A different pattern of results is found for component 5. The anomalous results mentioned previously are driven by the fact that the two most frequently occurring modes, two codes described as "broken," both have large conditional MTTF's.

An additional anomaly present in all of the parametric analyses is the value of the AD statistic calculated by Minitab. It is invariably very large and the values are identical for all fitted distributions. Although the value is calculated in all cases of fitting, the properties of this statistic for highly censored data with parameters estimated are not known and the calculated values shown are of no use in selecting one or more candidate distributions for further analysis.

In all of the cases analyzed, some results appear to be quite feasible, but many more do not. We conclude that the modeling efforts at the component level are inconclusive, at best. Some possible causes of this may be:

- The extreme censoring of the data, particularly at the component level, where there are typically a few hundred TTF values and 90,000 or more censored observations.
- None of the distributions and models tried adequately describe the claims/failure data. Since there are typically many failure modes, the true model may be much more complex than any of those used, e.g., mixtures with more than two components, competing risk with more than two, or some of the other models discussed in Chap. 13. These have not been tested.
- All EDF analyses give an estimated MTTF of about 75 weeks, which suggests that the estimation procedure used is not valid for data of this type.
- For almost all components, the conditional MTTF's differ significantly (see Table 17.11). This suggests that a lower level analysis than that at the component level may be necessary.
- The data may also be a problem. As is apparent from Table 17.6, in many instances very few claims were received (e.g., Months 6 and 13 on, and in many individual cells of the table). The use of only month of claim (with some four-week and some five-week months) may be a problem.
- Another data problem is the lag between production and actual use of the product. As noted, it is not possible to estimate the amount of the lag from the information available. Two possible approaches to dealing with this are: (1) analyze only the first 12 or 14 months of data, and (2) select reasonable possible values for the average length of the lag and reanalyze the data accordingly.

In any case, further analysis is needed in order to determine failure distributions and MTTF's. Additional data and information are required in this effort.

**Product Level** We look next at the analysis at the higher, product level. Here some of the anomalies may be expected to be mitigated by the large number of claims analyzed. We again consider nonparametric and parametric approaches.

*EDF* The item level survival function is shown in Fig. 17.13. Note that the estimated MTTF is given as 73.193 weeks, or 1.4 years, slightly less than those of the separate components, and again not feasible, perhaps for one or more of the reasons mentioned above.

*Parametric models* We again look at the basic distributions modeled in Minitab, plus the Weibull, exponential and lognormal with location parameters. The results are given in Fig. 17.14. Again, all values of the AD statistic are essentially identical and very large. Most graphs also indicate the presence of a possible outlier at the lower end. The same plots after removing the suspect point were obtained. The overall appearance of the plots was not much changed, nor was the value of the AD statistic. Since there is no evidence of an error in recording or of any other anomaly, this point will be retained in the analysis.

We also consider the mixture of two Weibull distributions and the Weibull competing risk models in an attempt to account for the apparent mixture shape in

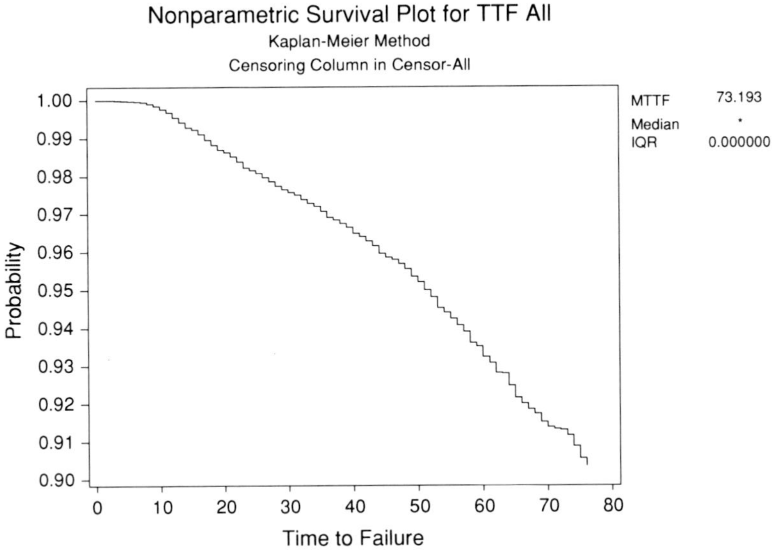


**Fig. 17.13** Empirical survival function based on all claims

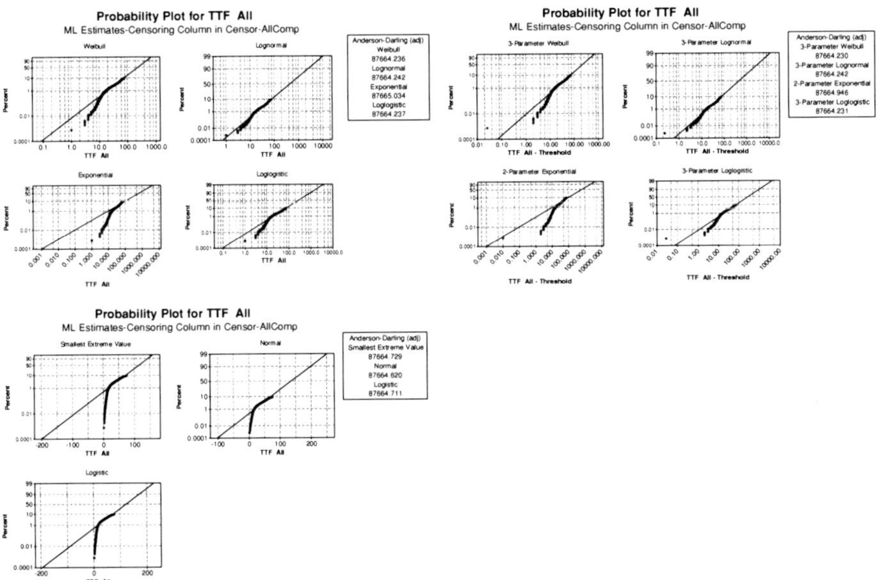


**Fig. 17.14** Distribution ID plots of TTF for all claims

the plots mentioned above. (These are not included in the figures below). None of the candidate models can be selected as a definitive best choice on the basis of the graphical analysis.

**Table 17.13** MLE's of MTTF's for failure models, all claims data

| Distribution | Estimated MTTF (years) |
|---|---|
| EDF | 1.41 |
| Two-parameter Weibull | 4.92 |
| Exponential | 18.37 |
| Normal | 2.60 |
| Extreme Value | 2.13 |
| Lognormal | 22.19 |
| Logistic | 2.34 |
| Loglogistic | 9.82 |
| Three-parameter Weibull | 5.14 |
| Three-parameter lognormal | 24.73 |
| Two-parameter exponential | 17.96 |
| Mixture of two Weibulls | 4.93 |
| Weibull–Weibull competing risk | 4.92 |

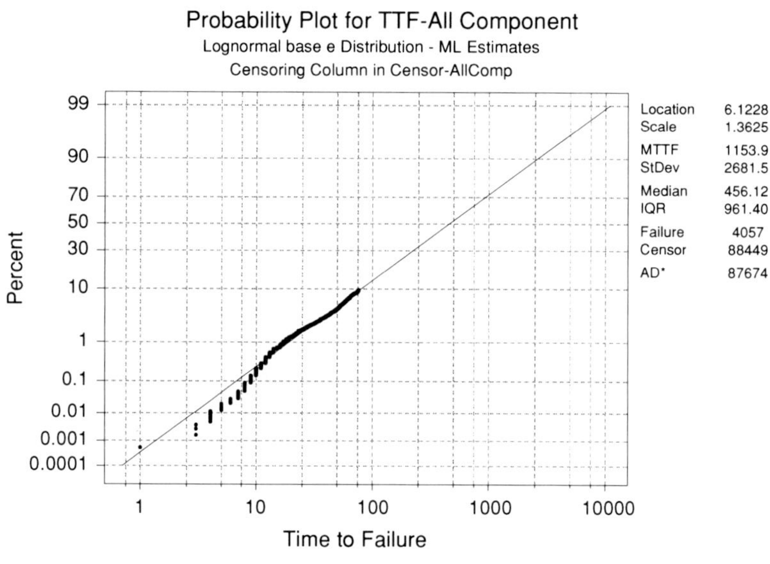


**Fig. 17.15** Fit of all claims data to the lognormal distribution

To obtain further insight into the validity of the various models, we again look at the estimated MTTF's for all of the models considered. These are given in Table 17.13. Again, the results show considerable variability, ranging from unrealistically low to very high. In short, the results for most models are feasible, though many are unlikely. In the analyses of the next sections, we will use several of the candidate models.

Graphically, the best fitting distribution is the lognormal. The detailed fit to this is shown in Fig. 17.15. The plot here again has the appearance of a mixture or

**Table 17.14** Parameter estimates and SE's for selected distributions

| Model | Parameters | MLE | SE |
|---|---|---|---|
| Weibull | Shape | 1.697 | 0.02255 |
| | Scale | 286.8 | 6.905 |
| Lognormal | Location | 6.123 | 0.03069 |
| | Scale | 1.362 | 0.01685 |
| Weibull–Weibull mixture | Shape 1 | 1.698 | 0.08078 |
| | Scale 1 | 300.0 | 92.461 |
| | Shape 2 | 1.696 | 0.05392 |
| | Scale 2 | 280.0 | 195.84 |
| | $p$ | 0.3616 | 5.697 |
| Weibull–Weibull Competing risk | Shape 1 | 1.697 | 0.04762 |
| | Scale 1 | 409.5 | 85.21 |
| | Shape 2 | 1.697 | 0.05540 |
| | Scale 2 | 456.9 | 111.81 |

other more complicated distribution and the MTTF is 1,154 weeks or 22.2 years, which is extremely large. The median, however, is 456 weeks or 8.8 years, which may be reasonable.

Finally, parameter estimates with estimated standard errors for the distributions discussed above are shown in Table 17.14. Note that the estimated SE's for the scale parameters of the Weibull mixture and competing risk models and for the mixture parameter $p$ of the mixture model are extremely large, so that the validity of the estimates is doubtful.

## 17.6 Estimation of Field Reliability

In estimating the field reliability of AC units, we use the basic two-parameter lognormal and Weibull models for TTF discussed above, as well as the Weibull mixture and competing risk models. The latter are an attempt to account for the various shapes apparent in fitting the standard distributions. In all cases, several values for the parameters will be used. Varying the parameter values will provide bounds of sorts on the results. The key information needed in the analysis is provided in Table 17.14. We do not use the three-parameter versions of the Weibull and lognormal distributions. It was thought that these might model the lag between production and sale, but the values of the threshold parameter in each case are relatively small and it is doubtful that they represent the actual lag. This must be taken into account when interpreting the results.

What we have chosen to do is basically a pragmatic approach, reflecting the uncertainty in the very tentative conclusions of the detailed analysis. This approach is acceptable as long as the results are understood in this light and the results are used with caution in arriving at management decisions.

**Table 17.15** Parameter values for reliability calculations

| Distribution | Parameter | Values | | | | | | |
|---|---|---|---|---|---|---|---|---|
| Weibull | Scale | 286.8 | 279.9 | 293.7 | 273.0 | 300.6 | 266.1 | 307.5 |
| | Shape | 1.697 | 1.674 | 1.720 | 1.652 | 1.742 | 1.629 | 1.765 |
| Lognormal | Location | 6.123 | 6.093 | 6.153 | 6.061 | 6.184 | 6.031 | 6.215 |
| | Scale | 1.362 | 1.346 | 1.379 | 1.329 | 1.396 | 1.312 | 1.413 |

In Sect. 17.3, the overall claims rate was estimated to be 4.41%. This was based on the observed 4,081 claims on the 92,530 units that had been sold during the observation period. This is a very low estimate of the *failure* rate.[5] It does not include failures under warranty that occurred after 6/30/98. It also does not include failures for which claims were not made, legitimate claims that may have been rejected, and so forth. This is important in interpreting the results given below.

As discussed in Chap. 3, reliability and failure rates are calculated as functions of time. In the context of warranty, a key measure of interest is the probability that an item does not fail during the warranty period. This is the major determinant of warranty cost. It is given by

$$R = R(W) = 1 - F(W), \tag{17.1}$$

where $F(.)$ is the CDF of time to failure $T$.

In principle, failure rates and reliability may be estimated parametrically or nonparametrically. The latter is based on the EDF. As we have seen, the EDF calculated from the data of this study leads to an unrealistically low estimate of the MTTF. As such, it will not be used here to estimate reliability.

For parametric estimates of reliability, we use the lognormal and Weibull distributions as well as the Weibull–Weibull mixture and competing risk models discussed in the previous section. Even though it is doubtful that any of these is the actual distribution of TTF, the analysis based on the data and these assumptions may provide some insight into future warranty costs.

Parameter values for the selected distributions are chosen as follows. As nominal values, we use those listed in Table 17.14. To investigate the sensitivity of the results to the parameters, we also use the nominal plus and minus one, two, and three times the standard errors given in Table 17.14 for the Weibull and lognormal distributions. For the Weibull mixture and competing risk models, we use only the nominal values because of the extreme values of SE's mentioned previously. The resulting sets of parameter values are given in Table 17.15 for the Weibull and lognormal distributions. The Weibull mixture and competing risk parameters are given in Table 17.14.

The parameter choices in Table 17.15 are the combinations that give the nominal, maximum and minimum estimates of reliabilities for the Weibull and

[5] In fact, as noted previously, claims rate is always an underestimate of failure rate, for a number of reasons, mostly connected with failures for which claims are not filed.

**Table 17.16** Estimated field reliability, Weibull distribution, $t$ = lifetime in years

| Parameter | | $R(t)$ | | | | | |
|---|---|---|---|---|---|---|---|
| $\alpha$ | $\beta$ | $t = 1.0$ | $t = 1.5$ | $t = 2.0$ | $t = 3.0$ | $t = 4.0$ | $t = 5.0$ |
| 286.8 | 1.697 | 0.9463 | 0.8961 | 0.8363 | 0.7006 | 0.5600 | 0.4289 |
| 279.9 | 1.674 | 0.9420 | 0.8889 | 0.8264 | 0.6867 | 0.5442 | 0.4132 |
| 293.7 | 1.720 | 0.9504 | 0.9028 | 0.8456 | 0.7140 | 0.5756 | 0.4445 |
| 276.0 | 1.652 | 0.9374 | 0.8814 | 0.8162 | 0.6725 | 0.5283 | 0.3975 |
| 300.6 | 1.742 | 0.9540 | 0.9090 | 0.8544 | 0.7269 | 0.5907 | 0.4599 |
| 266.1 | 1.629 | 0.9324 | 0.8733 | 0.8054 | 0.6577 | 0.5120 | 0.3818 |
| 307.5 | 1.765 | 0.9575 | 0.9150 | 0.8628 | 0.7394 | 0.6056 | 0.4754 |

Lognormal distributions, with confidence of roughly 50, 90 and 98% for the various combinations. In particular, the values in the first column of values (286.8 and 1.697) are the nominal values, the next two columns give nominal−SE and nominal + SE, and so forth.

Estimated reliabilities associated with these parameter values, defined to be $1 - P(T < t)$, are given in Table 17.16 for the Weibull distribution with parameters as listed in Table 17.15 ($\alpha$ is the scale parameter, $\beta$ the shape parameter) and various values of $t$ (in years). The failure probability is calculated as $1 - R(t)$.

Note that the estimated reliability of the product for the length of the warranty (1.5 years) at the nominal parameter values is 0.8961. Based on the results for parameter values plus or minus 2*SE*, the reliability is 88–91%. These results indicate that about 10% of the items can be expected to fail and lead to warranty claims within the warranty period, a rather high proportion of claims.

It is also seen from Table 17.16 that, as expected, the reliability decreases as $t$ increases, reaching less than 50% at $t = 5$ years

We next look at the corresponding results based on the assumption of a lognormal distribution of TTF. The parameter values ($\mu$ = location, $\sigma$ = scale) are those listed in Table 17.15. These include nominal values as well as estimated values plus or minus multiples of SE's, as for the Weibull. In this case, however, we look at combinations in which the positive deviation for one parameter is paired with the corresponding negative deviation for the other to assess their joint effect. The results are given in Table 17.17.

Both the Weibull and lognormal results indicate that the reliability of the AC unit, defined as the probability that the unit does not fail during the warranty period, is about 90%. Beyond the current warranty period, however, the estimated reliability based on the Weibull distribution decreases much more rapidly that does that based on the lognormal.

Finally, we look at the Weibull mixture and competing risk models at the nominal parameter values, given in Table 17.14. The results for the same values of $t$ are given in Table 17.18.

A comparison of these results with those of the previous tables shows that both the mixture and competing risk models give nearly identical results to those of the Weibull model with nominal parameters. It appears that one of the failure modes

Table 17.17 Estimated field reliability, lognormal distribution, $t$ = lifetime in years

| Parameter | | $R(t)$ | | | | | |
|---|---|---|---|---|---|---|---|
| $\mu$ | $\sigma$ | $t = 1.0$ | $t = 1.5$ | $t = 2.0$ | $t = 3.0$ | $t = 4.0$ | $t = 5.0$ |
| 6.123 | 1.362 | 0.9445 | 0.9025 | 0.8611 | 0.7845 | 0.7178 | 0.6600 |
| 6.092 | 1.379 | 0.9397 | 0.8958 | 0.8530 | 0.7751 | 0.7078 | 0.6500 |
| 6.153 | 1.346 | 0.9491 | 0.9090 | 0.8689 | 0.7939 | 0.7279 | 0.6702 |
| 6.061 | 1.396 | 0.9347 | 0.8889 | 0.8449 | 0.7656 | 0.6979 | 0.6401 |
| 6.184 | 1.329 | 0.9536 | 0.9155 | 0.8767 | 0.8033 | 0.7380 | 0.6805 |
| 6.031 | 1.413 | 0.9294 | 0.8819 | 0.9367 | 0.7562 | 0.6881 | 0.6303 |
| 6.215 | 1.312 | 0.9578 | 0.9217 | 0.8844 | 0.8127 | 0.7581 | 0.6910 |

Table 17.18 Estimated field reliability, Weibull mixture and competing risk models

| Distribution | $R(t)$ | | | | | |
|---|---|---|---|---|---|---|
| | $t = 1.0$ | $t = 1.5$ | $t = 2.0$ | $t = 3.0$ | $t = 4.0$ | $t = 5.0$ |
| Weibull Mixture | 0.9463 | 0.8961 | 0.8363 | 0.7007 | 0.5603 | 0.4293 |
| Weibull Competing Risk | 0.9463 | 0.8961 | 0.8362 | 0.7006 | 0.5600 | 0.4288 |

(risks) predominates. It is clear that if important decisions regarding reliability of the AC unit are to be made, additional data and further analysis are required.

## 17.7 Comparison of Warranty Policies

In evaluation of warranty policies, it is often useful to compare various policies that might be considered, both on new products and as alternative policies on existing products. The key concern is warranty cost. Modeling of warranty cost for 1-D warranties is discussed in Chap. 6 and for 2-D warranties in Chap. 7. Here we look briefly at a few alternatives to the nonrenewing FRW that covered the AC unit.

### 17.7.1 Nonrenewing FRW Alternatives

In considering alternative warranty terms, one of the first considerations is modifying the length of the existing warranty. Competitors' warranties play a role in this decision, but the most important initial concern is the cost to the company of such a modification. In the analysis of the previous section, we concluded that about 10%[6] of the units produced would result in warranty claims for the current warranty, for which $W = 1.5$ years. Since this is an excessive amount for products

[6] Note that this is the predicted percentage. The observed value of 4.1% is the percentage of units produced during the period of the study. Most of these units had not reached the warranty limit.

of this type, it is doubtful that a longer warranty would be considered unless the reliability of the product can be increased significantly.

If the market were such that a shorter warranty could be considered, the savings in warranty costs could be substantial. For example, it can be seen from Tables 17.16, 17.17, 17.18 that a one-year warranty would lead to about a 50% reduction in claims under all of the failure distributions considered.

Cost models for the nonrenewing FRW are discussed in Sect. 6.4. The models are based on the non-homogeneous Poisson Process (see Appendix B), with the expected cost of warranty given by (6.18). Application of the result requires estimation of the intensity function, which has not been pursued. For further discussion, see [1, 2, 3].

Note that a more detailed study of warranty costs must address the warranty as it is actually administered, which is more complex than the simple nonrenewing FRW in that the first failure is minimally repaired and the second leads to a replacement of the entire unit, which effectively treats the first failure as repairable and the second as nonrepairable. This has not been modeled, and new, more complex models are needed to carry out the analysis.

### 17.7.2 Alternative Policies

Here we comment briefly on a few alternatives that might be appropriate in principle for an AC unit. These are the nonrenewing PRW and the renewing FRW and PRW.

*Nonrenewing Pro Rata Warranty* The nonrenewing PRW is discussed in Sect. 6.4.3. This is basically a rebate policy, with a rebate that is a function of the amount of time in service of the unit prior to failure. The cost model for manufacturer is given in (6.26). Application of the model requires estimation of $F(W)$, given in the previous section, and of the partial expectation $\mu_W$. Estimation of $\mu_W$ for the Weibull distribution is discussed in detail in [1], Chap. 5. Because of the uncertainty in determining the distribution of TTF, we have not pursued the analysis in this case.

*Renewing FRW* The renewing free replacement warranty requires that replacements be supplied free of charge until one survives through the warranty period. This warranty is not often offered on items such as the room AC, since it is always more costly to the seller than the nonrenewing FRW. In this case, warranty costs are already high, and the renewing FRW would ordinarily not be considered. For comparison purposes, costs for the renewing FRW (assuming repair is "good-as-new") can be estimated as indicated in [1], Sect. 4.4.

*Renewing PRW* The renewing pro rata warranty may be a reasonable alternative to the FRW offered, except that it requires the buyer to purchase (at reduced cost) a replacement unit on failure of the original AC. This is not usually appropriate when the unit is repairable, unless the repair cost is substantial. This warranty is discussed in detail in [1], Sect. 5.3, and will not be considered further here.

## 17.8 Conclusions

We conclude with some comments on the results of this case, lessons learned, and suggestion for the practitioner regarding collection and analysis of 1-D warranty data. The following are a few observations regarding this case and analysis of 1-D data generally:

- The data of this study were relatively clean, with few extreme outliers and few obvious errors. In this sense, the data are of above-average quality of claims data.
- Pareto charts, as always, are useful in identifying key failure components and modes.
- Descriptive statistics and histograms of claims data are useful for summarizing and providing graphical representations of observed and reported failure times. They do not, however, provide an adequate description of the unconditional distribution of TTF.
- Detailed analyses of claims plus supplementary (censored) data in this case did not provide a solution to the problem of determining and estimating the distribution of TTF. Neither the EDF nor any of the distributions employed gave a useful and meaningful result. We conclude that (1) more complex models are needed to represent the AC unit failure times; (2) other data not available to the study (e.g., test data, follow-up information, etc.) are needed to more adequately estimate the field reliability of the unit and the cost of warranty.
- One of the key difficulties in analysis of the data is the somewhat extreme censoring encountered (about 95% of failure data are censored). This is not at all unusual for claims data, and it is essential to consider the censored data in fitting distributions. It does, however, invalidate the goodness-of-fit tests (e.g., the Anderson–Darling Test) and lead to difficulties in attempting to select a "best" fitting distribution.
- A data problem that was not addressed is the relative decrease in claims frequency during the last third of the observation period. It may be useful to reanalyze the data using only the first 12 months or some other period.
- In this study, limited data were available, with no possibility of obtaining additional information. In dealing with this, the approach taken here was to attempt to fit many models, look at various parameter values for each, calculate the corresponding reliability estimates, and compare the results. This provides a bound of sorts of the reliability, but the result should be used with caution because of the large areas of uncertainty.
- The models used in the study all predict that about 10% of the items will fail within the warranty period of $W = 1.5$ years. As $W$ increases, however, the predictions of the various models diverge. Additional data and further analysis is needed to characterize the reliability function.

- Conclusions regarding alternative warranties are very tentative. Based on this study alone, the only reasonable alternative appears to be to shorten the warranty period, if that is feasible within business objectives.
- There are many areas needing additional research. The key concern is the use of goodness-of-fit tests for censored data with unknown parameters. Estimation of parameters, reliability functions, and cost models using highly censored data also requires considerable further investigation.

## References

1. Blischke WR, Murthy DNP (1994) Warranty cost analysis. Marcel Dekker, New York
2. Blischke WR, Murthy DNP (eds) (1996) Product warranty handbook. Marcel Dekker, New York
3. Kalbfleisch JD, Lawless JF (1996) Statistical analysis of warranty claims data. In: Blischke WR, Murthy DNP (eds) Product warranty handbook, Chap 9. Marcel Dekker, New York

# Chapter 18
# Case Study 2: Analysis of Automobile Components Warranty Claims Data

## 18.1 Introduction

This chapter deals with a case study based on data collected in monitoring and dealing with warranty claims. The product is an automobile manufactured and sold in Asia in four different regions. As indicated in Example 1.2, an automobile is a complex system comprised of many components. We focus our attention on one component (which for reasons of commercial sensitivity we cannot name and call simply the "component"). We look at the warranty claims data for this component for automobiles sold over a 24-month period, and some supplementary data that the manufacturer provided for analysis.

The three approaches for analyzing 2-D data discussed in Chap. 14 are (1) 1-D conditional on usage, (2) composite scale, and (3) 2-D approach using bivariate distributions. Approaches 1 and 2 will be applied in this chapter. The outline of the chapter is as follows. Section 18.2 describes the objectives of the case study. Section 18.3 presents a description on data related issues, including the product and the component, the warranty, and the data used in this case study. In Sect. 18.4, we present the results of preliminary data analyses. Section 18.5 discusses Approach 1 for analyzing 2-D data, which is 1-D model conditional on usage rate. Section 18.6 discusses Approach 2 for modelling 2-D data, where the two scales—age and usage—are combined to define a single composite scale and to model the data based on the composite scale. In Sect. 18.7, we present the managerial implications for cost-benefit analysis, including improvement in reliability, reduction in warranty cost, and forecasting claims rates and costs. We conclude the case study with some suggestions for practitioners in Sect. 18.8.

W. R. Blischke et al., *Warranty Data Collection and Analysis*,
Springer Series in Reliability Engineering, DOI: 10.1007/978-0-85729-647-4_18,

## 18.2 Description of the Case

This section describes the case study, including (1) a description of the component and warranty, and (2) a discussion of the objectives of the study.

### *18.2.1 Description of Component and Warranty*

The product considered in this case study is an automobile component . Data are available for a 12-month period and are in batches consisting of total monthly production. The component is sold in Asia in four different regions. The operating environment in each region is different. The manufacturer has identified 14 different failure modes for the component. The component is non-repairable and the automobiles on which it is used are sold with a non-renewing 2-D FRW (Policy 5 of Chap. 2). The warranty region is a rectangle with $W = 18$ months (age limit) and $U = 100{,}000$ km (usage limit). The warranty ceases either due to the age or the usage limit being exceeded, whichever occurs first. The components were sold over 28 months and the data of the study were on claims that were filed over an observation interval of 29-months. The data provided are warranty claims data and supplementary data (see Chaps. 4 and 5) on a number of variables that are described in detail below.

### *18.2.2 Objectives of the Analysis*

In this study, we investigate questions of interest to the manufacturer regarding the reliability of the component in the context of the available data. The major objectives of the case are to:

- Focus on the reliability aspects of the component under consideration.
- Describe the preliminary analysis of data (see Chaps. 8, 9) to present summary statistics, graphs, etc., and to extract information to help in building statistical models.
- Investigate the relationship between age and usage of the component.
- Investigate the variation in component performance in different operating environments (from region to region), as discussed in Chap. 13.
- Model age conditional on usage rate and model the two variables age and usage by a composite scale variable.
- Discuss the effect of design changes—eliminating the dominant failure modes.
- Investigate production-related problems by looking at the reliability of items produced in different batches (monthly).

## 18.3 Data Provided for Analysis

This section presents a description of data related issues and describes the key features of the claims and supplementary data. It includes a list of available variables, units of measurement (transformed, if necessary, for confidentiality), censoring, number of censored and uncensored observations, and other pertinent information regarding the content and structure of the data.

### *18.3.1 Warranty Claims Data*

The structure of the claims data provided for analysis is similar to that of Scenario 5 of Structure 4 (see Fig. 14.2) discussed in Chap. 14. For each claim, the available data relating to component, customer, usage, etc., consisted of the following:

- Serial number of claim
- Date of production
- Date of sale
- Date of failure
- Age in days
- KM at failure
- Failure mode
- Region of use

The number of warranty claims over the 29-month period from the start of production was 2,230.

### *18.3.2 Supplementary Data*

The supplementary data provided for analysis has a structure similar to Structure 5 and Scenario 7 and discussed in Chap. 14 (see Fig. 14.2) and consisted of the following:

- Monthly production over 12 months
- Monthly sales over 28 months

For reasons of confidentiality, details of monthly production and sales are not disclosed. The total number of units produced over the 12-month period was 75,700. The total number sold was 75,666, implying that nearly all the items produced were sold, but with a lag[1] between production and sale dates.

[1] More on sales lag can be found in [6] and [5].

### *18.3.3 Data Problems*

It is important to note the following limitations of the data provided for analysis:

- Only product-related data (failure mode, age and usage at failure, etc.) are provided.
- There are several different agents servicing warranty claims. No information about the service agent or cost of servicing was provided.
- The monthly sales data were aggregated over all four regions.
- Data relating to usage are collected during the regular preventive maintenance servicing in 6-month or 12-month intervals. The manufacturer did not have access to these data and hence they are not available for analysis.

To overcome the above shortcomings, it is important to follow an adequate and effective data collection methodology that will provide information on all necessary variables.

## 18.4 Data Evaluation and Preliminary Analysis

The goals of this section are to evaluate the data and to deal with preliminary statistical analyses. As discussed in Chap. 8, preliminary analyses include basic descriptive statistics (mean, median, mode, standard deviation, etc.) and some graphical representations (histograms, scatter plots, Pareto charts, etc.) of important variables. The objective of the analysis of this section is to extract information to help in building models in Sects. 18.5 and 18.6.

Data analysis is done on the following three levels:

- *Level 1 (Aggregated Analysis)*: Analysis at the pooled level, aggregating over all regions and failure modes, to provide a single set of data.
- *Level 2 (Intermediate Analysis)*: Analysis at (1) failure mode level (aggregating over all regions but treating each failure mode separately—five sets of data) or at (2) region level (aggregating all failure modes but treating each region separately—four sets of data). The former would indicate the effect of failure modes on component life, ignoring region, and the latter indicate the effect of region on warranty claims, ignoring mode.
- *Level 3 (Detailed Analysis)*: Analysis at failure mode and region levels. This indicates the joint effect of different failure modes and regions. In this case we have 20 different data sets.

The preliminary analysis is done at all three levels. We begin the analysis at the pooled level (Level 1) and then analyze the data at the more detailed levels (Levels 2 and 3).[2]

---

[2] Note that as we move from Level 3 to Level 1, we are losing information. Level 3 is detailed whereas Levels 1 and 2 are aggregated.

## *18.4.1 Level 1—Preliminary Analysis*

### 18.4.1.1 Claims Data

Here the preliminary analyses of some variables are done based solely on claims data at Level 1 for 2,230 customers.[3] The data are Scenario 5 data and hence the results (based on claims data only) represent the conditional characteristics of the component. The heavy censoring of data makes the subsequent analysis more difficult. For the component, the main two variables that are important in field reliability studies are the age and usage at the time of warranty claim. The summary statistics for these two variables are given in Table 18.1.

Note that the means given in Table 18.1 are not the mean times to failure of all items, rather they are means of items that failed during the warranty period and led to claims.

Figure 18.1 is marginal plot of age and usage, that is, a scatter plot of age (in days) versus usage (kilometers driven) at failure under warranty with a histogram of the frequency of age and a histogram of the frequency of usage in the margins. This figure is useful in assessing the conditional distributions of two variables, age and usage, as well as the relationship between them.

The histogram of the age data indicates that the majority of cases are clustered at the upper end of the scale or toward the end of warranty age limit. The conditional distribution of age is negatively skewed. It is observed from Fig. 18.1 that after about 300 days, many components are likely to be in the highest usage group with the warranty for some ceasing due to usage limit.

The histogram of the usage data indicates that the majority of cases are clustered at the lower end of the scale, and the conditional distribution of usage is positive skewed. From the histograms, it can be seen that the number of claims is relatively higher during days 352–368 and usage 19,000–31,000 km.

The scatter plot shows a positive correlation between age and usage; that is, usage time increases (decreases) as the age increases (decreases), as expected. The Pearson correlation coefficient between age and usage is 0.564, with a $p$-value of 0.000 for the test of the correlation being zero (see Sect. 10.6.3). We conclude that, as would be expected, the correlation is not zero.

The results of this section are based solely on claims data and hence the results are of limited use. Fig. 18.1 is relevant for the 1-D, composite scale approach discussed in Sect. 18.6, since it provides information on usage rate needed for this analysis.

[3] There are a very small (negligible) number of components on which more than one claim was filed and this information is not given in the database. As a result, each new sale/replacement is identified with a unique customer number. We use the term "customer" to correspond to the identification number.

**Table 18.1** Summary statistics for the variables Age and Usage for failed items

| Statistics | Age (in Days) | Usage (in KM) |
|---|---|---|
| Mean | 326.01 | 29,488 |
| Standard error of Mean | 2.95 | 365 |
| Median | 346.5 | 27,194 |
| Standard Deviation | 139.46 | 17,257 |
| Coefficient of Variation | 42.78 | 58.52 |
| Minimum | 1 | 25 |
| Maximum | 540 | 96,110 |
| First quartile | 225 | 17,904 |
| Third quartile | 441.25 | 38,297 |
| Skewness | −0.44 | 0.97 |
| Kurtosis | −0.73 | 1.30 |

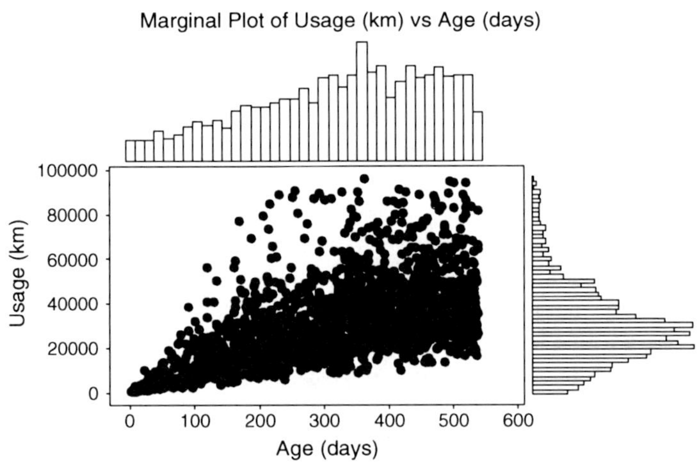


**Fig. 18.1** Age versus mileage plot of warranty claims data

#### 18.4.1.2 Claims Plus Supplementary Data

Here the preliminary analysis is done based on claims and supplementary data at Level 1. Data are Scenario 7 Data.

Censored data are important in fitting distributions and in estimating field reliability and related quantities, such as cost. The conditional analyses of previous sections give some idea of nature of the observed data, but are not adequate for these tasks. Including the censored data is especially important for two reasons: (1) the descriptive statistics (means, standard deviation's, etc.) and graphs (e.g., histograms) given are all descriptive only of the conditional distributions and can be misleading if interpreted as estimates applicable to the unconditional distribution of TTF; (2) in order to estimate reliability and costs, estimates of the parameters of the unconditional distribution are needed.

**Table 18.2** Number censored at each age

| Age at censoring $T$ (in months) | Number of units censored | Age at censoring $T$ (in months) | Number of units censored |
|---|---|---|---|
| 1 | 38 | 10 | 402 |
| 2 | 59 | 11 | 549 |
| 3 | 103 | 12 | 686 |
| 4 | 159 | 13 | 987 |
| 5 | 200 | 14 | 1,517 |
| 6 | 206 | 15 | 3,083 |
| 7 | 256 | 16 | 5,788 |
| 8 | 328 | 17 | 6,117 |
| 9 | 363 | 18 | 52,595 |

All of the data on items for which claims were not filed represent right-censored data, the censored values being the service times of the items. These are obtained (expressed in months) by tracking sales data through time. Thus they are available by MIS, since we have the sales data by month (see Sect. 11.5.1). The total number of units sold is 75,666 and total number of claims is 2,230. Thus the total number of units censored is 75,666 − 2,230 = 73,436. Table 18.2 shows censoring information, including number censored at each age. From the table, we see that for the component under study, the monthly sales amounts decrease with respect to calendar months and the age-based numbers of censored units increase.

The database includes information on the dates of production and sale. Therefore the sales lag is known. This was obtained from a MOP-MOS diagram. This diagram is not given for commercial reasons but we use the information for various kinds of analysis. The sales lag data indicate that a major percent of the components (76%) are sold within 3 months of production. For the sales lag (in months), the mean is 3.257, the median is 2.00, the standard deviation is 2.68, and the maximum is 21 months.

### *18.4.2 Level 2—Preliminary Analysis of Failure Modes*

The purpose of performing a failure mode analysis is to investigate the effects of several types of failures in a product or system. Pareto analysis (see Sect. 8.4.1) is an examination of failure mode frequency or relative frequency data to determine the most important failure modes that contribute to quality problems and to which quality improvement efforts should be directed. It is found that for the claims data of the units that are analyzed in this case study, fourteen failure modes were reported in the database. Figure 18.2 is a Pareto chart of failure modes, which are denoted FM01, FM02, …, FM14.

The Pareto chart indicates that the top 8 modes account for 98% of the total claims. The most common failure modes are FM01, FM02, FM03 and FM04,

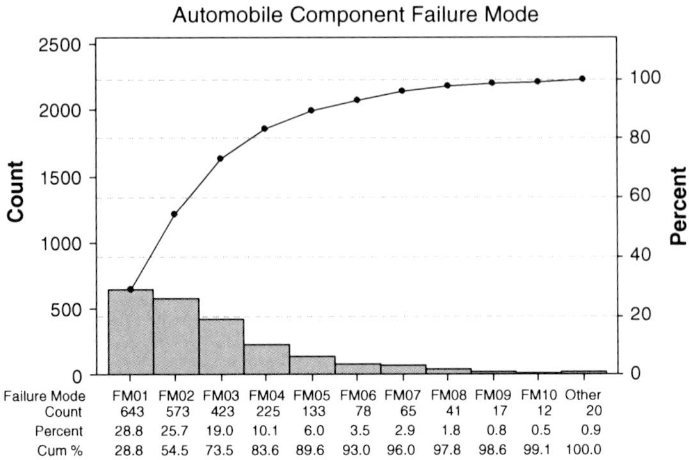


**Fig. 18.2** Pareto chart of automobile component failure modes

**Table 18.3** Summary statistics for the variables Age for main failure modes

| Failure Mode | $n$ | Mean | SE Mean | SD | CV | Min | Med | Max | Skew | Kurt |
|---|---|---|---|---|---|---|---|---|---|---|
| FM01 | 643 | 379.18 | 4.80 | 121.79 | 32.12 | 1 | 404 | 540 | −0.84 | 0.05 |
| FM02 | 573 | 285.30 | 5.69 | 136.15 | 47.72 | 1 | 298 | 538 | −0.15 | −0.89 |
| FM03 | 423 | 300.07 | 6.79 | 139.68 | 46.55 | 1 | 315 | 540 | −0.24 | −0.87 |
| FM04 | 225 | 342.12 | 8.79 | 131.86 | 38.54 | 28 | 357 | 539 | −0.57 | −0.44 |
| FM05–14 | 366 | 316.43 | 7.72 | 147.63 | 46.66 | 1 | 338 | 538 | −0.42 | −0.78 |

SE, SD, CV, Min, Max, Med, Skew and Kurt means standard error, standard deviation, coefficient of variation, minimum, maximum, median, skewness and kurtosis, respectively.

which account for 84% of the risk. Failure modes from FM05 to FM14 have considerably lower frequencies. Based on Fig. 18.2, we may conclude that efforts should be concentrated on failure modes FM01 through FM04 to eliminate them or to reduce the risks associated with them. That is, these failure modes should be the targets of investigation aimed at determining the root causes of failures and removing these root causes. We continue the analysis with special emphasis on these failure modes.

Summary statistics for the variables age and usage for the main failure modes are given in Tables 18.3 and 18.4, respectively. These statistics are also conditional on failures within the warranty period. Note that "FM05–14" means the aggregation of data for modes FM05 to FM10 plus "Other" in Fig. 18.2.

Tables 18.3 and 18.4 show that the mean and median failure times (with respect to both age and usage) are shortest for FM02 and longest for FM01. Figure 18.3 shows the histograms of the frequencies of age and usage by the five main failure modes. This figure indicates that for all of the five failure modes, the distributions

**Table 18.4** Summary statistics for the variables usage for main failure modes

| Failure mode | $n$ | Mean | SE Mean | SD | CV | Min | Med | Max | Skew | Kurt |
|---|---|---|---|---|---|---|---|---|---|---|
| FM01 | 643 | 32,886 | 634 | 16,082 | 48.9 | 29 | 30,222 | 94,732 | 1.06 | 1.78 |
| FM02 | 573 | 27,026 | 706 | 16,888 | 62.49 | 25 | 24,791 | 90,443 | 0.90 | 0.90 |
| FM03 | 423 | 29,003 | 862 | 17,719 | 61.09 | 38 | 27,051 | 88,321 | 0.91 | 0.86 |
| FM04 | 225 | 29,461 | 1,192 | 17,874 | 60.67 | 1,498 | 26,191 | 91,480 | 1.31 | 2.29 |
| FM05–14 | 366 | 27,947 | 943 | 18,037 | 64.54 | 33 | 24,973 | 96,110 | 1.06 | 1.48 |

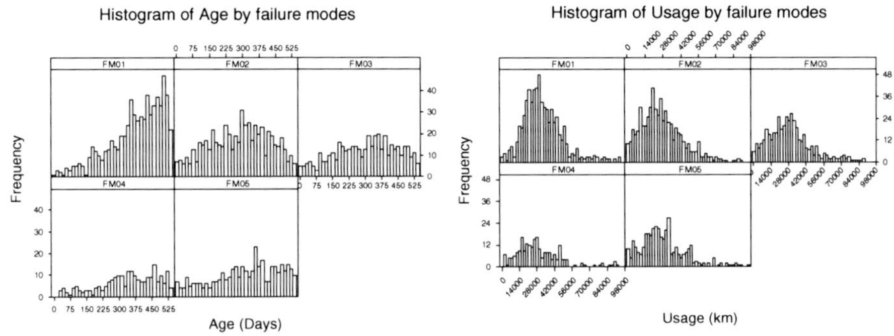


**Fig. 18.3** Histograms of age and usage for the main failure modes

of age are negatively skewed and the distributions for usage are positively skewed, as is also evident in Tables 18.3 and 18.4. The Pearson correlation coefficients between age and usage for FM01, FM02, FM03, FM04 and FM05–14 are 0.48, 0.57, 0.64, 0.49 and 0.59, respectively. In all cases, the $p$-values are 0.000 for the test of the correlations being zero.

To test equality of the means for the five failure modes and determine where significant differences exist, we use the ANOVA procedure with multiple comparisons (Sect. 10.5). Minitab outputs displaying the results for age and usage are shown in Figs. 18.4, 18.5, 18.6 and 18.7.

In Fig. 18.4 the $F$-value of 43.49 is significant at any level; the $p$-value given is 0.000. We conclude that the means of age for different failure modes differ. The output includes individual confidence intervals for the means, which give an indication of which population means may differ. To explore the differences among the means, the Minitab output for the Tukey's multiple comparison tests (see Sect. 10.5) is given in Fig. 18.5. Figure 18.5 indicates that the means for (FM02 and FM03), (FM03 and FM05–14) and (FM04 and FM05–14) are not statistically different because all of these the confidence intervals include 0. The means for the other seven combinations of failure modes are statistically different because the confidence intervals for these combinations of means exclude zero.

In the ANOVA table of Fig. 18.6, the $p$-value (0.000) for failure mode indicates that the means of usage for different failure modes differ. Tukey's multiple comparison tests given in Fig. 18.7 indicate that failure modes are not statistically

**One-way ANOVA: Age versus Failure mode**

```
Source          DF        SS      MS      F      P
Failure mode     4   3144011  786003  43.49  0.000
Error         2225  40208695   18071
Total         2229  43352706

S = 134.4   R-Sq = 7.25%   R-Sq(adj) = 7.09%


                                Individual 95% CIs For Mean Based on
                                Pooled StDev
Level       N   Mean  StDev  ---------+---------+---------+---------+
FM01      643  379.2  121.8                                (--*---)
FM02      573  285.3  136.1  (---*---)
FM03      423  300.1  139.7       (---*---)
FM04      225  342.1  131.9                  (-----*-----)
FM05-14   366  316.4  147.6          (---*----)
                             ---------+---------+---------+---------+
                                    300       330       360       390

Pooled StDev = 134.4
```

**Fig. 18.4** Minitab output of ANOVA and confidence intervals for means of age for failure mode

different with regard to average usage except for FM01 versus FM02, FM01 versus FM03, and FM01 versus FM05–14.

### *18.4.3 Level 2—Preliminary Analysis of Regions*

The component is used in four different regions, denoted R1, R2, R3 and R4. There are substantial differences among the usage environments (temperature, rainfall, road conditions, etc.) of these four regions that may effect the lifetime of the component. The purpose of the Level 2 analysis is to investigate the effect of region on the lifetime of the component.

Conditional summary statistics for the variables age and usage for the four regions are given in Tables 18.5 and 18.6, respectively. These tables show that the highest percent of the claims occurred in region R1 and the smallest percent occurred in region R2. Note that the region-wise sales amounts are not available, so the region-wise ratios of the numbers of failures to numbers sold are not given.

Table 18.5 shows that the mean and median failure ages are shortest for R2 and longest for R3. Table 18.6 shows that the mean and median failure kilometres are shortest for R2 and longest for R1.

We applied the ANOVA procedure to test equality of the means of age and usage for different regions. We found that the ANOVA tables with *p*-values (0.000) for both age and usage, indicating that the means of age and usage for different regions differ. Also the Tukey's multiple comparison tests indicate that all regions are statistically different from each other with regard to average age

```
Tukey 95% Simultaneous Confidence Intervals
All Pairwise Comparisons among Levels of Failure mode

Individual confidence level = 99.36%

Failure mode = FM01 subtracted from:

Failure
mode      Lower  Center  Upper  ---------+---------+---------+---------+
FM02     -115.0   -93.9  -72.8  (--*---)
FM03     -102.1   -79.1  -56.1    (---*---)
FM04      -65.5   -37.1   -8.6          (----*----)
FM05-14   -86.8   -62.7  -38.7       (---*---)
                                ---------+---------+---------+---------+
                                       -60         0        60       120

Failure mode = FM02 subtracted from:

Failure
mode     Lower  Center  Upper  ---------+---------+---------+---------+
FM03      -8.7    14.8   38.3                    (--*---)
FM04      28.0    56.8   85.7                          (---*----)
FM05-14    6.6    31.1   55.7                      (---*---)
                               ---------+---------+---------+---------+
                                      -60         0        60       120

Failure mode = FM03 subtracted from:

Failure
mode     Lower  Center  Upper  ---------+---------+---------+---------+
FM04      11.8    42.0   72.3                       (----*----)
FM05-14   -9.8    16.4   42.6                   (----*---)
                               ---------+---------+---------+---------+
                                      -60         0        60       120

Failure mode = FM04 subtracted from:

Failure
mode     Lower  Center  Upper  ---------+---------+---------+---------+
FM05-14  -56.8   -25.7    5.4            (----*----)
                               ---------+---------+---------+---------+
                                      -60         0        60       120
```

**Fig. 18.5** Minitab output of Tukey's multiple comparison tests for means of age for failure mode

except for R1 versus R3 and that all regions are not statistically different from each other with regard to average usage except for R1 versus R2, R1 versus R3, and R1 versus R4.[4]

The marginal plots of the failure age and usage for the four regions are shown in Fig. 18.8. The histograms of the frequencies of age and usage in Fig. 18.8 indicate that for all regions the distributions of age are negatively skewed and the distributions for usage are positively skewed. The scatter plots in Fig. 18.8 indicate the relationships between failure age and usage and show positive correlations between age and usage for the four regions (for R1, R2, R3 and R4 the values are

---

[4] The detailed results of ANOVA and Tukey's multiple comparison tests are not shown here.

**One-way ANOVA: Usage versus Failure mode**

```
Source         DF           SS          MS      F      P
Failure mode    4  11869202816  2967300704  10.13  0.000
Error        2225  6.51974E+11   293022173
Total        2229  6.63844E+11

S = 17118   R-Sq = 1.79%   R-Sq(adj) = 1.61%


                              Individual 95% CIs For Mean Based on
                              Pooled StDev
Level      N   Mean  StDev  --------+---------+---------+---------+-
FM01     643  32886  16082                            (-----*----)
FM02     573  27026  16888  (-----*-----)
FM03     423  29003  17719         (------*------)
FM04     225  29461  17874         (---------*---------)
FM05-14  366  27947  18037     (------*------)
                            --------+---------+---------+---------+-
                                27500     30000     32500     35000

Pooled StDev = 17118
```

**Fig. 18.6** Minitab output of ANOVA and confidence intervals for means of usage for failure mode

0.55, 0.61, 0.51 and 0.59, respectively). For all regions, the *p*-values are 0.000 for the test of the correlations being zero. The scatter plots also can be used to indicate that the usage rates differ from region to region. Both the mean and median usage rates are high for region R1 and low for region R3.

### *18.4.4 Level 3—Preliminary Analysis of Joint Effect*

The purpose of Level 3 analysis is to investigate the joint effect of different failure modes and regions. The summary statistics for the variables age and usage are given, respectively, in Tables 18.7 and 18.8 for Level 3 with 20 different data sets (5 FM $\times$ 4 R). In these tables, "FM*x*R*y*" means the combination of failure mode *x* and region *y*. The highest frequency occurs for FM01R1—failure mode FM01 and region R1.

The sample statistics given in Tables 18.7 and 18.8 for the different combinations of failure modes and regions provide useful information on the differences between failure times for failed items based on failure mode and/or region of use. In particular, one can look for interaction between these two factors. This involves comparing failure modes within regions, or, equivalently, comparing regions for each failure mode.

Interaction plots among failure modes and regions for both age and usage are given in Fig. 18.9. In these interaction plots, the failure mode by region for both age and usage plots show nonparallel lines, indicating that there is evidence of interaction in all of the plots. Those for usage are much more extreme.

```
Tukey 95% Simultaneous Confidence Intervals
All Pairwise Comparisons among Levels of Failure mode

Individual confidence level = 99.36%

Failure mode = FM01 subtracted from:

Failure
mode        Lower   Center   Upper    -+---------+---------+---------+--------
FM02        -8545    -5861   -3177    (-----*------)
FM03        -6809    -3884    -959        (------*-------)
FM04        -7044    -3425     194       (--------*--------)
FM05-14     -7999    -4940   -1880     (-------*------)
                                      -+---------+---------+---------+--------
                                   -8000     -4000         0      4000

Failure mode = FM02 subtracted from:

Failure
mode        Lower   Center   Upper    -+---------+---------+---------+--------
FM03        -1018     1977    4972                      (-------*------)
FM04        -1241     2435    6111                      (---------*--------)
FM05-14     -2205      921    4047                   (-------*-------)
                                      -+---------+---------+---------+--------
                                   -8000     -4000         0      4000

Failure mode = FM03 subtracted from:

Failure
mode        Lower   Center   Upper    -+---------+---------+---------+--------
FM04        -3397      458    4313                (--------*---------)
FM05-14     -4391    -1056    2279             (-------*--------)
                                      -+---------+---------+---------+--------
                                   -8000     -4000         0      4000

Failure mode = FM04 subtracted from:

Failure
mode        Lower   Center   Upper    -+---------+---------+---------+--------
FM05-14     -5472    -1514    2444          (---------*---------)
                                      -+---------+---------+---------+--------
                                   -8000     -4000         0      4000
```

**Fig. 18.7** Minitab output of Tukey's multiple comparison tests for means of usage for failure mode

**Table 18.5** Summary statistics for Age (in days) for different regions

| Region | $n$ | Mean | SE Mean | SD | CV | Min | Med | Max | Skew | Kurt |
|---|---|---|---|---|---|---|---|---|---|---|
| R1 | 1,410 | 337.66 | 3.62 | 136.10 | 40.31 | 1 | 357 | 540 | −0.54 | −0.56 |
| R2 | 110 | 259.50 | 12.80 | 133.80 | 51.55 | 1 | 255 | 517 | −0.04 | −0.87 |
| R3 | 146 | 342.70 | 10.80 | 130.80 | 38.17 | 7 | 364.5 | 539 | −0.51 | −0.47 |
| R4 | 564 | 305.55 | 6.11 | 145.18 | 47.52 | 1 | 315 | 540 | −0.25 | −0.96 |

**Table 18.6** Summary statistics for Usage (in km) for different regions

| Region | $n$ | Mean | SE Mean | SD | CV | Min | Med | Max | Skew | Kurt |
|---|---|---|---|---|---|---|---|---|---|---|
| R1 | 1,410 | 31,939 | 477 | 17,899 | 56.04 | 25 | 29,391 | 94,732 | 0.90 | 1.00 |
| R2 | 110 | 23,219 | 1,446 | 15,167 | 65.32 | 29 | 21,577 | 85,655 | 1.15 | 2.83 |
| R3 | 146 | 26,790 | 1,153 | 13,935 | 52.02 | 407 | 26,038 | 84,698 | 1.21 | 3.18 |
| R4 | 564 | 25,280 | 653 | 15,506 | 61.34 | 82 | 23,229 | 96,110 | 1.00 | 1.61 |

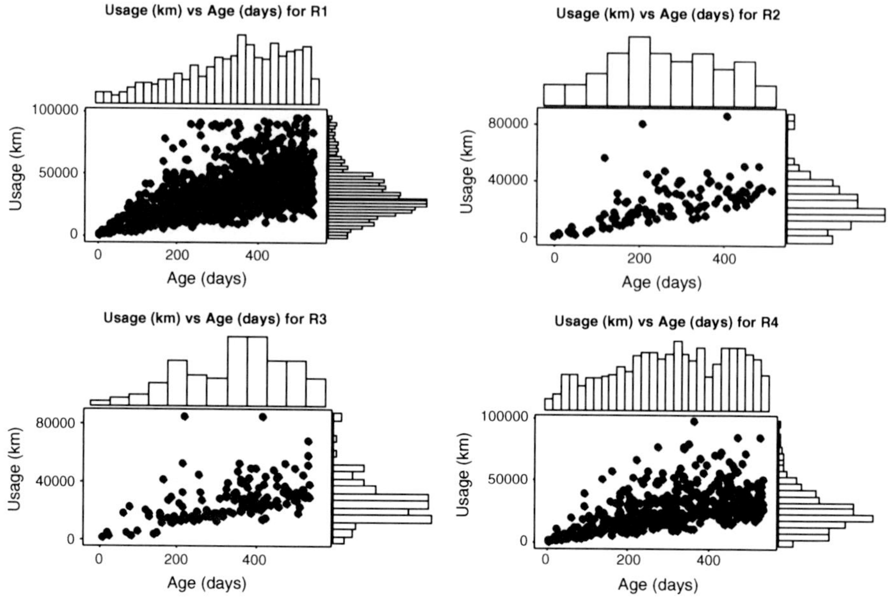


**Fig. 18.8** Age versus mileage plot of warranty claims data for different regions

**Table 18.7** Summary statistics for Age for different combinations of failure modes and regions

| Failure Mode-Region | $n$ | Mean | SE Mean | SD | CV | Min | Med | Max | Skew | Kurt |
|---|---|---|---|---|---|---|---|---|---|---|
| FM01R1 | 517 | 381.90 | 5.25 | 119.47 | 31.28 | 13 | 402 | 540 | −0.83 | 0.09 |
| FM01R2 | 22 | 353.80 | 26.80 | 125.70 | 35.52 | 1 | 378.5 | 517 | −1.22 | 1.53 |
| FM01R3 | 36 | 385.50 | 20.10 | 120.40 | 31.22 | 63 | 412.5 | 537 | −1.20 | 0.99 |
| FM01R4 | 68 | 363.30 | 16.80 | 138.30 | 38.07 | 59 | 409 | 539 | −0.63 | −0.76 |
| FM02R1 | 333 | 293.29 | 7.10 | 129.58 | 44.18 | 1 | 305 | 538 | −0.22 | −0.77 |
| FM02R2 | 37 | 231.60 | 19.20 | 116.50 | 50.31 | 1 | 221 | 493 | 0.23 | −0.30 |
| FM02R3 | 33 | 302.10 | 24.50 | 140.70 | 46.57 | 7 | 332 | 529 | −0.49 | −0.58 |
| FM02R4 | 170 | 278.10 | 11.40 | 149.20 | 53.64 | 3 | 273.5 | 534 | −0.05 | −1.10 |
| FM03R1 | 267 | 304.38 | 8.56 | 139.83 | 45.94 | 5 | 318 | 540 | −0.29 | −0.84 |
| FM03R2 | 24 | 257.30 | 28.60 | 140.20 | 54.50 | 1 | 266.5 | 485 | −0.26 | −0.69 |
| FM03R3 | 32 | 309.50 | 21.60 | 122.10 | 39.44 | 26 | 340.5 | 534 | −0.21 | −0.27 |
| FM03R4 | 100 | 295.80 | 14.40 | 144.40 | 48.82 | 20 | 310.5 | 540 | −0.11 | −1.09 |
| FM04R1 | 104 | 361.90 | 12.50 | 127.60 | 35.25 | 35 | 380 | 538 | −0.82 | −0.10 |
| FM04R2 | 9 | 208.80 | 44.50 | 133.40 | 63.90 | 41 | 153 | 428 | 0.44 | −1.16 |
| FM04R3 | 9 | 431.40 | 31.40 | 94.20 | 21.83 | 283 | 470 | 539 | −0.56 | −1.38 |
| FM04R4 | 103 | 326.00 | 12.70 | 129.30 | 39.67 | 28 | 322 | 539 | −0.40 | −0.37 |
| FM05R1 | 189 | 328.50 | 10.90 | 150.20 | 45.73 | 1 | 359 | 538 | −0.62 | −0.58 |
| FM05R2 | 18 | 230.10 | 29.80 | 126.60 | 55.03 | 42 | 204 | 479 | 0.43 | −0.38 |
| FM05R3 | 36 | 344.30 | 21.50 | 128.80 | 37.41 | 69 | 358.5 | 534 | −0.14 | −1.00 |
| FM05R4 | 123 | 302.40 | 13.30 | 147.30 | 48.70 | 1 | 302 | 538 | −0.28 | −0.85 |

**Table 18.8** Summary statistics for Usage for different combinations of failure modes and regions

| Failure Mode-Region | $n$ | Mean | SE Mean | SD | CV | Min | Med | Max | Skew | Kurt |
|---|---|---|---|---|---|---|---|---|---|---|
| FM01R1 | 517 | 33,775 | 727 | 16,528 | 48.94 | 767 | 30,611 | 94,732 | 1.02 | 1.51 |
| FM01R2 | 22 | 31,891 | 3,721 | 17,455 | 54.73 | 29 | 31,757 | 85,655 | 1.12 | 3.44 |
| FM01R3 | 36 | 31,355 | 2,497 | 14,980 | 47.78 | 1,825 | 30,159 | 84,440 | 1.08 | 3.25 |
| FM01R4 | 68 | 27,265 | 1,339 | 11,039 | 40.49 | 5,281 | 26,133 | 59,648 | 0.50 | 0.07 |
| FM02R1 | 333 | 28,885 | 954 | 17,408 | 60.27 | 25 | 26,122 | 90,443 | 0.92 | 0.89 |
| FM02R2 | 37 | 20,296 | 2,020 | 12,284 | 60.53 | 53 | 18,646 | 47,224 | 0.57 | −0.30 |
| FM02R3 | 33 | 21,075 | 1,554 | 8,925 | 42.35 | 407 | 22,136 | 38,982 | −0.51 | 0.14 |
| FM02R4 | 170 | 26,003 | 1,324 | 17,266 | 66.40 | 329 | 24,055 | 84,168 | 0.70 | 0.19 |
| FM03R1 | 267 | 30,417 | 1,118 | 18,264 | 60.05 | 475 | 28,256 | 88,321 | 0.83 | 0.66 |
| FM03R2 | 24 | 23,776 | 2,526 | 12,375 | 52.05 | 38 | 26,040 | 42,362 | −0.49 | −0.65 |
| FM03R3 | 32 | 29,563 | 2,778 | 15,713 | 53.15 | 2,416 | 27,691 | 67,722 | 0.49 | −0.32 |
| FM03R4 | 100 | 26,301 | 1,759 | 17,595 | 66.90 | 1,201 | 23,034 | 83,386 | 1.25 | 1.68 |
| FM04R1 | 104 | 35,932 | 1,994 | 20,335 | 56.59 | 2,774 | 34,035 | 91,480 | 1.08 | 1.11 |
| FM04R2 | 9 | 14,263 | 3,547 | 10,641 | 74.60 | 2,311 | 12,410 | 30,939 | 0.55 | −1.10 |
| FM04R3 | 9 | 25,773 | 2,263 | 6,790 | 26.35 | 15,468 | 26,770 | 34,343 | −0.58 | −0.83 |
| FM04R4 | 103 | 24,577 | 1,333 | 13,533 | 55.06 | 1,498 | 21,687 | 74,136 | 0.99 | 1.47 |
| FM05R1 | 189 | 32,248 | 1,417 | 19,484 | 60.42 | 33 | 29,550 | 89,227 | 0.71 | 0.40 |
| FM05R2 | 18 | 22,364 | 4,465 | 18,945 | 84.71 | 3,010 | 17,943 | 79,453 | 1.97 | 4.39 |
| FM05R3 | 36 | 25,253 | 2,450 | 14,699 | 58.21 | 3,300 | 24,071 | 84,698 | 2.05 | 6.96 |
| FM05R4 | 123 | 22,942 | 1,324 | 14,685 | 64.01 | 82 | 22,181 | 96,110 | 1.36 | 4.49 |

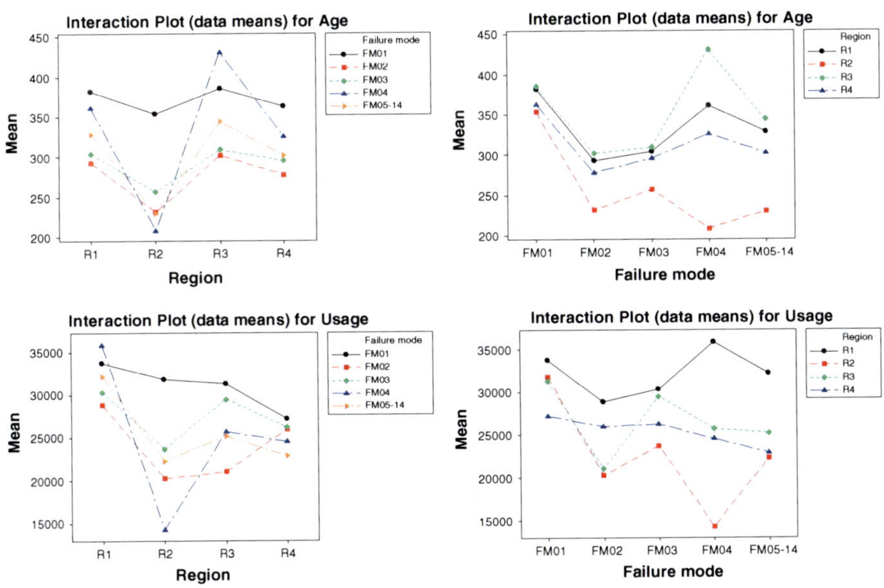


**Fig. 18.9** Interaction plots among failure modes and regions for age and usage

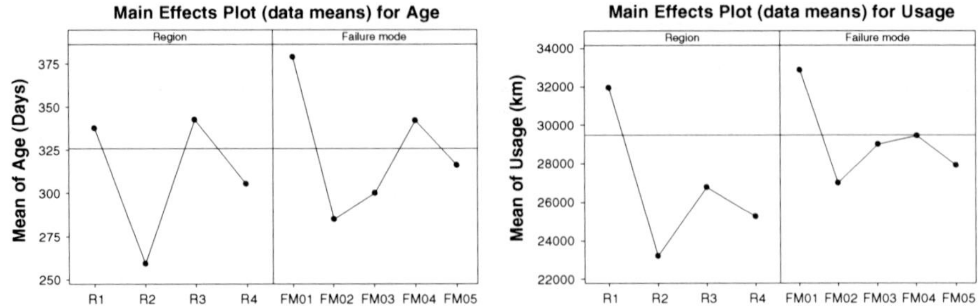


**Fig. 18.10** Main effects plots of failure modes and regions for age and usage

**General Linear Model: Age versus Failure mode, Region**

```
Factor         Type   Levels  Values
Failure mode   fixed       5  FM01, FM02, FM03, FM04, FM05-14
Region         fixed       4  R1, R2, R3, R4

Analysis of Variance for Age, using Adjusted SS for Tests

Source                 DF    Seq SS    Adj SS  Adj MS      F      P
Failure mode            4   3144011   1046349  261587  14.65  0.000
Region                  3    566034    674690  224897  12.60  0.000
Failure mode*Region    12    183911    183911   15326   0.86  0.590
Error                2210  39458750  39458750   17855
Total                2229  43352706

S = 133.621   R-Sq = 8.98%   R-Sq(adj) = 8.20%
```

**Fig. 18.11** Minitab output of GLM of age versus failure mode and region

These interactions are tested statistically by use of a model-fitting procedure such as GLM (General Linear Model), available in Minitab.

The main effects plots of failure mode and region for age and usage are shown in Fig. 18.10. Figure 18.10 shows the effects of failure mode and region upon age and usage, in each case ignoring the other variable. These are basically averages of the values in the plots of Fig. 18.9 and show roughly the same overall pattern, indicating that the interaction is not too severe, so that comparison of main effect levels is meaningful.

We next look at the statistical tests of main effects and interaction. The Minitab output of GLM of age versus failure mode and region and usage versus failure mode and region are given in Figs. 18.11 and 18.12, respectively. Figure 18.11 indicates that the interaction effects of failure mode and region on age are not statistically significant ($p = 0.590$). But in Fig. 18.12, the significance value for each term is less than 0.05, indicating that interaction is significant at 5% (but not at 1%) for usage. As a result, the main effects plots of Fig. 18.10 are fairly reasonable (i.e., not misleading) for age, but less so for usage.

**General Linear Model: Usage versus Failure mode, Region**

```
Factor        Type   Levels  Values
Failure mode  fixed       5  FM01, FM02, FM03, FM04, FM05-14
Region        fixed       4  R1, R2, R3, R4

Analysis of Variance for Usage, using Adjusted SS for Tests

Source                 DF       Seq SS       Adj SS      Adj MS      F      P
Failure mode            4  11869202816   5528759756  1382189939   4.87  0.001
Region                  3  18489109494  21876621865  7292207288  25.71  0.000
Failure mode*Region    12   6690594537   6690594537   557549545   1.97  0.024
Error                2210  6.26795E+11  6.26795E+11   283617480
Total                2229  6.63844E+11

S = 16840.9   R-Sq = 5.58%   R-Sq(adj) = 4.77%
```

**Fig. 18.12** Minitab output of GLM of usage versus failure mode and region

Based on the results of preliminary data analyses, the following managerial implications may be drawn:

- The main failure mode is FM02 which has the shortest mean and median failure times for both age and usage. To reduce its effect on warranty costs, efforts should be concentrated on eliminating or minimizing this failure mode.
- Extensive investigation on the environmental impact of region R2 on the lifetime of the component should be conducted (as the mean and median failure times for both age and usage are shortest for this region). Design changes may be needed to protect the component from the harshest environmental effects encountered in this region.
- Testing at the 5% level, (1) we cannot claim different mean ages between FM02 and FM03, FM03 and FM05–14, and FM04 and FM05–14; (2) we can claim different mean usages between FM01 and FM02, FM01 and FM03, and FM01 and FM05–14; (3) all regions are statistically different from each other with regard to average age except for R1 versus R3; and (4) all regions are not statistically different from each other with regard to average usage except for R1 versus R2, R1 versus R3, and R1 versus R4.
- At 5% level of significance, the interaction effects of failure mode and region are not statistically significant for age but significant for usage. This implies that the effect of failure mode on usage varies as a function of region.

## 18.5 Analysis Based on Conditional Usage Rate

This Section applies Approach 1 discussed in Sect. 14.4 for modelling age at failure as a function of usage rate , which is calculated as (thousands of km)/(age in months).

**Table 18.9** Anderson-Darling (Adj.) values for various distributions for usage-rate

| Distribution | AD (adj.) |
|---|---|
| Weibull | 66.33 |
| Lognormal | 14.38 |
| Exponential | 311.36 |
| Loglogistic | 12.85 |
| 3-Parameter Weibull | 38.80 |
| 3-Parameter Lognormal | 8.90 |
| 2-Parameter Exponential | 174.39 |
| 3-Parameter Loglogistic | 6.98 |
| Smallest Extreme Value | 256.02 |
| Normal | 103.70 |
| Logistic | 60.60 |

## *18.5.1 Level 1 Analysis*

### 18.5.1.1 Usage-rate Distribution based on Claims Data

For the 2,230 customers (all with a single failure), usage rates are calculated as $UR_i = z_i$ in units of 1,000 km per month (see Sect. 14.4.1). Here the usage rate is estimated based solely on claims data at Level 1. For the usage rate variable, the sample mean is 2.871, the standard deviation is 1.598, the minimum is 0.611, the median is 2.418, and the maximum is 14.000.

Eleven parametric distributions, listed in Table 18.9, are applied for modeling the usage rate data. The adjusted Anderson–Darling values for these distributions are shown in Table 18.9.

According to AD, the first, second, third and fourth choices of models are: 3-parameter loglogistic, 3-parameter lognormal, 2-parameter loglogistic and 2-parameter lognormal. Figure 18.13 shows the probability plots for usage rates for these four distributions. The 3-parameter loglogistic distribution has the smallest AD value; however, probability plots indicate that the 2-parameter lognormal model also fits the data nearly as well. We choose the 2-parameter lognormal model for the usage-rate variable $Z$ because of its simplicity in applications.

The lognormal distribution overview plot shown in Fig. 18.14, where the ML estimates of the location and scale parameters are given, respectively, as $\hat{\mu} = 0.9358$ and $\hat{\sigma} = 0.4708$, and the mean mileage accumulation rate as 2.848 (or 2,848 km/month).

Additional details on the distribution of usage rate may be obtained from follow-up surveys or recall data. Examples of these approaches are given in [8] and [19, 20].

### 18.5.1.2 Claims Plus Supplementary Data

The goal of this section is to derive an estimate of the PDF $f(t|z)$ based on claims plus supplementary (censored) data. To find these distributions, the usage-rates

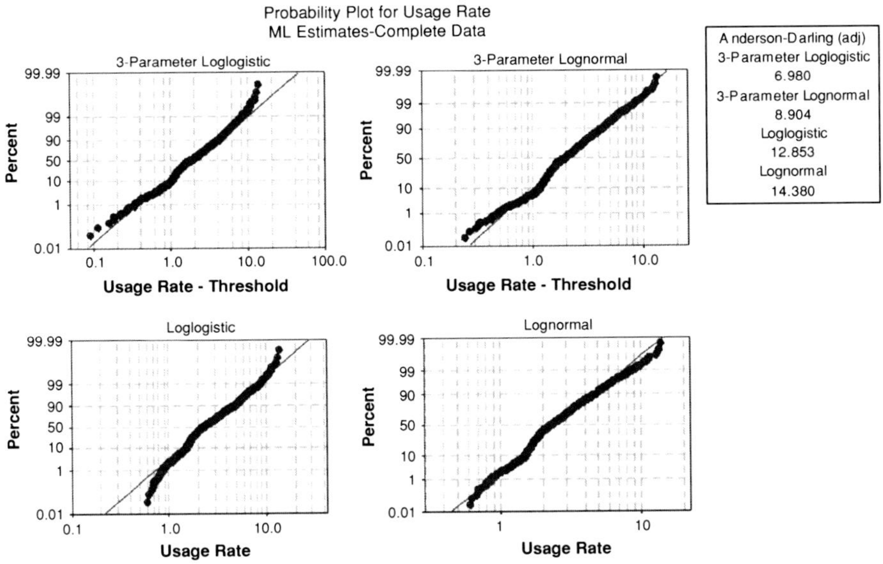


**Fig. 18.13** Four distributions probability plots for usage-rate (claims data only)

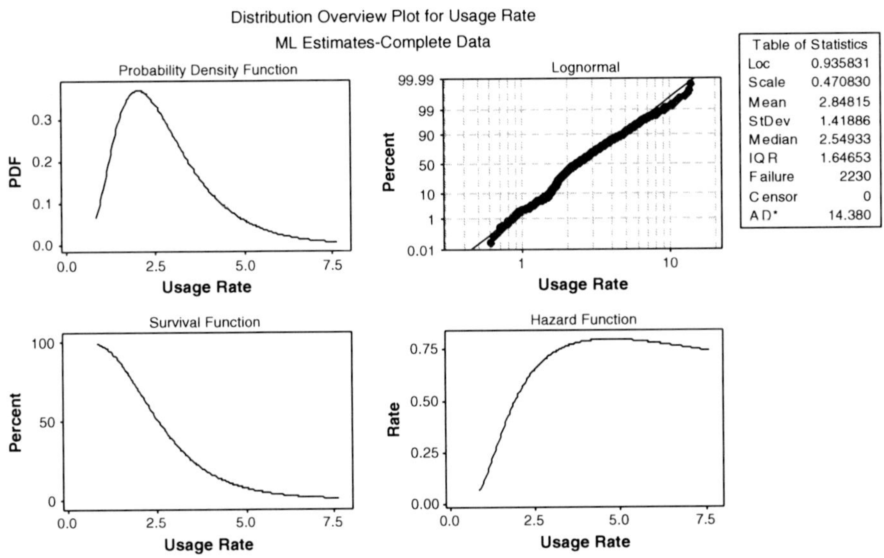


**Fig. 18.14** Lognormal distribution overview plot for usage-rate (claims data only)

(minimum = 0.6105, maximum = 14.00) are grouped into five groups. The groups and the number of customers in each group are given in Table 18.10.

The age-based failures are classified according to the five usage rate (UR) groups given in Table 18.10 and analyses are done by individual group. When

**Table 18.10** Usage-rate groups

| Group | Usage-rate | Mid-value (z) | Frequency | Percent |
|---|---|---|---|---|
| 1 | 0–2 | 1.00 | 783 | 35.11 |
| 2 | 2–3 | 2.50 | 698 | 31.30 |
| 3 | 3–4 | 3.50 | 385 | 17.26 |
| 4 | 4–6 | 5.00 | 257 | 11.52 |
| 5 | 6–14 | 10.00 | 107 | 4.80 |
| | Total | | 2,230 | 100.00 |

doing the analysis for a given UR group for a particular age, the other data are considered as censored for that age (originally censored items at that age are also included in the analysis as censored data). It was found that the lognormal distribution provides the best fit for age (in months) for individual groups, with the Weibull distribution as an alternative. Here we consider the lognormal distribution and estimate the MLEs of the parameters for different UR groups.

Using the same procedure applied in Example 14.1, we can express the location and scale parameters as functions of mid-values of usage-rate groups $z$ (given in Table 18.10) as follows[5]:

$$\mu_z = 5.305 + 0.8846z \tag{18.1}$$

and

$$\sigma_z = 1.270 + 0.2538z \tag{18.2}$$

Equations (18.1, 18.2) may be used to compare mean ages for a given $z$. For the Level 1 analysis with claims and supplementary data, the age-based lifetime ($T$) for given usage-rate ($z$) can be assumed to be a lognormal distribution with location parameter $\mu_z$ and scale parameter $\sigma_z$, defined, respectively, in (18.1) and (18.2). Thus,

$$f(t|z) = \frac{1}{\sigma_z t\sqrt{2\pi}} \exp\left[-\frac{(\log t - \mu_z)^2}{2\sigma_z^2}\right], \quad t \geq 0 \tag{18.3}$$

This is a one-dimensional model for $T$ conditional on $z$.

A comparison of the empirical distribution function (EDF)[6] and the estimated CDF based on (18.3) for given usage rate $z = 3.50$ is shown in Fig. 18.15. Figure 18.15 shows that the MLE of the CDF overestimates the observed at early ages (1–12 months), but at the later ages (13–18 months) it is very close to the nonparametric estimate given by the EDF.

[5] Regression analysis is used to find these relationships.

[6] Calculation of the empirical distribution function (EDF) is discussed in Sect. 8.5.1.

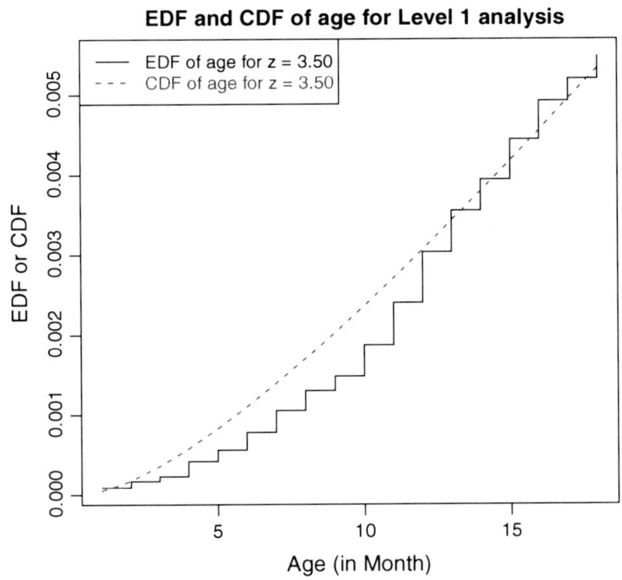


**Fig. 18.15** EDF and CDF of age for $z = 3.50$ for Level 1 analysis

## *18.5.2 Level 2 Analysis by Failure Mode*

This section discusses the analysis by individual failure mode.

### 18.5.2.1 Claims Plus Supplementary Data

In this section, we use Level 2 analysis to find suitable parametric forms for $f(t|z)$ for the five failure modes, FM01, FM02, …, FM05-14, of the component. When we looking at a single failure mode, all of the remaining items, including those that failed by another mode, are right-censored, with censoring times being TTF for items that failed by one of the other failure modes, and service times as above for the rest. The models for individual failure modes for different UR groups are selected based on the minimum AD values from a set of 4 models, Weibull, normal, lognormal and exponential. For each failure mode, the respective parameters are then expressed approximately as functions of usage rate $z$. The selected models with their parameters are summarized in Table 18.11.

Figure 18.16 shows the CDFs for the main failure modes for $z = 3.50$, constructed based on the results of Table 18.11. This figure compares the CDFs for different failure modes for $t = 1, 2, \ldots, 18$ months.

Figure 18.16 indicates that the probability of failure (survival) increases (decreases) rapidly with respect to age for FM01 compared with other failure modes.

**Table 18.11** Models for age conditional on usage rate for different failure modes based on claims and censored data

| Failure Mode | Model for $f(t\|z)$ | Parameters (Approximate) $[0 < z \leq 10]$ |
|---|---|---|
| FM01 | Lognormal | $\mu_z = 6.028 - 0.2373z + 0.1346z^2$ |
| | | $\sigma_z = 1.196 - 0.09447z + 0.03929z^2$ |
| FM02 | Lognormal | $\mu_z = 8.066 + 0.8291z$ |
| | | $\sigma_z = 2.123 + 0.2004z$ |
| FM03 | Lognormal | $\mu_z = 8.127 + 0.8987z$ |
| | | $\sigma_z = 1.971 + 0.2198z$ |
| FM04 | Lognormal | $\mu_z = 7.215 + 0.6567z$ |
| | | $\sigma_z = 1.526 + 0.1494z$ |
| FM05-14 | Lognormal | $\mu_z = 6.705 + 1.496z$ |
| | | $\sigma_z = 1.543 + 0.3820z$ |

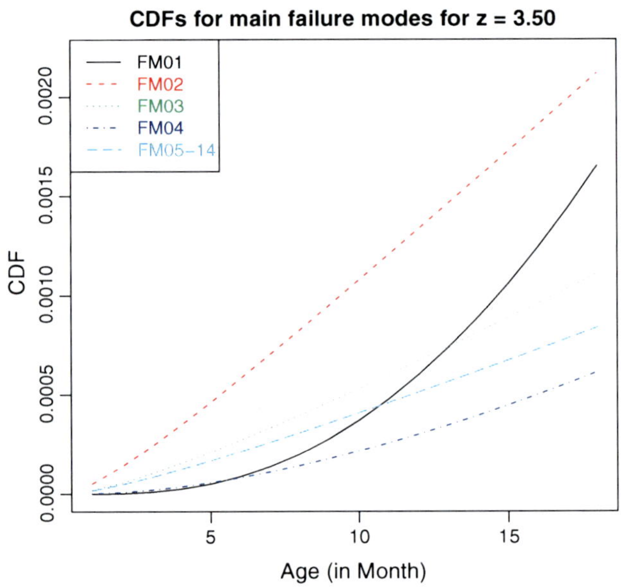


**Fig. 18.16** Comparison of CDFs for different failure modes for $z = 3.50$

Some observations regarding the results of this section:

- The usage rate distribution is a conditional distribution, as it is based only on failures that occurred within the warranty period and censored data are not taken into account. A usage rate distribution using supplemental data consisting of both failed and non-failed items (obtained from follow-up survey data or recall data[7]) would be more informative and useful.

[7] See, for example, [8, 17, 19, 20, 21].

- In this analysis, a two-dimensional problem is reduced to one-dimensional by the conditional distribution function of the time to failure, given the usage rate. For a given usage rate, the conditional distribution function of the time to first failure can be obtained.
- The derived relationships for the parameters and the distributions of $f(t|z)$ are approximate relationships and distributions. It is important to note that one should use caution when using such approximate results, especially when derived from data having an extreme number of censoring observations.

## 18.6 Analysis Based on Composite Scale Model Formulation

The composite scale model formulation (GK approach) discussed in Sect. 14.5 reduces the two variables Age (*T*) and Usage (*X*) to a single variable *V*, given in (14.17), for the *i*th observation, as a linear combination, namely

$$V_i = (1 - \varepsilon)T_i + \varepsilon X_i \tag{18.4}$$

where the weighting factor $\varepsilon$ is determined so as to minimize the sample coefficient of variation $cv = s_V/\bar{v}$ with $S_V$ and $\bar{v}$ being the sample standard deviation and mean of the transformed variable. $\varepsilon$ is obtained by a simple search over the region $0 \le \varepsilon \le 1$. *V* is called the composite scale. Given the value of $\varepsilon$, failures are characterized by the univariate distribution function $F_V(v)$.

For convenience, in the remainder of this section units are changed from days to months and km to 1,000 km. Thus the warranty period becomes $(W, U) = (18, 100)$. The data are Scenario 7 with claims plus supplementary (censored) data. For the data, we know ages at censoring times, but not usage. A number of approaches may be taken to approximating usage at censoring (see Chap. 14 and [3]). We use two approaches, one based on the mean usage of the component, and the other on the median usage. For the data, the average usage rate was found to be 2,871.4 km/month; the median was 2417.9 km/month. In this section, the result based on the mean usage is called Approach (i); that based on median usage is called Approach (ii).

The ages ($T_i$) and usages ($X_i$) at time of censoring as indicated above, are given in Table 18.12 for $i = 1, \ldots, 18$, for the two approaches.

The estimation of $\varepsilon$ and the model parameters involves a four-step procedure, as discussed in Sect. 14.5. Here we apply this procedure except that in Step 2, instead of nonparametric estimation of the CV, we fit the *V*'s to a hypothetical distribution using maximum likelihood and estimate the CV based on the fitted distribution [4].

The above procedure is applied to Levels 1 and 2 analyses[8] in the following sub-sections.

---

[8] Level 3 analysis is not given because there are very few numbers of failures for some cases (see Tables 18.7 and 18.8) compared with huge numbers of censored observations.

**Table 18.12** Ages and usages at censoring times for Approaches (i) and (ii)

| Age at censoring $T$ (in months) | Number of unit censored | Usage at censoring, $X$ (in 1,000 km) | |
|---|---|---|---|
| | | Approach (i) | Approach (ii) |
| 1 | 38 | 2.871 | 2.418 |
| 2 | 59 | 5.743 | 4.836 |
| 3 | 103 | 8.614 | 7.254 |
| 4 | 159 | 11.486 | 9.672 |
| 5 | 200 | 14.357 | 12.090 |
| 6 | 206 | 17.228 | 14.507 |
| 7 | 256 | 20.100 | 16.925 |
| 8 | 328 | 22.971 | 19.343 |
| 9 | 363 | 25.843 | 21.761 |
| 10 | 402 | 28.714 | 24.179 |
| 11 | 549 | 31.585 | 26.597 |
| 12 | 686 | 34.457 | 29.015 |
| 13 | 987 | 37.328 | 31.433 |
| 14 | 1,517 | 40.200 | 33.851 |
| 15 | 3,083 | 43.071 | 36.269 |
| 16 | 5,788 | 45.942 | 38.686 |
| 17 | 6,117 | 48.814 | 41.104 |
| 18 | 52,595 | 51.685 | 43.522 |

### *18.6.1 Level 1 Modeling*

Level 1 analysis is done based on Approaches (i) and (ii) by using the above procedure assuming the two-parameter Weibull distribution for the variable $V$. Figure 18.17 shows the estimated CV for given a set of values of $\varepsilon$ for Approach (ii). For this Approach, the estimate of $\varepsilon$ is 0.509 and the corresponding value of the CV is 0.6036. For Approach (i) the estimates of $\varepsilon$ and CV are 0.0017 and 0.6555.

The estimates of the scale parameter $\alpha$ and shape parameter $\beta$ of the Weibull distribution for $F_V(v)$ based on the two approaches, along with estimated standard errors and MTTFs are given in Table 18.13.

The resulting asymptotic confidence intervals for $\alpha$ and $\beta$ are given in Table 18.14.

For Approach (i), which uses the mean usages, $\varepsilon$ is very close to zero, meaning that $V$ is very close to $T$. Thus it is expected that the results for $V$ would be very close to those for $T$. To verify this, we fit the Weibull distribution to Age ($T$) for Level 1 data.[9] The MLEs of Weibull parameters are $\hat{\alpha} = 163.110$ and $\hat{\beta} = 1.5585$ and MTTF = 146.617 months. The results for Approach (i) given in Table 18.13 are very close to these estimates.

[9] Another choice can be the lognormal model for age for level 1 analysis.

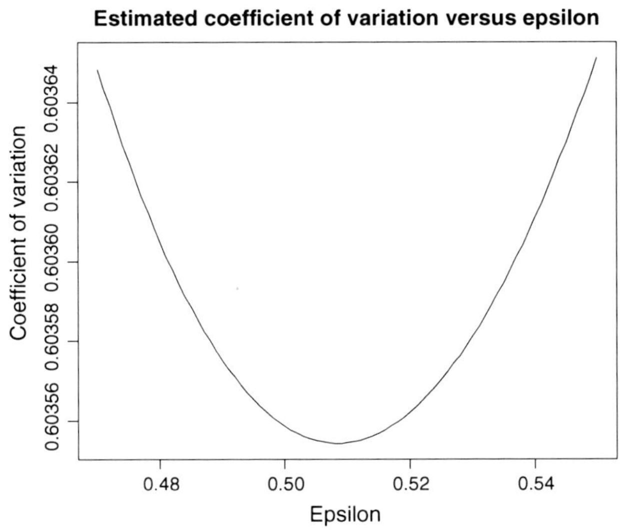


**Fig. 18.17** Estimated coefficient of variation as a function of $\varepsilon$ for Approach (ii)

**Table 18.13** Numerical results for the Weibull distribution for $V$

| Approach | $\hat{\alpha}$ | $\hat{\beta}$ | SE($\alpha$) | SE($\beta$) | MTTF |
|---|---|---|---|---|---|
| (i) | 163.626 | 1.558 | 7.973 | 0.0325 | 147.080 |
| (ii) | 231.699 | 1.706 | 10.284 | 0.0355 | 206.683 |

**Table 18.14** Confidence intervals for Weibull parameters

| Approach | 95% C.I for $\alpha$ | | 95% C.I for $\beta$ | |
|---|---|---|---|---|
| | Lower | Upper | Lower | Upper |
| (i) | 148.721 | 180.024 | 1.496 | 1.624 |
| (ii) | 212.394 | 252.758 | 1.638 | 1.777 |

The variations between the results of Approaches (i) and (ii) suggest that the outcome of the composite model depends highly on the assumed censoring values for usage that were given in Table 18.12. We recommend that both approaches by applied in analysis of a data set with the results then compare with previous results or results of similar components if available.

### *18.6.2 Level 2 Modeling*

As before, the failure modes are assumed to be competing risks, that is, when considering a particular failure mode, failures by other modes are considered to be censored and are treated the same as actual censored data.

**Table 18.15** Sample statistics for usage rate for various failure modes

| Failure Mode | $n$ | Mean | Minimum | Median | Maximum |
|---|---|---|---|---|---|
| FM01 | 643 | 2.7325 | 0.6105 | 2.3217 | 11.2601 |
| FM02 | 573 | 3.0158 | 0.6355 | 2.5051 | 13.6953 |
| FM03 | 423 | 3.0268 | 0.7382 | 2.6111 | 9.2815 |
| FM04 | 225 | 2.7030 | 0.7970 | 2.0960 | 12.7850 |
| FM05-14 | 366 | 2.8132 | 0.7018 | 2.2492 | 14.0000 |

**Table 18.16** Numerical results for the Weibull distribution for $V$ for Level 2

| Failure Mode | Approach | $\varepsilon$ | CV | $\hat{\alpha}$ | $\hat{\beta}$ | MTTF |
|---|---|---|---|---|---|---|
| FM01 | (i) | 0.0005 | 0.4399 | 122.571 | 2.42409 | 108.678 |
| | (ii) | 0.650 | 0.3831 | 172.465 | 2.82878 | 153.632 |
| FM02 | (i) | 0.0004 | 0.8062 | 848.262 | 1.24812 | 790.329 |
| | (ii) | 0.475 | 0.7507 | 1,094.72 | 1.3462 | 1004.35 |
| FM03 | (i) | 0.0003 | 0.7520 | 805.723 | 1.3437 | 739.459 |
| | (ii) | 0.510 | 0.7046 | 1,131.69 | 1.44106 | 1027.02 |
| FM04 | (i) | 0.00006 | 0.5626 | 400.42 | 1.84341 | 355.709 |
| | (ii) | 0.999 | 0.4472 | 413.964 | 2.37967 | 366.916 |
| FM05-14 | (i) | 0.00025 | 0.7496 | 885.22 | 1.34831 | 811.919 |
| | (ii) | 0.815 | 0.6654 | 1,110.93 | 1.53334 | 1000.34 |

Descriptive statistics for the usage rate variable $Z$ for various failure modes are given in Table 18.15. We note that the sample mean and median usage rates by failure mode given in Table 18.15 were used in Table 18.12 for estimating the approximate usage at censoring by Approaches (i) and (ii).

Again the two-parameter Weibull distribution is assumed for the composite scale variable $V$ for all failure modes. The estimates of $\varepsilon$, CV, and the scale parameter $\alpha$ and shape parameter $\beta$ of the Weibull distribution for $F_V(v)$ based on the two approaches, along with estimated MTTFs, are given in Table 18.16.

Approaches (i) and (ii) give very different results with respect to the estimates of parameters and MTTFs. With respect to the MTTFs, the conservative choice is Approach (i), which uses the mean usage at failure as the guessed value for usage of unfailed items. In Table 18.16, for Approach (i) and for all failure modes, $\varepsilon$ is very close to zero, meaning that the results for $V$ are very close to those for $T$. For this purpose, Table 18.16 can be compared with Fig. 18.18. Figure 18.18 shows the Weibull probability plots for Age for the main failure modes. The "Table of Statistics" given in this figure gives the MLEs of Weibull distributions for individual failure modes, $F$ and $C$ indicate, respectively, the number of failure and censored observations. The results of Approach (i) in Table 18.16 are very close to the output given in Fig. 18.18.

Results of Approach (i) and Approach (ii) highly depend on the assumed values of the censored usage. A comparison between Approaches (i) and (ii) in another application can be found in [3].

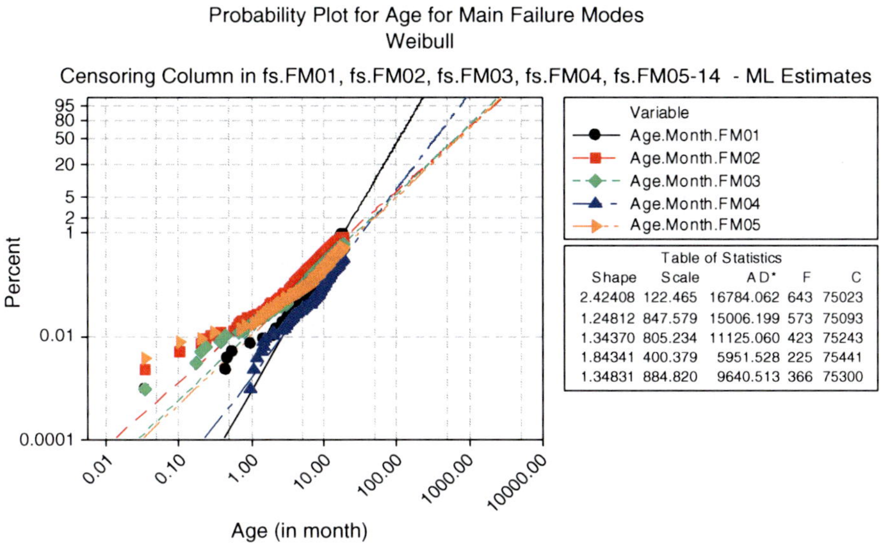


**Fig. 18.18** Weibull probability plots for age for main failure modes

## 18.7 Managerial Implications

First we summarize what we have found in the above analyses. The principal findings, in short, are as follows[10]:

- The components that failed during the warranty period have conditional means of 326 days for age and 29,488 km for usage. The Pearson correlation coefficient between failure age and usage is 0.564, and there is evidence at $\alpha = 0.01$ that the correlation is not zero.
- The component has four main failure modes, denoted FM01, FM02, FM03 and FM04, which jointly account for 84% of the risk. The preliminary results show that the mean and median failure times (with respect to both age and usage) are shortest for FM02. For all of the main failure modes, the distributions of age are negatively skewed and the distributions of usage are positively skewed. The outputs of ANOVA indicate that the means of age and usage for different failure modes differ.
- The component is used in four different regions, denoted R1, R2, R3 and R4. The mean and median failure times with respect to both age and usage are shortest for R2. The outputs of ANOVA indicate that the means of age and usage for different regions differ.
- The distribution of age conditional on usage rate is found to be lognormal, but the Weibull would be a reasonable alternative choice. The composite scale

[10] Some findings and managerial implications are also pointed out in Sect. 18.4.

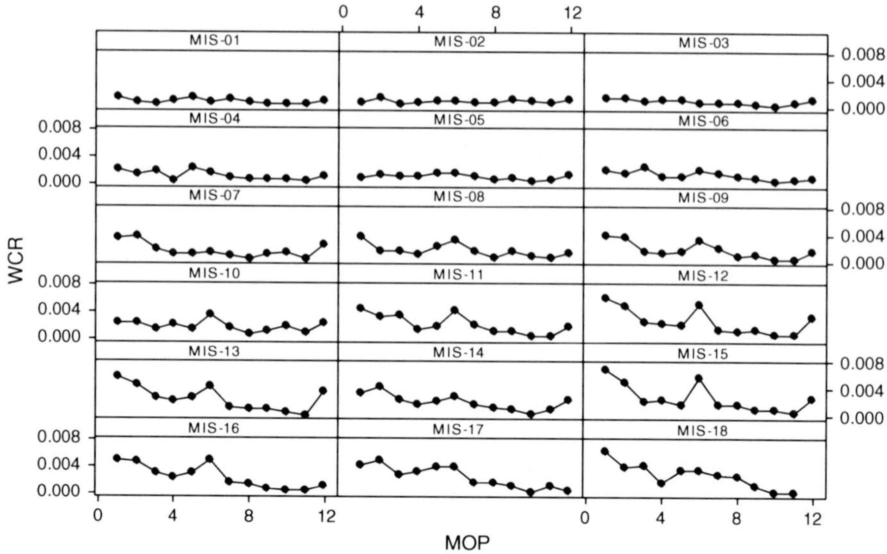


**Fig. 18.19** MOP-MIS chart of WCR for Level 1 analysis

variable is modeled by the Weibull distribution. Both methods reduce the analysis of the two-dimensional warranty to a one-dimensional problem and provide some useful information regarding the lifetime distribution of the component at level 1 and level 2.

Next we discuss the managerial implications that focus on the practical use of the information regarding product reliability and warranty for making decisions, e.g., (i) how the reliability varies with respect to MOP and MIS, (ii) how the reliability improves and warranty costs come down by eliminating the main failure modes (by changing design), and (iii) what is the role of the base warranty and extended warranty with respect to cost and customer satisfaction, etc. Based on the above information, managerial decision can be made concerning whether or not the potential gain is worth the improvement effort.

### *18.7.1 MOP-MIS Diagrams*

The MOP-MIS diagram, discussed in Chap. 11, can be used to compare the quality of the component from warranty claim rates (WCR) by each MOP and MIS. Here we construct a MOP-MIS diagram based on the method given in Sects. 11.5.2 and 11.5.3. Figure 18.19 is the MOP-MIS diagram for Level 1 analysis, which can help to determine if the failures are related to MIS and also to MOP.

As can be seen from Fig. 18.19, there is not much variation in WCR for MIS from Months 01–06 for all MOP. For some MIS, the WCR are significantly high

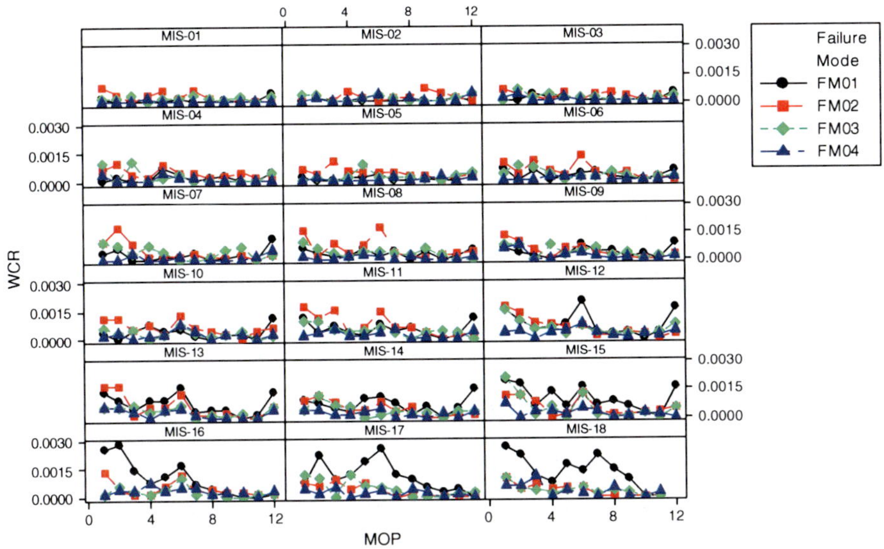


**Fig. 18.20** MOP-MIS chart of WCR for Level 2 (failure mode wise) data

for the first two MOP, for the sixth MOP, and for the last MOP. This indicates that there are some problems with the January, February, June and December MOP and these MOP need further investigation.

Figure 18.20 is the MOP-MIS diagram for Level 2 failure mode analysis. This figure can be used to investigate the WCR with respect to failure mode, MIS and MOP.

Figure 18.20 indicates that for the failure mode FM01 the WCR are very high compared with other failure modes for MIS 15–18 for all most all MOP. For some MIS and MOP, the WCR are high for the failure mode FM02. Therefore, to increase the reliability of the component, effort should be concentrated on failure modes FM01 and FM02. Elimination of these or reducing the risks associated with them would significantly decrease warranty claims.

The MOP-MIS charts given above are useful in determining whether or not there are problems in production. If the WCR for a MOP is unexpectedly high, this indicates that there may be production-related problems in that MOP.

### 18.7.2 Elimination of the Dominant Failure Mode

In this section, we look at modeling through elimination of the main failure modes one at a time. This enables us to investigate how the reliability of the component improves and the warranty costs are reduced by successively removing failure modes. Figure 18.21 shows the conditional reliability function $R(t|z)$ for various values of $z$ (1.0, 2.5, 3.5, 5.0, 10.0 and 100/18) for different ages $t = 1, 2, \ldots,$ 18 months. In the plots, “All FM included” means that $R(t|z)$ is given by

$$R(t|z) = (1 - F_{2FM01}(t|z)) \times (1 - F_{2FM02}(t|z)) \times (1 - F_{2FM03}(t|z)) \times (1 - F_{2FM04}(t|z)) \times (1 - F_{2FM05-14}(t|z)), \quad (18.5)$$

FM01 eliminated means the first term of the right side of (18.5) equals 1, and so on for other failure modes. $F_{ij}(.)$ is the CDF for Level $i$ analysis with failure mode $j$.

Figure 18.21 indicates that the effect of a failure mode on the component reliability does not vary much with respect to usage rate. This suggests that if we design out failure mode FM02, the reliability of the component improves vastly within the warranty period. This investigation is important not only for assessing warranty costs, but also for assuring customer satisfaction and product reputation.

### *18.7.3 Forecasting Claims and Costs*

Table 18.17 gives the conditional CDF of TTF given $z$, denoted by $F(t|z)$. In this table, $W^* = 18, 36$ and $72$ months and $z = 1.00, 2.50, 3.50, 5.00$ and $10.00$ (in 1,000 km per month), the mid-values of the usage rate groups. For $W^* = 18$ months, Pr(Claim|$W^*$, $z$) denotes the probability of claim occurrence within the each usage rate group, which is calculated by dividing the number of claims in that group, given in Table 18.10, by the total number of sales, 75,666. For example, for UR Group 1, Pr(Claim|$W^*$, $z$) = 783/75,666 = 0.0103. As the claim frequencies in usage rate groups 1–5 decrease (Table 18.10), Pr(Claim|$W^*$, $z$) also decrease from Group 1 to Group 5. Note that there are relatively far fewer in Group 5 than in Group 1.

$F_1(\text{Claim}|W^*, z)$ and $F_{2FM}(\text{Claim}|W^*, z)$ denote the conditional CDFs for Level 1 and for Level 2 (FM-wise), respectively (see Figs. 18.15, 18.16). For $W^* = 18$ months, $F_1(\text{Claim}|W^*, z)$ and $F_{2FM}(\text{Claim}|W^*, z)$ can be compared with Pr(Claim|$W^*$, $z$). One can see from the table that $F_1(\text{Claim}|W^*, z)$ and $F_{2FM}(\text{Claim}|W^*, z)$ give comparable values of the empirical estimate of Pr(Claim|$W^*$, $z$).

For the Level 1 analysis, the approximated total CDFs for 18, 36 and 72 months are 0.0332, 0.0864 and 0.1978. These results indicate that by Level 1 analysis the forecasted claims for warranty periods 18, 36 and 72 months are approximately 3.32, 8.64 and 19.78%, respectively. In using this method, it is assumed that the distribution of usage rate would be same in the various forecasting periods. The forecasted claims by Level 2 analysis for the same warranty period are also given in Table 18.17. Level 1 analysis provides conservative estimates of predicted CDFs for all prediction months compared with the Level 2 analysis.

## 18.8 Concluding Comments

In this case, we have attempted to analyze 2-D warranty claims data on a subsystem of an automobile. Two approaches, (1) 1-D model conditional on usage rate and (2) composite scale (Gertsbakh–Kordonsky method) model formulation,

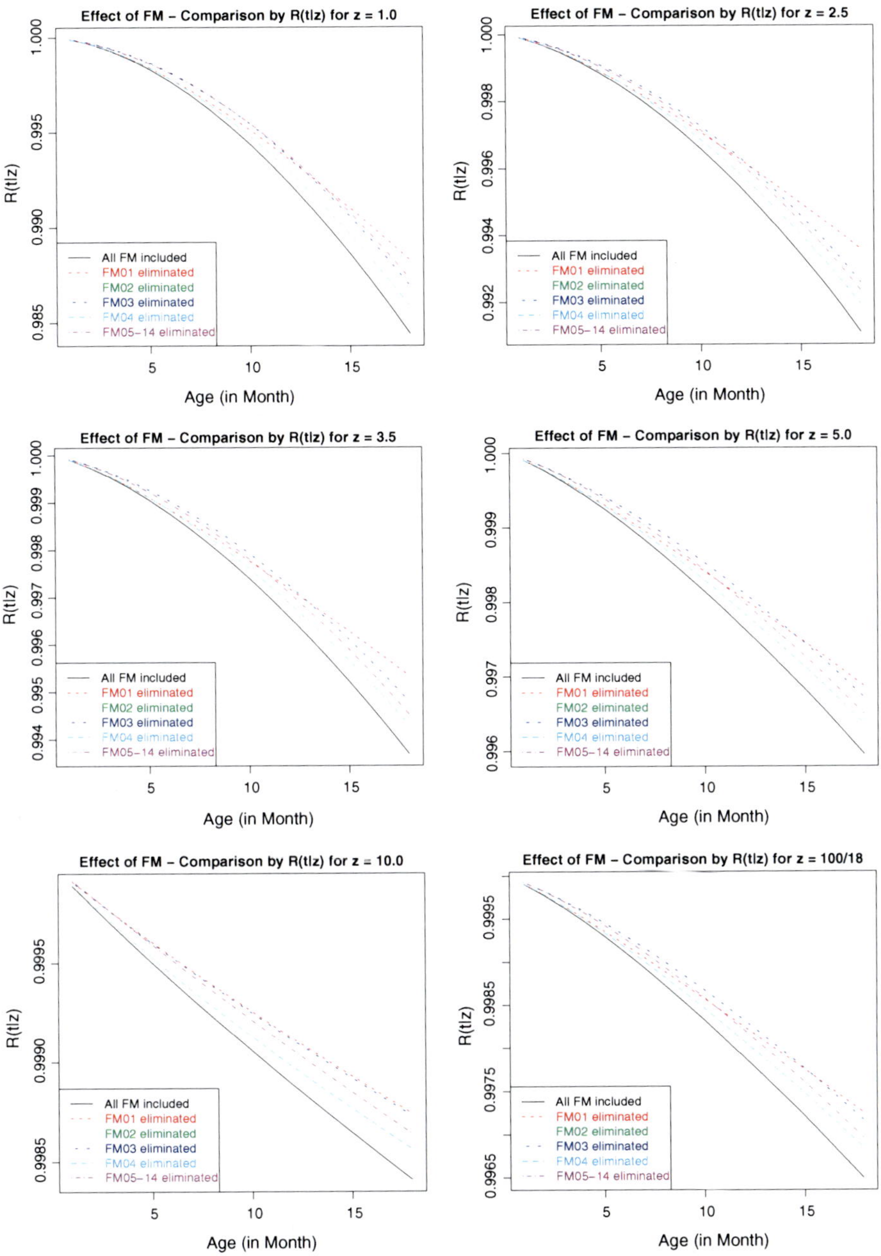


**Fig. 18.21** Conditional reliability functions $R(t|z)$ with failure modes successively eliminated

**Table 18.17** Prediction of conditional CDF $F(t|z)$

| $W^*$ | Estimates of | Usage Rate Groups | | | | | Total |
|---|---|---|---|---|---|---|---|
| | | Group1 | Group2 | Group3 | Group4 | Group5 | |
| 18 | $\Pr(\text{Claim}\|W^*, z)$ | 0.0103 | 0.0092 | 0.0051 | 0.0034 | 0.0014 | 0.0295 |
| | $F_1(\text{Claim}\|W^*, z)$ | 0.0152 | 0.0076 | 0.0053 | 0.0035 | 0.0016 | 0.0332 |
| | $F_{2FM}(\text{Claim}\|W^*, z)$ | 0.0156 | 0.0089 | 0.0063 | 0.0040 | 0.0016 | 0.0365 |
| 36 | $F_1(\text{Claim}\|W^*, z)$ | 0.0436 | 0.0195 | 0.0128 | 0.0078 | 0.0028 | 0.0864 |
| | $F_{2FM}(\text{Claim}\|W^*, z)$ | 0.0502 | 0.0280 | 0.0183 | 0.0101 | 0.0029 | 0.1095 |
| 72 | $F_1(\text{Claim}\|W^*, z)$ | 0.1047 | 0.0445 | 0.0280 | 0.0159 | 0.0048 | 0.1978 |
| | $F_{2FM}(\text{Claim}\|W^*, z)$ | 0.1419 | 0.0817 | 0.0508 | 0.0244 | 0.0052 | 0.3040 |

including preliminary analyses, were employed for analyzing the warranty data. Some findings and recommendations based on the case study are as follows:

- For this case, the results of Approach 1 depend on the assumed distribution and the estimated parameters that are functions of usage rate. In Approach 2, the assumed method of dealing with censoring affects the results substantially.
- Warranty claims data are generally error-prone, incomplete and highly censored. Therefore, special attention should be given by warranty practitioners to collecting, editing and properly analyzing warranty claims data.
- For this case study, the main limitations of the data are that (1) the information on service agent and the cost of servicing are not provided, (2) the region-wise sales amounts are not given, and there is no supplementary data collected during the mandatory 6 or 12 month servicing of the automobile. To overcome these limitations, it is important to follow a good data collection methodology with all necessary variables. Also, additional information on usage rate collected by means of a customer survey would be useful for the case study.
- The analysis based solely on claims data is used to extract information to help in building models, but it underestimates reliability and gives a biased picture of component reliability, whereas the results based on claims plus supplementary data gives a more accurate picture of component reliability.
- Since the region-wise sales amounts are not known, the region-wise age-based number of censored units are also unknown. As a result, for Level 2 analysis, only failure mode-wise modelling by Approaches 1 and 2 are given.
- The problem of characterizing the bivariate joint failure distribution of age and usage discussed in Sect. 14.6 has not been addressed. Several bivariate models for 2-D warranty policies can be found in [14, 15, 2, 7, 16, 1] and [11–13]. In addition, the estimation of age-based claims rates discussed in Sect. 12.5 has not applied in detail in this case study. Further applications of these methods would enrich the analysis of the data.
- If management were to decide to eliminate the dominant failure mode, the problem of whether it is due to manufacturing or design must be addressed. An analysis based on MOP is useful in this regard. This should give an idea of the possible improvement in reliability and reduction in warranty costs attainable by fixing this defect. Whether or not to go ahead with the necessary design

and/or production changes would be determined on the basis of a cost-benefit analysis. The process can then be repeated, addressing additional failure modes.

- This is a case in which the data are in the form of a highly censored sample, which is typical of warranty claims data. As a result, one should use caution when using the results. Statistics using only the reported failure data, ignoring censoring, can be misleading.
- Additional details on the analysis of automobile component warranty claims data can be found in [5, 8–10, 17–20, 22].

## References


1. Baik J, Murthy DNP, Jack N (2004) Two-dimensional failure modelling and minimal repair. Nav Res Logist 51:345–362
2. Blischke WR, Murthy DNP (eds) (1996) Product warranty handbook. Marcel dekker, Inc., New york
3. Iskandar BP, Blischke WR (2003) Reliability and warranty analysis of a motorcycle based on claims data. In: Blischke WR, Murthy DNP (eds) Case studies in reliability and maintenance. wiley, New york, pp 623–656
4. Jiang R, Jardine AKS (2006) Composite scale modeling in the presence of censored data. Reliab Eng Sys Saf 91:756–764
5. Karim MR (2008) Modelling sales lag and reliability of an automobile component from warranty database. Int J Reliab Saf 2:234–247
6. Karim MR, Suzuki K (2004) Analysis of field failure warranty data with sales lag. Pakistan J Statist 20:93–102
7. Kim HG, Rao BM (2000) Expected warranty cost of a two-attribute free-replacement warranties based on a bi-variate exponential distribution. Comput Ind Eng 38:425–434
8. Lawless JF, Hu XJ, Cao J (1995) Methods for the estimation of failure distributions and rates from automobile warranty data. Lifetime Data Anal 1:227–240
9. Lu MW (1998) Automotive reliability prediction based on early field failure warranty data. Qual Reliab Eng Int 14:103–108
10. Majeske KD (2003) A mixture model for automobile warranty data. Reliab Eng Sys Saf 81:71–77
11. Manna DK, Pal S, Sinha S (2006) Optimal determination of warranty region for 2D policy: A customers' perspective. Comput Ind Eng 50:161–174
12. Manna DK, Pal S, Sinha S (2007) A use-rate based failure model for two-dimensional warranty. Comput Ind Eng 52:229–240
13. Manna DK, Pal S, Sinha S (2008) A note on calculating cost of two-dimensional warranty policy. Comput Ind Eng 54:1071–1077
14. Moskowitz H, Chun YH (1994) A Poisson regression model for two-attribute warranty policies. Nav Res Logist 41:355–376
15. Murthy DNP, Iskandar BP, Wilson RJ (1995) Two-dimensional failure free warranties: Two-dimensional point process models. Oper Res 43:356–366
16. Pal S, Murthy GSR (2003) An application of Gumbel's bivariate exponential distribution in estimation of warranty cost of motorcycles. Int J Qual Reliab Manag 20:488–502
17. Rai B, Singh N (2003) Hazard rate estimation from incomplete and unclean warranty data. Reliab Eng Sys Saf 81:79–92
18. Rai B, Singh N (2004) Modeling and analysis of automobile warranty data in presence of bias due to customer-rush near warranty expiration limit. Reliab Eng Sys Saf 86:83–94

19. Rai B, Singh N (2005) A modeling framework for assessing the impact of new time/mileage warranty limits on the number and cost of automotive warranty claims. Reliab Eng Sys Saf 88:157–169
20. Rai B, Singh N (2006) Customer-rush near warranty expiration limit and nonparametric hazard rate estimation from the known mileage accumulation rates. IEEE Trans Reliab 55:480–489
21. Suzuki K, Karim MR, Wang L (2001) Statistical analysis of reliability warranty data. In: Balakrishnan N, Rao CR (eds) Handbook of statistics: advances in reliability–Volume 20. Elsevier Science, New York, pp 585–609
22. Yang G, Zaghati Z (2002) Two-dimensional reliability modeling from warranty data. Proc Annu Reliab Maintainab Symp 272-278

# Appendix A: Basic Concepts from Probability Theory

In this Appendix, we give a brief introduction to elementary probability theory, which is the basis of the mathematical approach to modeling failures. The presentation is non-rigorous. The objective is to develop an intuitive feel for the topic that forms the foundation for most models used in solving reliability related problems.

## A.1 Scalar Random Variables

A scalar random variable $X$ is useful in representing the outcome of an uncertain event. It can be either discrete or continuous. A discrete random variable takes on at most a countable number of values (for example, the set of nonnegative integers) and a continuous random variable can take on values from a set of possible values which is uncountable (for example, values in the interval $(-\infty, \infty)$).

Because the outcomes are uncertain, the value assumed by $X$ is uncertain before the event occurs. Once the event occurs, $X$ assumes a certain value. The standard convention used is as follows: $X$ (upper case) represents the random variable before the event and the value it assumes after the event is represented by $x$ (lower case).

### A.1.1 Distribution and Density Functions

The distribution function $F(x; \theta)$ is defined as the probability that $X \leq x$ and is given by

$$F(x;\theta) = P\{X \leq x\} \tag{A.1}$$

W. R. Blischke et al., *Warranty Data Collection and Analysis*,
Springer Series in Reliability Engineering, DOI: 10.1007/978-0-85729-647-4,

The domain of $F(x;\,\theta)$ is $(-\infty, \infty)$, the range is [0,1], and $\theta$ denotes the set of parameters of the distribution function. Often the parameters are omitted for notational ease, so that one uses $F(x)$ instead of $F(x;\,\theta)$. We will do this in the remainder of the appendix.

$F(x)$ has the following properties:

- $F(x)$ is a non decreasing function in $x$.
- $F(-\infty) = 0$ and $F(\infty) = 1$
- For $x_1 < x_2$, $P\{x_1 < X \leq x_2\} = F(x_2) - F(x_1)$

When $X$ is continuous valued and $F(x)$ is differentiable, the density function $f(x)$ is given by

$$f(x) = \frac{dF(x)}{dx} \tag{A.2}$$

$f(x)$ may be interpreted as

$$P\{x < X \leq x + \delta x\} \approx f(x)\delta x + O(\delta x^2). \tag{A.3}$$

When $X$ takes on only values in a set $(x_1, x_2, \ldots, x_n)$, with $n$ being finite or infinite, the probability that $X = x_i$ is given by

$$p_i = P\{X = x_i\}, \quad i = 1, 2, \ldots, n \tag{A.4}$$

In this case, $X$ is called a *discrete* random variable, and the CDF is a step function with steps of height $p_i$ at each of the possible values $x_i$.[1] $p_i$ has the following properties:

- $p_i \geq 0$ is a non decreasing function in $x$.
- $\sum_{i=1}^{n} p_i = 1$

#### A.1.1.1 Moments of Random Variables

The $j$th *moment* of the random variable $X, M_j(\theta)$ is given by[2]

$$M_j(\theta) = E[X^j] = \begin{cases} \int_0^{\infty} x^j f(x)dx, & \text{if } X \text{ is continuous} \\ \sum_x x^j P\{X = x\}, & \text{if } X \text{ is discrete} \end{cases} \tag{A.5}$$

The first moment of $X$ is called the *mean* and is usually denoted $\mu$, so that

---

[1] As before, the parameters may be omitted for notational ease, so that $p_i$ is often used instead of $p_i(\theta)$.

[2] The parameters are omitted for notational ease, so that one uses $M_j$ instead of $M_j(\theta)$. The same is true for $\mu_j(\theta)$.

$$\mu = E[X] \tag{A.6}$$

The *j*th *central moment* of the random variable $X$, $\mu_j$, is given by

$$\mu_j = E[(X - \mu)^j] \tag{A.7}$$

The second central moment of $X$ is called the *variance* and is usually denoted $\sigma^2$, so that

$$\sigma^2 = E[(X - \mu)^2] \tag{A.8}$$

$\sigma$ is called the *standard deviation*.

#### A.1.1.2 Fractiles of Distributions

For a continuous distribution, the *a*-fractile, $x_\alpha$, for a given $\alpha$, $0<\alpha<1$, is a number such that

$$P\{X \leq x_\alpha\} = F(x_\alpha) = \alpha \tag{A.9}$$

The fractiles for $\alpha = 0.25$ and 0.75 are called *first* and *third quartiles*, respectively, of the distribution, and the 0.50-fractile is called the *median*.

### A.1.2 Discrete Distributions

The following are some well known discrete distributions that are useful in failure modeling[3]:

*Bernoulli Distribution* Here $X$ takes on two possible values, 0 and 1, with probabilities given by

$$p_0 = p \quad \text{and} \quad p_1 = (1 - p) \tag{A.10}$$

The parameter set is $\theta = \{p\}$, with $0 \leq p \leq 1$. The mean and variance are

$$\mu = p \quad \text{and} \quad \sigma^2 = p(1 - p) \tag{A.11}$$

*Binomial Distribution* $X$ assumes integer values from 0 to $n$, where $n$ is a positive integer and $p_i$, $0 \leq i \leq n$, is given by

$$p_i = \frac{n!}{i!(n-i)!} p^i (1 - p)^{(n-i)} \tag{A.12}$$

[3] Most basic books on statistics and probability discuss some of the well-known distributions. References [14] and [15] give a more comprehensive coverage of many discrete distributions.

The parameter set is $\theta = \{n, p\}$ with $0 \le p \le 1$ and $0 < n < \infty$. The mean and variance are

$$\mu = np \quad \text{and} \quad \sigma^2 = np(1-p) \tag{A.13}$$

*Geometric Distribution* $X$ assumes integer values from 0 to $\infty$, with probabilities $p_i$, $0 \le i < \infty$, given by

$$p_i = (1-p)^i p \tag{A.14}$$

The parameter set is $\theta = \{p\}$ with $0 \le p \le 1$. The mean and variance are

$$\mu = \frac{(1-p)}{p} \quad \text{and} \quad \sigma^2 = \frac{(1-p)}{p^2} \tag{A.15}$$

*Hypergeometric Distribution* $X$ assumes integer values in the interval $(\max\{0, n - N + D\}, \min\{n, D\})$, where $N$, $D$ and $n$ are the three parameters of the distribution, with $N$, $D$ and $n$ positive integers satisfying $n \le N$ and $D \le N$. The $p_i$ are given by

$$p_i = P(X = i) = \frac{\binom{D}{i}\binom{N-D}{n-i}}{\binom{N}{n}} \tag{A.16}$$

The mean and variance are given by

$$\mu = \frac{nD}{N} \quad \text{and} \quad V(X) = \frac{(N-n)n}{N-1}\left(\frac{D}{N}\right)\left(1 - \frac{D}{N}\right) \tag{A.17}$$

*Poisson Distribution* $X$ assumes integer values from 0 to $\infty$. $p_i$, $0 \le i < \infty$, is given by

$$p_i = \frac{e^{-\lambda}\lambda^i}{i!} \tag{A.18}$$

The parameter set is $\theta = \{\lambda\}$, with $\lambda > 0$. The mean and variance are given by

$$\mu = \lambda \quad \text{and} \quad \sigma^2 = \lambda \tag{A.19}$$

*Multinomial Distribution* This is an extension of the binomial distribution to the case where there are $k$ possible outcomes, with corresponding probabilities of occurrence $p_1, p_2, \ldots, p_k$ (with $\sum_{i=1}^{k} p_i = 1$). The probability of observing $n_i$ items of type $i$ $(i = 1, 2, \ldots, k)$ in a sample of size $n$ from an infinite population is given by

$$P(n_1, n_2, \ldots n_k) = \frac{n!}{n_1! n_2! \ldots n_k!} p_1^{n_1} p_2^{n_2} \ldots p_k^{n_k}, \quad n_i \ge 0, \quad \sum_{i=1}^{k} n_i = n, \tag{A.20}$$

## A.1.3 Continuous Distribution and Density Functions

Some basic continuous distribution functions useful in failure modeling and statistical analysis are the following[4]:

### A.1.3.1 Basic Distributions and Density Functions

*Exponential Distribution* The distribution function for the exponential distribution is given by

$$F(x;\theta) = 1 - e^{-\lambda x}, \quad x \geq 0, \tag{A.21}$$

The parameter set is $\theta = \{\lambda\}$, with $\lambda > 0$. The density function is

$$f(x;\theta) = \lambda e^{-\lambda x} \tag{A.22}$$

The first two moments are given by

$$\mu = \frac{1}{\lambda} \quad \text{and} \quad \sigma^2 = \frac{1}{\lambda^2} \tag{A.23}$$

*Gamma Distribution* The gamma density function is given by

$$f(x;\theta) = \frac{x^{\alpha-1} e^{-x/\beta}}{\beta^{\alpha}\Gamma(\alpha)}, \quad x \geq 0, \tag{A.24}$$

The parameter set is $\theta = \{\alpha, \beta\}$, with $\alpha > 0$ and $\beta > 0$.

The distribution and failure rate functions are complicated functions involving confluent hyper-geometric functions [2]. The mean and variance are

$$\mu = \alpha\beta \quad \text{and} \quad \sigma^2 = \alpha\beta^2 \tag{A.25}$$

*Normal Gaussian Distribution* The density function for the normal distribution is given by

$$f(x;\theta) = \frac{e^{-(x-\mu)^2/2\sigma^2}}{\sigma\sqrt{2\pi}}, \quad -\infty < x < \infty, \tag{A.26}$$

The parameter set is $\theta = \{\mu, \sigma^2\}$, with $\sigma > 0$ and $-\infty < \mu < \infty$. It is not possible to give analytical expressions for the distribution function. The mean and variance, $\mu$ and $\sigma^2$, are also the parameters of the distribution.

4 Most basic books on statistics and probability discuss some of the well-known distributions. References [16, 17] give a more comprehensive coverage of many continuous distributions.

*Uniform (Rectangular) Distribution* The density function is given by

$$f(x;\theta) = \frac{1}{b-a}, \quad a \leq x \leq b. \tag{A.27}$$

The parameter set is $\theta = \{a, b\}$, with $a < b$. The distribution function is given by

$$F(x;\theta) = \frac{x-a}{b-a} \tag{A.28}$$

The mean and variance are

$$\mu = (a+b)/2 \quad \text{and} \quad \sigma^2 = (b-a)^2/12 \tag{A.29}$$

*Weibull Distribution* The two-parameter Weibull distribution function is given by

$$F(x;\theta) = 1 - e^{-(x/\alpha)^\beta}, \quad x \geq 0. \tag{A.30}$$

The parameter set is $\theta = \{\alpha, \beta\}$, with $\alpha > 0$ and $\beta > 0$. The failure density function is given by

$$f(x;\theta) = \frac{\beta x^{(\beta-1)} e^{-(x/\alpha)^\beta}}{\alpha^\beta} \tag{A.31}$$

The mean and variance are

$$\mu = \Gamma\left(1 + \frac{1}{\beta}\right)\alpha \quad \text{and} \quad \sigma^2 = \left[\Gamma\left(1 + \frac{2}{\beta}\right) - \left\{\Gamma\left(1 + \frac{1}{\beta}\right)\right\}^2\right]\alpha^2 \tag{A.32}$$

Here $\Gamma(\cdot)$ is the Gamma-function. Extensive table can be found in [2].

*Smallest Extreme Value Distribution* The distribution function of smallest extreme value (SEV) distribution is given by

$$F(x;\mu,\sigma) = 1 - \exp[-\exp\{(x-\mu)/\sigma\}], \quad -\infty < x < \infty. \tag{A.33}$$

The parameter set is $\theta = \{\mu, \sigma\}$, where $\mu$ $(-\infty < \mu < \infty)$ is the location parameter and $\sigma > 0$ is the scale parameter. The density function is given by

$$f(x;\mu,\sigma) = \frac{1}{\sigma}\exp[(x-\mu)/\sigma - \exp\{(x-\mu)/\sigma\}], \quad -\infty < x < \infty. \tag{A.34}$$

The mean and variance are

$$E(X) = \mu - \sigma\gamma \quad \text{and} \quad V(X) = \frac{\sigma^2\pi^2}{6} \tag{A.35}$$

where $\gamma = 0.5772$ is Euler's constant.

It can be shown that the smallest extreme value distribution reduces to a Weibull distribution (A.30) under the transformation

$$\mu = \ln(\alpha) \quad \text{and} \quad \sigma = \frac{1}{\beta} \tag{A.36}$$

*Largest Extreme Value Distribution* The distribution function of largest extreme value (LEV) distribution is given by

$$F(x;\mu,\sigma)=\exp[-\exp\{-(x-\mu)/\sigma\}],\quad -\infty<x<\infty. \tag{A.37}$$

The parameter set is $\theta=\{\mu,\sigma\}$, where $\mu$ $(-\infty<\mu<\infty)$ is a location parameter and $\sigma>0$ is a scale parameter. The density function is given by

$$f(x;\mu,\sigma)=\frac{1}{\sigma}\exp[-(x-\mu)/\sigma-\exp\{-(x-\mu)/\sigma\}],\quad -\infty<x<\infty. \tag{A.38}$$

The mean and variance are

$$E(X)=\mu+\sigma\gamma \quad \text{and} \quad V(X)=\frac{\sigma^2\pi^2}{6} \tag{A.39}$$

where $\gamma=0.5772$ is Euler's constant.

The smallest and largest extreme value distributions have a simple relationship. If $T\sim\text{LEV}(\mu,\sigma)$, then $X=-T\sim\text{SEV}(-\mu,\sigma)$.

*Fréchet Distribution* The distribution function of the Fréchet distribution is given by

$$F(x;\mu,\sigma)=\exp[-(\mu/x)^{\sigma}],\quad x>0. \tag{A.40}$$

The parameter set is $\theta=\{\mu,\sigma\}$, where $\mu>0$ is a location parameter and $\sigma>0$ is a scale parameter. The density function is given by

$$f(x;\mu,\sigma)=\frac{\sigma}{\mu}\left(\frac{\mu}{x}\right)^{\sigma+1}\exp[-(\mu/x)^{\sigma}],\quad x>0. \tag{A.41}$$

The mean and variance are

$$E(X)=\Gamma\left(1-\frac{1}{\sigma}\right) \quad \text{and} \quad V(X)=\Gamma\left(1-\frac{2}{\sigma}\right)-\left\{\Gamma\left(1-\frac{1}{\sigma}\right)\right\}^2 \tag{A.42}$$

The mean and variance exist only if $\sigma>1$ and $\sigma>2$, respectively.

### A.1.3.2 Derived Continuous Distribution and Density Functions

The derived distributions given below are obtained by (i) transformation of the random variable from a basic distribution (for example, the log normal distribution), (ii) modification of the form of a basic distribution by introducing additional parameters (for example, the exponentiated Weibull distribution) and, (iii) devising forms that involve two or more basic distribution functions (for example, mixtures of distribution, competing risk models). We present a few of each form of derived distribution.[5]

---

[5] For additional details with regard to the three types, see [4, 34].

*Inverse Gaussian (Wald) Distribution* The density function is given by

$$f(x) = \left(\frac{\lambda}{2\pi x^3}\right)^{1/2} \exp\left(\frac{-\lambda(x-\mu)^2}{2\mu^2 x}\right), \quad x > 0. \tag{A.43}$$

The parameter set is $\theta = \{\mu, \lambda\}$, with $\mu > 0$ and $\lambda > 0$. The mean is $\mu$ and the variance is $\mu^3/\lambda$.

*Lognormal Distribution* The density function is given by

$$f(x;\theta) = \frac{e^{-\{(\log(x)-\mu)^2/2\sigma^2\}}}{\sigma x\sqrt{2\pi}}, \quad x \geq 0. \tag{A.44}$$

The parameter set is $\theta = \{\mu, \sigma\}$ with $\sigma > 0$ and $-\infty < \mu < \infty$. It is not possible to give an analytical expression for the distribution function. The mean and variance are

$$E(X) = e^{(\mu+\sigma^2/2)} \quad \text{and} \quad V(X) = \omega(\omega-1)e^{2\mu} \tag{A.45}$$

where $\omega = e^{\sigma^2}$. The distribution is related to the normal in that if $X$ is lognormal $(\mu, \sigma)$, then $Y = \log(X)$ is $N(\mu, \sigma)$.

*Three Parameter Weibull Distribution* This is an extension of the two-parameter Weibull distribution (A.30), given by

$$F(x;\theta) = 1 - e^{-(\{x-\tau\}/\alpha)^\beta}, \quad x \geq \tau. \tag{A.46}$$

The additional parameter is the location parameter $\tau > 0$. The mean and variance are given by

$$\mu = \tau + \Gamma\left(1 + \frac{1}{\beta}\right)\alpha \quad \text{and} \quad \sigma^2 = \left[\Gamma\left(1 + \frac{2}{\beta}\right) - \left\{\Gamma\left(1 + \frac{1}{\beta}\right)\right\}^2\right]\alpha^2 \tag{A.47}$$

*Extended Weibull Distribution* [29] The distribution function is given by

$$F(x) = 1 - \frac{ve^{-(x/\alpha)^\beta}}{1-(1-v)e^{-(x/\alpha)^\beta}}, \quad x \geq 0 \tag{A.48}$$

with $0 \leq v \leq 1$. The distribution reduces to the two-parameter Weibull (A.30) when $v = 1$.

*Modified Weibull Distribution* [21] The distribution function is given by

$$F(x) = 1 - \exp(-\{x/\alpha\}^\beta e^{vx}), \quad x \geq 0, \tag{A.49}$$

with $v \geq 0$. The distribution reduces to the two-parameter Weibull (A.30) when $v = 0$.

*Exponentiated Weibull Distribution* [32] The distribution function is given by

$$F(x) = [1 - \exp\{-(x/\alpha)^{\beta}\}]^{\nu}, \quad x \geq 0, \tag{A.50}$$

with $\nu \geq 0$. The distribution reduces to the two-parameter Weibull (A.30) when $\nu = 1$.

*Four parameter Weibull Distribution* [19] The distribution function is given by

$$F(x) = 1 - \exp\left\{-\lambda\left(\frac{x-a}{b-x}\right)^{\beta}\right\}, \quad 0 \leq a \leq x \leq b < \infty, \tag{A.51}$$

with $\lambda > 0$ and $\beta > 0$. Note that the support is a finite interval.

*Mixtures of Distributions* A *finite mixture* of distributions is a weighted average of distribution functions given by

$$F(x) = \sum_{i=1}^{K} p_i F_i(x) \tag{A.52}$$

with $p_i \geq 0, \; i = 1, 2, \ldots, K, \sum_{i=1}^{K} p_i = 1$ and $F_i(x) \geq 0, \; i = 1, 2, \ldots, K$ distribution functions (called the *components* of the mixture). If the components are differentiable, then the density function is given by

$$f(x) = \sum_{i=1}^{K} p_i f_i(x) \tag{A.53}$$

*Competing Risk* The distribution function is given by

$$F(x) = 1 - \prod_{i=1}^{K} (1 - F_i(x)) \tag{A.54}$$

The density function is

$$f(x) = \sum_{i=1}^{K} \left[ \prod_{\substack{k=1 \\ k \neq i}}^{K} \{1 - F_k(x)\} \right] f_i(x) \tag{A.55}$$

*Multiplicative* The distribution function is given by

$$F(x) = \prod_{i=1}^{K} F_i(x), \quad x \geq 0 \tag{A.56}$$

The density function is given by

$$f(x) = \sum_{i=1}^{K} \prod_{\substack{k=1 \\ k \neq i}}^{K} F_k(x) f_i(x) \tag{A.57}$$

#### A.1.3.3 Distributions of Importance in Statistical Inference

The following distributions are used extensively in data analysis. They are employed in many important applications in estimation and hypothesis testing.

*Chi-Square Distribution* The Chi-Square ($\chi^2$) distribution is related to the distribution of the sum of squares of normal random variables. The density function is

$$f(x) = \frac{x^{(v-2)/2} e^{-x/2}}{2^{v/2} \Gamma(v/2)}, \quad x > 0, \tag{A.58}$$

where $\Gamma(\cdot)$ is the gamma function. The distribution function is an incomplete gamma function [16]. The parameter is $v$, a positive integer called *degrees of freedom*. This density is a gamma distribution with shape parameter $\alpha = v/2$ and scale parameter $\beta = 2$. The mean is $v$ and the variance is $2v$.

*Student-t Distribution* The density is

$$f(x) = \frac{\Gamma[(v+1)/2]}{\sqrt{\pi v}\Gamma(v/2)[1 + x^2/v]^{(v+1)/2}}, \quad -\infty < x < \infty \tag{A.59}$$

The parameter is $v$; $v$ is a positive integer and is called *degrees of freedom*. The CDF is a complex expression [17]. The mean is infinite if $v = 1$, and zero if $v \geq 2$. The variance is infinite if $v = 1$ or 2 and $v/(v - 2)$ if $v > 2$.

*F Distribution* The density function of the F distribution, also called the "variance ratio" or the "Fisher-Snedecor" distribution, is given by

$$f(x) = \frac{\Gamma[(v_1 + v_2)/2](v_1/v_2)^{v_1/2} x^{(v_1-2)/2}}{\Gamma(v_1/2)\Gamma(v_2/2)[1 + v_1 x/v_2]^{(v_1+v_2)/2}}, \quad x > 0. \tag{A.60}$$

The parameter set is $\theta = \{v_1, v_2\}$. Both parameters are positive integers called *degrees of freedom*. The mean is $v_2/(v_2 - 2)$ if $v_2 > 2$ and infinite otherwise. The variance is infinite if $v_2 \leq 4$ and $[2v_2^2(v_1 + v_2 - 2)]/[v_1(v_2 - 2)^2(v_2 - 4)]$ if $v_2 > 4$.

## A.2 Two or More Random Variables

We first consider distributions in the case of two variables and then the general case *of more than two*.

## A.2.1 Two Random Variables

We shall confine our discussion to two continuous random variables, denoted $X$ and $Y$.

### A.2.1.1 Joint, Marginal and Conditional Distribution and Density Functions

The *joint distribution function* $F(x, y)$ is given by

$$F(x,y) = P\{X \leq x, Y \leq y\} \tag{A.61}$$

The random variables are said to be jointly continuous if there exists a function $f(x, y)$, called the *joint probability density function*, such that

$$f(x,y) = \frac{\partial^2 F(x,y)}{\partial x \partial y} \tag{A.62}$$

The *marginal distribution functions* $F_X(x)$ and $F_Y(y)$ are given by

$$F_X(x) = F(x,\infty) \quad \text{and} \quad F_Y(y) = F(\infty,y) \tag{A.63}$$

The two marginal density functions are given by

$$f_X(x) = \frac{dF_X(x)}{dx} \quad \text{and} \quad f_Y(y) = \frac{dF_Y(y)}{dy}. \tag{A.64}$$

The *conditional distribution* of $X$ given that $Y = y$ is denoted $F(x|y)$ and given by

$$F(x|y) = P\{X \leq x | Y = y\} \tag{A.65}$$

The conditional distribution of $Y$ given that $X = x$, $F(y|x)$, is defined similarly.

For jointly continuous random variables with a joint density function $f(x, y)$, the *conditional probability density* function of $X$, given $Y = y$, is given by

$$f(x|y) = \frac{f(x,y)}{f_Y(y)} \tag{A.66}$$

Similarly,

$$f(y|x) = \frac{f(x,y)}{f_X(x)} \tag{A.67}$$

The random variables $X$ and $Y$ are said to be *independent* (or *statistically independent*) if and only if

$$F(x,y) = F_X(x)\, F_Y(y) \tag{A.68}$$

for all $x$ and $y$.

The results are similar for discrete random variables, with summation replacing integration.

#### A.2.1.2 Moments of Two Random Variables

The covariance of $X$ and $Y$ is defined as

$$Cov(X, Y) = E[\{X - E[X]\}\{Y - E[Y]\}] = E[X\,Y] - E[X]\,E[Y] \tag{A.69}$$

The correlation $\rho_{XY}$ is defined as

$$\rho_{XY} = \frac{Cov(X, Y)}{\sigma_X \sigma_Y}, \tag{A.70}$$

where $\sigma_X$ and $\sigma_Y$ are the standard deviations of $X$ and $Y$, respectively. The random variables $X$ and $Y$ are said to be *uncorrelated* if $\rho_{XY} = 0$. Note that independent random variables are uncorrelated but that the converse is not necessarily true.

#### A.2.1.3 Conditional Expectation

$E[X|Y = y]$ is called the conditional expectation of $X$ given that $Y = y$. The unconditional expectation of $X$, given by

$$E[X] = \int_{-\infty}^{\infty} x\,f_X(x)\,dx, \tag{A.71}$$

is related to the conditional expectation by the relation

$$E[X] = \int_{-\infty}^{\infty} E[X|Y = y]\,f_Y(y)\,dy. \tag{A.72}$$

This is written symbolically as

$$E[X] = E[E[X|Y]] \tag{A.73}$$

#### A.2.1.4 Bivariate Distribution and Density Functions

There are many multi-dimensional distributions.[6] We list a few that are useful in reliability applications and data analysis.

*Bivariate Normal Distribution* The joint distribution function is given by

---

[6] Reference [18] discusses several multivariate distributions; Reference [12] deals with several bivariate distributions.

$$F(x,y) = \frac{1}{2\pi\sigma_X\sigma_Y\sqrt{1-\rho^2}} \times \exp\left[-\frac{1}{2(1-\rho^2)}\left(\left(\frac{x-\mu_X}{\sigma_X}\right)^2 - 2\rho\left(\frac{x-\mu_X}{\sigma_X}\right)\left(\frac{y-\mu_Y}{\sigma_Y}\right) + \left(\frac{y-\mu_Y}{\sigma_Y}\right)^2\right)\right] \tag{A.74}$$

where $\rho = \rho_{XY}$ is the correlation coefficient and the remaining parameters are the means and standard deviations of the marginal distributions of $X$ and $Y$. Both marginal distributions are normal, as are the conditional distributions. In the latter case, the means and standard deviations are functions of the condition [18].

*Bivariate Exponential Distributions* A variety of bivariate exponential distributions have been proposed in the literature. We list two of these.

1. Marshall and Olkin [29]

$$\bar{F}(x,y) = \exp\{-[\lambda_1 x + \lambda_2 y + \lambda_{12}\max(x,y)]\} \tag{A.75}$$

where

$$\bar{F}(x,y) = P\{X > x, Y > y\} \tag{A.76}$$

The marginal distributions are given by

$$F_X(x) = 1 - \exp\{-(\lambda_1 + \lambda_{12})x\} \tag{A.77}$$

and

$$F_Y(y) = 1 - \exp\{-(\lambda_2 + \lambda_{12})y\} \tag{A.78}$$

respectively. It is easily shown that

$$P(X > Y) = \frac{\lambda_2}{\lambda_1 + \lambda_2 + \lambda_{12}}, \tag{A.79}$$

$$P(X < Y) = \frac{\lambda_1}{\lambda_1 + \lambda_2 + \lambda_{12}}, \tag{A.80}$$

and

$$P(X = Y) = \frac{\lambda_{12}}{\lambda_1 + \lambda_2 + \lambda_{12}} \tag{A.81}$$

2. Gumbel [9]

$$\bar{F}(x,y) = \exp(-x/\theta_1) + \exp(-y/\theta_2) - \exp\{-[(x/\theta_1)^m + (y/\theta_2)^m]^{1/m}\} \tag{A.82}$$

---

[7] See [34] for additional details on these as well as other bivariate Weibull distributions.

*Bivariate Weibull Distributions* A variety of bivariate Weibull distributions have been proposed in the literature. We list a few of these.[7]

1. Marshall and Olkin [29]

$$\bar{F}(x,y) = \exp\{-[\lambda_1 x^{\beta_1} + \lambda_2 y^{\beta_2} + \lambda_{12}\max(x^{\beta_1}, y^{\beta_2})]\} \tag{A.83}$$

2. Lee [23]

$$\bar{F}(x,y) = \exp\{-[\lambda_1 c_1^{\beta} x^{\beta} + \lambda_2 c_2^{\beta} y^{\beta} + \lambda_{12}\max(c_1^{\beta} x^{\beta}, c_2^{\beta} y^{\beta})]\} \tag{A.84}$$

3. Lu and Bhattacharyya [27]

$$\bar{F}(x,y) = \exp\{-(x/\alpha_1)^{\beta_1} - (y/\alpha_2)^{\beta_2} - \delta h(x,y)\} \tag{A.85}$$

Different forms for the function of $h(x, y)$ yield a family of models. One form for $h(x, y)$ is the following:

$$h(x,y) = [(x/\alpha_1)^{\beta_1/m} + (y/\alpha_2)^{\beta_2/m}]^m \tag{A.86}$$

This results in

$$\bar{F}(x,y) = \exp\left\{-(x/\alpha_1)^{\beta_1} - (y/\alpha_2)^{\beta_2} - \delta\left[(x/\alpha_1)^{\beta_1/m} + (y/\alpha_2)^{\beta_2/m}\right]^m\right\} \tag{A.87}$$

## A.2.2 General Case

The $k$ (>2) random variables may be represented by the vector $(X_1, X_2, \ldots, X_k)$. The approach is similar to the two random variable case, but involving an $k$-dimensional distribution function $F(x_1, x_2, \ldots, x_k)$. We have $k$ marginal distributions and several different conditional distributions, depending how the $k$-variables are divided into two sets, with the distribution of the first set conditioned on the values of the variables in the second. Similarly, there are many different correlation coefficients. Details can be found in [18].

# Appendix B: Introduction to Point Processes

One-dimensional [two-dimensional] point processes are useful for modeling random events involving warranty (e.g., number of warranty claims for products sold with 1-D [2-D] warranty). In this appendix, we discuss a few such processes and present some results (without proof) that will be used in the modeling and analysis of warranties.[8]

## B.1 One-dimensional Point Processes

A one-dimensional point process is a continuous-time stochastic process characterized by events that occur randomly along the time continuum.

### *B.1.1 Counting Processes*

A point process $\{N(t),\ t \geq 0\}$ is a counting process if it represents the number of events that have occurred until time $t$. It must satisfy:

- $N(t) \geq 0$.
- $N(t)$ integer valued.
- If $s < t$, then $N(s) \leq N(t)$
- For $s < t$, $\{N(t) - N(s)\}$ is the number of events in the interval $(s,t]$.

It is assumed that $N(0) = 0$.

[8] Proofs of the results can be found in many books on probability, for example, [43, 44]. For more on point processes, see [6].

W. R. Blischke et al., *Warranty Data Collection and Analysis*,
Springer Series in Reliability Engineering, DOI: 10.1007/978-0-85729-647-4,

#### B.1.1.1 Non-Stationary Poisson Process

A counting process $\{N(t),\ t \geq 0\}$ is a non-stationary Poisson process if

- $N(0) = 0$.
- $\{N(t),\ t \geq 0\}$ has independent increments.
- $P\{N(t + \delta t) - N(t) = 1\} = \lambda(t)\delta t + o(\delta t)$.
- $P\{N(t + \delta t) - N(t) \geq 2\} = o(\delta t)$.

$\lambda(t)$ is called the intensity function and is nonnegative. The function

$$\Lambda(t) = \int_0^t \lambda(x)\, dx \tag{B.1}$$

is called the cumulative intensity function.
*Distribution and Moments of N(t)* The probability that $N(t) = j$ is given by

$$p_j(t) = P\{N(t) = j\} = \frac{e^{-\Lambda(t)}\{\Lambda(t)\}^j}{j!} \tag{B.2}$$

for $j \geq 0$. The mean of $N(t)$ is given by

$$M(t) = E[N(t)] = \Lambda(t) \tag{B.3}$$

The variance of $N(t)$ is given by

$$V[N(t)] = E[\{N(t) - \Lambda(t)\}^2] = \Lambda(t) \tag{B.4}$$

*Comment*: If $\lambda(t) = \lambda$, a constant, then the process is a *stationary Poisson process*.

#### B.1.1.2 Renewal Processes

A counting process $\{N(t),\ t \geq 0\}$ is an ordinary renewal process if

- $N(0) = 0$.
- $\tilde{X}_1$, the time to occurrence of the first event (counting from time $t = 0$) and $\tilde{X}_j, j \geq 2$, the time between the $(j - 1)$st and $j$th events, are a sequence of independent and identically distributed random variables with distribution function $F(x)$.
- $N(t) = \text{Sup}\ \{n : S_n \leq t\}$, where

$$S_0 = 0, \quad S_n = \sum_{i=1}^{n} \tilde{X}_i,\ n \geq 1 \tag{B.5}$$

*Distribution and Moments of* $N(t)$ The probability that $N(t) = j$ is given by

$$p_j(t) = P\{N(t) = j\} = F^{(n)}(t) - F^{(n+1)}(t), \tag{B.6}$$

where $F^{(n)}(t)$ is the $n$-fold convolution of $F(t)$ with itself. This is obtained in a recursive manner as follows:

$$F^{(j+1)}(t) = \int_0^\infty F^{(j)}(t - t')f(t')dt', \tag{B.7}$$

with $F^{(0)}(t) = 1$.

The expected value of $N(t),\ t \geq 0$, denoted $M(t)$, is given by

$$M(t) = \sum_{j=1}^{\infty} F^{(j)}(t) \tag{B.8}$$

$M(t)$ may also be obtained as the solution of the integral equation

$$M(t) = F(t) + \int_0^t M(t - x)f(x)dx \tag{B.9}$$

This is called the *renewal integral equation* and $M(t)$ is called the *renewal function* associated with the distribution function $F(t)$.

The variance of $N(t),\ t \geq 0$, is given by

$$V(t) = \sum_{n=1}^{\infty} (2n - 1)F^{(n)}(t) - [M(t)]^2. \tag{B.10}$$

For large $t$, an approximation of $M(t)$ involving the first two moments of $\tilde{X}_i$ is given by

$$M(t) \approx t/\mu + \sigma^2/(2\mu^2) - 1/2 \tag{B.11}$$

### B.1.1.3 Delayed Renewal Process

A counting process $\{N(t), t \geq 0\}$ is a delayed renewal process if

- $N(0) = 0$.
- $\tilde{X}_1$, the time to the first event, is a non-negative random variable with distribution function $F(x)$.
- $\tilde{X}_j, j \geq 2$, the time intervals between the $j$th and $(j - 1)$st events, are independent and identically distributed random variables with a distribution function $F_1(x)$ that is different from $F(x)$.
- $N(t) = \text{Sup}\ \{n : S_n \leq t\}$, where $S_n$ is given by (B.5).

*Comment*: When $F_1(x)$ equals $F(x)$, the delayed renewal process reduces to an ordinary renewal process.
*First Moment* $M_d(t)$, the expected number of renewals over [0, *t*) for the delayed renewal process, is given by

$$M_d(t) = F(t) + \int_0^t M_1(t-x)\, f(x)\, dx \tag{B.12}$$

## B.1.2 Mean Function of a Point Process

The mean function of a point process $N(t)$, often referred to as the *mean cumulative function* (MCF), is defined as the expected value of $N(t)$. This is given by $\mu(t) = E[N(t)]$. If $\mu(t)$ is differentiable, then

$$v(t) = \frac{dE[N(t)]}{dt} = \frac{d\mu(t)}{dt} \tag{B.13}$$

$v(t)$ is called the *recurrence rate* or *intensity function.* In the context of reliability, where $N(t)$ denotes the number of failures, it is also referred to as the rate of occurrence of failures (ROCOF).[9]
*Comment*: $\mu(t) = \Lambda(t)$ (given by (B.1)) in the case of a non-stationary Poisson process and $\mu(t) = M(t)$ (given by (B.8) or (B.9)) in the case of a renewal process.

## B.1.3 Other Processes

### B.1.3.1 Alternating Renewal Process

In an ordinary renewal process, the inter-event times are independent and identically distributed. In an alternating renewal process, the inter-event times are independent, but not identically distributed. More specifically, the odd numbered inter-event times (i.e., $\tilde{X}_1, \tilde{X}_3, \tilde{X}_5, \ldots$) are from a common distribution function $F(x)$ and the even numbered (i.e. $\tilde{X}_2, \tilde{X}_4, \tilde{X}_6, \ldots$) are from a common a distribution function $G(x)$, that is different from $F(x)$.

### B.1.3.2 Marked Point Process

A marked point process is a point process with an auxiliary variable, called a mark, associated with each event. Let $\tilde{Y}_i, i \geq 1$, denote the mark attached to the *i*th event.

[9] For more details on MCF and ROCOF, see [42].

A *simple marked* point process is characterized by

- $\{N(t), t \geq 0\}$, a stationary Poisson process with intensity $\lambda$, and
- a sequence of independent and identically distributed random variables $\{\tilde{Y}_i\}$, called marks, which are independent of the Poisson process.

#### B.1.3.3 Cumulative Process

A cumulative process, $\psi(t)$, is given by

$$\psi(t) = \sum_{i=1}^{N(t)} \tilde{Y}_i \tag{B.14}$$

with $N(t)$ a marked point process and $\tilde{Y}_i$ the mark attached to event $i$. The cumulative process is sometimes also called a *compound Poisson process.*

## B.2 Two-dimensional Point Processes

Two-dimensional point processes deal with random events on a two-dimensional plane, with one axis representing time and the other representing usage. In the ensuing, $(T_i, X_i)$, $i \geq 1$, denotes the time and usage of an item at the time at which the $i$th event occurs and $T_0 = X_0 = 0$. $N(t, x)$ denotes the number of events occurring over the interval $[0, t) \times [0, x)$.

### *B.2.1 Two-dimensional Renewal Processes*

A two-dimensional renewal process is characterized by

- $N(0, 0) = 0$.
- $(\tilde{T}_i, \tilde{X}_i)$, $i \geq 1$, are a sequence of independent and identically distributed random variables with bivariate distribution function $F(t, x)$, where $\tilde{T}_i = T_i - T_{i-1}$ and $\tilde{X}_i = X_i - X_{i-1}$, $i \geq 1$.
- $N(t, x) = \text{Sup}\ \{n : T_n \leq t, X_n \leq x\}$

*Distribution and Moments*[10] The probability that $N(t, x) = n$ is given by

$$p_n(t, x) = F^{(n)}(t, x) - F^{(n+1)}(t, x), \quad n \geq 0, \tag{B.15}$$

where $F^{(n)}(t, x)$ is the $n$-fold bivariate convolution of $F(t, x)$ with itself.

[10] For details, see [13].

The expected number of failures over $[0,t) \times [0,x)$ is obtained by solution of the two-dimensional integral equation

$$M(t,x) = F(t,x) + \int_0^t \int_0^x M(t-u, x-v) f(u,v) dv du. \tag{B.16}$$

# Appendix C: Probability Plots

In this appendix, we look at various theoretical and empirical plots. Which empirical plot is appropriate in any given application depends on the type of data available. These plots help in deciding whether or not a given data set can be modeled by a specified distribution function. We first look at the empirical distribution function (EDF) plot. This does not involve a transformation of the data. We then look at the WPP plot, which may be used to decide whether one or more of the many Weibull-derived distributions can be used to model a given data set. We conclude with a brief discussion of other probability plots.[11]

## C.1 Types of Data

The types of data on which the plots are based may be (i) complete data, (ii) incomplete data, and (iii) grouped data.

- Complete Data: The data comprise solely the failure times for the $n$ units and are given by the set $\{t_1, t_2, \ldots, t_n\}$, where $t_i$ is the $i$th observation.
- Censored Data: The data consist of failure times of failed items and the ages of units that have not yet failed at the time of observation. For unit $i$, the observation is the age at failure $t_i$ if the unit has failed and $\tilde{t}_i$, the age of the unit, if it is still working.
- Grouped Data: The data available are not the failure times (as in the case complete data) but only the number of failures that occurred in different time intervals. Let the number of observations in interval $j$, $1 \leq j \leq J$, be $d_j$, with the interval given by $[a_{j-1}, a_j)$, where $J$ is the number of intervals, $a_0 = 0$, and $a_J = \infty$. Note that $n = \sum_{j=1}^{J} d_j$.

[11] We outline the procedures for other plots. For details, see [8, 22, 35].


Springer Series in Reliability Engineering, DOI: 10.1007/978-0-85729-647-4,

## C.2 Plot of the Empirical Distribution Function (EDF)

The EDF $\hat{F}(t)$ is a step function. The calculation of $\hat{F}(t)$ depends on the type of data available.

### C.2.1 Complete Data

In this case, the data are given by $t_1, t_2, \ldots, t_n$. The EDF is obtained as follows:

1. Reorder the data from the smallest to the largest, obtaining the ordered observations $t_{(1)} \leq t_{(2)} \leq \cdots \leq t_{(n)}$
2. Compute $\hat{F}(t_i) = \frac{i}{n+1}$ for $1 \leq i \leq n$.

The EDF for complete data is given by

$$\hat{F}(t) = \begin{cases} 0, & 0 \leq t < t_1 \\ \hat{F}(t_i), & t_i \leq t < t_{i+1}, \; 1 \leq i \leq (n-1) \end{cases} \tag{C.1}$$

### C.2.2 Right Censored Data

In this case, the procedure is as follows:

1. Order the observations from the smallest to the largest.
2. For each uncensored observation, compute $I_j$ and $N_j$ as follows:

$$I_j = \frac{n+1-N_p}{1+C} \quad \text{and} \quad N_j = N_p + I_p \tag{C.2}$$

   where $I_j$ is the increment for the $j$th uncensored datum, $N_p$ is the order of the previous uncensored observation, and $C$ is the number of data points remaining in the data set, including the current data point, with $N_p = 0$ for the first uncensored observation.
3. For each uncensored observation $t_{(j)}$, compute the EDF as

$$\hat{F}(t_{(j)}) = \frac{N_j - 0.3}{n + 0.4} \tag{C.3}$$

The complete function is given by (C.1), using $\hat{F}(t_{(j)})$ given by (C.3) instead of $\hat{F}(t_i)$.

### C.2.3 Grouped Data

In this case, let the number of observations in interval $j$ $(1 \leq j \leq J)$ be $d_j$, with the interval given by $[a_{j-1}, a_j)$, with $a_0 = 0$ and $a_J = \infty$.

**Table C.1** Classification of shapes for the WPP

| Type | Description |
|---|---|
| A | Straight line |
| B | Concave |
| B1 | Concave with left asymptote vertical |
| C | Convex |
| C1 | Convex with right asymptote vertical |
| D | Single inflection point (S-shaped) with parallel asymptotes |
| D1 | Single inflection point (S-shaped) with vertical asymptotes |
| E1 | Bell shaped |
| E(n) | Multiple inflection points ($n \geq 1$ and odd) |

The EDF is calculated as

$$\hat{F}(a_j) = \frac{\sum_{i=1}^{j} d_i}{n}, \tag{C.4}$$

with $n = \sum_{j=1}^{J} d_j$.

## C.3 WPP Plots

### *C.3.1 Theoretical Plot*

For a failure distribution $F(t)$, the Weibull transformation is given by

$$y = \log(-\log(1 - F(t))) \quad \text{and} \quad x = \log(t) \tag{C.5}$$

A plot of $y$ versus $x$ is called the theoretical WPP Plot.[12]

The two-parameter Weibull distribution given by (A.30) is transformed into a linear relationship

$$y = \beta[x - \log(\alpha)] \tag{C.6}$$

under the Weibull transformation. For other Weibull derived distributions, the relationship is non-linear. Murthy et al. [34] characterizes the different possible shapes. These are listed in Table C.1

[12] In the early 1970's, a special paper was developed for plotting data under this transformation. The plotting paper was referred to as Weibull Probability Paper (WPP) and the plot called the WPP plot. At present, most computer reliability software packages and many statistical program packages contain programs to produce these plots automatically, but the plot continues to be called a WPP plot.

**Table C.2** Shapes of WPP Plots for various Weibull-derived distributions

| Distribution[a] | Shapes |
|---|---|
| Two-parameter Weibull (A.30) | A |
| Three-parameter Weibull (A.46) | B1 |
| Extended Weibull (A.48) | B when $\nu < 1$; C when $\nu > 1$ |
| Modified Weibull (A.49) | B when $\nu > 1$; C when $\nu < 1$ |
| Exponentiated Weibull (A.50) | C |
| Four-parameter Weibull (A.51) | D1 |
| Mixture (A.52) with $K = 2$ | D when $\beta_1 = \beta_2$; E(3) when $\beta_1 \neq \beta_2$ |
| Competing Risk (A.54) with $K = 2$ | C |
| Multiplicative (A.56) with $K = 2$ | B |

[a] The numbers refer to equation numbers in Appendix A

Shapes for the various Weibull-derived models discussed in Appendix A are indicated in Table C.2.[13]

## C.3.2 Empirical WPP Plots

An empirical WPP plot is a plot on Weibull probability paper of an empirical distribution function (EDF) instead of the true distribution function. The procedure for plotting the empirical WPP for various data sets is as follows:

### C.3.2.1 Complete Data

The first two steps are as in Sect. C.2.1. The remaining steps are as follows:

3. Compute $y_i = \log(-\log(1 - \hat{F}(t_{(i)})))$ for $1 \leq i \leq n$.
4. Compute $x_i = \log(t_{(i)})$ for $1 \leq i \leq n$.
5. Plot $y_i$ versus $x_i$ for $1 \leq i \leq n$.

A smooth curve to fit the plotted data yields the empirical WPP Plot.

### C.3.2.2 Right Censored Data

The first three steps are as in Sect. C.2.2. The remaining step is as follows:

4. Plot $y_j = \log(-\log(1 - \hat{F}(t_{(j)})))$ versus $x_j = \log(t_{(j)})$ for each uncensored observation.

A smooth curve to fit the plotted data yields the empirical WPP Plot.

[13] See [34] for details of the different shapes for other distributions that are either derived from or linked to the Weibull distribution.

#### C.3.2.3 Grouped Data

The first step is as in Sect. C.2.3. The remaining step is as follows:

2. Plot $y_j = \log(-\log(1 - \hat{F}(a_j)))$ versus $x_j = \log(a_j)$ for $1 \le j \le J$.

A smooth curve to fit the plotted data yields the empirical WPP Plot.

### *C.3.3 Model Selection*

The selection of a distribution to model a given data set based on WPP plots involves comparing the empirical plot with each theoretical plot to see whether or not the shapes of the two are similar. If the shapes are similar, the theoretical distribution is a candidate model. This issue is discussed further in [33, 34].

## C.4 Other Plots

Plots have been proposed to determine if a given data set can be modeled by distributions other than Weibull or Weibull-derived distributions. Many software packages for reliability data modeling and statistical analysis have plots to determine if a given data set can be modeled by an exponential, normal or log-normal distribution, and other distributions. We discuss the basis for the theoretical plots of some of these below. The empirical plots follow along the lines indicated in the above section, using the appropriate transformation.

### *C.4.1 Exponential Probability Plot*

For a failure distribution $F(t)$, the exponential probability plot is plot of $y$ versus $x$ under the transformation

$$y = -\log(1 - F(t)) \quad \text{and} \quad x = t \tag{C.7}$$

If $F(t)$ is the exponential distribution (given by (A.21)), then (C.7) reduces to a straight line given by

$$y = \lambda x \tag{C.8}$$

### *C.4.2 Normal Distribution Plot*

For a failure distribution $F(t)$, the normal probability plot is plot of $y$ versus $x$ under the transformation

$$y = F^{-1}(p) \quad \text{and} \quad x = t_p \tag{C.9}$$

If $F(t)$ is the normal distribution (given by (A.26)), then (C.9) reduces to a straight line given by

$$x = \mu + \sigma y \tag{C.10}$$

### C.4.3 Lognormal Probability Plot

For a failure distribution $F(t)$, the lognormal probability plot is plot of $y$ versus $x$ under the transformation

$$y = \Phi^{-1}(p) \quad \text{and} \quad x = \log(t_p), \tag{C.11}$$

where the function $\Phi^{-1}(\cdot)$ is the inverse of the standard normal distribution function.

If $F(t)$ is the lognormal distribution (given by (A.44)), then (C.11) reduces to a straight line given by

$$x = \mu + \sigma y \tag{C.12}$$

### C.4.4 Smallest Extreme Value Probability Plot

For a failure distribution $F(t)$, the smallest extreme value (SEV) probability plot is plot of $y$ versus $x$ under the transformation

$$y = \Phi_{sev}^{-1}(p) \quad \text{and} \quad x = t_p \tag{C.13}$$

where $\Phi_{sev}^{-1}(p) = \log[-\log(1-p)]$.

If $F(t)$ is the smallest extreme value distribution (given by (A.33)), then (C.13) reduces to a straight line given by

$$x = \mu + \sigma y \tag{C.14}$$

### C.4.5 Largest Extreme Value Probability Plot

For a failure distribution $F(t)$, the largest extreme value (LEV) probability plot is plot of $y$ versus $x$ under the transformation

$$y = \Phi_{lev}^{-1}(p) \quad \text{and} \quad x = t_p \tag{C.15}$$

where $\Phi_{lev}^{-1}(p) = -\log[-\log(p)]$.

If $F(t)$ is the largest extreme value distribution (given by (A.33)), then (C.15) reduces to a straight line given by

$$x = -\mu + \sigma y \tag{C.16}$$

## C.4.6 Fréchet Probability Plot

Taking natural logs twice of the Fréchet distribution function (A.40) yields a linear relation in $\ln(t_p)$ given by

$$-\log[-\log(p)] = -\sigma \log(\mu) + \sigma \log(t_p) \tag{C.17}$$

# Appendix D: Statistical Theory

In this appendix, we provide a brief introduction to some aspects of theoretical statistics that are important in understanding some of the topics covered in the text. Included are comments on selected items from the theory of estimation, a brief development of maximum likelihood estimation, estimation of functions of parameters, and use of the EM algorithm in analysis of incomplete data.

## D.1 Estimation

### *D.1.1 Introduction*

The objective of statistical inference generally is to use sample information to make statements about populations. In estimation, the specific objective is to provide methods of *estimating* (providing numerical values for) unknown population parameters or other population characteristics. Implicit in this approach is that the form of the CDF[14] is known or assumed. We consider estimation of a parameter $\theta$, which may be a scalar or vector parameter.

We also assume that the inference is to be based on a sample of $n$ independent and identically distributed (*iid*) observations. These are indicated by capital letters if they are considered to be random variables, with lower case used to indicate numerical values. An *estimator* of a parameter $\theta$ is a function $\hat{\theta} = \hat{\theta}(Y_1, \ldots, Y_n)$, which is a random variable. A numerical value of the estimator, $\hat{\theta} = \hat{\theta}(y_1, \ldots, y_n)$, is called an *estimate*. In general, a caret placed over a symbol will mean that the quantity is an estimator or an estimate. If it is not clear from the context which of these is meant, the more complete notation will be used.

[14] See Appendix A for many examples of CDFs.

W. R. Blischke et al., *Warranty Data Collection and Analysis*,

Springer Series in Reliability Engineering, DOI: 10.1007/978-0-85729-647-4,

The remainder of this appendix will be devoted to a discussion of topics in estimation theory. The inference procedures discussed above are *point* estimators and estimates, and the derivation of these will be the focus of the discussion. These estimators are the basis of much of statistical inference, including *confidence interval* estimation, i.e., construction of a set of values along with a statement of the likelihood that the true value of the parameter is contained in the set, and the related area of hypothesis testing.

Confidence intervals and other inferences may be based on exact results; these require knowledge of the exact distribution of the estimator. In many cases, it is not possible to obtain this, but it is possible to obtain an *asymptotic* distribution, which may be used as the basis of, for example, an approximate confidence interval.

### *D.1.2 Approaches to Estimation*

There are many approaches to the construction of estimators. A very commonly used procedure is maximum likelihood (ML). Maximum likelihood estimators (MLEs) are "best" in a number of senses. These will be discussed in the next sections.

Other methods of estimation[15] include:

- Moment estimation: Express a set of $k$ population moments in terms of the $k$ unknown parameters. Solve the resulting equations to express the parameters in terms of the moments. Substitute sample moments for population moments.
- Least Squares (LS) estimation: Express the observations in terms of a model (often a linear model) involving the parameters. Sum the squares of deviations of the observations from the model. LS estimators are those values that minimize this expression.
- Bayes estimation: Determine a prior distribution of the parameters. Express the joint distribution of the data in terms of this and the assumed distribution of the observations. Use Bayes' Theorem to determine a posterior distribution and base the estimator on this.
- Minimum Chi-Square estimation: Form a Chi-Square statistic based on the difference between the observations (often grouped) and a model. The estimators are those values that minimize this quantity.
- Best Asymptotically Normal (BAN) estimation: A large class of estimators that are asymptotically normally distributed. If certain conditions are met, the MLE, minimum Chi-Square, and LS estimators are BAN.

[15] See books on theoretical statistics such as [5, 11, 24, 46] for details on these and other methods.

### *D.1.3 Properties of Estimators*

In practice, it is necessary to choose among the many possible estimators. This is done on the base of how they perform when applied to the many data sets that may occur. The objective is to use a "best" procedure, i.e., one that is optimal in one or more senses. Criteria of optimality in this context include the following[16]:

*Sufficiency* An estimator $\hat{\theta}$ is *sufficient* if the conditional distribution of $Y_1$, $Y_2$, … , $Y_n$ given $\hat{\theta}$ does not depend on $\theta$. This implies that $\hat{\theta}$ contains all of the sample information about $\theta$.

*Unbiasedness* An estimator $\hat{\theta}$ is *unbiased* if $\mathrm{E}(\hat{\theta}) = \theta$, for all values of $\theta$.

*Asymptotic unbiasedness* An estimator $\hat{\theta}$ is *asymptotically unbiased* if $\mathrm{E}(\hat{\theta}) \to \theta$, as $n \to \infty$, for all values of $\theta$.

*Consistency* An estimator $\hat{\theta}$ is *consistent* if for all $\theta$, $P(|\hat{\theta} - \theta| > \varepsilon) \to 0$ as $n \to \infty$ for any $\varepsilon > 0$.

*Efficiency* An estimator is *efficient* within a class of estimators (e.g., unbiased) if no other estimator in the class has smaller variance.

*Asymptotic efficiency* An estimator is asymptotically efficient if it is efficient as $n \to \infty$.

There are many other such criteria. See [4] and the references cited. All of the above and some others are desirable properties of estimators. It is very rare, however, that an estimator can be found that satisfies all, or even many, of the optimality criteria.

An important result that is useful in evaluating estimators is the *Cramér-Rao Inequality*, which gives a lower bound on the variance of any unbiased estimator. The result is as follow:

*Cramér-Rao Inequality* Suppose that $T = T(Y_1, Y_2, \ldots, Y_n)$ is an unbiased estimator of a function $\tau(\theta)$, with $k = 1$. Then under certain regularity conditions on $\tau$ and the distribution of $Y_1, Y_2, \ldots, Y_n$ [46],

$$V(T) \geq \frac{[\tau'(\theta)]^2}{E\left\{\frac{\partial \log[L(Y_1,\ldots,Y_n;\theta]}{\partial\theta}\right\}^2} \tag{D.1}$$

for all $\theta$, where

$$L(Y_1, \ldots, Y_n; \theta) = \prod_{i=1}^{n} f(Y_i; \theta) \tag{D.2}$$

[16] Note that the definitions given are not completely rigorous, but are intended to give the reader a sense of these criteria. For mathematically precise definitions see books on theoretical statistics such as [5, 11, 46].

The denominator of (D.1) is known as the *Fisher Information*, and is denoted $I(\theta)$. If $\tau(\theta) = \theta$, the lower bound on the variance of an unbiased estimator is $1/I(\theta)$.

If the variance of an estimator achieves this bound, it is an efficient, unbiased estimator. If it achieves the bound as $n \to \infty$, the estimator is asymptotically efficient.

If an estimator $\hat{\theta}$ of $\theta$ is biased, with bias given by $b(\theta) = E(\hat{\theta}) - \theta$, the bound becomes

$$V(\hat{\theta}) \geq \frac{[1 + b'(\theta)]^2}{I(\theta)} \tag{D.3}$$

For $k > 1$, the results extend to bounds on the covariance matrix of a vector of estimators of the elements of $\theta$. The bounds are based on the $k \times k$ *Fisher information matrix*, with elements

$$I_{ij} = E\left\{\frac{\partial}{\partial\theta_i}\log[f(y;\theta)]\frac{\partial}{\partial\theta_j}\log[f(y;\theta)]\right\} \tag{D.4}$$

See the references on theoretical statistics cited above for additional details.

## D.2 Method of Maximum Likelihood

### *D.2.1 Complete Data*

The *maximum likelihood* estimator (MLE) is obtained by maximizing the *likelihood function*, which is defined to be the joint distribution of the observations in a random sample. The likelihood function is given in (D.2) for complete data. For ease of computation, we ordinarily maximize the natural log of the likelihood function.[17] Maximization is with regard to the components of $\theta$, and, under the assumption of differentiability, is accomplished by equating the derivatives to zero and solving the resulting *likelihood equations*. If necessary, the ML equations may be solved by numerical methods. Solutions for a number of life distributions are given in most statistical packages.

The rationale for use of the MLE is the *Maximum Likelihood Principle*, which essentially states that one should choose as an estimator the values of the parameters that make the data actually observed most likely to have occurred. In practice, the MLE is used because it is optimal in many ways.

The optimality of the MLE depends on certain *regularity* conditions. These are: (1) the first two derivatives of the log-likelihood with respect to the components of

---

[17] Thus we deal with a sum, and the resulting equations are simpler. Since log is a one-to-one function, the solutions are identical.

$\theta$ must be defined; and (2) $I(\theta)$ must not be zero and must be a continuous of the components of $\theta$. Under these conditions, the MLEs are consistent, asymptotically unbiased, asymptotically efficient, and asymptotically normally distributed.

These results apply to incomplete as well as complete data. In the next two sections we look at the likelihood functions for censored data and grouped data.

## D.2.2 Censored Data

The likelihood function for censored data depends on the type of censoring. We consider the types ordinarily found in reliability and warranty claims data.

*Type I censoring* Censoring as a function of time is called Type I censoring. We are concerned with right censoring. In reliability testing, this occurs when testing is stopped at a specified time $T$. For claims data, censoring occurs at the end of the warranty period, i.e., at $T = W$.

Suppose that all $n$ items are put on test or sold at time 0 and that $r$ items have failed by time $T$. The data may be written as ordered observations, in which case they consist of failure times for the first $r$ items, say $Y_1, \ldots, Y_r$, and the value $T$ for the remaining (censored) items. The likelihood function is

$$L(Y_1, \ldots, Y_n; \theta) = \left\{ \prod_{i=1}^{r} f(Y_i) \right\} [1 - F(T; \theta)]^{n-r} \tag{D.5}$$

Note that here $r$ is a random variable. The implication of this is that maximization may not be approached simply by differentiation of the likelihood or log-likelihood, and alternate methods, e.g., search routines, are required.

In practice, warranty claims data are often multiply censored. This occurs in warranty data when items are sold at different times. In this case, the likelihood function becomes

$$L(Y_1, \ldots, Y_n; \theta) = \prod_{i=1}^{r} f(Y_i) \prod_{i=r+1}^{n} [F(W; \theta) - F(Y_i; \theta)] \tag{D.6}$$

where $Y_{r+1}, \ldots, Y_n$ are the times of sale of the unfailed items.

*Type II censoring* Testing continues until a predetermined number $r$ of failures occur. The data are as above. The likelihood function is

$$L(Y_1, \ldots, Y_n; \theta) = \prod_{i=1}^{r} f(Y_i) \prod_{i=r+1}^{n} [1 - F(Y_i; \theta)] \tag{D.7}$$

The MLEs are obtained by minimization of (D.7). The properties of the MLEs are as indicated above. For additional results, including the likelihood function for other types of censoring, see [30].

### D.2.3 Grouped Data

We assume that the observations are grouped into $k$ intervals defined by endpoints $y'_0, y'_1, \ldots, y'_k$. The number of observations falling into the $i$th interval is $n_i$, where $\sum_{i=1}^{k} n_i = n$. The likelihood function is given by

$$L(n_1, \ldots, n_k; \theta) = \frac{n!}{n_1! \ldots n_k!} \prod_{i=1}^{k} [F(y'_i; \theta) - F(y'_{i-1}; \theta)]^{n_i} \tag{D.8}$$

For data that are censored as well as grouped, the likelihood function is modified as in (D.6). Let $r_i$ denote the number of observations censored at $i$th interval. The likelihood function can be written as

$$L(\theta) \propto \prod_{i=1}^{k} [F(y'_i; \theta) - F(y'_{i-1}; \theta)]^{n_i} [1 - F(y'_i; \theta)]^{r_i} \tag{D.9}$$

### D.2.4 Asymptotic Confidence Intervals and Tests

The asymptotic normality of the MLEs may be used to obtain asymptotic confidence intervals and asymptotic test of hypotheses. Asymptotic variances and covariances of the estimators are obtained as the elements of the inverse of the information matrix with elements given by (D.4). These may be estimated by substituting MLEs of the parameters involved. The confidence intervals and tests are then done by use of the procedures based on the normal distribution.

## D.3 Estimation of Functions of Random Variables

In reliability and warranty analyses, we often encounter problems in which estimators of functions of parameters are needed. Here we look briefly at some approaches to problems of this type and illustrate the methodology by considering sums, products and ratios of random variables. Most of the results given are asymptotic results. These provide the means of obtaining asymptotic confidence intervals and tests.

### D.3.1 Asymptotic Mean and Variance of a Function

Assume that $Y$ is a random variable with mean $\mu$ and finite variance $V(Y)$ and let $\tau(Y)$ be a twice-differentiable function of $Y$. Then

$$E[\tau(Y)] \approx \tau(\mu) + 0.5\,\tau''(\mu)V(Y) \tag{D.10}$$

and

$$V((\tau(Y)) \approx [\tau'(\mu)]^2 V(Y). \tag{D.11}$$

In practice, only the first-order approximation of the expectation is used, in which case the result is $E[\tau(Y)] \approx \tau(\mu)$.

This extends to $k$ random variables as follows: Let $Y_1$, $Y_2$, ..., $Y_k$ be random variables with respective means $\mu_i$, variances $\sigma_i^2 = V(Y_i)$ and covariances $\sigma_{ij} = \mathrm{Cov}(Y_i, Y_j) = E[(Y_i - \mu_i)(Y_j - \mu_j)]$, and let $\tau(Y_1, Y_2, \ldots, Y_k)$ be a function such that all second-order derivatives exist. Then

$$E[\tau(Y_l, \ldots, Y_k)] \approx \tau(\mu_l, \ldots, \mu_k) + \sum_{i=1}^{k} \sigma_i^2 \left(\frac{\partial^2 \tau}{\partial Y_i^2}\right)\bigg|_{\mu_l,\ldots,\mu_k} + 2\sum_{i<j} \sigma_{ij} \left(\frac{\partial^2 \tau}{\partial Y_i \partial Y_j}\right)\bigg|_{\mu_l,\ldots,\mu_k} \tag{D.12}$$

and

$$V(\tau(Y_i, \ldots, Y_k)) \approx \sum_{i=1}^{k} \sigma_i^2 \left(\frac{\partial \tau}{\partial Y_i}\right)^2\bigg|_{\mu_1,\ldots,\mu_k} + 2\sum_{i<j} \sigma_{ij} \left(\frac{\partial \tau}{\partial Y_i}\right)\left(\frac{\partial \tau}{\partial Y_j}\right)\bigg|_{\mu_1,\ldots,\mu_k} \tag{D.13}$$

Again, in practice only the first order approximation to the expectation is used.

It was noted above that the MLE is asymptotically normally distributed. Under fairly general conditions, this is true of functions of the MLE as well, with means and variances are given in (D.10–D.13). It follows that asymptotic confidence intervals and tests may be constructed based on these results. These are appropriate for large $n$ and are obtained by substitution of MLE's into the formulas for the asymptotic variance and use of fractiles of the standard normal distribution.

### *D.3.2 Sums of Random Variables*

Assume that $Y_1, Y_2, \ldots, Y_n$ are random variables with respective means $\mu_i$ and finite variances $\sigma_i^2$ and covariances $\sigma_{ij}$. Let $c_0, c_1, \ldots, c_n$ be a sequence of constants and $Y = c_0 + \sum_{i=1}^{n} c_i Y_i$. Then the mean and variance of $Y$ are given by

$$E(Y) = c_0 + \sum_{i=1}^{n} c_i \mu_i \tag{D.14}$$

and

$$V(Y) = \sum_{i=1}^{n} c_i^2 \sigma_i^2 + 2\sum_{i<j} c_i c_j \sigma_{ij}. \tag{D.15}$$

It follows from these results that if the $Y_i$'s are independent and identically distributed, then the sample mean $\bar{Y}$ has expectation $\mu$ and variance $\sigma^2/n$. Another important result is that if, in addition, the $Y_i$'s are normally distributed, then $\bar{Y}$ is also normal. Finally, by the Central Limit Theorem, under fairly general conditions $\bar{Y}$ is asymptotically normally distributed.

## D.3.3 Products of Random Variables

Estimation of products is of important in many reliability applications. We consider $k$ independent random variables $Y_1, Y_2, \ldots, Y_k$ with respective means $\mu_i$ and finite variances $\sigma_i^2$. Let $Y = Y_1 \cdot Y_2 \cdot \ldots \cdot Y_k$. Then $E(Y) = \mu_1 \cdot \mu_2 \cdot \ldots \cdot \mu_k$. The variance is more complicated. If $k = 2$,

$$V(Y) = \mu_1^2\sigma_2^2 + \mu_2^2\sigma_1^2 + \sigma_1^2\sigma_2^2. \tag{D.16}$$

For $k = 3$,

$$V(Y) = \mu_1^2\mu_2^2\sigma_3^2 + \mu_1^2\mu_3^2\sigma_2^2 + \mu_2^2\mu_3^2\sigma_1^2 + \mu_1^2\sigma_2^2\sigma_3^2 + \mu_2^2\sigma_1^2\sigma_3^2 + \mu_3^2\sigma_1^2\sigma_2^2 + \sigma_1^2\sigma_2^2\sigma_3^2. \tag{D.17}$$

The general result involves $2^k - 1$ terms, involving all combinations of products of squares of means and variances except the term involving only squares of means.

## D.3.4 Ratios of Random Variables

If $Y_1$ and $Y_2$ are random variables with respective means $\mu_i$ and finite variances $\sigma_i^2$ and covariances $\sigma_{12}$, then the mean and variance of $Y = Y_1/Y_2$ are approximated by

$$E(Y) \approx \frac{\mu_1}{\mu_2} - \frac{\sigma_{12}}{\mu_2^2} + \frac{\mu_1\sigma_2^2}{\mu_2^3} \tag{D.18}$$

and

$$V(Y) \approx \left(\frac{\mu_1}{\mu_2}\right)^2 \left[\frac{\sigma_1^2}{\mu_2^2} + \frac{\sigma_2^2}{\mu_1^2} - \frac{2\sigma_{12}}{\mu_1\mu_2}\right]. \tag{D.19}$$

## D.4 MLE for Incomplete Data using the EM Algorithm

The *Expectation-Maximization* (EM) algorithm is a broadly applicable iterative procedure for computing maximum likelihood estimates in problems with incomplete data. The EM algorithm consists of two conceptually distinct steps at each iteration: the *expectation* or E-step and the *maximization* or M-step.[18]

Suppose we have a model for a set of complete data $Y$, with associated density $f(Y|\theta)$, where $\theta = (\theta_1, \theta_2, \ldots, \theta_d)'$ is a vector of unknown parameters with parameter space $\Omega$ We write $Y = (Y_{obs}, Y_{mis})$ where $Y_{obs}$ represent the observed part of $Y$ and $Y_{mis}$ denotes the missing values. The EM algorithm is designed to find the value the value of $\theta$, denoted $\theta^*$, that maximizes the incomplete data log-likelihood $\log L(\theta) = \log f(Y_{obs}|\theta)$, that is, the MLE of $\theta$ based on the observed data $Y_{obs}$.

The EM algorithm starts with an initial value $\theta^{(0)} \in \Omega$. Suppose that $\theta^{(k)}$ denotes the estimate of $\theta$ at the $k$th iteration; then the $(k + 1)$st iteration can be described in two steps as follows:

*E-step*: Find the conditional expected complete-data log-likelihood given observed data and $\theta = \theta^{(k)}$:

$$\begin{aligned} Q(\theta|\theta^k) &= E(log\, L(Y|Y_{obs}, \theta = \theta^{(k)})) \\ &= \int \log\, L(\theta|Y) f(Y_{mis}|Y_{obs}, \theta = \theta^k) dY_{mis} \end{aligned} \tag{D.20}$$

which, in the case of linear exponential family, amounts to estimating the sufficient statistics for the complete data.

*M-step*: Determine $\theta^{(k+1)}$ to be a value of $\theta \in \Omega$ that maximizes $Q(\theta|\theta^{(k)})$.

The MLE of $\theta$ is found by iterating between the E and M steps until a convergence criterion is met. In some cases, it may not be numerically feasible to find the value of $\theta$ that globally maximizes the function $Q(\theta|\theta^{(k)})$ in the M-step. In such situations, a *Generalized EM* (GEM) *algorithm* [7] is used to choose $\theta^{(k+1)}$ in the M-step such that the condition

$$Q(\theta^{(k+1)}|\theta^{(k)}) \geq Q(\theta^{(k)}|\theta^{(k)}) \tag{D.21}$$

holds. For any EM or GEM algorithm, the change from $\theta^{(k)}$ to $\theta^{(k+1)}$ increases the likelihood; that is,

$$\log L(\theta^{(k+1)}) \geq \log L(\theta^{(k)}) \tag{D.22}$$

which follows from the definition of GEM and Jensen's inequality.[19] This fact implies that the log-likelihood, log $L(\theta)$, increases monotonically on any iteration

---

[18] For details, see [7, 10, 25, 28].

[19] See p. 47 of [40].

sequence generated by the EM algorithm, which is the fundamental property for the convergence of the algorithm.[20] Meng and Rubin [31], Louis [26] and Oakes [38] derived methods for obtaining the asymptotic variance-covariance matrix of the EM-computed estimator.

[20] Detailed properties of the algorithm, including the convergence properties, are given in [7, 28, 41, 48].

# Appendix E: Statistical Tables

Percentiles of statistical distributions and related tables are needed for many purposes in statistical inference. Even though most, if not all, of these can be obtained on-line or from statistical program packages, it is often useful to have the tables at hand in books such as this. We provide the following statistical tables:

E.1 Fractiles $z_p$ of the standard normal distribution
E.2 Fractiles of the Student-t distribution
E.3 Fractiles of the Chi-Square distribution shapes.
E.4 Fractiles of the F distribution
E.5 Factors for two-sided normal tolerance limits
E.6 Factors for one-sided normal tolerance limits
E.7 Factors for two-sided nonparametric tolerance limits

**Table E.1** Fractiles $z_p$ of the Standard Normal Distribution. ($P(Z \leq z_p) = p$)

| $p$ | $z_p$ | $p$ | $z_p$ |
|---|---|---|---|
| 0.0005 | −3.291 | 0.8000 | 0.842 |
| 0.0010 | −3.091 | 0.8500 | 1.036 |
| 0.0025 | −2.807 | 0.9000 | 1.282 |
| 0.0050 | −2.576 | 0.9500 | 1.645 |
| 0.0100 | −2.326 | 0.9750 | 1.960 |
| 0.0200 | −2.054 | 0.9800 | 2.054 |
| 0.0250 | −1.960 | 0.9900 | 2.326 |
| 0.0500 | −1.645 | 0.9950 | 2.576 |
| 0.1000 | −1.282 | 0.9975 | 2.807 |
| 0.1500 | −1.036 | 0.9990 | 3.091 |
| 0.2000 | −0.842 | 0.9995 | 3.291 |

W. R. Blischke et al., *Warranty Data Collection and Analysis*,
Springer Series in Reliability Engineering, DOI: 10.1007/978-0-85729-647-4,

**Table E.2** Fractiles of the Student-t distribution

| *df* | *p* | | | | |
|---|---|---|---|---|---|
| | 0.900 | 0.950 | 0.975 | 0.990 | 0.995 |
| 1 | 3.078 | 6.314 | 12.706 | 31.821 | 63.657 |
| 2 | 1.886 | 2.920 | 4.303 | 6.965 | 9.925 |
| 3 | 1.638 | 2.353 | 3.182 | 4.541 | 5.841 |
| 4 | 1.533 | 2.132 | 2.776 | 3.747 | 4.604 |
| 5 | 1.476 | 2.015 | 2.571 | 3.365 | 4.032 |
| 6 | 1.440 | 1.943 | 2.447 | 3.143 | 3.707 |
| 7 | 1.415 | 1.895 | 1.365 | 2.998 | 3.499 |
| 8 | 1.397 | 1.860 | 2.306 | 2.896 | 2.355 |
| 9 | 1.383 | 1.833 | 2.262 | 2.821 | 3.250 |
| 10 | 1.372 | 1.812 | 2.228 | 2.764 | 3.169 |
| 11 | 1.363 | 1.796 | 2.201 | 2.718 | 3.106 |
| 12 | 1.356 | 1.782 | 2.179 | 2.681 | 3.055 |
| 13 | 1.350 | 1.771 | 2.160 | 2.650 | 3.012 |
| 14 | 1.345 | 1.761 | 2.145 | 2.624 | 2.977 |
| 15 | 1.341 | 1.753 | 2.131 | 2.602 | 2.947 |
| 16 | 1.337 | 1.746 | 2.120 | 2.583 | 2.921 |
| 17 | 1.333 | 1.740 | 2.110 | 2.567 | 2.898 |
| 18 | 1.330 | 1.734 | 2.101 | 2.552 | 2.878 |
| 19 | 1.328 | 1.729 | 2.093 | 2.539 | 2.861 |
| 20 | 1.325 | 1.725 | 2.086 | 2.528 | 2.845 |
| 21 | 1.323 | 1.721 | 2.080 | 2.518 | 2.831 |
| 22 | 1.321 | 1.717 | 2.074 | 2.508 | 2.819 |
| 23 | 1.319 | 1.714 | 2.069 | 2.500 | 2.807 |
| 24 | 1.318 | 1.711 | 2.064 | 2.492 | 2.797 |
| 25 | 1.316 | 1.708 | 2.060 | 2.485 | 1.787 |
| 26 | 1.315 | 1.706 | 2.056 | 2.479 | 2.779 |
| 27 | 1.314 | 1.703 | 2.052 | 2.473 | 2.771 |
| 28 | 1.313 | 1.701 | 2.048 | 2.467 | 2.763 |
| 29 | 1.311 | 1.699 | 2.045 | 2.462 | 2.756 |
| 30 | 1.310 | 1.697 | 2.042 | 2.457 | 2.750 |
| 35 | 1.306 | 1.690 | 2.030 | 2.438 | 2.715 |
| 40 | 1.303 | 1.684 | 2.021 | 2.423 | 2.704 |
| 45 | 1.301 | 1.679 | 2.014 | 2.412 | 2.690 |
| 50 | 1.299 | 1.676 | 2.009 | 2.403 | 2.678 |
| 55 | 1.297 | 1.673 | 2.004 | 2.396 | 2.668 |
| 60 | 1.296 | 1.671 | 2.000 | 2.390 | 2.660 |
| 65 | 1.295 | 2.669 | 1.997 | 2.385 | 2.654 |
| 70 | 1.294 | 1.667 | 1.994 | 2.381 | 2.648 |
| 75 | 1.293 | 2.665 | 1.992 | 2.377 | 2.643 |
| 80 | 1.292 | 1.664 | 1.990 | 2.374 | 2.639 |
| 85 | 1.292 | 1.663 | 1.988 | 2.371 | 2.635 |
| 90 | 1.291 | 1.662 | 1.987 | 2.369 | 2.632 |
| 95 | 1.291 | 1.661 | 1.985 | 2.366 | 2.629 |
| 100 | 1.290 | 1.660 | 1.984 | 2.364 | 2.626 |
| 200 | 1.286 | 1.653 | 1.972 | 2.345 | 2.601 |
| 500 | 1.283 | 1.648 | 1.965 | 2.334 | 2.586 |
| $\infty$ | 1.282 | 1.645 | 1.960 | 2.326 | 2.576 |

**Table E.3** Fractiles of the Chi-Square distribution

| *df* | *p* | | | | | | | | | |
|---|---|---|---|---|---|---|---|---|---|---|
| | 0.005 | 0.010 | 0.025 | 0.050 | 0.100 | 0.900 | 0.950 | 0.975 | 0.990 | 0.995 |
| 1 | $0.0^4393$ | $0.0^31571$ | $0.0^39821$ | $0.0^2393$ | 0.01579 | 2.706 | 3.841 | 5.024 | 6.635 | 7.879 |
| 2 | 0.01003 | 0.02010 | 0.05064 | 0.1026 | 0.2107 | 4.605 | 5.991 | 7.378 | 9.210 | 10.60 |
| 3 | 0.07172 | 0.1148 | 0.2158 | 0.3519 | 0.5844 | 6.251 | 7.815 | 9.348 | 11.345 | 12.84 |
| 4 | 0.2070 | 0.2971 | 0.4844 | 0.7107 | 1.064 | 7.779 | 9.488 | 11.14 | 13.28 | 14.86 |
| 5 | 0.4117 | 0.5543 | 0.8312 | 1.145 | 1.610 | 9.236 | 11.07 | 12.83 | 15.09 | 16.75 |
| 6 | 0.6757 | 0.8721 | 1.237 | 1.635 | 2.204 | 10.64 | 12.59 | 14.45 | 16.81 | 18.55 |
| 7 | 0.9893 | 1.239 | 1.690 | 2.167 | 2.833 | 12.02 | 14.07 | 16.01 | 18.48 | 20.28 |
| 8 | 1.344 | 1.647 | 2.180 | 2.733 | 3.490 | 13.36 | 15.51 | 17.53 | 20.09 | 21.96 |
| 9 | 1.735 | 2.088 | 2.700 | 3.325 | 4.168 | 14.68 | 16.92 | 19.02 | 21.67 | 23.59 |
| 10 | 2.156 | 2.558 | 3.247 | 3.940 | 4.865 | 15.99 | 18.31 | 20.48 | 23.21 | 25.19 |
| 11 | 2.603 | 3.054 | 3.816 | 4.575 | 5.578 | 17.28 | 19.68 | 21.92 | 24.73 | 26.76 |
| 12 | 3.074 | 3.571 | 4.404 | 5.226 | 6.304 | 18.55 | 21.03 | 23.34 | 26.22 | 28.30 |
| 13 | 3.565 | 4.107 | 5.009 | 5.892 | 7.042 | 19.81 | 22.36 | 24.74 | 27.69 | 29.82 |
| 14 | 4.075 | 4.660 | 5.629 | 6.571 | 7.790 | 21.06 | 23.68 | 26.12 | 29.14 | 31.32 |
| 15 | 4.601 | 5.229 | 6.262 | 7.261 | 8.547 | 22.31 | 25.00 | 27.49 | 30.58 | 32.80 |
| 16 | 5.142 | 5.812 | 6.908 | 7.962 | 9.312 | 23.54 | 26.30 | 28.85 | 32.00 | 34.27 |
| 17 | 5.697 | 6.408 | 7.564 | 8.672 | 10.09 | 24.77 | 27.59 | 30.19 | 33.41 | 35.72 |
| 18 | 6.265 | 7.015 | 8.231 | 9.391 | 10.86 | 25.99 | 28.87 | 31.53 | 34.81 | 37.16 |
| 19 | 6.844 | 7.633 | 8.907 | 10.12 | 11.65 | 27.20 | 30.14 | 32.85 | 36.19 | 38.58 |
| 20 | 7.434 | 8.260 | 9.591 | 10.85 | 12.44 | 28.41 | 31.41 | 34.17 | 37.57 | 40.00 |
| 21 | 8.034 | 8.897 | 10.28 | 11.59 | 13.24 | 29.62 | 32.67 | 35.48 | 38.93 | 41.40 |
| 22 | 8.643 | 9.542 | 10.98 | 12.34 | 14.04 | 30.81 | 33.92 | 36.78 | 40.29 | 42.80 |
| 23 | 9.260 | 10.20 | 11.69 | 13.09 | 14.85 | 32.01 | 35.17 | 38.08 | 41.64 | 44.18 |
| 24 | 9.886 | 10.86 | 12.40 | 13.85 | 15.66 | 33.20 | 36.42 | 39.36 | 42.98 | 45.56 |
| 25 | 10.52 | 11.52 | 13.12 | 14.61 | 16.47 | 34.38 | 37.65 | 40.65 | 44.31 | 46.93 |
| 26 | 11.16 | 12.20 | 13.84 | 15.38 | 17.29 | 35.56 | 38.89 | 41.92 | 45.64 | 48.29 |
| 27 | 11.81 | 12.88 | 14.57 | 16.15 | 18.11 | 36.74 | 40.11 | 43.19 | 46.96 | 49.64 |
| 28 | 12.46 | 13.56 | 15.31 | 16.93 | 18.94 | 37.92 | 41.34 | 44.46 | 48.28 | 50.99 |
| 29 | 13.12 | 14.26 | 16.05 | 17.71 | 19.77 | 39.09 | 42.56 | 45.72 | 49.59 | 52.34 |
| 30 | 13.79 | 14.95 | 16.79 | 18.49 | 20.60 | 40.26 | 43.77 | 46.98 | 50.89 | 53.67 |
| 40 | 20.71 | 22.16 | 24.43 | 26.51 | 29.05 | 51.81 | 55.76 | 59.34 | 63.69 | 66.77 |
| 50 | 27.99 | 29.71 | 32.36 | 34.76 | 37.69 | 63.17 | 67.50 | 71.42 | 76.15 | 79.49 |
| 60 | 35.53 | 37.48 | 40.48 | 43.19 | 46.46 | 74.40 | 79.08 | 83.30 | 88.38 | 91.95 |
| 70 | 43.28 | 45.44 | 48.76 | 51.74 | 55.33 | 85.53 | 90.53 | 95.02 | 100.43 | 104.21 |
| 80 | 51.17 | 53.54 | 57.15 | 60.39 | 64.28 | 96.58 | 101.88 | 106.63 | 112.33 | 116.32 |
| 90 | 59.20 | 61.75 | 65.65 | 69.13 | 73.29 | 107.57 | 113.15 | 118.14 | 124.12 | 128.30 |
| 100 | 67.33 | 70.06 | 74.22 | 77.93 | 82.36 | 118.50 | 124.34 | 129.56 | 135.81 | 140.17 |

**Table E.4** *F* distribution ($p$ = upper-tail probability, $n_1$: denominator *df*; $n_2$: numerator *df*).

| $n_2$ | $p$ | $n_1$ 1 | 2 | 3 | 4 | 5 | 6 | 7 | 8 | 9 | 10 |
|---|---|---|---|---|---|---|---|---|---|---|---|
| | 0.10 | 4.06 | 3.78 | 3.62 | 3.52 | 3.45 | 3.40 | 3.37 | 3.34 | 3.32 | 3.30 |
| 5 | 0.05 | 6.61 | 5.79 | 5.41 | 5.19 | 5.05 | 4.95 | 4.88 | 4.82 | 4.77 | 4.74 |
| | 0.01 | 16.26 | 13.27 | 12.06 | 11.39 | 10.97 | 10.67 | 10.46 | 10.29 | 10.16 | 10.05 |
| | 0.10 | 3.78 | 3.46 | 3.29 | 3.18 | 3.11 | 3.05 | 3.01 | 2.98 | 2.96 | 2.94 |
| 6 | 0.05 | 5.99 | 5.14 | 4.76 | 4.53 | 4.39 | 4.28 | 4.21 | 4.15 | 4.10 | 4.06 |
| | 0.01 | 13.75 | 10.92 | 9.78 | 9.15 | 8.75 | 8.47 | 8.26 | 8.10 | 7.98 | 7.87 |
| | 0.10 | 3.59 | 3.26 | 3.07 | 2.96 | 2.88 | 2.83 | 2.78 | 2.75 | 2.72 | 2.70 |
| 7 | 0.05 | 5.59 | 4.74 | 4.35 | 4.12 | 3.97 | 3.87 | 3.79 | 3.73 | 3.68 | 3.64 |
| | 0.01 | 12.25 | 9.55 | 8.45 | 7.85 | 7.46 | 7.19 | 6.99 | 6.84 | 6.72 | 6.62 |
| | 0.10 | 3.46 | 3.11 | 2.92 | 2.81 | 2.73 | 2.67 | 2.62 | 2.59 | 2.56 | 2.54 |
| 8 | 0.05 | 5.32 | 4.46 | 4.07 | 3.84 | 3.69 | 3.58 | 3.50 | 3.44 | 3.39 | 3.35 |
| | 0.01 | 11.26 | 8.65 | 7.59 | 7.01 | 6.63 | 6.37 | 6.18 | 6.03 | 5.91 | 5.81 |
| | 0.10 | 3.36 | 3.01 | 2.81 | 2.69 | 2.61 | 2.55 | 2.51 | 2.47 | 2.44 | 2.42 |
| 9 | 0.05 | 5.12 | 4.26 | 3.86 | 3.63 | 3.48 | 3.37 | 3.29 | 3.23 | 3.18 | 3.14 |
| | 0.01 | 10.56 | 8.02 | 6.99 | 6.42 | 6.06 | 5.80 | 5.61 | 5.47 | 5.35 | 5.26 |
| | 0.10 | 3.29 | 2.92 | 2.73 | 2.61 | 2.52 | 2.46 | 2.41 | 2.38 | 2.35 | 2.32 |
| 10 | 0.05 | 4.96 | 4.10 | 3.71 | 3.48 | 3.33 | 3.22 | 3.14 | 3.07 | 3.02 | 2.98 |
| | 0.01 | 10.04 | 7.56 | 6.55 | 5.99 | 5.64 | 5.39 | 5.20 | 5.06 | 4.94 | 4.85 |
| | 0.10 | 3.23 | 2.86 | 2.66 | 2.54 | 2.45 | 2.39 | 2.34 | 2.30 | 2.27 | 2.25 |
| 11 | 0.05 | 4.84 | 3.98 | 3.59 | 3.36 | 3.20 | 3.09 | 3.01 | 2.95 | 2.90 | 2.85 |
| | 0.01 | 9.65 | 7.21 | 6.22 | 5.67 | 5.32 | 5.07 | 4.89 | 4.74 | 4.63 | 4.54 |
| | 0.10 | 3.18 | 2.81 | 2.61 | 2.48 | 2.39 | 2.33 | 2.28 | 2.24 | 2.21 | 2.19 |
| 12 | 0.05 | 4.75 | 3.89 | 3.49 | 3.26 | 3.11 | 3.00 | 2.91 | 2.85 | 2.80 | 2.75 |
| | 0.01 | 9.33 | 6.93 | 5.95 | 5.41 | 5.06 | 4.82 | 4.64 | 4.50 | 4.39 | 4.30 |
| | 0.10 | 3.14 | 2.76 | 2.56 | 2.43 | 2.35 | 2.28 | 2.23 | 2.20 | 2.16 | 2.14 |
| 13 | 0.05 | 4.67 | 3.81 | 3.41 | 3.18 | 3.03 | 2.92 | 2.83 | 2.77 | 2.71 | 2.67 |
| | 0.01 | 9.07 | 6.70 | 5.74 | 5.21 | 4.86 | 4.62 | 4.44 | 4.30 | 4.19 | 4.10 |
| | 0.10 | 3.10 | 2.73 | 2.52 | 2.39 | 2.31 | 2.24 | 2.19 | 2.15 | 2.12 | 2.10 |
| 14 | 0.05 | 4.60 | 3.74 | 3.34 | 3.11 | 2.96 | 2.85 | 2.76 | 2.70 | 2.65 | 2.60 |
| | 0.01 | 8.86 | 6.51 | 5.56 | 5.04 | 4.69 | 4.46 | 4.28 | 4.14 | 4.03 | 3.94 |
| | 0.10 | 3.07 | 2.70 | 2.49 | 2.36 | 2.27 | 2.21 | 2.16 | 2.12 | 2.09 | 2.06 |
| 15 | 0.05 | 4.54 | 3.68 | 3.29 | 3.06 | 2.90 | 2.79 | 2.71 | 2.64 | 2.59 | 2.54 |
| | 0.01 | 8.68 | 6.36 | 5.42 | 4.89 | 4.56 | 4.32 | 4.14 | 4.00 | 3.89 | 3.80 |
| | 0.10 | 3.05 | 2.67 | 2.46 | 2.33 | 2.24 | 2.18 | 2.13 | 2.09 | 2.06 | 2.03 |
| 16 | 0.05 | 4.49 | 3.63 | 3.24 | 3.01 | 2.85 | 2.74 | 2.66 | 2.59 | 2.54 | 2.49 |
| | 0.01 | 8.53 | 6.23 | 5.29 | 4.77 | 4.44 | 4.20 | 4.03 | 3.89 | 3.78 | 3.69 |

| $n_2$ | $p$ | $n_1$ 12 | 15 | 20 | 25 | 30 | 40 | 50 | 60 | 120 | 1000 |
|---|---|---|---|---|---|---|---|---|---|---|---|
| | 0.10 | 3.27 | 3.24 | 3.21 | 3.19 | 3.17 | 3.16 | 3.15 | 3.14 | 3.12 | 3.11 |
| 5 | 0.05 | 4.68 | 4.62 | 4.56 | 4.52 | 4.50 | 4.46 | 4.44 | 4.43 | 4.40 | 4.37 |
| | 0.01 | 9.89 | 9.72 | 9.55 | 9.45 | 9.38 | 9.29 | 9.24 | 9.20 | 9.11 | 9.03 |
| | 0.10 | 2.90 | 2.87 | 2.84 | 2.81 | 2.80 | 2.78 | 2.77 | 2.76 | 2.74 | 2.72 |
| 6 | 0.05 | 4.00 | 3.94 | 3.87 | 3.83 | 3.81 | 3.77 | 3.75 | 3.74 | 3.70 | 3.67 |
| | 0.01 | 7.72 | 7.56 | 7.40 | 7.30 | 7.23 | 7.14 | 7.09 | 7.06 | 6.97 | 6.89 |
| | 0.10 | 2.67 | 2.63 | 2.59 | 2.57 | 2.56 | 2.54 | 2.52 | 2.51 | 2.49 | 2.47 |
| 7 | 0.05 | 3.57 | 3.51 | 3.44 | 3.40 | 3.38 | 3.34 | 3.32 | 3.30 | 3.27 | 3.23 |
| | 0.01 | 6.47 | 6.31 | 6.16 | 6.06 | 5.99 | 5.91 | 5.86 | 5.82 | 5.74 | 5.66 |

(continued)

**Table E.4** (continued)

| $n_2$ | $p$ | $n_1$ 12 | 15 | 20 | 25 | 30 | 40 | 50 | 60 | 120 | 1000 |
|---|---|---|---|---|---|---|---|---|---|---|---|
| | 0.10 | 2.50 | 2.46 | 2.42 | 2.40 | 2.38 | 2.36 | 2.35 | 2.34 | 2.32 | 2.30 |
| 8 | 0.05 | 3.28 | 3.22 | 3.15 | 3.11 | 3.08 | 3.04 | 3.02 | 3.01 | 2.97 | 2.93 |
| | 0.01 | 5.67 | 5.52 | 5.36 | 5.26 | 5.20 | 5.12 | 5.07 | 5.03 | 4.95 | 4.87 |
| | 0.10 | 2.38 | 2.34 | 2.30 | 2.27 | 2.25 | 2.23 | 2.22 | 2.21 | 2.18 | 2.16 |
| 9 | 0.05 | 3.07 | 3.01 | 2.94 | 2.89 | 2.86 | 2.83 | 2.80 | 2.79 | 2.75 | 2.71 |
| | 0.01 | 5.11 | 4.96 | 4.81 | 4.71 | 4.65 | 4.57 | 4.52 | 4.48 | 4.40 | 4.32 |
| | 0.10 | 2.28 | 2.24 | 2.20 | 2.17 | 2.16 | 2.13 | 2.12 | 2.11 | 2.08 | 2.06 |
| 10 | 0.05 | 2.91 | 2.85 | 2.77 | 2.73 | 2.70 | 2.66 | 2.64 | 2.62 | 2.58 | 2.54 |
| | 0.01 | 4.71 | 4.56 | 4.41 | 4.31 | 4.25 | 4.17 | 4.12 | 4.08 | 4.00 | 3.92 |
| | 0.10 | 2.21 | 2.17 | 2.12 | 2.10 | 2.08 | 2.05 | 2.04 | 2.03 | 2.00 | 1.98 |
| 11 | 0.05 | 2.79 | 2.72 | 2.65 | 2.60 | 2.57 | 2.53 | 2.51 | 2.49 | 2.45 | 2.41 |
| | 0.01 | 4.40 | 4.25 | 4.10 | 4.01 | 3.94 | 3.86 | 3.81 | 3.78 | 3.69 | 3.61 |
| | 0.10 | 2.15 | 2.10 | 2.06 | 2.03 | 2.01 | 1.99 | 1.97 | 1.96 | 1.93 | 1.91 |
| 12 | 0.05 | 2.69 | 2.62 | 2.54 | 2.50 | 2.47 | 2.43 | 2.40 | 2.38 | 2.34 | 2.30 |
| | 0.01 | 4.16 | 4.01 | 3.86 | 3.76 | 3.70 | 3.62 | 3.57 | 3.54 | 3.45 | 3.37 |
| | 0.10 | 2.10 | 2.05 | 2.01 | 1.98 | 1.96 | 1.93 | 1.92 | 1.90 | 1.88 | 1.85 |
| 13 | 0.05 | 2.60 | 2.53 | 2.46 | 2.41 | 2.38 | 2.34 | 2.31 | 2.30 | 2.25 | 2.21 |
| | 0.01 | 3.96 | 3.82 | 3.66 | 3.57 | 3.51 | 3.43 | 3.38 | 3.34 | 3.25 | 3.18 |
| | 0.10 | 2.05 | 2.01 | 1.96 | 1.93 | 1.91 | 1.89 | 1.87 | 1.86 | 1.83 | 1.80 |
| 14 | 0.05 | 2.53 | 2.46 | 2.39 | 2.34 | 2.31 | 2.27 | 2.24 | 2.22 | 2.18 | 2.14 |
| | 0.01 | 3.80 | 3.66 | 3.51 | 3.41 | 3.35 | 3.27 | 3.22 | 3.18 | 3.09 | 3.02 |
| | 0.10 | 2.02 | 1.97 | 1.92 | 1.89 | 1.87 | 1.85 | 1.83 | 1.82 | 1.79 | 1.76 |
| 15 | 0.05 | 2.48 | 2.40 | 2.33 | 2.28 | 2.25 | 2.20 | 2.18 | 2.16 | 2.11 | 2.07 |
| | 0.01 | 3.67 | 3.52 | 3.37 | 3.28 | 3.21 | 3.13 | 3.08 | 3.05 | 2.96 | 2.88 |
| | 0.10 | 1.99 | 1.94 | 1.89 | 1.86 | 1.84 | 1.81 | 1.79 | 1.78 | 1.75 | 1.72 |
| 16 | 0.05 | 2.42 | 7.35 | 2.28 | 2.23 | 2.19 | 2.15 | 2.12 | 2.11 | 2.06 | 2.02 |
| | 0.01 | 3.55 | 3.41 | 3.26 | 3.16 | 3.10 | 3.02 | 2.97 | 2.93 | 2.84 | 2.76 |

| $n_2$ | $p$ | $n_1$ 1 | 2 | 3 | 4 | 5 | 6 | 7 | 8 | 9 | 10 |
|---|---|---|---|---|---|---|---|---|---|---|---|
| | 0.10 | 3.03 | 2.64 | 2.44 | 2.31 | 2.22 | 2.15 | 2.10 | 2.06 | 2.03 | 2.00 |
| 17 | 0.05 | 4.45 | 3.59 | 3.20 | 2.96 | 2.81 | 2.70 | 2.61 | 2.55 | 2.49 | 2.45 |
| | 0.01 | 8.40 | 6.11 | 5.19 | 4.67 | 4.34 | 4.10 | 3.93 | 3.79 | 3.68 | 3.59 |
| | 0.10 | 3.01 | 2.62 | 2.42 | 2.29 | 2.20 | 2.13 | 2.08 | 2.04 | 2.00 | 1.98 |
| 18 | 0.05 | 4.41 | 3.55 | 3.16 | 2.93 | 2.77 | 2.66 | 2.58 | 2.51 | 2.46 | 2.41 |
| | 0.01 | 8.29 | 6.01 | 5.09 | 4.58 | 4.25 | 4.01 | 3.84 | 3.71 | 3.60 | 3.51 |
| | 0.10 | 2.99 | 2.61 | 2.40 | 2.27 | 2.18 | 2.11 | 2.06 | 2.02 | 1.98 | 1.96 |
| 19 | 0.05 | 4.38 | 3.52 | 3.13 | 2.90 | 2.74 | 2.63 | 2.54 | 2.48 | 2.42 | 2.38 |
| | 0.01 | 8.18 | 5.93 | 5.01 | 4.50 | 4.17 | 3.94 | 3.77 | 3.63 | 3.52 | 3.43 |
| | 0.10 | 2.97 | 2.59 | 2.38 | 2.25 | 2.16 | 2.09 | 2.04 | 2.00 | 1.96 | 1.94 |
| 20 | 0.05 | 4.35 | 3.49 | 3.10 | 2.87 | 2.71 | 2.60 | 2.51 | 2.45 | 2.39 | 2.35 |
| | 0.01 | 8.10 | 5.85 | 4.94 | 4.43 | 4.10 | 3.87 | 3.70 | 3.56 | 3.46 | 3.37 |
| | 0.10 | 2.96 | 2.57 | 2.36 | 2.23 | 2.14 | 2.08 | 2.02 | 1.98 | 1.95 | 1.92 |
| 21 | 0.05 | 4.32 | 3.47 | 3.07 | 2.84 | 2.68 | 2.57 | 2.49 | 2.42 | 2.37 | 2.32 |
| | 0.01 | 8.02 | 5.78 | 4.87 | 4.37 | 4.04 | 3.81 | 3.64 | 3.51 | 3.40 | 3.31 |
| | 0.10 | 2.95 | 2.56 | 2.35 | 2.22 | 2.13 | 2.06 | 2.01 | 1.97 | 1.93 | 1.90 |
| 22 | 0.05 | 4.30 | 3.44 | 3.05 | 2.82 | 2.66 | 2.55 | 2.46 | 2.40 | 2.34 | 2.30 |
| | 0.01 | 7.95 | 5.72 | 4.82 | 4.31 | 3.99 | 3.76 | 3.59 | 3.45 | 3.35 | 3.26 |
| | 0.10 | 2.94 | 2.55 | 2.34 | 2.21 | 2.11 | 2.05 | 1.99 | 1.95 | 1.92 | 1.89 |
| 23 | 0.05 | 4.28 | 3.42 | 3.03 | 2.80 | 2.64 | 2.53 | 2.44 | 2.37 | 2.32 | 2.27 |
| | 0.01 | 7.88 | 5.66 | 4.76 | 4.26 | 3.94 | 3.71 | 3.54 | 3.41 | 3.30 | 3.21 |

(continued)

**Table E.4** (continued)

| $n_2$ | $p$ | $n_1$ 1 | 2 | 3 | 4 | 5 | 6 | 7 | 8 | 9 | 10 |
|---|---|---|---|---|---|---|---|---|---|---|---|
| | 0.10 | 2.93 | 2.54 | 2.33 | 2.19 | 2.10 | 2.04 | 1.98 | 1.94 | 1.91 | 1.88 |
| 24 | 0.05 | 4.26 | 3.40 | 3.01 | 2.78 | 2.62 | 2.51 | 2.42 | 2.36 | 2.30 | 2.25 |
| | 0.01 | 7.82 | 5.61 | 4.72 | 4.22 | 3.90 | 3.67 | 3.50 | 3.36 | 3.26 | 3.17 |
| | 0.10 | 2.92 | 2.53 | 2.32 | 2.18 | 2.09 | 2.02 | 1.97 | 1.93 | 1.89 | 1.87 |
| 25 | 0.05 | 4.24 | 3.39 | 2.99 | 2.76 | 2.60 | 2.49 | 2.40 | 2.34 | 2.28 | 2.24 |
| | 0.01 | 7.77 | 5.57 | 4.68 | 4.18 | 3.85 | 3.63 | 3.46 | 3.32 | 3.22 | 3.13 |
| | 0.10 | 2.88 | 2.49 | 2.28 | 2.14 | 2.05 | 1.98 | 1.93 | 1.88 | 1.85 | 1.82 |
| 30 | 0.05 | 4.17 | 3.32 | 2.92 | 2.69 | 2.53 | 2.42 | 2.33 | 2.27 | 2.21 | 2.16 |
| | 0.01 | 7.56 | 5.39 | 4.51 | 4.02 | 3.70 | 3.47 | 3.30 | 3.17 | 3.07 | 2.98 |
| | 0.10 | 2.84 | 2.44 | 2.23 | 2.09 | 2.00 | 1.93 | 1.87 | 1.83 | 1.79 | 1.76 |
| 40 | 0.05 | 4.08 | 3.23 | 2.84 | 2.61 | 2.45 | 2.34 | 2.25 | 2.18 | 2.12 | 2.08 |
| | 0.01 | 7.31 | 5.18 | 4.31 | 3.83 | 3.51 | 3.29 | 3.12 | 2.99 | 2.89 | 2.80 |
| | 0.10 | 2.81 | 2.41 | 2.20 | 2.06 | 1.97 | 1.90 | 1.84 | 1.80 | 1.76 | 1.73 |
| 50 | 0.05 | 4.03 | 3.18 | 2.79 | 2.56 | 2.40 | 2.29 | 2.20 | 2.13 | 2.07 | 2.03 |
| | 0.01 | 7.17 | 5.06 | 4.20 | 3.72 | 3.41 | 3.19 | 3.02 | 2.89 | 2.78 | 2.70 |

| $n_2$ | $p$ | $n_1$ 12 | 15 | 20 | 25 | 30 | 40 | 50 | 60 | 120 | 1000 |
|---|---|---|---|---|---|---|---|---|---|---|---|
| | 0.10 | 1.96 | 1.91 | 1.86 | 1.83 | 1.81 | 1.78 | 1.76 | 1.75 | 1.72 | 1.69 |
| 17 | 0.05 | 2.38 | 2.31 | 2.23 | 2.18 | 2.15 | 2.10 | 2.08 | 2.06 | 2.01 | 1.97 |
| | 0.01 | 3.46 | 3.31 | 3.16 | 3.07 | 3.00 | 2.92 | 2.87 | 2.83 | 2.75 | 2.66 |
| | 0.10 | 1.93 | 1.89 | 1.84 | 1.80 | 1.78 | 1.75 | 1.74 | 1.72 | 1.69 | 1.66 |
| 18 | 0.05 | 2.34 | 2.27 | 2.19 | 2.14 | 2.11 | 2.06 | 2.04 | 2.02 | 1.97 | 1.92 |
| | 0.01 | 3.37 | 3.23 | 3.08 | 2.98 | 2.92 | 2.84 | 2.78 | 2.75 | 2.66 | 2.58 |
| | 0.10 | 1.91 | 1.86 | 1.81 | 1.78 | 1.76 | 1.73 | 1.71 | 1.70 | 1.67 | 1.64 |
| 19 | 0.05 | 2.31 | 2.23 | 2.16 | 2.11 | 2.07 | 2.03 | 2.00 | 1.98 | 1.93 | 1.88 |
| | 0.01 | 3.30 | 3.15 | 3.00 | 2.91 | 2.84 | 2.76 | 2.71 | 2.67 | 2.58 | 2.50 |
| | 0.10 | 1.89 | 1.84 | 1.79 | 1.76 | 1.74 | 1.71 | 1.69 | 1.68 | 1.64 | 1.61 |
| 20 | 0.05 | 2.28 | 2.20 | 2.12 | 2.07 | 2.04 | 1.99 | 1.97 | 1.95 | 1.90 | 1.85 |
| | 0.01 | 3.23 | 3.09 | 2.94 | 2.84 | 2.78 | 2.69 | 2.64 | 2.61 | 2.52 | 2.43 |
| | 0.10 | 1.87 | 1.83 | 1.78 | 1.74 | 1.72 | 1.69 | 1.67 | 1.66 | 1.62 | 1.59 |
| 21 | 0.05 | 2.25 | 2.18 | 2.10 | 2.05 | 2.01 | 1.96 | 1.94 | 1.92 | 1.87 | 1.82 |
| | 0.01 | 3.17 | 3.03 | 2.88 | 2.79 | 2.72 | 2.64 | 2.58 | 2.55 | 2.46 | 2.37 |
| | 0.10 | 1.86 | 1.81 | 1.76 | 1.73 | 1.70 | 1.67 | 1.65 | 1.64 | 1.60 | 1.57 |
| 22 | 0.05 | 2.23 | 2.15 | 2.07 | 2.02 | 1.98 | 1.94 | 1.91 | 1.89 | 1.84 | 1.79 |
| | 0.01 | 3.12 | 2.98 | 2.83 | 2.73 | 2.67 | 2.58 | 2.53 | 2.50 | 2.40 | 2.32 |
| | 0.10 | 1.84 | 1.80 | 1.74 | 1.71 | 1.69 | 1.66 | 1.64 | 1.62 | 1.59 | 1.55 |
| 23 | 0.05 | 2.20 | 2.13 | 2.05 | 2.00 | 1.96 | 1.91 | 1.88 | 1.86 | 1.81 | 1.76 |
| | 0.01 | 3.07 | 2.93 | 2.78 | 2.69 | 2.62 | 2.54 | 2.48 | 2.45 | 2.35 | 2.27 |
| | 0.10 | 1.83 | 1.78 | 1.73 | 1.70 | 1.67 | 1.64 | 1.62 | 1.61 | 1.57 | 1.54 |
| 24 | 0.05 | 2.18 | 2.11 | 2.03 | 1.97 | 1.94 | 1.89 | 1.86 | 1.84 | 1.79 | 1.74 |
| | 0.01 | 3.03 | 2.89 | 2.74 | 2.64 | 2.58 | 2.49 | 2.44 | 2.40 | 2.31 | 2.22 |
| | 0.10 | 1.82 | 1.77 | 1.72 | 1.68 | 1.65 | 1.63 | 1.61 | 1.58 | 1.56 | 1.52 |
| 25 | 0.05 | 2.16 | 2.05 | 2.01 | 1.96 | 1.92 | 1.87 | 1.84 | 1.82 | 1.77 | 1.72 |
| | 0.01 | 2.99 | 2.85 | 2.70 | 2.60 | 2.54 | 2.45 | 2.40 | 2.36 | 2.27 | 2.18 |
| | 0.10 | 1.77 | 1.72 | 1.67 | 1.63 | 1.60 | 1.57 | 1.55 | 1.54 | 1.50 | 1.46 |
| 30 | 0.05 | 2.09 | 2.01 | 1.93 | 1.88 | 1.84 | 1.79 | 1.76 | 1.74 | 1.68 | 1.63 |
| | 0.01 | 2.84 | 2.70 | 2.55 | 2.45 | 2.39 | 2.30 | 2.25 | 2.21 | 2.11 | 2.02 |

(continued)

**Table E.4** (continued)

| $n_2$ | $p$ | $n_1$ 12 | 15 | 20 | 25 | 30 | 40 | 50 | 60 | 120 | 1000 |
|---|---|---|---|---|---|---|---|---|---|---|---|
| | 0.10 | 1.71 | 1.66 | 1.61 | 1.57 | 1.54 | 1.51 | 1.48 | 1.47 | 1.42 | 1.38 |
| 40 | 0.05 | 2.00 | 1.92 | 1.84 | 1.78 | 1.74 | 1.69 | 1.66 | 1.64 | 1.58 | 1.52 |
| | 0.01 | 2.66 | 2.52 | 2.37 | 2.27 | 2.20 | 2.11 | 2.06 | 2.02 | 1.92 | 1.82 |
| | 0.10 | 1.68 | 1.63 | 1.57 | 1.53 | 1.50 | 1.46 | 1.44 | 1.42 | 1.38 | 1.33 |
| 50 | 0.05 | 1.95 | 1.87 | 1.78 | 1.73 | 1.69 | 1.63 | 1.60 | 1.55 | 1.51 | 1.45 |
| | 0.01 | 2.56 | 2.42 | 2.27 | 2.17 | 2.10 | 2.01 | 1.95 | 1.91 | 1.80 | 1.70 |

| $n_2$ | $p$ | $n_1$ 1 | 2 | 3 | 4 | 5 | 6 | 7 | 8 | 9 |
|---|---|---|---|---|---|---|---|---|---|---|
| | 0.10 | 2.79 | 2.39 | 2.18 | 2.04 | 1.95 | 1.87 | 1.82 | 1.77 | 1.74 |
| 60 | 0.05 | 4.00 | 3.15 | 2.76 | 2.53 | 2.37 | 2.25 | 2.17 | 2.10 | 2.04 |
| | 0.01 | 7.08 | 4.98 | 4.13 | 3.65 | 3.34 | 3.12 | 2.95 | 2.82 | 2.72 |
| | 0.10 | 2.76 | 2.36 | 2.14 | 2.00 | 1.91 | 1.83 | 1.78 | 1.73 | 1.69 |
| 100 | 0.05 | 3.94 | 3.09 | 2.70 | 2.46 | 2.31 | 2.19 | 2.10 | 2.03 | 1.97 |
| | 0.01 | 6.90 | 4.82 | 3.98 | 3.51 | 3.21 | 2.99 | 2.82 | 2.69 | 2.59 |
| | 0.10 | 2.73 | 2.33 | 2.11 | 1.97 | 1.88 | 1.80 | 1.75 | 1.70 | 1.66 |
| 200 | 0.05 | 3.89 | 3.04 | 2.65 | 2.42 | 2.26 | 2.14 | 2.06 | 1.98 | 1.93 |
| | 0.01 | 6.76 | 4.71 | 3.88 | 3.41 | 3.11 | 2.89 | 2.73 | 2.60 | 2.50 |
| | 0.10 | 2.71 | 2.31 | 2.09 | 1.95 | 1.85 | 1.78 | 1.72 | 1.68 | 1.64 |
| 1000 | 0.05 | 3.85 | 3.00 | 2.61 | 2.38 | 2.22 | 2.11 | 2.02 | 1.95 | 1.89 |
| | 0.01 | 6.66 | 4.63 | 3.80 | 3.34 | 3.04 | 2.82 | 2.66 | 2.53 | 2.43 |

| $n_2$ | $p$ | $n_1$ 10 | 12 | 15 | 20 | 25 | 30 | 40 | 50 | 60 | 120 | 1000 |
|---|---|---|---|---|---|---|---|---|---|---|---|---|
| | 0.10 | 1.71 | 1.66 | 1.60 | 1.54 | 1.50 | 1.48 | 1.44 | 1.41 | 1.40 | 1.35 | 1.30 |
| 60 | 0.05 | 1.99 | 1.92 | 1.84 | 1.75 | 1.69 | 1.65 | 1.59 | 1.56 | 1.53 | 1.47 | 1.40 |
| | 0.01 | 2.63 | 2.50 | 2.35 | 2.20 | 2.10 | 2.03 | 1.94 | 1.88 | 1.84 | 1.73 | 1.62 |
| | 0.10 | 1.66 | 1.61 | 1.56 | 1.49 | 1.45 | 1.42 | 1.38 | 1.35 | 1.34 | 1.28 | 1.22 |
| 100 | 0.05 | 1.93 | 1.85 | 1.77 | 1.68 | 1.62 | 1.57 | 1.52 | 1.48 | 1.45 | 1.38 | 1.30 |
| | 0.01 | 2.50 | 2.37 | 2.22 | 2.07 | 1.97 | 1.89 | 1.80 | 1.74 | 1.69 | 1.57 | 1.45 |
| | 0.10 | 1.63 | 1.58 | 1.52 | 1.46 | 1.41 | 1.38 | 1.34 | 1.31 | 1.29 | 1.23 | 1.16 |
| 200 | 0.05 | 1.88 | 1.80 | 1.72 | 1.62 | 1.56 | 1.52 | 1.46 | 1.41 | 1.39 | 1.30 | 1.21 |
| | 0.01 | 2.41 | 2.27 | 2.13 | 1.97 | 1.87 | 1.79 | 1.69 | 1.63 | 1.58 | 1.45 | 1.30 |
| | 0.10 | 1.61 | 1.55 | 1.49 | 1.43 | 1.38 | 1.35 | 1.30 | 1.27 | 1.25 | 1.18 | 1.08 |
| 1000 | 0.05 | 1.84 | 1.76 | 1.68 | 1.58 | 1.52 | 1.47 | 1.41 | 1.36 | 1.33 | 1.24 | 1.11 |
| | 0.01 | 2.34 | 2.20 | 2.06 | 1.90 | 1.79 | 1.72 | 1.61 | 1.54 | 1.50 | 1.35 | 1.16 |

**Table E.5** Factors for two-sided tolerance intervals, normal distribution (confidence $\gamma$, coverage $p$)

| $n \backslash p$ | $\gamma = 0.90$ | | | | $\gamma = 0.95$ | | | | $\gamma = 0.99$ | | | |
|---|---|---|---|---|---|---|---|---|---|---|---|---|
| | 0.75 | 0.90 | 0.95 | 0.99 | 0.75 | 0.90 | 0.95 | 0.99 | 0.75 | 0.90 | 0.95 | 0.99 |
| 2 | 11.407 | 15.978 | 18.800 | 24.167 | 22.858 | 32.019 | 37.674 | 48.430 | 114.363 | 160.193 | 188.491 | 242.300 |
| 3 | 4.132 | 5.847 | 6.919 | 8.974 | 5.922 | 8.380 | 9.916 | 12.861 | 13.378 | 18.930 | 22.401 | 29.055 |
| 4 | 2.932 | 4.166 | 4.943 | 6.440 | 3.779 | 5.369 | 6.370 | 8.299 | 6.614 | 9.398 | 11.150 | 14.527 |
| 5 | 2.454 | 3.494 | 4.152 | 5.423 | 3.002 | 4.275 | 5.079 | 6.634 | 4.643 | 6.612 | 7.855 | 10.260 |
| 6 | 2.196 | 3.131 | 3.723 | 4.870 | 2.604 | 3.712 | 4.414 | 5.775 | 3.743 | 5.337 | 6.345 | 8.301 |
| 7 | 2.034 | 2.902 | 3.452 | 4.521 | 2.361 | 3.369 | 4.007 | 5.248 | 3.233 | 4.613 | 5.488 | 7.187 |
| 8 | 1.921 | 2.743 | 3.264 | 4.278 | 2.197 | 3.136 | 3.732 | 4.891 | 2.905 | 4.147 | 4.936 | 6.468 |
| 9 | 1.839 | 2.626 | 3.125 | 4.098 | 2.078 | 2.967 | 3.532 | 4.631 | 2.677 | 3.822 | 4.550 | 5.966 |
| 10 | 1.775 | 2.535 | 3.018 | 3.959 | 1.987 | 2.839 | 3.379 | 4.433 | 2.508 | 3.582 | 4.265 | 5.594 |
| 11 | 1.724 | 2.463 | 2.933 | 3.849 | 1.916 | 2.737 | 3.259 | 4.277 | 2.378 | 3.397 | 4.045 | 5.308 |
| 12 | 1.683 | 2.404 | 2.863 | 3.758 | 1.858 | 2.655 | 3.162 | 4.150 | 2.274 | 3.250 | 3.870 | 5.079 |
| 13 | 1.648 | 2.355 | 2.805 | 3.682 | 1.810 | 2.587 | 3.081 | 4.044 | 2.190 | 3.130 | 3.727 | 4.893 |
| 14 | 1.619 | 2.314 | 2.756 | 3.618 | 1.770 | 2.529 | 3.012 | 3.955 | 2.120 | 3.029 | 3.608 | 4.737 |
| 15 | 1.594 | 2.278 | 2.713 | 3.562 | 1.735 | 2.480 | 2.954 | 3.878 | 2.060 | 2.945 | 3.507 | 4.605 |
| 16 | 1.572 | 2.246 | 2.676 | 3.514 | 1.705 | 2.437 | 2.903 | 3.812 | 2.009 | 2.872 | 3.421 | 4.492 |
| 17 | 1.552 | 2.219 | 2.643 | 3.471 | 1.679 | 2.400 | 2.858 | 3.754 | 1.965 | 2.808 | 3.345 | 4.393 |
| 18 | 1.535 | 2.194 | 2.614 | 3.433 | 1.655 | 2.366 | 2.819 | 3.702 | 1.926 | 2.753 | 3.279 | 4.307 |
| 19 | 1.520 | 2.172 | 2.588 | 3.399 | 1.635 | 2.337 | 2.784 | 3.656 | 1.891 | 2.703 | 3.221 | 4.230 |
| 20 | 1.506 | 2.152 | 2.564 | 3.368 | 1.616 | 2.310 | 2.752 | 3.615 | 1.860 | 2.659 | 3.168 | 4.161 |
| 21 | 1.493 | 2.135 | 2.543 | 3.340 | 1.599 | 2.286 | 2.723 | 3.577 | 1.833 | 2.620 | 3.121 | 4.100 |
| 22 | 1.482 | 2.118 | 2.524 | 3.315 | 1.584 | 2.264 | 2.697 | 3.543 | 1.808 | 2.584 | 3.078 | 4.044 |
| 23 | 1.471 | 2.103 | 2.506 | 3.292 | 1.570 | 2.244 | 2.673 | 3.512 | 1.785 | 2.551 | 3.040 | 3.993 |
| 24 | 1.462 | 2.089 | 2.489 | 3.270 | 1.557 | 2.225 | 2.651 | 3.483 | 1.764 | 2.522 | 3.004 | 3.947 |
| 25 | 1.453 | 2.077 | 2.474 | 3.251 | 1.545 | 2.208 | 2.631 | 3.457 | 1.745 | 2.494 | 2.972 | 3.904 |
| 26 | 1.444 | 2.065 | 2.460 | 3.232 | 1.534 | 2.193 | 2.612 | 3.432 | 1.727 | 2.469 | 2.941 | 3.865 |

(continued)

**Table E.5** (continued)

| n\p | $\gamma = 0.90$ | | | | $\gamma = 0.95$ | | | | $\gamma = 0.99$ | | | |
|---|---|---|---|---|---|---|---|---|---|---|---|---|
| | 0.75 | 0.90 | 0.95 | 0.99 | 0.75 | 0.90 | 0.95 | 0.99 | 0.75 | 0.90 | 0.95 | 0.99 |
| 27 | 1.437 | 2.054 | 2.447 | 3.215 | 1.523 | 2.178 | 2.595 | 3.409 | 1.711 | 2.446 | 2.914 | 3.828 |
| 30 | 1.417 | 2.025 | 2.413 | 3.170 | 1.497 | 2.140 | 2.549 | 3.350 | 1.668 | 2.385 | 2.841 | 3.733 |
| 35 | 1.390 | 1.988 | 2.368 | 3.112 | 1.462 | 2.090 | 2.490 | 3.272 | 1.613 | 2.306 | 2.748 | 3.611 |
| 40 | 1.370 | 1.959 | 2.334 | 3.066 | 1.435 | 2.052 | 2.445 | 3.213 | 1.571 | 2.247 | 2.677 | 3.518 |
| 45 | 1.354 | 1.935 | 2.306 | 3.030 | 1.414 | 2.021 | 2.408 | 3.165 | 1.539 | 2.200 | 2.621 | 3.444 |
| 50 | 1.340 | 1.916 | 2.284 | 3.001 | 1.396 | 1.996 | 2.379 | 3.126 | 1.512 | 2.162 | 2.576 | 3.385 |
| 55 | 1.329 | 1.901 | 2.265 | 2.976 | 1.382 | 1.976 | 2.354 | 3.094 | 1.490 | 2.130 | 2.538 | 3.335 |
| 60 | 1.320 | 1.887 | 2.248 | 2.955 | 1.369 | 1.958 | 2.333 | 3.066 | 1.471 | 2.103 | 2.506 | 3.293 |
| 65 | 1.312 | 1.875 | 2.235 | 2.937 | 1.359 | 1.943 | 2.315 | 3.042 | 1.455 | 2.080 | 2.478 | 3.257 |
| 70 | 1.304 | 1.865 | 2.222 | 2.920 | 1.349 | 1.929 | 2.299 | 3.021 | 1.440 | 2.060 | 2.454 | 3.225 |
| 75 | 1.298 | 1.856 | 2.211 | 2.906 | 1.341 | 1.917 | 2.285 | 3.002 | 1.428 | 2.042 | 2.433 | 3.197 |
| 80 | 1.292 | 1.848 | 2.202 | 2.894 | 1.334 | 1.907 | 2.272 | 2.986 | 1.417 | 2.026 | 2.414 | 3.173 |
| 85 | 1.287 | 1.841 | 2.193 | 2.882 | 1.327 | 1.897 | 2.261 | 2.971 | 1.407 | 2.012 | 2.397 | 3.150 |
| 90 | 1.283 | 1.834 | 2.185 | 2.872 | 1.321 | 1.889 | 2.251 | 2.958 | 1.398 | 1.999 | 2.382 | 3.130 |
| 95 | 1.278 | 1.828 | 2.178 | 2.863 | 1.315 | 1.881 | 2.241 | 2.945 | 1.390 | 1.987 | 2.368 | 3.112 |
| 100 | 1.275 | 1.822 | 2.172 | 2.854 | 1.311 | 1.874 | 2.233 | 2.934 | 1.383 | 1.977 | 2.355 | 3.096 |
| 110 | 1.268 | 1.813 | 2.160 | 2.839 | 1.302 | 1.861 | 2.218 | 2.915 | 1.369 | 1.958 | 2.333 | 3.066 |
| 120 | 1.262 | 1.804 | 2.150 | 2.826 | 1.294 | 1.850 | 2.205 | 2.898 | 1.358 | 1.942 | 2.314 | 3.041 |
| 130 | 1.257 | 1.797 | 2.141 | 2.814 | 1.288 | 1.841 | 2.194 | 2.883 | 1.349 | 1.928 | 2.298 | 3.019 |
| 140 | 1.252 | 1.791 | 2.134 | 2.804 | 1.282 | 1.833 | 2.184 | 2.870 | 1.340 | 1.916 | 2.283 | 3.000 |
| 150 | 1.248 | 1.785 | 2.127 | 2.795 | 1.277 | 1.825 | 2.175 | 2.859 | 1.332 | 1.905 | 2.270 | 2.983 |
| 160 | 1.245 | 1.780 | 2.121 | 2.787 | 1.272 | 1.819 | 2.167 | 2.848 | 1.326 | 1.896 | 2.259 | 2.968 |
| 170 | 1.242 | 1.775 | 2.116 | 2.780 | 1.268 | 1.813 | 2.160 | 2.839 | 1.320 | 1.887 | 2.248 | 2.955 |
| 180 | 1.239 | 1.771 | 2.111 | 2.774 | 1.264 | 1.808 | 2.154 | 2.831 | 1.314 | 1.879 | 2.239 | 2.942 |
| 190 | 1.236 | 1.767 | 2.106 | 2.768 | 1.261 | 1.803 | 2.148 | 2.823 | 1.309 | 1.872 | 2.230 | 2.931 |
| 200 | 1.234 | 1.764 | 2.102 | 2.762 | 1.258 | 1.798 | 2.143 | 2.816 | 1.304 | 1.865 | 2.222 | 2.921 |

(continued)

**Table E.5** (continued)

| $n \backslash p$ | $\gamma = 0.90$ | | | | $\gamma = 0.95$ | | | | $\gamma = 0.99$ | | | |
|---|---|---|---|---|---|---|---|---|---|---|---|---|
| | 0.75 | 0.90 | 0.95 | 0.99 | 0.75 | 0.90 | 0.95 | 0.99 | 0.75 | 0.90 | 0.95 | 0.99 |
| 250 | 1.224 | 1.750 | 2.085 | 2.740 | 1.245 | 1.780 | 2.121 | 2.788 | 1.286 | 1.839 | 2.191 | 2.880 |
| 300 | 1.217 | 1.740 | 2.073 | 2.725 | 1.236 | 1.767 | 2.106 | 2.767 | 1.273 | 1.820 | 2.169 | 2.850 |
| 400 | 1.207 | 1.726 | 2.057 | 2.703 | 1.223 | 1.749 | 2.084 | 2.739 | 1.255 | 1.794 | 2.138 | 2.809 |
| 500 | 1.201 | 1.717 | 2.046 | 2.689 | 1.215 | 1.737 | 2.070 | 2.721 | 1.243 | 1.777 | 2.117 | 2.783 |
| 600 | 1.196 | 1.710 | 2.038 | 2.678 | 1.209 | 1.729 | 2.060 | 2.707 | 1.234 | 1.764 | 2.102 | 2.763 |
| 700 | 1.192 | 1.705 | 2.032 | 2.670 | 1.204 | 1.722 | 2.052 | 2.697 | 1.227 | 1.755 | 2.091 | 2.748 |
| 800 | 1.189 | 1.701 | 2.027 | 2.663 | 1.201 | 1.717 | 2.046 | 2.688 | 1.222 | 1.747 | 2.082 | 2.736 |
| 900 | 1.187 | 1.697 | 2.023 | 2.658 | 1.198 | 1.712 | 2.040 | 2.682 | 1.218 | 1.741 | 2.075 | 2.726 |
| 1000 | 1.185 | 1.695 | 2.019 | 2.654 | 1.195 | 1.709 | 2.036 | 2.676 | 1.214 | 1.736 | 2.068 | 2.718 |
| $\infty$ | 1.150 | 1.645 | 1.960 | 2.576 | 1.150 | 1.645 | 1.960 | 2.576 | 1.150 | 1.645 | 1.960 | 2.576 |

**Table E.6** Factors for one-sided tolerance intervals, normal distribution (confidence $\gamma$, coverage $p$)

| $n \backslash p$ | $\gamma = 0.90$ | | | | | $\gamma = 0.95$ | | | | |
|---|---|---|---|---|---|---|---|---|---|---|
| | 0.900 | 0.950 | 0.975 | 0.990 | 0.999 | 0.900 | 0.950 | 0.975 | 0.990 | 0.999 |
| 2 | 10.253 | 13.090 | 15.586 | 18.500 | 24.582 | 20.581 | 26.260 | 31.257 | 37.094 | 49.276 |
| 3 | 4.258 | 5.311 | 6.244 | 7.340 | 9.651 | 6.155 | 7.656 | 8.986 | 10.553 | 13.857 |
| 4 | 3.188 | 3.957 | 4.637 | 5.438 | 7.129 | 4.162 | 5.144 | 6.015 | 7.042 | 9.214 |
| 5 | 2.744 | 3.401 | 3.983 | 4.668 | 6.113 | 3.413 | 4.210 | 4.916 | 5.749 | 7.509 |
| 6 | 2.494 | 3.093 | 3.621 | 4.243 | 5.556 | 3.008 | 3.711 | 4.332 | 5.065 | 6.614 |
| 7 | 2.333 | 2.893 | 3.389 | 3.972 | 5.201 | 2.756 | 3.401 | 3.971 | 4.643 | 6.064 |
| 8 | 2.219 | 2.754 | 3.227 | 3.783 | 4.955 | 2.582 | 3.188 | 3.724 | 4.355 | 5.689 |
| 9 | 2.133 | 2.650 | 3.106 | 3.641 | 4.771 | 2.454 | 3.032 | 3.543 | 4.144 | 5.414 |
| 10 | 2.066 | 2.568 | 3.011 | 3.532 | 4.628 | 2.355 | 2.911 | 3.403 | 3.981 | 5.204 |
| 11 | 2.012 | 2.503 | 2.936 | 3.444 | 4.515 | 2.275 | 2.815 | 3.291 | 3.852 | 5.036 |
| 12 | 1.966 | 2.448 | 2.872 | 3.371 | 4.420 | 2.210 | 2.736 | 3.201 | 3.747 | 4.900 |
| 13 | 1.928 | 2.403 | 2.820 | 3.310 | 4.341 | 2.155 | 2.670 | 3.125 | 3.659 | 4.787 |
| 14 | 1.895 | 2.363 | 2.774 | 3.257 | 4.274 | 2.108 | 2.614 | 3.060 | 3.585 | 4.690 |
| 15 | 1.866 | 2.329 | 2.735 | 3.212 | 4.215 | 2.068 | 2.566 | 3.005 | 3.520 | 4.607 |
| 16 | 1.842 | 2.299 | 2.700 | 3.172 | 4.164 | 2.032 | 2.523 | 2.956 | 3.463 | 4.534 |
| 17 | 1.819 | 2.272 | 2.670 | 3.137 | 4.118 | 2.002 | 2.486 | 2.913 | 3.414 | 4.471 |
| 18 | 1.800 | 2.249 | 2.643 | 3.106 | 4.078 | 1.974 | 2.453 | 2.875 | 3.370 | 4.415 |
| 19 | 1.781 | 2.228 | 2.618 | 3.078 | 4.041 | 1.949 | 2.423 | 2.840 | 3.331 | 4.364 |
| 20 | 1.765 | 2.208 | 2.597 | 3.052 | 4.009 | 1.926 | 2.396 | 2.809 | 3.295 | 4.319 |
| 21 | 1.750 | 2.190 | 2.575 | 3.028 | 3.979 | 1.905 | 2.371 | 2.781 | 3.262 | 4.276 |
| 22 | 1.736 | 2.174 | 2.557 | 3.007 | 3.952 | 1.887 | 2.350 | 2.756 | 3.233 | 4.238 |
| 23 | 1.724 | 2.159 | 2.540 | 2.987 | 3.927 | 1.869 | 2.329 | 2.732 | 3.206 | 4.204 |
| 24 | 1.712 | 2.145 | 2.525 | 2.969 | 3.904 | 1.853 | 2.309 | 2.711 | 3.181 | 4.171 |
| 25 | 1.702 | 2.132 | 2.510 | 2.952 | 3.882 | 1.838 | 2.292 | 2.691 | 3.158 | 4.143 |
| 30 | 1.657 | 2.080 | 2.450 | 2.884 | 3.794 | 1.778 | 2.220 | 2.608 | 3.064 | 4.022 |
| 35 | 1.623 | 2.041 | 2.406 | 2.833 | 3.730 | 1.732 | 2.166 | 2.548 | 2.994 | 3.934 |
| 40 | 1.598 | 2.010 | 2.371 | 2.793 | 3.679 | 1.697 | 2.126 | 2.501 | 2.941 | 3.866 |
| 45 | 1.577 | 1.986 | 2.344 | 2.762 | 3.638 | 1.669 | 2.092 | 2.463 | 2.897 | 3.811 |
| 50 | 1.560 | 1.965 | 2.320 | 2.735 | 3.604 | 1.646 | 2.065 | 2.432 | 2.863 | 3.766 |
| 60 | 1.532 | 1.933 | 2.284 | 2.694 | 3.552 | 1.609 | 2.022 | 2.384 | 2.807 | 3.695 |
| 70 | 1.511 | 1.909 | 2.257 | 2.663 | 3.513 | 1.581 | 1.990 | 2.348 | 2.766 | 3.643 |
| 80 | 1.495 | 1.890 | 2.235 | 2.638 | 3.482 | 1.560 | 1.965 | 2.319 | 2.733 | 3.601 |
| 90 | 1.481 | 1.874 | 2.217 | 2.618 | 3.456 | 1.542 | 1.944 | 2.295 | 2.706 | 3.567 |
| 100 | 1.470 | 1.861 | 2.203 | 2.601 | 3.435 | 1.527 | 1.927 | 2.276 | 2.684 | 3.539 |
| 120 | 1.452 | 1.841 | 2.179 | 2.574 | 3.402 | 1.503 | 1.899 | 2.245 | 2.649 | 3.495 |
| 145 | 1.436 | 1.821 | 2.158 | 2.550 | 3.371 | 1.481 | 1.874 | 2.217 | 2.617 | 3.455 |
| 300 | 1.386 | 1.765 | 2.094 | 2.477 | 3.280 | 1.417 | 1.800 | 2.133 | 2.522 | 3.335 |
| 500 | 1.362 | 1.736 | 2.062 | 2.442 | 3.235 | 1.385 | 1.763 | 2.092 | 2.475 | 3.277 |
| $\infty$ | 1.282 | 1.645 | 1.960 | 2.326 | 3.090 | 1.282 | 1.645 | 1.960 | 2.326 | 3.090 |

**Table E.7** Two-sided nonparametric tolerance intervals

| $n\backslash p$ | $\gamma = 0.75$ | | | | $\gamma = 0.90$ | | | |
|---|---|---|---|---|---|---|---|---|
| | 0.75 | 0.90 | 0.95 | 0.99 | 0.75 | 0.90 | 0.95 | 0.99 |
| 50 | 5,5 | 2,1 | – | – | 5,4 | 1,1 | – | – |
| 55 | 6,6 | 2,2 | 1,1 | – | 5,5 | 2,1 | – | – |
| 60 | 7,6 | 2,2 | 1,1 | – | 6,5 | 2,1 | – | – |
| 65 | 7,7 | 3,2 | 1,1 | – | 6,6 | 2,2 | – | – |
| 70 | 8,7 | 3,2 | 1,1 | – | 7,6 | 2,2 | – | – |
| 75 | 8,8 | 3,3 | 1,1 | – | 7,7 | 2,2 | – | – |
| 80 | 9,8 | 3,3 | 2,1 | – | 8,7 | 3,2 | 1,1 | – |
| 85 | 10,9 | 4,3 | 2,1 | – | 8,8 | 3,2 | 1,1 | – |
| 90 | 10,10 | 4,3 | 2,1 | – | 9,8 | 3,2 | 1,1 | – |
| 95 | 11,10 | 4,3 | 2,1 | – | 9,9 | 3,3 | 1,1 | – |
| 100 | 11,11 | 4,4 | 2,1 | – | 10,10 | 3,3 | 1,1 | – |
| 110 | 12,12 | 5,4 | 2,2 | – | 11,11 | 4,3 | 2,1 | – |
| 120 | 14,13 | 5,5 | 2,2 | – | 12,12 | 4,4 | 2,1 | – |
| 130 | 15,14 | 6,5 | 3,2 | – | 13,13 | 5,4 | 2,1 | – |
| 140 | 16,15 | 6,6 | 3,2 | – | 14,14 | 5,5 | 2,2 | – |
| 150 | 17,17 | 6,6 | 3,3 | – | 16,15 | 5,5 | 2,2 | – |
| 170 | 20,19 | 7,7 | 4,3 | – | 18,17 | 6,6 | 3,2 | – |
| 200 | 23,23 | 9,8 | 4,4 | – | 21,21 | 8,7 | 3,3 | – |
| 300 | 35,35 | 13,13 | 6,6 | 1,1 | 33,32 | 12,11 | 5,5 | – |
| 400 | 47,47 | 18,18 | 9,8 | 2,1 | 45,44 | 16,16 | 8,7 | 1,1 |
| 500 | 59,59 | 23,22 | 11,11 | 2,1 | 57,56 | 23,20 | 10,9 | 1,1 |
| 600 | 72,71 | 28,27 | 13,13 | 2,2 | 68,68 | 26,25 | 12,11 | 2,1 |
| 700 | 84,83 | 33,32 | 16,15 | 3,2 | 80,80 | 30,30 | 14,14 | 2,2 |
| 800 | 96,96 | 37,27 | 18,18 | 3,3 | 92,92 | 35,34 | 16,16 | 3,2 |
| 900 | 108,108 | 42,42 | 21,20 | 4,3 | 104,104 | 40,39 | 19,18 | 3,2 |
| 1000 | 121,120 | 47,47 | 23,22 | 4,4 | 117,116 | 44,44 | 21,20 | 3,3 |
| $n\backslash p$ | $\gamma = 0.95$ | | | | $\gamma = 0.99$ | | | |
| | 0.75 | 0.90 | 0.95 | 0.99 | 0.75 | 0.90 | 0.95 | 0.99 |
| 50 | 4,4 | 1,1 | – | – | 3,3 | – | – | – |
| 55 | 5,4 | 1,1 | – | – | 4,3 | – | – | – |
| 60 | 5,5 | 1,1 | – | – | 4,4 | – | – | – |
| 65 | 6,5 | 2,1 | – | – | 5,4 | 1,1 | – | – |
| 70 | 6,6 | 2,1 | – | – | 5,5 | 1,1 | – | – |
| 75 | 7,6 | 2,1 | – | – | 5,5 | 1,1 | – | – |
| 80 | 7,7 | 2,2 | – | – | 6,5 | 1,1 | – | – |
| 85 | 8,7 | 2,2 | – | – | 6,6 | 2,1 | – | – |
| 90 | 8,8 | 3,2 | – | – | 7,6 | 2,1 | – | – |
| 95 | 9,8 | 3,2 | 1,1 | – | 7,7 | 2,1 | – | – |
| 100 | 9,9 | 3,2 | 1,1 | – | 8,7 | 2,2 | – | – |
| 110 | 10,10 | 3,3 | 1,1 | – | 9,8 | 2,2 | – | – |
| 120 | 11,11 | 4,3 | 1,1 | – | 10,9 | 3,2 | – | – |
| 130 | 13,12 | 4,4 | 2,1 | – | 11,10 | 3,3 | 1,1 | — |

(continued)

**Table E.7** (continued)

| $n \backslash p$ | $\gamma = 0.95$ | | | | $\gamma = 0.99$ | | | |
|---|---|---|---|---|---|---|---|---|
| | 0.75 | 0.90 | 0.95 | 0.99 | 0.75 | 0.90 | 0.95 | 0.99 |
| 140 | 14,13 | 4,4 | 2,1 | – | 12,11 | 3,3 | 1,1 | – |
| 150 | 15,14 | 5,4 | 2,1 | – | 13,13 | 4,3 | 1,1 | – |
| 170 | 17,16 | 6,5 | 2,2 | – | 15,15 | 5,4 | 2,1 | – |
| 200 | 20,20 | 7,6 | 3,2 | – | 18,18 | 6,5 | 2,2 | – |
| 300 | 32,31 | 11,11 | 5,4 | – | 29,29 | 10,9 | 4,3 | – |
| 400 | 43,43 | 15,15 | 7,6 | – | 40,40 | 14,13 | 6,5 | – |
| 500 | 55,54 | 20,19 | 9,8 | 1,1 | 52,51 | 18,17 | 7,7 | – |
| 600 | 67,66 | 24,24 | 11,10 | 1,1 | 63,63 | 22,22 | 9,9 | – |
| 700 | 78,78 | 29,28 | 13,13 | 2,1 | 75,74 | 26,26 | 11,11 | 1,1 |
| 800 | 90,90 | 33,33 | 15,15 | 2,2 | 86,86 | 31,30 | 13,13 | 1,1 |
| 900 | 102,102 | 38,37 | 18,17 | 2,2 | 98,97 | 35,35 | 15,15 | 2,1 |
| 1000 | 114,114 | 43,42 | 20,19 | 3,2 | 110,109 | 40,39 | 18,17 | 2,1 |

Values $(r, s)$ such that we may assert with confidence at least $\gamma$ that 100P percent of a population lies between the $r$th smallest and the $s$th largest of a random sample of $n$ from that population (no assumption of normality required).

When the values of $r$ and $s$ given in the table are not equal, they are interchangeable; i.e., for $n = 120$ with confidence at least 0.75 we may assert that 75% of the population lies between the 14th smallest and the 13th largest values, or between the 13th smallest and the 14th largest values.

This table is based on [45].

# Appendix F: Data Sets

This Appendix contains sixteen data sets (or partial sets in cases where the full set is large) obtained from companies that collected various types of information in the process of servicing warranty claims or assessing product reliability. Many of these are used as examples or cases in one or more chapters of the book.

## F.1 Data Set 1 [Home Air Conditioners-I]

The data consist of 729 failures of "Split type" and "Window type" home air conditioners (AC's) during the year 2004. The total number of units sold was 95,320. This and the number of failure are broken down by AC type. In addition, the data are classified into 13 different failure modes (called "Problem"), but individual failure dates are not given, nor is it indicated how many in each failure group are of each type of AC. Counts of failures within failure modes are given by "MODEL." There are over 100 different models listed in the original data.

The data on failure mode are given in Table F.1. Model is not defined in the data description, and is not included in the table. No further information regarding the warranty claims is available.

## F.2 Data Set 2 [Four-Wheel-Drive Automobiles]

Claims data involving engine problems on four-wheel-drive vehicles imported into Australia are given in Table F.2. The total number of vehicles sold was approximately 5000. The warranty was an FRW with W = 40,000 km. There were a total of 329 warranty claims on these vehicles, of which 32 involved engine problems. The data for these, including odometer reading at failure and cost of repair, are given in the table.

W. R. Blischke et al., *Warranty Data Collection and Analysis*,
Springer Series in Reliability Engineering, DOI: 10.1007/978-0-85729-647-4,

**Table F.1** Air-conditioner failure modes

| Mode | Problem | Number of failures |
|---|---|---|
| 1 | Bearing Case Defective | 1 |
| 2 | Fan Blade Hitting/Damaged | 43 |
| 3 | Ventilation Lever Damaged | 81 |
| 4 | Front Panel/Intake Grille Damaged | 152 |
| 5 | Grille Door Damaged | 158 |
| 6 | Control Panel Damaged | 38 |
| 7 | Print Circuit Board Damaged | 23 |
| 8 | Fan Blade Damaged | 3 |
| 9 | Compressor Noisy | 40 |
| 10 | Gas Leakage | 66 |
| 11 | Receiver Not Functioning | 27 |
| 12 | Air Vane Panel Damaged | 39 |
| 13 | PC BOARD & Fan Motor Not Operating | 58 |

**Table F.2** Warranty claims data for automobile engines

| Auto | Km at failure (000) | Cost of repair ($) |
|---|---|---|
| 1 | 13.1 | 24.60 |
| 2 | 29.2 | 5037.28 |
| 3 | 13.2 | 253.50 |
| 4 | 10.0 | 26.35 |
| 5 | 21.4 | 1712.47 |
| 6 | 14.5 | 127.20 |
| 7 | 12.6 | 918.53 |
| 8 | 27.4 | 34.68 |
| 9 | 35.5 | 1007.27 |
| 10 | 15.1 | 658.36 |
| 11 | 17.0 | 42.96 |
| 12 | 27.8 | 77.22 |
| 13 | 2.4 | 77.57 |
| 14 | 38.6 | 831.61 |
| 15 | 17.5 | 432.89 |
| 16 | 14.0 | 60.35 |
| 17 | 15.3 | 48.05 |
| 18 | 19.2 | 388.30 |
| 19 | 4.4 | 149.36 |
| 20 | 19.0 | 7.75 |
| 21 | 32.4 | 29.91 |
| 22 | 23.7 | 27.58 |
| 23 | 16.8 | 1101.90 |
| 24 | 2.3 | 27.78 |
| 25 | 26.7 | 1638.73 |
| 26 | 5.3 | 11.70 |
| 27 | 29.0 | 98.90 |
| 28 | 10.1 | 77.24 |

(continued)

**Table F.2** (continued)

| Auto | Km at failure (000) | Cost of repair ($) |
|---|---|---|
| 29 | 18.0 | 42.71 |
| 30 | 4.5 | 1546.75 |
| 31 | 18.7 | 556.93 |
| 32 | 31.6 | 78.42 |

## F.3 Data Set 3 [Battery Failures]

Incomplete failure data on a sample of 54 batteries are given in Table F.3. The data include failure times for 39 items that failed under warranty and service times for 15 items that had not failed at the time of observation.

## F.4 Data Set 4 [Bond Strength of Adhesive]

In a study of the reliability of a key component of many of its systems, a manufacturer of audio equipment performed a number of strength tests on an adhesive used to bond two metallic parts. The study was done during the development phase of one of its largest selling home sound systems. Two subsets of the data obtained in the study are given in Tables F.4 and F.5 (obtained from [4]). The first is from a study of the strength of the bond under four different environmental conditions, ranging from the temperature and relative humidity (RH) of a warm room, to hot, extremely humid conditions. The second is from a series of tests of items stored for varying periods of time under normal warehouse conditions.

## F.5 Data Set 5 [Hydraulic Systems]

Hydraulic components are essential subsystems of large load-haul-dump (LHD) machines used to move ore and rock in mining operations. Table F.6 lists times (operating hours) between successive failures of hydraulic systems in 4 LHD's used in underground mines in Sweden [20]. Operational data such as these are useful in formulating maintenance policies, devising engineering modifications, developing new products, and selecting warranty policies.

## F.6 Data Set 6 [Jet Engine Failure]

Failures of a jet engine on a fleet of military aircraft at a particular airbase are shown in Table F.7 [1]. The fleet consisted of 31 aircraft, of which six had experienced engine failures. The table gives time to failure for these and service

**Table F.3** Battery failure data

| Time to failure | | | Service time |
|---|---|---|---|
| 64 | 599 | 852 | 131 |
| 66 | 619 | 929 | 162 |
| 164 | 631 | 948 | 163 |
| 178 | 639 | 973 | 202 |
| 185 | 645 | 977 | 232 |
| 299 | 656 | 1084 | 245 |
| 319 | 681 | 1100 | 286 |
| 383 | 722 | 1100 | 302 |
| 385 | 727 | 1350 | 315 |
| 405 | 738 | | 337 |
| 482 | 761 | | 845 |
| 492 | 765 | | 983 |
| 506 | 788 | | 1259 |
| 548 | 801 | | 1384 |
| 589 | 848 | | 1421 |

**Table F.4** Bond strength (pounds) under various test conditions

| Test conditions | | | |
|---|---|---|---|
| 27°C, 50% RH | 27°C, 70% RH | 32°C, 70% RH | 27°C, 100% RH |
| 345 | 210 | 378 | 45 |
| 230 | 272 | 254 | 4 |
| 26 | 210 | 278 | 12 |
| 91 | 247 | 253 | 4 |
| 222 | 223 | 359 | 3 |
| 325 | 263 | 276 | 132 |
| 251 | 9 | 245 | 8 |
| 131 | 265 | 282 | 1 |
| 322 | 214 | 308 | 48 |
| 237 | 32 | 126 | 4 |
| 8 | 282 | 265 | |
| 306 | 1 | 245 | |
| 45 | 276 | 266 | |
| 272 | 202 | 289 | |
| 264 | 231 | 176 | |
| 277 | 251 | 273 | |
| 332 | 5 | 253 | |
| 100 | 75 | 320 | |
| 7 | 325 | 139 | |
| 254 | 50 | 266 | |
| 157 | 240 | | |
| 260 | 68 | | |
| 204 | 233 | | |

**Table F.5** Bond strength (pounds) versus length of storage (days)

| Days | Strength | Days | Strength | Days | Strength |
|---|---|---|---|---|---|
| 3 | 296 | 11 | 278 | 52 | 264 |
| 3 | 337 | 11 | 271 | 52 | 306 |
| 3 | 266 | 11 | 316 | 52 | 290 |
| 4 | 197 | 12 | 312 | 68 | 212 |
| 4 | 312 | 12 | 308 | 68 | 268 |
| 4 | 317 | 12 | 309 | 68 | 290 |
| 5 | 309 | 13 | 289 | 75 | 318 |
| 5 | 297 | 13 | 212 | 75 | 260 |
| 5 | 296 | 13 | 249 | 75 | 246 |
| 6 | 307 | 14 | 298 | 80 | 290 |
| 6 | 298 | 14 | 252 | 80 | 271 |
| 6 | 320 | 14 | 326 | 80 | 360 |
| 7 | 323 | 17 | 332 | 82 | 294 |
| 7 | 287 | 17 | 247 | 82 | 356 |
| 7 | 344 | 17 | 311 | 82 | 284 |
| 10 | 352 | 49 | 237 | | |
| 10 | 235 | 49 | 278 | | |
| 10 | 290 | 49 | 217 | | |

**Table F.6** Time (hours) between failures of hydraulic systems in LHD's

| LHD1 | LHD3 | LHD11 | LHD17 |
|---|---|---|---|
| 327 | 637 | 353 | 401 |
| 125 | 40 | 96 | 36 |
| 7 | 397 | 49 | 18 |
| 6 | 36 | 211 | 159 |
| 107 | 54 | 82 | 341 |
| 277 | 53 | 175 | 171 |
| 54 | 97 | 79 | 24 |
| 332 | 63 | 117 | 350 |
| 510 | 216 | 26 | 72 |
| 110 | 118 | 4 | 303 |
| 10 | 125 | 5 | 34 |
| 9 | 25 | 60 | 45 |
| 85 | 4 | 39 | 324 |
| 27 | 101 | 35 | 2 |
| 59 | 184 | 258 | 70 |
| 16 | 167 | 97 | 57 |
| 8 | 81 | 59 | 103 |
| 34 | 46 | 3 | 11 |
| 21 | 18 | 37 | 5 |
| 152 | 32 | 8 | 3 |
| 158 | 219 | 245 | 144 |
| 44 | 405 | 79 | 80 |
| 18 | 20 | 49 | 53 |
| | 248 | 31 | 84 |
| | 140 | 259 | 218 |
| | | 283 | 122 |
| | | 150 | |
| | | 24 | |

**Table F.7** Failure data for jet engines (flight hours)

| Failure times | Service times (Non-failures) | |
|---|---|---|
| 684 | 350 | 1350 |
| 701 | 650 | 1450 |
| 770 | 750 | 1550 |
| 812 | 850 | 1550 |
| 821 | 850 | 1650 |
| 845 | 950 | 1750 |
| | 950 | 1850 |
| | 1050 | 1850 |
| | 1050 | 1950 |
| | 1150 | 2050 |
| | 1150 | 2050 |
| | 1250 | 2050 |
| | 1250 | |

times for the remaining 25 engines that did not experience failures. (The data for service times were grouped; the table shows midpoints of the groups.) Data of this type are essential for administration of a Reliability Improvement Warranty [3], which has been widely used in military procurement in the USA.

## F.7 Data Set 7 [Fan Failures]

Nelson [35, p. 133] reported the hours to fan failure on 12 diesel generators and the censoring hours on 58 generators without a fan failure. Assume that these data represent a complete reporting of failures during the warranty period. To make a new data set under the warranty system, [47] randomly selected $(100\,p^*)\%$ of a total of 70 fans to comprise the follow-up study. Here the fraction of items that are followed up is $p^* = 30/70$. In the warranty period, all failures will be reported even if they are not being followed up. For a non-failure observation that is not followed up, the real time during the warranty period is unknown. Table F.8 shows the result of a random selection of followed-up observations. The values in parentheses are unobserved data, that is, non-failure observations that are not followed up.

## F.8 Data Set 8 [Construction Machine Failure and Follow-up Data]

Suzuki [47] presents the result of an observational study of a construction machine throughout the one-year period of its warranty. The results indicate that $N = 77$, $p^* = 20/77$, $n_u = 9$, $n_c = 17$, and $n_l = 51$. Here $n_u$ = the number of items that failed in the warranty period, $n_c$ = the number of items without failure in the warranty period but for which usage was determined through follow-up, $n_l$ = the number of items

**Table F.8** Fan failure data with a random selection of follow-up observation [47]

| $Z_i$ | $\delta_i$ | $Z_i$ | $\delta_i$ | $Z_i$ | $\delta_i$ | $Z_i$ | $\delta_i$ |
|---|---|---|---|---|---|---|---|
| 450 | 1 | 2,200 | 0 | 4,600 | 1 | 7,800 | 0 |
| (460) | 0 | 3,000 | 0 | 4,850 | 0 | 8,100 | 0 |
| 1,150 | 1 | 3,000 | 0 | 4,850 | 0 | (8,100) | 0 |
| 1,150 | 1 | (3,000) | 0 | (4,850) | 0 | (8,200) | 0 |
| (1,560) | 0 | (3,000) | 0 | 4,850 | 0 | (8,500) | 0 |
| 1,600 | 1 | 3,100 | 1 | (5,000) | 0 | 8,500 | 0 |
| (1,660) | 0 | (3,200) | 0 | (5,000) | 0 | 8,500 | 0 |
| (1,850) | 0 | 3,450 | 1 | (5,000) | 0 | 8,750 | 1 |
| 1,850 | 0 | 3,750 | 0 | 6,100 | 1 | (8,750) | 0 |
| (1,850) | 0 | 3,750 | 0 | 6,100 | 0 | (8,750) | 0 |
| 1,850 | 0 | (4,150) | 0 | 6,100 | 0 | 9,400 | 0 |
| 1,850 | 0 | (4,150) | 0 | 6,100 | 0 | (9,900) | 0 |
| 2,030 | 0 | (4,150) | 0 | 6,300 | 0 | (10,100) | 0 |
| (2,030) | 0 | (4,150) | 0 | (6,450) | 0 | (10,100) | 0 |
| (2,030) | 0 | (4,300) | 0 | 6,450 | 0 | (10,100) | 0 |
| 2,070 | 1 | 4,300 | 0 | (6,700) | 0 | 11,500 | 0 |
| 2,070 | 1 | (4,300) | 0 | 7,450 | 0 | | |
| 2,080 | 1 | (4,300) | 0 | (7,800) | 0 | | |

**Table F.9** Failure and follow-up data of a construction machine [47]

| $n_u$ | $n_c$ | |
|---|---|---|
| 70 | 537 | 1156 |
| 149 | 704 | 1168 |
| 190 | 757 | 1258 |
| 247 | 908 | 1280 |
| 283 | 964 | 1362 |
| 442 | 1072 | 1413 |
| 779 | 1100 | 1602 |
| 1373 | 1117 | 1771 |
| 1590 | 1124 | |

without failure that have not been followed up in the warranty period (the usages for these items have not been observed) and $N = n_u + n_c + n_l$ is the total number of items. The real operating hours for $n_u = 9$ and $n_c = 17$ machines are shown in Table F.9.

## F.9 Data Set 9 [Aircraft Air Conditioning Units]

The failure data of Table F.10 are a partial set of data [39] that have been used extensively in the statistical and reliability literature for illustration of various concepts and techniques. The observations are times between failures of air

**Table F.10** Time between failures of AC units

| Failure | Aircraft | | |
|---|---|---|---|
| | TBF-7909 | TBF-7912 | TBF-7913 |
| 1 | 90 | 23 | 97 |
| 2 | 10 | 261 | 51 |
| 3 | 60 | 87 | 11 |
| 4 | 186 | 7 | 4 |
| 5 | 61 | 120 | 141 |
| 6 | 49 | 14 | 18 |
| 7 | 14 | 62 | 142 |
| 8 | 24 | 47 | 68 |
| 9 | 56 | 225 | 77 |
| 10 | 20 | 71 | 80 |
| 11 | 79 | 246 | 1 |
| 12 | 84 | 21 | 16 |
| 13 | 44 | 42 | 106 |
| 14 | 59 | 20 | 206 |
| 15 | 29 | 5 | 82 |
| 16 | 118 | 12 | 54 |
| 17 | 25 | 120 | 31 |
| 18 | 156 | 11 | 216 |
| 19 | 310 | 3 | 46 |
| 20 | 76 | 14 | 111 |
| 21 | 26 | 71 | 39 |
| 22 | 44 | 11 | 63 |
| 23 | 23 | 14 | 18 |
| 24 | 62 | 11 | 191 |
| 25 | 130 | 16 | 18 |
| 26 | 208 | 90 | 163 |
| 27 | 70 | 1 | 24 |
| 28 | 101 | 16 | |
| 29 | 208 | 52 | |

conditioning units on commercial aircraft. The original data set reported by Proschan included data on thirteen aircraft. The results for three of these are given in Table F.10.

## F.10 Data Set 10 [Valve Seat for Diesel Engines]

Table F.11 shows engine age (in days) at the time of a valve seat replacement for a fleet of 41 diesel engines. These data on a sample of systems appeared in [36, 37] and also in [30], p. 635.

**Table F.11** Diesel engine age at time of replacement of valve seats

| System ID | Days observed | Engine age at replacement time | | | |
|---|---|---|---|---|---|
| 251 | 761 | | | | |
| 252 | 759 | | | | |
| 327 | 667 | 98 | | | |
| 328 | 667 | 326 | 653 | 653 | |
| 329 | 665 | | | | |
| 330 | 667 | 84 | | | |
| 331 | 663 | 87 | | | |
| 389 | 653 | 646 | | | |
| 390 | 653 | 92 | | | |
| 391 | 651 | | | | |
| 392 | 650 | 258 | 328 | 377 | 621 |
| 393 | 648 | 61 | 539 | | |
| 394 | 644 | 254 | 276 | 298 | 640 |
| 395 | 642 | 76 | 538 | | |
| 396 | 641 | 635 | | | |
| 397 | 649 | 349 | 404 | 561 | |
| 398 | 631 | | | | |
| 399 | 596 | | | | |
| 400 | 614 | 120 | 479 | | |
| 401 | 582 | 323 | 449 | | |
| 402 | 589 | 139 | 139 | | |
| 403 | 593 | | | | |
| 404 | 589 | 573 | | | |
| 405 | 606 | 165 | 408 | 604 | |
| 406 | 594 | 249 | | | |
| 407 | 613 | 344 | 497 | | |
| 408 | 595 | 265 | 586 | | |
| 409 | 389 | 166 | 206 | 348 | |
| 410 | 601 | | | | |
| 411 | 601 | 410 | 581 | | |
| 412 | 611 | | | | |
| 413 | 608 | | | | |
| 414 | 587 | | | | |
| 415 | 603 | 367 | | | |
| 416 | 585 | 202 | 563 | 570 | |
| 417 | 587 | | | | |
| 418 | 578 | | | | |
| 419 | 578 | | | | |
| 420 | 586 | | | | |
| 421 | 585 | | | | |
| 422 | 582 | | | | |

**Table F.12** Aggregated warranty claims data for an automobile component

| Age ($t$) | No. of failures ($n_{\bullet t}$) | No. censored ($\tilde{n}_t$) | No. at risk ($N_t$) |
|---|---|---|---|
| 1 | 9 | 631 | 104005 |
| 2 | 22 | 1506 | 103365 |
| 3 | 64 | 3229 | 101837 |
| 4 | 80 | 3774 | 98544 |
| 5 | 61 | 3677 | 94690 |
| 6 | 47 | 3695 | 90952 |
| 7 | 28 | 3689 | 87210 |
| 8 | 36 | 3656 | 83493 |
| 9 | 22 | 3351 | 79801 |
| 10 | 31 | 3242 | 76428 |
| 11 | 32 | 3873 | 73155 |
| 12 | 25 | 4298 | 69250 |
| 13 | 13 | 4270 | 64927 |
| 14 | 7 | 3600 | 60644 |
| 15 | 4 | 3091 | 57037 |
| 16 | 4 | 3117 | 53942 |
| 17 | 8 | 3208 | 50821 |
| 18 | 5 | 47600 | 47605 |

## F.11 Data Set 11 [Automobile Component-I]

Table F.12 shows the aggregated warranty claims and censored data for a specific automobile component[21] (unit) under the warranty period of 18 months.

## F.12 Data Set 12 [Automobile Component-II]

The item is a component of an automobile sold in Asia with an 18 month warranty. The monthly sales data ($S_{ij}$) and failures as a function of MIS $t$ (age) and MOS ($j$) for a particular MOP (September, $i = 9$) are given in Table F.13.

## F.13 Data Set 13 [Tensile Strength of Fibers]

Composite materials consist of a matrix material and reinforcing elements. The latter are the predominant factor in determining the strength of the material. The data of Table F.14 are measurements of fiber strength as measured by stress

[21] The information regarding the names of the component and manufacturing company are not disclosed to protect proprietary nature of the information.

**Table F.13** Monthly sales ($S_{ij}$) and failures ($n_{it}$) indexed by MOS ($j$) and MIS ($t$) for a particular MOP ($i = 9$)

| $j$ | $S_{ij}$ | Failures $\{n_{it}\}$ in MIS ($t$) under warranty | | | | | | | | | | | | | | | | | | Tot |
|---|---|---|---|---|---|---|---|---|---|---|---|---|---|---|---|---|---|---|---|---|
| | | 1 | 2 | 3 | 4 | 5 | 6 | 7 | 8 | 9 | 10 | 11 | 12 | 13 | 14 | 15 | 16 | 17 | 18 | |
| 1 | 1064 | | | 1 | | | 1 | | 2 | 1 | 1 | 2 | 1 | | 1 | 1 | | | 1 | 12 |
| 2 | 3600 | 4 | 5 | | 1 | 1 | 3 | 5 | 6 | 4 | 5 | 4 | 3 | 4 | 3 | 3 | 5 | 7 | 5 | 68 |
| 3 | 2113 | | | 1 | 1 | 2 | | 3 | 2 | 1 | 1 | 2 | 5 | 2 | 3 | 2 | 1 | | 3 | 29 |
| 4 | 720 | 1 | 1 | | | | 1 | | 1 | 1 | | 1 | 1 | | 1 | 1 | | 1 | 1 | 11 |
| 5 | 442 | | | | | | | | | | | 1 | 3 | 2 | | 1 | | 1 | | 8 |
| 6 | 235 | | | | 1 | | 2 | 1 | 1 | 1 | | | | | | 1 | | | | 7 |
| 7 | 168 | | | | | | | | | | | | | | | | | | | 0 |
| 8 | 94 | | | | | | | | | | | | | | | | | | | 0 |
| 9 | 74 | | | | | 1 | | | | | 2 | | | | | | | | | 3 |
| 10 | 90 | | | 1 | 1 | | | | | | | | | | | | | | | 2 |
| 11 | 51 | | 1 | | | | | | | | 1 | | | | | | | | | 2 |
| 12 | 16 | | | | | | | | | | | | | | | | | | | 0 |
| 13 | 63 | | | | | | | | | | | | | | | | | | | 0 |
| 14 | 82 | | 1 | | 1 | 2 | | | | | | | | | | | | | | 4 |
| 15 | 51 | | | | | | | | | | | | | | | | | | | 0 |
| 16 | 27 | | | | | | | | | | | | | | | | | | | 0 |
| 17 | 8 | | | | | | | | | | | | | | | | | | | 0 |
| 18 | 20 | | | | | | | | | | | | | | | | | | | 0 |
| 19 | 12 | | | | | | | | | | | | | | | | | | | 0 |
| 20 | 8 | 1 | | | | | | | | | | | | | | | | | | 1 |
| Tot | 8938 | 6 | 8 | 3 | 5 | 6 | 7 | 9 | 12 | 8 | 10 | 10 | 13 | 8 | 8 | 9 | 6 | 9 | 10 | 147 |

applied until fracture failure of silicon carbide fibers after extraction from a ceramic matrix [49]. One of the objectives of the experiment was to compare strengths of fibers of varying lengths. Fiber lengths used in the study were 5, 12.5, 25.4, and 265 mm. Sample sizes of 50 were used for all but the 25.4 mm fibers, for which a sample of size 64 was used.

## F.14 Data Set 14 [Automobile Component-III]

Table F.15 shows a part of the warranty claims data for an automobile component (20 observations out of 498).[22] Failure modes, type of automobile used the component and auto-used Zone/Region are shown in codes.

[22] The information regarding the names of the component and manufacturing company are not disclosed to protect proprietary nature of the information.

**Table F.14** Tensile Strength of SiC Fibers

| Fiber length (mm) | | | | | | | |
|---|---|---|---|---|---|---|---|
| 5 | 12.7 | 25.4 | 265 | 5 | 12.7 | 25.4 | 265 |
| 2.36 | 1.96 | 1.25 | 0.36 | 3.81 | 3.29 | 2.81 | 1.93 |
| 2.40 | 1.98 | 1.50 | 0.50 | 3.88 | 3.30 | 2.82 | 1.96 |
| 2.54 | 2.06 | 1.57 | 0.57 | 3.93 | 3.36 | 2.90 | 1.97 |
| 2.67 | 2.07 | 1.85 | 0.95 | 3.94 | 3.39 | 2.92 | 1.99 |
| 2.68 | 2.07 | 1.92 | 0.99 | 3.94 | 3.39 | 2.93 | 2.04 |
| 2.69 | 2.11 | 1.94 | 1.09 | 3.94 | 3.41 | 3.02 | 2.06 |
| 2.70 | 2.22 | 2.00 | 1.09 | 3.98 | 3.41 | 3.11 | 2.06 |
| 2.77 | 2.25 | 2.02 | 1.33 | 4.04 | 3.43 | 3.11 | 2.08 |
| 2.77 | 2.39 | 2.13 | 1.33 | 4.07 | 3.52 | 3.14 | 2.11 |
| 2.79 | 2.42 | 2.17 | 1.37 | 4.08 | 3.72 | 3.20 | 2.26 |
| 2.83 | 2.63 | 2.17 | 1.38 | 4.08 | 3.96 | 3.20 | 2.27 |
| 2.91 | 2.67 | 2.20 | 1.38 | 4.16 | 4.07 | 3.22 | 2.27 |
| 3.04 | 2.75 | 2.23 | 1.39 | 4.18 | 4.09 | 3.26 | 2.38 |
| 3.05 | 2.75 | 2.24 | 1.41 | 4.22 | 4.13 | 3.29 | 2.39 |
| 3.06 | 2.75 | 2.30 | 1.42 | 4.24 | 4.13 | 3.30 | 2.47 |
| 3.24 | 2.89 | 2.33 | 1.42 | 4.35 | 4.14 | 3.34 | 2.48 |
| 3.27 | 2.93 | 2.42 | 1.45 | 4.37 | 4.15 | 3.35 | 2.73 |
| 3.28 | 2.95 | 2.43 | 1.49 | 4.50 | 4.29 | 3.37 | 2.74 |
| 3.34 | 2.96 | 2.45 | 1.50 | | | 3.43 | |
| 3.36 | 2.97 | 2.49 | 1.56 | | | 3.43 | |
| 3.39 | 3.00 | 2.51 | 1.57 | | | 3.47 | |
| 3.51 | 3.03 | 2.54 | 1.57 | | | 3.61 | |
| 3.53 | 3.04 | 2.57 | 1.75 | | | 3.61 | |
| 3.59 | 3.05 | 2.62 | 1.78 | | | 3.62 | |
| 3.63 | 3.07 | 2.66 | 1.79 | | | 3.64 | |
| 3.64 | 3.08 | 2.68 | 1.79 | | | 3.72 | |
| 3.64 | 3.13 | 2.71 | 1.82 | | | 3.79 | |
| 3.66 | 3.20 | 2.72 | 1.83 | | | 3.84 | |
| 3.71 | 3.22 | 2.76 | 1.86 | | | 3.93 | |
| 3.73 | 3.23 | 2.79 | 1.89 | | | 4.03 | |
| 3.75 | 3.26 | 2.79 | 1.90 | | | 4.07 | |
| 3.78 | 3.27 | 2.80 | 1.92 | | | 4.13 | |

## F.15 Data Set 15 [Aircraft Windshield]

The windshield on a large aircraft is a complex piece of equipment, comprised basically of several layers of material, including a very strong outer skin with a heated layer just beneath it, all laminated under high temperature and pressure. Failures of these items are not structural failures. Instead, they typically involve damage or delamination of the non-structural outer ply, or failure of the heating system. These failures do not result in damage to the aircraft, but do result in replacement of the windshield because of decreased visibility.

**Table F.15** Detailed warranty claims data of an automobile component

| Serial no. | Date of production | Date of sale | Date of failure | Age in days | Used KM at failure | Failure mode | Used region | Type of auto used the unit |
|---|---|---|---|---|---|---|---|---|
| 1 | 01-Aug-01 | 01-Sep-01 | 15-Jan-02 | 136 | 36487 | M2 | R1 | A2 |
| 2 | 01-May-01 | 02-Sep-01 | 15-Dec-01 | 104 | 2381 | M2 | R4 | A2 |
| 3 | 01-Feb-01 | 07-Sep-01 | 15-Dec-01 | 99 | 14507 | M2 | R3 | A2 |
| 4 | 10-Jan-01 | 12-Sep-01 | 15-Dec-01 | 94 | 7377 | M1 | R4 | A2 |
| 5 | 01-May-01 | 12-Sep-01 | 15-Dec-01 | 94 | 10790 | M1 | R4 | A1 |
| 6 | 05-Jul-01 | 12-Sep-01 | 15-Feb-02 | 156 | 47312 | M3 | R1 | A1 |
| 7 | 05-Apr-01 | 13-Sep-01 | 05-Jul-02 | 295 | 56943 | M2 | R4 | A1 |
| 8 | 10-Jun-01 | 15-Sep-01 | 12-Jul-02 | 300 | 45292 | M1 | R3 | A2 |
| 9 | 05-Apr-01 | 22-Sep-01 | 15-Nov-01 | 54 | 5187 | M2 | R3 | A1 |
| 10 | 01-Aug-00 | 24-Sep-01 | 15-Dec-01 | 82 | 4512 | M3 | R1 | A1 |
| 11 | 01-Dec-00 | 27-Sep-01 | 15-Mar-02 | 169 | 18175 | M1 | R3 | A1 |
| 12 | 10-Jan-01 | 27-Sep-01 | 15-Mar-02 | 169 | 18106 | M1 | R3 | A1 |
| 13 | 01-Apr-01 | 27-Sep-01 | 26-Sep-02 | 364 | 27008 | M1 | R3 | A2 |
| 14 | 05-Apr-01 | 28-Sep-01 | 15-Dec-01 | 78 | 11600 | M1 | R4 | A1 |
| 15 | 01-May-01 | 28-Sep-01 | 15-Dec-01 | 78 | 7900 | M1 | R4 | A1 |
| 16 | 10-Jun-01 | 29-Sep-01 | 15-Dec-01 | 77 | 17620 | M1 | R4 | A1 |
| 17 | 10-Jun-01 | 29-Sep-01 | 15-Oct-01 | 16 | 7762 | M1 | R4 | A1 |
| 18 | 10-Jun-01 | 01-Oct-01 | 15-Mar-02 | 165 | 39487 | M1 | R3 | A1 |
| 19 | 01-Jun-00 | 02-Oct-01 | 15-Nov-01 | 44 | 6420 | M1 | R3 | A2 |
| 20 | 01-Mar-01 | 07-Oct-01 | 10-Apr-02 | 185 | 45121 | M2 | R4 | A1 |

**Table F.16** Windshield failure and censored data

| Failure times (thousand hours) | | | | Service times (thousand hours) | | |
|---|---|---|---|---|---|---|
| 0.040 | 1.866 | 2.385 | 3.443 | 0.046 | 1.436 | 2.592 |
| 0.301 | 1.876 | 2.481 | 3.467 | 0.140 | 1.492 | 2.600 |
| 0.309 | 1.899 | 2.610 | 3.478 | 0.150 | 1.580 | 2.670 |
| 0.557 | 1.911 | 2.625 | 3.578 | 0.248 | 1.719 | 2.717 |
| 0.943 | 1.912 | 2.632 | 3.595 | 0.280 | 1.794 | 2.819 |
| 1.070 | 1.914 | 2.646 | 3.699 | 0.313 | 1.915 | 2.820 |
| 1.124 | 1.981 | 2.661 | 3.779 | 0.389 | 1.920 | 2.878 |
| 1.248 | 2.010 | 2.688 | 3.924 | 0.487 | 1.963 | 2.950 |
| 1.281 | 2.038 | 2.823 | 4.035 | 0.622 | 1.978 | 3.003 |
| 1.281 | 2.085 | 2.890 | 4.121 | 0.900 | 2.053 | 3.102 |
| 1.303 | 2.089 | 2.902 | 4.167 | 0.952 | 2.065 | 3.304 |
| 1.432 | 2.097 | 2.934 | 4.240 | 0.996 | 2.117 | 3.483 |
| 1.480 | 2.135 | 2.962 | 4.255 | 1.003 | 2.137 | 3.500 |
| 1.505 | 2.154 | 2.964 | 4.278 | 1.010 | 2.141 | 3.622 |
| 1.506 | 2.190 | 3.000 | 4.305 | 1.085 | 2.163 | 3.665 |
| 1.568 | 2.194 | 3.103 | 4.376 | 1.092 | 2.183 | 3.695 |
| 1.615 | 2.223 | 3.114 | 4.449 | 1.152 | 2.240 | 4.015 |
| 1.619 | 2.224 | 3.117 | 4.485 | 1.183 | 2.341 | 4.628 |
| 1.652 | 2.229 | 3.166 | 4.570 | 1.244 | 2.435 | 4.806 |
| 1.652 | 2.300 | 3.344 | 4.602 | 1.249 | 2.464 | 4.881 |
| 1.757 | 2.324 | 3.376 | 4.663 | 1.262 | 2.543 | 5.140 |
| 1.795 | 2.349 | 3.385 | 4.694 | 1.360 | 2.560 | – |

**Table F.17** Photocopier's service history data

| Counter | Day | Component | Counter | Day | Component |
|---|---|---|---|---|---|
| 60152 | 29 | Cleaning Web | 365075 | 397 | Toner Filter |
| 60152 | 29 | Toner Filter | 365075 | 397 | Drum Claws |
| 60152 | 29 | Feed Rollers | 365075 | 397 | Ozone Filter |
| 132079 | 128 | Cleaning Web | 370070 | 468 | Feed Rollers |
| 132079 | 128 | Drum Cleaning Blade | 378223 | 492 | Drum |
| 132079 | 128 | Toner Guide | 390459 | 516 | Upper Fuser Roller |
| 220832 | 227 | Toner Filter | 427056 | 563 | Cleaning Web |
| 220832 | 227 | Cleaning Blade | 427056 | 563 | Upper Fuser Roller |
| 220832 | 227 | Dust Filter | 449928 | 609 | Toner Filter |
| 220832 | 227 | Drum Claws | 449928 | 609 | Feed Rollers |
| 252491 | 276 | Drum Cleaning Blade | 449928 | 609 | Upper Roller Claws |
| 252491 | 276 | Cleaning Blade | 472320 | 677 | Feed Rollers |
| 252491 | 276 | Drum | 472320 | 677 | Cleaning Blade |
| 252491 | 276 | Toner Guide | 472320 | 677 | Upper Roller Claws |
| 365075 | 397 | Cleaning Web | 501550 | 722 | Cleaning Web |
| 501550 | 722 | Dust Filter | 933637 | 1410 | Feed Rollers |
| 501550 | 722 | Drum | 933637 | 1410 | Dust Filter |
| 501550 | 722 | Toner Guide | 933637 | 1410 | Ozone Filter |
| 533634 | 810 | TS Block Front | 933785 | 1412 | Cleaning Web |

(continued)

**Table F.17** (continued)

| Counter | Day | Component | Counter | Day | Component |
|---|---|---|---|---|---|
| 533634 | 810 | Charging Wire | 936597 | 1436 | Drive Gear D |
| 583981 | 853 | Cleaning Blade | 938100 | 1448 | Cleaning Web |
| 597739 | 916 | Cleaning Web | 944235 | 1460 | Dust Filter |
| 597739 | 916 | Drum Claws | 944235 | 1460 | Ozone Filter |
| 597739 | 916 | Drum | 984244 | 1493 | Feed Rollers |
| 597739 | 916 | Toner Guidc | 984244 | 1493 | Charging Wire |
| 624578 | 956 | Charging Wire | 994597 | 1514 | Cleaning Web |
| 660958 | 996 | Lower Roller | 994597 | 1514 | Ozone Filter |
| 675841 | 1016 | Cleaning Web | 994597 | 1514 | Optics PS Felt |
| 675841 | 1016 | Feed Rollers | 1005842 | 1551 | Upper Fuser Roller |
| 684186 | 1074 | Toner Filter | 1005842 | 1551 | Upper Roller Claws |
| 684186 | 1074 | Ozone Filter | 1005842 | 1551 | Lower Roller |
| 716636 | 1111 | Cleaning Web | 1014550 | 1560 | Feed Rollers |
| 716636 | 1111 | Dust Filter | 1014550 | 1560 | Drive Gear D |
| 716636 | 1111 | Upper Roller Claws | 1045893 | 1583 | Cleaning Web |
| 769384 | 1165 | Feed Rollers | 1045893 | 1583 | Toner Guide |
| 769384 | 1165 | Upper Fuser Roller | 1057844 | 1597 | Cleaning Blade |
| 769384 | 1165 | Optics PS Felt | 1057844 | 1597 | Drum |
| 787106 | 1217 | Cleaning Blade | 1057844 | 1597 | Charging Wire |
| 787106 | 1217 | Drum Claws | 1068124 | 1609 | Cleaning Web |
| 787106 | 1217 | Toner Guide | 1068124 | 1609 | Toner Filter |
| 840494 | 1266 | Feed Rollers | 1068124 | 1609 | Ozone Filter |
| 840494 | 1266 | Ozone Filter | 1072760 | 1625 | Feed Rollers |
| 851657 | 1281 | Cleaning Blade | 1072760 | 1625 | Dust Filter |
| 851657 | 1281 | Toner Guide | 1072760 | 1625 | Ozone Filter |
| 872523 | 1312 | Drum Claws | 1077537 | 1640 | Cleaning Web |
| 872523 | 1312 | Drum | 1077537 | 1640 | Optics PS Felt |
| 900362 | 1356 | Cleaning Web | 1077537 | 1640 | Charging Wire |
| 900362 | 1356 | Upper Fuser Roller | 1099369 | 1650 | TS Block Front |
| 900362 | 1356 | Upper Roller Claws | 1099369 | 1650 | Charging Wire |

Data on failure and service times for a particular model windshield are given in Table F.16 (from [4]). The data consist of 153 observations, of which 88 are classified as failed windshields and the remaining 65 are service times of windshields that had not failed at the time of observation.

## F.16 Data Set 16 [Photocopier]

The data recorded from the photocopier's service history are given in Table F.17. Each row describes a part that was replaced, giving the number of copies made at the time of replacement, the age of the machine in days, and the component replaced. Most services involved replacing multiple components.

## References


1. Abernethy RB, Breneman JR, Medlin CH, Reinman GL (1983) Weibull analysis handbook. Report No. AFWAL-TR-83-2079, Aero Propulsion Laboratory, USAF, Wright-Patterson AFB, Ohio
2. Abramowitz M, Stegun IA (1964) Handbook of mathematical functions. Applied Mathematics Series No. 55, National Bureau of Standards, Washington, D.C.
3. Blischke WR, Murthy DNP (1994) Warranty cost analysis. Dekker, New York
4. Blischke WR, Murthy DNP (2000) Reliability. Wiley, New York
5. Casella G, Berger RL (2001) Statistical inference. Duxbury, New York
6. Cox DR, Isham V (1980) Point processes. Chapman and Hall, London
7. Dempster AP, Laird NM, Rubin DB (1977) Maximum likelihood from incomplete data via the EM algorithm (with discussion). J Royal Statist Soc B 39:1–38
8. Dodson B (1994) Weibull analysis. ASQ Quality Press, Milwaukee, Wisconsin
9. Gumbel EJ (1960) Bivariate exponential distributions. J Am Statist Assoc 55:698–707
10. Hartley HQ (1958) Maximum likelihood estimation from incomplete data. Biometrics 14:174–194
11. Hogg RV, Craig A, McKean JW (2004) Introduction to mathematical statistics, 6th edn. Prentice Hall, New York
12. Hutchinson TP, Lai CD (1990) Continuous bivariate distributions–emphasising applications. Rumsby Scientific, Adelaide, Australia
13. Hunter JJ (1974) Renewal theory in two dimensions: basic results. Adv App Probab 6:376–391
14. Johnson NL, Kotz S (1969) Discrete distributions. Houghton Mifflin Co., Boston
15. Johnson NL, Kotz S (1969) Distributions in statistics: discrete distributions. Wiley, New York
16. Johnson NL, Kotz S (1970) Distributions in statistics: continuous univariate distributions–I. Wiley, New York
17. Johnson NL, Kotz S (1970) Distributions in statistics: continuous univariate distributions–II. Wiley, New York
18. Johnson NL, Kotz S (1972) Distributions in statistics: continuous multivairate distributions. Wiley, New York
19. Kies JA (1958) The strength of glass, Naval Res-Lab. Report No. 5093, Washington D.C.
20. Kumar U, Klefsjö B (1992) Reliability analysis of hydraulic systems of LHD machines using the power law process model. Reliab Eng Sys Saf 35:217–224
21. Lai CD, Xie M, Murthy DNP (2003) Modified Weibull model. IEEE Tran Rel 52:33–37
22. Lawless JF (1982) Statistical methods for lifetime data. Wiley, New York
23. Lee L (1979) Multivariate distributions having Weibull properties. J Multivar Analysis 9:267–277
24. Lehman EL, Casella G (1998) Theory of point estimation. Springer, New York
25. Little RJA, Rubin DB (1987) Statistical analysis with missing data. Wiley, New York
26. Louis TA (1982) Finding the observed information matrix when using the EM algorithm. J Royal Statist Soc B 44:226–233
27. Lu JC, Bhattacharyya GK (1990) Some new constructions of bivariate Weibull Models. Ann Inst Statist Math 42:543–559
28. McLachlan GJ, Krishnan T (1997) The EM algorithm and extensions. Wiley, New York
29. Marshall AW, Olkin I (1967) A multivariate exponential distribution. J Am Statist Assoc 62:30–44
30. Meeker WQ, Escobar LA (1998) Statistical methods for reliability data. Wiley, New York
31. Meng XL, Rubin DB (1991) Using EM to obtain asymptotic variance-covariance matrices: the SEM algorithm. J Am Statist Assoc 86:899–909

32. Mudholkar GS, Kollia GD (1994) Generalized Weibull family: A structural analysis. Comm in Stat Ser A: Theory and Methods 23:1149–1171
33. Murthy DNP, Bulmer M, Eccleston JE (2004) Weibull model selection. Reliab Eng Sys Saf 86:257–267
34. Murthy DNP, Xie M, Jiang R (2003) Weibull models. Wiley, New York
35. Nelson W (1982) Applied life data analysis. Wiley, New York
36. Nelson W (1995) Confidence limits for recurrence data—applied to cost or number of product repairs. Technometrics 37:147–157
37. Nelson W, Doganaksoy N (1989) A Computer program for an estimate and confidence limits for the mean cumulative function for cost or number of repairs of repairable products. TIS report 89CRD239, General Electric Company Research and Development, Schenectady, NY
38. Oakes D (1999) Direct calculation of the information matrix via the EM algorithm. J Royal Statist Soc B 61:479–482
39. Proschan F (1963) Theoretical explanation of observed decreasing failure rate. Technometrics 5:375–383
40. Rao CR (1972) Linear statistical inference and its applications. Wiley, New York
41. Redner RA, Walker HF (1984) Mixture densities, maximum likelihood and the EM algorithm. SIAM Rev 26:195–239
42. Rigdon SE, Basu AP (2000) Statistical methods for the reliability of repairable systems. Wiley, New York
43. Ross SM (1970) Applied probability models with optimization applications. Holden–Day San, Francisco
44. Ross SM (1980) Stochastic processes. Wiley, New York
45. Somerville PN (1958) Tables for obtaining non-parametric tolerance limits. Ann Math Stat 29:599–601
46. Stuart A, Ord JK (1991) Kendall's advanced theory of statistics, vol 2, 5th edn. Oxford University Press, New York
47. Suzuki K (1985) Estimation of lifetime parameters from incomplete field data. Technometrics 27:263–271
48. Wu CFJ (1983) On the convergence properties of the EM algorithm. Ann Statist 11:95–103
49. Zok FW, Chen X, Weber DH (1995) Tensile strength of SiC fibers. J Am Ceram Soc 78:1965–1968

# Index



W. R. Blischke et al., *Warranty Data Collection and Analysis*,
Springer Series in Reliability Engineering, DOI: 10.1007/978-0-85729-647-4,

**Q**